Biodiversity Studies

A Bibliographic Review

Charles H. Smith

The Scarecrow Press, Inc.
Lanham, Maryland, and London
2000

SCARECROW PRESS, INC.

Published in the United States of America
by Scarecrow Press, Inc.
4720 Boston Way, Lanham, Maryland 20706
http://www.scarecrowpress.com

4 Pleydell Gardens, Folkestone
Kent CT20 2DN, England

Copyright © 2000 by Charles H. Smith

All rights reserved. No part of this publication may be reproduced, stored in a retrieval system, or transmitted in any form or by any means, electronic, mechanical, photocopying, recording, or otherwise, without the prior permission of the publisher.

British Library Cataloguing in Publication Information Available

Library of Congress Cataloging-in-Publication Data

Smith, Charles H. (Charles Hyde), 1950–
 Biodiversity studies : a bibliographic review / Charles H. Smith.
  p. cm.
 Includes bibliographic references and indexes.
 ISBN 0-8108-3754-4 (alk. paper)
  1. Biological diversity—Bibliography. 2. Biological diversity conservation—Bibliography. I. Title.

Z7405.N38 S54   2000
[QH75]
016.33395'15—dc21   99-053779

∞™ The paper used in this publication meets the minimum requirements of American National Standard for Information Sciences—Permanence of Paper for Printed Library Materials, ANSI/NISO Z39.48–1992.
Manufactured in the United States of America.

Contents

Preface	vii
Scope	
Acknowledgments	
Introduction	ix
Biodiversity	
How to Use This Work	
Abbreviations	xiii
The Bibliographies	
Introduction	3
Bibliography I: Books and Articles	5
Bibliography II: Special Issues	393
The Indexes	
Introduction	403
Index I: General Subjects	405
Index II: Geographical Subjects	443
Index III: Organism-Centered Subjects	451
About the Author	461

Preface

Scope

Although biodiversity has existed as a recognized concept for less than fifteen years, its literature has already swollen to an extent making any thoughts of a fully comprehensive one-volume bibliographic treatment unrealistic. Nevertheless, this is a good time—before the first studies have become "history," and the volume of writings becomes so great as to defy more than selective sampling—to make *some* effort in that direction. With this in mind, I have endeavored to provide a study which, while largely ignoring the *facts* of biodiversity, attempts to reasonably thoroughly review its many facets of study.

My emphasis here is not, therefore, on "the" biodiversity itself. Instead, I have attempted to canvass the literature reflecting the *investigation* of biodiversity—that is, the written record of efforts to measure, understand, monitor, and preserve it. In consequence, one should not look to this work for help in coming to grips with the world's countless local or regional floras or faunas, or indeed for analogous coverage of the many facets of the basic biology of individual living things. Many other sources exist which deal with information of this sort, and I recommend consulting with a librarian as a first step in finding them. Similarly, this is not the place for obtaining directory-type information on individuals or organizations involved in biodiversity-related study.

Even with this stated focus a number of further concessions to the space available had to be made. First and foremost, it was necessary to exclude from consideration the individual papers appearing in edited collections. Bibliography I lists several hundred such collections, and a full second volume of size equal to the present one would have been needed to survey the many thousands of works of this type. Second, it should be noted that almost every entry in the list is in the English language, and that there are no references to electronic sources. Further, the list contains very little if any mention of dissertations or theses, paper abstracts, book reviews, news or newspaper stories, reference works (dictionaries, encyclopedias, bibliographies, etc.), systematic or faunal/floral revisions, or transcripts of government hearings-related testimony. Last, no juvenile literature is cited.

What *is* featured is the primary literature of biodiversity-directed natural science, social science, and humanities subjects, plus a fair amount of science journalism. Relevant items were located through a search of about one hundred electronic databases, and the hand examination of an approximately equal number of print bibliographies and references cited lists. Inclusion in the collection was ultimately decided on the basis of (1) subject content and (2) bibliometric comparisons.

Acknowledgments

I would especially like to acknowledge graduate assistant support made possible by a Western Kentucky University Faculty Research Grant; the work was additionally facilitated by a Faculty Research Leave awarded me by the Department of Library Public Services at WKU. Also, my thanks to four excellent student workers who made contributions to the study's progress: Kelly Haston, Jill Arrance, David Spence, and Lu Yinghua.

Introduction

Biodiversity

The term "biodiversity" is a shortened form of the phrase "biological diversity," and derives historically almost entirely from a single event: a Fall 1986 conference organized under the auspices of the National Academy of Sciences and the Smithsonian Institution. The main papers presented at the conference, provocatively titled the "National Forum on BioDiversity," were later (1988) committed to a collection titled *Biodiversity* (item 5551 of the present work) under the editorship of biologist Edward O. Wilson. The word caught on immediately.

The neologism's rather straightforward history would seemingly imply a correspondingly abrupt revolution as regards the emergence of the *concept* of biodiversity, but here the story becomes more complicated. The first place to turn for perspective on the matter ordinarily would be the term's formal definition, but in this direction one runs into a wall of truism. Writing recently on the subject, Wilson has drawn attention to the question as follows:

> So what is it? Biologists are inclined to agree that it is, in one sense, everything. Biodiversity is defined as all hereditarily based variation at all levels of organization, from genes within a single local population or species, to the species composing all or part of a local community, and finally to the communities themselves that compose the living parts of the multifarious ecosystems of the world. The key to effective analysis of biodiversity is the precise definition of each level of organization when it is being addressed. (item 4199, p. 1)

This is all well and good, but one could certainly not be blamed for thinking that a concept which seeks to describe "everything," even if only "in one sense," must have limited practical value—especially if it is truly necessary to revert to a "level of organization" approach to apply it in analysis. This criticism is buoyed by the fact that the "biodiversity movement" itself has actually been rather long in the making. Scientific interest in the diversity of life and its reasons for being can be traced back at least as far as Aristotle; neither should one ignore the more recent revolutions begun by figures such as Linnæus and Darwin, and the many ways their work has changed our perceptions of natural interrelations over the past two hundred years.

In the same essay quoted from above, Wilson notes that as regards biodiversity "what finally has given it such extraordinarily widespread attention is the realization that it is disappearing" (item 4199, p. 1). But even the history of worry over the disappearance of species and environments goes back well beyond the current era, to at least the mid-nineteenth century. In 1864 the American writer George Perkins Marsh brought out a highly successful and influential book entitled *Man and Nature* (item 3144) which fully explored this very subject. By the mid-twentieth century and the efforts of such naturalists as Rachel Carson and Aldo Leopold, a full-blown "environmental movement" had begun. Pesticide use and endangered species were the first beneficiaries of attention by investigators in the 1950s and 1960s, but by the 1970s the plights of degraded communities, habitats, and ecosystems were also being taken under full consideration. Meanwhile, investigators in the fields of wildlife conservation, community ecology, biogeography, genetics, systematics, environmental economics, and evolutionary biology were examining a variety of subjects that remain at the core of most of the work undertaken by present day proponents of biodiversity study.

Is there then indeed a distinct and delimitable "biodiversity movement" at all, or have scientists merely been acting out previous roles under a new and catchy banner? To this question I would reply, "Yes, the movement is a new and distinct one, but it has more to do with human aspirations than it does with biological focus." While it is true that a search for interdependencies within the natural world increasingly dominates our approach to environmental studies, the real measure of success of the biodiversity concept is the degree to which it has united natural, social, and moral/ethical objectives. There are thus two distinct sides to biodiversity studies, one in which the natural basis of diversity is emphasized, and another in which the social, economic, and ethical consequences of the conservation of life are central. The first set of studies dwells on the main defining characteristics of diversity *per se*: those evolutionary, biogeographical, and ecological forces that conspire to create and maintain heterogeneity in natural populations, communities, and environments. The second concerns itself with the ways human beings influence such goings-on, and how we can aspire to a form of management of our biological environment that is beneficial both to it and to ourselves. Biodiversity *matters*, and in the final analysis it is societal recognition of the fact (and the implications of *that* fact) that gives its study both currency and urgency.

Not that even this perspective is an *entirely* new one, of course. The practical and ethical rationale for studying the diversity of life was stated in timeless fashion by the great English naturalist Alfred Russel Wallace in a paper published as long ago as 1863:

> [The naturalist] looks upon every species of animal and plant now living as the individual letters which go to make up one of the volumes of our earth's

history; and, as a few lost letters may make a sentence intelligible, so the extinction of the numerous forms of life which the progress of cultivation invariably entails will necessarily render obscure this invaluable record of the past. It is, therefore, an important object, which governments and scientific institutions should immediately take steps to secure, that in all tropical countries colonized by Europeans the most perfect collections possible in every branch of natural history should be made and deposited in national museums, where they may be available for study and interpretation. If this is not done, future ages will certainly look back upon us as a people so immersed in the pursuit of wealth as to be blind to higher considerations. They will charge us with having culpably allowed the destruction of some of those records of Creation which we had it in our power to preserve; and while professing to regard every living thing as the direct handiwork and best evidence of a Creator, yet, with a strange inconsistency, seeing many of them perish irrecoverably from the face of the earth, uncared for and unknown. (*J. of the Royal Geographical Society* 33: 217-234, on p. 234)

Wallace, history's pre-eminent naturalist-collector, did not live in an era that much concerned itself with goals of conservation. We do; it remains to be seen, however, whether his "not be blind to higher considerations" admonition can become a rallying call, or end up a verdict.

How to Use This Work

In keeping with the notion that the "biodiversity movement" is defined by interdependent natural processes and social agenda, I have endeavored in this bibliographic review to include literature bearing on both the biological study of diversity itself, and the socio-natural science of diversity conservation. And, while its emphasis on primary sources argues for its primary use by scientists, educators, and professionals-to-be (especially, graduate students), it cites a fair amount of material that can be understood by undergraduates or even high school students.

The work includes a number of novel construction features, and I strongly recommend that the reader first peruse the Introductions to the Bibliographies and Indexes sections for explanations before diving in. These features are designed to facilitate item selection, and indeed to stimulate thought by exposing emphases and connections inherent in the study of the subject.

A number of abbreviations of journal titles, organizations, and terms have been employed in both the Bibliographies and the Indexes. The most significant of these (especially, "bd" for biodiversity, and "bdc" for biodiversity conservation) are listed on page xiii.

Abbreviations

ARES: *Annual Review of Ecology and Systematics*
A.T.: Added Terms
bd: biodiversity
bdc: biodiversity conservation
CB: *Conservation Biology*
CBD: Convention on Biological Diversity
CUP: Cambridge University Press
GATT: General Agreement on Tariffs and Trade
GEF: Global Environment Facility
INBio: Instituto Nacional de Biodiversidad (Costa Rica)
IPGRI: International Plant Genetic Resources Institute
IUBS: International Union of Biological Sciences
IUCN: International Union for Conservation of Nature and Natural Resources
IUDZG: International Union of Directors of Zoological Gardens
MAB: Man and the Biosphere Program
OECD: Organization for Economic Cooperation and Development
OUP: Oxford University Press
PNAS: *Proceedings of the National Academy of Sciences of the U.S.A.*
PRSL: *Proceedings of the Royal Society of London*
PTRSL: *Philosophical Transactions of the Royal Society of London*
SCOPE: Special Committee on Problems of the Environment
SLOSS: "single large, or several small?" [concerning reserve design]
TREE: *Trends in Ecology & Evolution*
UNCED: United Nations Conference on Environment and Development
UNDP: United Nations Development Programme
UNEP: United Nations Environment Programme
USAID: United States Agency for International Development
WRI: World Resources Institute
WWF: World Wildlife Fund/WorldWide Fund for Nature

The Bibliographies

The Bibliographies: Introduction

Two bibliographies are included in this work. Bibliography I, much the longer of the two, lists the approximately 1,200 monographs and 4,500 shorter articles featured here, whereas Bibliography II provides a selective accounting of "special issues" of serial publications whose main theme is biodiversity studies-related.

I have already indicated the kinds of literature that are (and are not) surveyed here in the Preface and Introduction; we now may turn to a brief discussion of the main aspects of the bibliographic form. My basic goal in this respect was to provide for each entry as much information as possible in as little space as possible. To achieve this optimum I have blended certain elements of standard bibliographic citation form with some additional information. Each entry contains a slightly abbreviated (see below) form of the usual citation format, plus (as relevant) bibliometric ratings, short annotations, and/or a list of descriptive "added terms."

Journal articles are bibliometrically rated along a scale of 1 to 3 according to the average number of times they have been listed as cited over a several- to ten- or twelve-year period (i.e., since 1986) in the combined *Citation Index* portions of *Science Citation Index*, *Social Sciences Citation Index*, and *Humanities Citation Index*. The rating appears directly after the pagination data in brackets [ ]; a '1' rating corresponds to about 0 to 4 citations per year, a '2,' about 5 to 20, and a '3,' over 20. Items published in 1997 and 1998 are usually not rated, for obvious reasons.

Monographs are rated along the same 1 to 3 scale, but on three different factors: (1) number of OCLC network libraries (out of a total of over 30,000 member institutions) whose collections hold the item; (2) number of times the work was reviewed after publication (as indicated by records in *Book Review Index* and the electronic databases *Periodical Abstracts* and *Expanded Academic Index*); and (3) number of citations it has generated (as described above for journal articles). This information is also displayed within brackets; thus, and for example, a '[3/3/1]' display describes a book that is held by, at the very least, most college and university libraries, that generated a good deal of reviewing interest (i.e., extending well beyond specialized circles), and that has not been much cited as contributing to the march of professional knowledge. Again, some of the newest items are not rated.

3

Many of the citations have in addition been supplemented with brief annotations and/or one or more "added terms" to help the reader assess main subject content. Any given entry in the list may contain any combination of none to all of these three kinds of citation enhancement (i.e., ratings, annotation, and added terms). Note that not all "added terms" have corresponding entries in the Indexes.

The citations themselves are designed to convey only the minimum amount of information needed to identify the work for retrieval. I list all authors when there are three or fewer, but only the first two (plus "et al.") when there are four or more.

Bibliography I is arranged alphabetically by author. Bibliography II is arranged alphabetically by journal title and, it should be noted, contains citation information only (i.e., no bibliometric ratings or additional descriptive material).

About one-quarter of the 5,700 items in Bibliography I have been distinguished in the Indexes on the basis of their high bibliometric ratings (see explanation on page 403). The following are the ten book publishers most frequently represented on this "honors" list (publisher names are followed by the number of titles involved): Cambridge (36), Island (28), Oxford (24), Columbia (17), Chapman & Hall (16), Princeton (13), Springer (12), Smithsonian (11), Wiley (10), and Yale (10). Correspondingly, the honors list for the twelve serial publications most frequently represented in this way is as follows: *Science* (79), *Conservation Biology* (73), *American Naturalist* (68), *Ecology* (62), *Nature* (41), *Bioscience* (37), *Biological Conservation* (31), *Annual Review of Ecology and Systematics* (29), *Oikos* (24), *Trends in Ecology & Evolution* (23), *Ecological Applications* (21), and *Proceedings of the National Academy of Sciences of the U.S.A.* (19).

It may be of some aid to specially single out those works that have been influential enough to achieve bibliometrically a '3' citation impact score here. The item numbers of these works are:

33, 103, 134, 135, 306, 345, 419, 477, 503, 519, 588, 601, 634, 636, 637, 639, 711, 803, 820, 835, 838, 1003, 1007, 1009, 1039, 1048, 1132, 1138, 1182, 1251, 1259, 1281, 1282, 1328, 1336, 1460, 1466, 1505, 1573, 1642, 1663, 1707, 1725, 1743, 1750, 1756, 1882, 1889, 1890, 1962, 1970, 2010, 2025, 2047, 2050, 2059, 2065, 2083, 2098, 2100, 2114, 2121, 2177, 2204, 2245, 2249, 2250, 2277, 2318, 2322, 2327, 2332, 2337, 2346, 2392, 2425, 2485, 2662, 2665, 2702, 2801, 2882, 2907, 2993, 3002, 3006, 3014, 3032, 3033, 3035, 3037, 3080, 3130, 3206, 3243, 3397, 3451, 3494, 3548, 3557, 3558, 3596, 3597, 3600, 3680, 3684, 3790, 3803, 3813, 3814, 3881, 3883, 3893, 3899, 3955, 3968, 4062, 4077, 4218, 4285, 4287, 4302, 4325, 4337, 4368, 4425, 4450, 4470, 4551, 4560, 4596, 4618, 4651, 4652, 4668, 4692, 4693, 4706, 4725, 4754, 4806, 4817, 4858, 5081, 5087, 5088, 5090, 5094, 5122, 5160, 5234, 5245, 5258, 5286, 5304, 5334, 5414, 5450, 5456, 5457, 5468, 5469, 5489, 5551, 5554, 5610, 5619, 5680.

Bibliography I: Books and Articles

1. Abate, Tom, 1993. Big computers tackle the job of saving little birds. *Bioscience* 43: 514-516. [1] How supercomputers are being used to help protect the habitats of threatened species. *A.T.:* California.
2. Abbott, Ian, 1980. Theories dealing with the ecology of land birds on islands. *Advances in Ecological Research* 11: 329-371. [2].
3. Abe, Takuya, S.A. Levin, & Masahiko Higashi, eds., 1997. *Biodiversity: An Ecological Perspective* [conference papers]. New York and London: Springer. 294p. [1/1/-].
4. Abel, N.O.J., & P.M. Blaikie, 1986. Elephants, people, parks and development: the case of the Luangwa Valley, Zambia. *Environmental Management* 10: 735-751. [1] *A.T.:* wildlife conservation.
5. Abele, Lawrence G., 1976. Comparative species richness in fluctuating and constant environments: coral-associated decapod crustaceans. *Science* 192: 461-463. [1] *A.T.:* environmental disturbance.
6. Abele, Lawrence G., & K. Walters, 1979. Marine benthic diversity: a critique and alternative explanation. *J. of Biogeography* 6: 115-126. [1].
7. Abensperg-Traun, Max, & D. Steven, 1997. Latitudinal gradients in the species richness of Australian termites (Isoptera). *Australian J. of Ecology* 22: 471-476.
8. Abensperg-Traun, Max, G.W. Arnold, et al., 1996. Biodiversity indicators in semi-arid, agricultural Western Australia. *Pacific Conservation Biology* 2: 375-389. [1].
9. Ablett, E.M., & J.S. Mattick, 1994. Conservation of invertebrate biodiversity: the role of *ex situ* preservation of genetic material. *Memoirs of the Queensland Museum* 36: 3-6. [1] *A.T.:* gene banks.
10. Abrahamson, Dean E., ed., 1989. *The Challenge of Global Warming*. Washington, DC: Island Press. 358p. [3/2/2] *A.T.:* greenhouse effect.
11. Abramovitz, Janet N., 1989. *A Survey of U.S.-based Efforts to Research and Conserve Biological Diversity in Developing Countries*. Washington, DC: Center for International Development and Environment, WRI. 71p. [1/1/1].
12. ———, 1991. Biodiversity: inheritance from the past, investment in the future. *Environmental Science & Technology* 25: 1816-1818. [1] *A.T.:* bd loss.

13. ———, 1991. *Investing in Biological Diversity: U.S. Research and Conservation Efforts in Developing Countries.* Washington, DC: WRI. 94p. [1/1/1].

14. ———, 1994. *Trends in Biodiversity Investments: U.S.-based Funding for Research and Conservation in Developing Countries, 1987-1991.* Washington, DC: WRI. 93p. [1/1/1].

15. ———, 1995. Freshwater failures: the crises on five continents. *World Watch* 8(5): 27-35. [1] *A.T.:* freshwater environments; environmental management.

16. ———, 1996. *Imperiled Waters, Impoverished Future: The Decline of Freshwater Ecosystems.* Washington, DC: Worldwatch Paper 128. 80p. [2/1/1] *A.T.:* water pollution; environmental degradation.

17. Abramovitz, Janet N., & Roberta Nichols, 1992. Women and biodiversity: ancient reality, modern imperative. *Development* (Rome) (2): 85-90. [1] *A.T.:* sustainable development; rural areas.

18. Abrams, Peter A., 1988. Resource productivity-consumer species diversity: simple models of competition in spatially heterogeneous environments. *Ecology* 69: 1418-1433. [1] *A.T.:* patch dynamics.

19. ———, 1995. Monotonic or unimodal diversity-productivity gradients: what does competition theory predict? *Ecology* 76: 2019-2027. [2].

20. Abrams, Robert H., 1990. Statutory protection of biodiversity in the national forests. *Michigan Academician* 22: 319-335. [1] *A.T.:* environmental legislation.

21. Abramsky, Z., & M.L. Rosenzweig, 1984. Tilman's predicted productivity-diversity relationship shown by desert rodents. *Nature* 309: 150-151. [1] *A.T.:* rainfall gradients.

22. Abranches, J., P. Valente, et al., 1998. Yeast diversity and killer activity dispersed in fecal pellets from marsupials and rodents in a Brazilian tropical habitat mosaic. *FEMS Microbiology Ecology* 26: 27-33.

23. Accaputo-Gendron, Josephine, & Morris Goldner, 1993. Roger Stanier: diversity as the key to a new era for biology. *Perspectives in Biology and Medicine* 37: 48-54. [1].

24. Acharya, Rohini, 1991. Patenting of biotechnology: GATT and the erosion of the world's biodiversity. *J. of World Trade* 25: 71-87. [1] *A.T.:* intellectual property.

25. Ackerman, Diane, 1995. *The Rarest of the Rare: Vanishing Animals, Timeless Worlds.* New York: Random House. 184p. [3/3/1] *A.T.:* threatened and endangered species.

26. Action for World Development, 1995. *Rice Beyond the "Green Revolution": Seeking Equity, Sustainable Farming and Biodiversity: Focus: the Philippines.* Surry Hills, New South Wales: Action for World Development NSW. 71p. [1/1/1].

27. Adam, Paul, 1992. *Australian Rainforests.* New York: OUP. 308p.

[1/1/1] *A.T.:* rainforest conservation.

28. Adams, C.E., S.L. Jester, & J.K. Thomas, 1995. National overview of regulations to conserve amphibians and reptiles. *Wildlife Society Bull.* 23: 391-396. [1] *A.T.:* state regulations.

29. Adams, J.M., 1989. Species diversity and productivity of trees. *Plants Today* 2: 183-187. [1] *A.T.:* North America; biomass accumulation.

30. Adams, J.M., & F.I. Woodward, 1989. Patterns in tree species richness as a test of the glacial extinction hypothesis. *Nature* 339: 699-701. [2] *A.T.:* Europe; North America; Eastern Asia.

31. Adams, Lowell W., 1994. *Urban Wildlife Habitats: A Landscape Perspective.* Minneapolis: University of Minnesota Press. 186p. [2/1/1].

32. Adams, Lowell W., & L.E. Dove, 1989. *Wildlife Reserves and Corridors in the Urban Environment: A Guide to Ecological Landscape Planning and Resource Conservation.* Columbia, MD: National Institute for Urban Wildlife. 91p. [2/1/1].

33. Adams, Mark D., & A.R. Kerlavage, 1995. Initial assessment of human gene diversity and expression patterns based upon 83 million nucleotides of cDNA sequence. *Nature* 377, Suppl.: 3-174. [3].

34. Adams, Robert McC., March 1992. Smithsonian horizons [editorial]. *Smithsonian* 22(12): 10. [1] Concerning the tradeoff between bdc and loss of livelihood.

35. Adams, Robert P., & J.E. Adams, eds., 1992. *Conservation of Plant Genes: DNA Banking and In Vitro Biotechnology* [organizational meeting papers]. San Diego: Academic Press. 345p. [1/1/1] *A.T.:* plant germplasm resources; plant gene banks.

36. Adams, Robert P., J.S. Miller, et al., eds., 1994. *Conservation of Plant Genes II: Utilization of Ancient and Modern DNA* [conference papers]. St. Louis: Missouri Botanical Garden. 276p. [1/1/1] *A.T.:* biotechnology; plant germplasm resources; fossil DNA.

37. Adams, Sean, Dec. 1993. Genetic fingerprinting helps sort out look-alikes. *Agricultural Research* 41(12): 8-11. [1] *A.T.:* germplasm resources; crop genetics.

38. ———, Nov. 1994. Roots: returning to the apple's birthplace. *Agricultural Research* 42(11): 18-21. [1] *A.T.:* germplasm resources; Kazakhstan; Kyrgyzstan.

39. ———, March 1995. Rhizobial magic: collection preserves valuable nitrogen-fixing bacteria. *Agricultural Research* 43(3): 16-17. [1].

40. ———, Nov. 1996. Special legumes may be an unopened medicine chest. *Agricultural Research* 44(11): 12-15. [1] *A.T.:* seed collections; germplasm resources.

41. Adams, William M., 1990. *Green Development: Environment and Sustainability in the Third World.* London and New York: Routledge. 257p. [2/2/2] *A.T.:* Green movement.

42. Adamus, P.R., 1995. Validating a habitat evaluation method for predicting avian richness. *Wildlife Society Bull.* 23: 743-749. [1] *A.T.:* Colorado Plateau; riparian habitats.

43. Adis, J., 1990. Thirty million arthropod species—too many or too few? *J. of Tropical Ecology* 6: 115-118. [1].

44. Adjeroud, Mehdi, & Bernard Salvat, 1996. Spatial patterns in biodiversity of a fringing reef community along Opunohu Bay, Moorea, French Polynesia. *Bull. of Marine Science* 59: 175-187. [1].

45. Adlard, R.D., & P.J. O'Donoghue, 1998. Perspectives on the biodiversity of parasitic protozoa in Australia. *International J. for Parasitology* 28: 887-897. *A.T.:* species boundaries; systematics.

46. Adler, G.H., 1994. Avifaunal diversity and endemism on tropical Indian Ocean islands. *J. of Biogeography* 21: 85-95. [1] *A.T.:* species turnover; isolation.

47. Adrain, Jonathan M., R.A. Fortey, & S.R. Westrop, 1998. Post-Cambrian trilobite diversity and evolutionary faunas. *Science* 280: 1922-1925. *A.T.:* Ordovician.

48. Aengst, Peter, Jeremy Anderson, et al., July-Aug. 1997. Introduction to habitat conservation planning. *Endangered Species Update* 14(7-8): 5-9. [1] *A.T.:* habitat conservation plans (HCPs); private lands.

49. Agard, J.B.R., R.H. Hubbard, & J.K. Griffith, 1996. The relation between productivity, disturbance and the biodiversity of Caribbean phytoplankton: applicability of Huston's dynamic equilibrium model. *J. of Experimental Marine Biology and Ecology* 202: 1-17. [1].

50. Agardy, M. Tundi, 1994. Advances in marine conservation: the role of marine protected areas. *TREE* 9: 267-270. [1] *A.T.:* land use zoning.

51. ———, 1997. *Marine Protected Areas and Ocean Conservation.* Austin, TX: R.G. Landes. 244p. [1/-/-].

52. Agudo, M., & D. Rodriguez, 1993. A proposal for the establishment of the National Center for Phytogeographic Resource Conservation in the Venezuelan Republic. *Science of the Total Environment* 139-140: 193-196. [1].

53. Ahearn, Sean C., J.L.D. Smith, & Catherine Wee, 1990. Framework for a geographically referenced conservation database: case study Nepal. *Photogrammetric Engineering and Remote Sensing* 56: 1477-1481. [1] *A.T.:* GIS.

54. Aiken, S. Robert, & C.H. Leigh, 1992. *Vanishing Rain Forests: The Ecological Transition in Malaysia.* New York: OUP. 194p. [1/1/1] *A.T.:* deforestation; rainforest ecology.

55. Airaudi, D., & V.F. Marchisio, 1997. Fungal biodiversity in the air of Turin. *Mycopathologia* 136: 95-102. *A.T.:* airborne microorganisms; Italy.

56. Akçakaya, H. Resit, Nov. 1994. GIS enhances endangered species conservation efforts. *GIS World* 7(11): 36-40. [1] *A.T.:* population viability

analysis.

57. Akerele, Olayiwola, V.H. Heywood, & Hugh Synge, eds., 1991. *The Conservation of Medicinal Plants: Proceedings of an International Consultation.* Cambridge, U.K., and New York: CUP. 362p. [1/1/1].

58. Alard, D., J.-F. Bance, & P.-N. Frileux, 1994. Grassland vegetation as an indicator of the main agro-ecological factors in a rural landscape: consequences for biodiversity and wildlife conservation in central Normandy (France). *J. of Environmental Management* 42: 91-109. [1].

59. Alberch, Pere, 1993. Museums, collections, and biodiversity inventories. *TREE* 8: 372-375. [1].

60. Albritton, Claude C., Jr., 1989. *Catastrophic Episodes in Earth History.* London and New York: Chapman & Hall. 221p. [2/2/1] *A.T.:* extinction; evolution.

61. Alcorn, Janis B., B.M. Beehler, & J.F. Swartzendruber, eds., 1992. *Papua New Guinea Conservation Needs Assessment.* Washington, DC: Biodiversity Support Program. 3 vols. [1/1/1] *A.T.:* environmental policy.

62. Aldebert, Y., 1997. Demersal resources of the Gulf of Lions (NW Mediterranean). Impact of exploitation on fish diversity. *Vie et Milieu* 47: 275-284. *A.T.:* France.

63. Alderson, Lawrence, ed., 1990. *Genetic Conservation of Domestic Livestock* [conference proceedings]. Wallingford, Oxon, U.K.: CAB International. 242p. [1/1/1] *A.T.:* animal germplasm resources; livestock breeds.

64. Aldhous, Peter, 1993. Tropical deforestation: not just a problem in Amazonia. *Science* 259: 1390. [1].

65. Aldrich, P.R., & J.L. Hamrick, 1998. Reproductive dominance of pasture trees in a fragmented tropical forest mosaic. *Science* 281: 103-105. *A.T.:* genetic markers.

66. Alexander, Donald, 1990. Bioregionalism: science or sensibility? *Environmental Ethics* 12: 161-171. [1].

67. Algeo, Thomas J., R.A. Berner, et al., March 1995. Late Devonian oceanic anoxic events and biotic crises: "rooted" in the evolution of vascular land plants? *GSA Today* 5(3): 45, 64-66. [1] *A.T.:* biogeochemical cycles.

68. Allan, David G., J.A. Harrison, et al., 1997. The impact of commercial afforestation on bird populations in Mpumalanga Province, South Africa—ir.sights from bird-atlas data. *Biological Conservation* 79: 173-185.

69. Allan, J. David, & A.S. Flecker, 1993. Biodiversity conservation in running waters. *Bioscience* 43: 32-43. [2] *A.T.:* riverine species; habitat loss.

70. Allan, R.J., & M.A. Zarull, 1995. Impacts on biodiversity and species changes in the lower Great Lakes. *Lakes & Reservoirs: Research and Management* 1: 157-162. [1] *A.T.:* environmental degradation.

71. Allard, R.W., 1990. The genetics of host-pathogen coevolution: implications for genetic resource conservation. *Heredity* 81: 1-6. [1] *A.T.:* genetic variation.

72. Allem, Antonio C., 1997. Roadside habitats: a missing link in the conservation agenda. *Environmentalist* 17: 7-10. [1] *A.T.:* Brazil.
73. Allen, Douglas C., 1992. Biodiversity and forestry. *Proc. of the Society of American Foresters National Convention 1992*: 26-34. [1].
74. Allen, Edith B., M.F. Allen, et al., 1995. Patterns and regulation of mycorrhizal plant and fungal diversity. *Plant and Soil* 170: 47-62. [2].
75. Allen, Glover M., 1942. *Extinct and Vanishing Mammals of the Western Hemisphere, With the Marine Species of All the Oceans*. Cambridge, MA: Special Publication No. 11, American Committee for International Wild Life Protection. 620p. [2/-/1].
76. Allen, Joel A., 1876. Geographical variation among North American mammals, especially in respect to size. *Bull. of the United States Geological and Geographical Survey of the Territories* 2: 309-344. [1] The basis of Allen's rule.
77. Allen, J.C., W.M. Schaffer, & D. Rosko, 1993. Chaos reduces species extinction by amplifying local population noise. *Nature* 364: 229-232. [2].
78. Allen, Keith C., & D.E.G. Briggs, eds., 1989. *Evolution and the Fossil Record*. London: Belhaven Press. 265p. [2/2/1] *A.T.:* paleobiology.
79. Allen, Leslie, Sept.-Oct. 1992. Plugging the gaps. *Nature Conservancy* 42(5): 8-9. [1] *A.T.:* GIS; gap analysis.
80. Allen, T.F.H., & T.W. Hoekstra, 1990. The confusion between scale-defined levels and conventional levels of organization in ecology. *J. of Vegetation Science* 1: 5-12. [1] *A.T.:* plants; population ecology.
81. ———, 1992. *Toward a Unified Ecology*. New York: Columbia University Press. 384p. [2/2/2] *A.T.:* philosophy of ecology.
82. Allen, William H., 1988. Biocultural restoration of a tropical forest: architects of Costa Rica's emerging Guanacaste National Park plan to make it an integral part of local culture. *Bioscience* 38: 156-161. [1].
83. ———, 1994. Reintroduction of endangered plants. *Bioscience* 44: 65-68. [1].
84. ———, April 1995. The reintroduction myth. *American Horticulturist* 74(4): 33-37. [1] Discusses the effectiveness of reintroduction operations.
85. Allen-Wardell, G., P. Bernhardt, et al., 1998. The potential consequences of pollinator declines on the conservation of biodiversity and stability of food crop yields. *CB* 12: 8-17. *A.T.:* honey bees.
86. Allison, G.W., Jane Lubchenco, & M.H. Carr, 1998. Marine reserves are necessary but not sufficient for marine conservation. *Ecological Applications* 8: S79-S92.
87. Allison, Peter A., & D.E.G. Briggs, 1993. Paleolatitudinal sampling bias, Phanerozoic species diversity, and the end-Permian extinction. *Geology* 21: 65-68. [1] *A.T.:* latitudinal diversity gradients.
88. Allkin, B., & P. Winfield, 1993. Cataloguing biodiversity: new approaches to old problems. *Biologist* 40: 179-183. [1] *A.T.:* bd databases;

information systems.

89. Allner, Thomas, Robert Backhaus, et al., 1992. Biodiversity indication by Earth observation data. In *Environment Observation and Climate Modelling Through International Space Projects* (Paris: European Space Agency), Vol. 2: 863-867. [1] *A.T.:* bd monitoring; remote sensing.

90. Allsopp, Dennis, R.R. Colwell, & D.L. Hawksworth, eds., 1995. *Microbial Diversity and Ecosystem Function* [workshop proceedings]. Wallingford, Oxon, U.K.: CAB International. 482p. [1/1/1].

91. Almeda, Frank, & C.M. Pringle, eds., 1988. *Tropical Rainforests: Diversity and Conservation* [symposium papers]. San Francisco: California Academy of Sciences and AAAS. 306p. [1/1/1].

92. Alpert, Peter, 1993. Conserving biodiversity in Cameroon. *Ambio* 22: 44-49. [1] *A.T.:* international cooperation.

93. ———, 1993. Support for biodiversity research from the US Agency for International Development. *Bioscience* 43: 628-631. [1].

94. ———, 1996. Integrated conservation and development projects: examples from Africa. *Bioscience* 46: 845-855. [1] *A.T.:* ICDPs; bdc; developing countries.

95. Alroy, John, 1992. Conjunction among taxonomic distributions and the Miocene mammalian biochronology of the Great Plains. *Paleobiology* 18: 326-343. [1].

96. Altieri, Miguel A., 1991. How best can we use biodiversity in agroecosystems? *Outlook on Agriculture* 20: 15-23. [1] *A.T.:* insect pests; Latin America; traditional agriculture.

97. ———, 1992. Sustainable agricultural development in Latin America: exploring the possibilities. *Agriculture, Ecosystems & Environment* 39: 1-21. [1] Examines issues confronting attainment of sustainable agriculture in Latin America. *A.T.:* developing countries.

98. ———, 1993. Ethnoscience and biodiversity: key elements in the design of sustainable pest management systems for small farmers in developing countries. *Agriculture, Ecosystems & Environment* 46: 257-272. [1] *A.T.:* sustainable agriculture; integrated pest management.

99. ———, 1994. *Biodiversity and Pest Management in Agroecosystems.* New York: Food Products Press. 185p. [1/1/1] *A.T.:* agricultural ecology; insect pests.

100. ———, 1995. *Agroecology: The Science of Sustainable Agriculture* (2nd ed.). Boulder, CO: Westview Press. 433p. [2/1/2].

101. Altieri, Miguel A., & L.C. Merrick, 1987. *In situ* conservation of crop genetic resources through maintenance of traditional farming systems. *Economic Botany* 41: 86-96. [1] *A.T.:* developing countries; rural development projects.

102. Altieri, Miguel A., M.K. Anderson, & L.C. Merrick, 1987. Peasant agriculture and the conservation of crop and wild plant resources. *CB* 1: 49-

58. [1] *A.T.:* traditional agriculture.

103. Alvarez, Luis W., Walter Alvarez, et al., 1980. Extraterrestrial cause for the Cretaceous-Tertiary extinction. *Science* 208: 1095-1108. [3] The study that propelled the asteroid collision hypothesis of dinosaur extinction into prominence.

104. Alvarez, Walter, 1997. *T. Rex and the Crater of Doom.* Princeton, NJ: Princeton University Press. 185p. [3/3/-] *A.T.:* geological catastrophes; extinction.

105. Alvarez, Walter, & Frank Asaro, Oct. 1990. What caused the mass extinction: an extraterrestrial impact. *Scientific American* 263(4): 78-84. [1].

106. Alverson, William S., Walter Kuhlmann, & D.M. Waller, 1994. *Wild Forests: Conservation Biology and Public Policy.* Washington, DC: Island Press. 300p. [2/1/1] *A.T.:* forest management.

107. Amann, R.I., 1995. Fluorescently labeled, ribosomal RNA-targeted oligonucleotide probes in the study of microbial ecology. *Molecular Ecology* 4: 543-553. [2].

108. Ameziane, N., & M. Roux, 1997. Biodiversity and historical biogeography of stalked crinoids (Echinodermata) in the deep sea. *Biodiversity and Conservation* 6: 1557-1570.

109. Amos, Bill, & A.R. Hoelzel, 1992. Applications of molecular genetic techniques to the conservation of small populations. *Biological Conservation* 61: 133-144. [1] *A.T.:* DNA fingerprinting; genetic variability.

110. Amos, Nevil, J.B. Kirkpatrick, & Melissa Giese, 1993. *Conservation of Biodiversity, Ecological Integrity and Ecologically Sustainable Development: A Discussion Paper.* Fitzroy, Victoria, Australia: Australian Conservation Foundation. 116p. [1/1/1] *A.T.:* Australia.

111. Anadu, P.A., P.O. Elamah, & J.F. Oates, 1988. The bushmeat trade in southwestern Nigeria: a case study. *Human Ecology* 16: 199-208. [1] *A.T.:* hunting.

112. Andersen, Alan N., 1992. Regulation of "momentary" diversity by dominant species in exceptionally rich ant communities of the Australian seasonal tropics. *American Naturalist* 140: 401-420. [2].

113. ———, 1995. Measuring more of biodiversity: genus richness as a surrogate for species richness in Australian ant faunas. *Biological Conservation* 73: 39-43. [1] *A.T.:* higher taxa.

114. Andersen, R.A., 1992. Diversity of eukaryotic algae. *Biodiversity and Conservation* 1: 267-292. [2].

115. Anderson, Anthony B., ed., 1990. *Alternatives to Deforestation: Steps Toward Sustainable Use of the Amazon Rain Forest* [conference papers]. New York: Columbia University Press. 281p. [2/2/1].

116. Anderson, H. Michael, winter 1994. Reforming national-forest policy. *Issues in Science and Technology* 10(2): 40-47. [1] *A.T.:* forest management; Forest Service.

117. Anderson, I., 25 Nov. 1995. Oceans plundered in the name of medicine [editorial]. *New Scientist* 148(2005): 5. [1] *A.T.:* bioprospecting.

118. Anderson, John, Heidi Anderson, et al., 1996. The Triassic explosion(?): a statistical model for extrapolating biodiversity based on the terrestrial Molteno Formation. *Paleobiology* 22: 318-328. [1] Estimates floral and faunal diversities in Late Triassic South Africa.

119. Anderson, M.J., & A.J. Underwood, 1997. Effects of gastropod grazers on recruitment and succession of an estuarine assemblage: a multivariate and univariate approach. *Oecologia* 109: 442-453. *A.T.:* New South Wales.

120. Anderson, M. Kat., 1997. California's endangered peoples and endangered ecosystems. *American Indian Culture and Research Journal* 21(3): 7-31. [1] Describes the relation between Native American cultures and Californian bd.

121. Anderson, Sydney, 1985. *The Theory of Range Size (RS) Distributions.* New York: American Museum of Natural History Novitates No. 2833. 20p. [1/1/1] *A.T.:* zoogeography.

122. ———, 1994. Area and endemism. *Quarterly Review of Biology* 69: 451-471. [1] *A.T.:* zoogeography.

123. Anderson, Sydney, & L.F. Marcus, 1993. Effect of quadrat size on measurements of species density. *J. of Biogeography* 20: 421-428. [1] *A.T.:* Australia; mammals.

124. Anderson, Terry L., & P.J. Hill, eds., 1995. *Wildlife in the Marketplace: The Political Economy Forum.* Lanham, MD: Rowman & Littlefield. 191p. [2/2/2] *A.T.:* wild animal trade.

125. Andersson, Lennart, 1996. An ontological dilemma: epistemology and methodology of historical biogeography [editorial]. *J. of Biogeography* 23: 269-277. [1] *A.T.:* vicariance.

126. Ando, A., J. Camm, et al., 1998. Species distributions, land values, and efficient conservation. *Science* 279: 2126-2128. *A.T.:* ecological economics.

127. Andow, D.A., 1991. Vegetational diversity and arthropod population response. *Annual Review of Entomology* 36: 561-586. [2] *A.T.:* diversity-stability hypothesis; herbivory; polyculture.

128. Andow, D.A., & O. Imura, 1994. Specialization of phytophagous arthropod communities on introduced plants. *Ecology* 75: 296-300. [1] *A.T.:* Japan; biological invasions.

129. Andow, D.A., P.M. Kareiva, et al., 1990. Spread of invading organisms. *Landscape Ecology* 4: 177-188. [2] *A.T.:* biological invasions; diffusion modelling; range expansion.

130. André, Henri M., François Brechignac, & Pierre Thibault, 1994. Biodiversity in model ecosystems [letter]. *Nature* 371: 565. [1]. *A.T.:* microcosms.

131. André, Henri M., P. Lebrun, & M.-I. Noti, 1992. Biodiversity in Africa: a plea for more data [editorial]. *J. of African Zoology* 106: 3-16. [1] *A.T.:* research programs.

132. André, Henri M., M.-I. Noti, & P. Lebrun, 1994. The soil fauna: the other last biotic frontier. *Biodiversity and Conservation* 3: 45-56. [1] *A.T.:* microarthropods; coastal dunes; France.

133. Andreev, Alexander, 1995. Bird fauna of northeast Asia: a summary of the unique biodiversity and the priorities for conservation. *Ibis* 137, Suppl. 1: S195-S197. [1] *A.T.:* Russia.

134. Andrén, Henrik, 1994. Effects of habitat fragmentation on birds and mammals in landscapes with different proportions of suitable habitat: a review. *Oikos* 71: 355-366. [3] *A.T.:* landscape ecology.

135. Andrewartha, Herbert G., & L.C. Birch, 1954. *The Distribution and Abundance of Animals*. Chicago: University of Chicago Press. 782p. [2/-/3] *A.T.:* animal ecology.

136. Andrews, C., 1990. The ornamental fish trade and fish conservation. *J. of Fish Biology* 37, Suppl. A: 53-59. [1] *A.T.:* captive breeding; wild animal trade.

137. Andrews, P.R., R. Borris, et al., 1996. Preservation and utilization of natural biodiversity in context of search for economically valuable medicinal biota. *Pure and Applied Chemistry* 68: 2325-2332. [1] IUPAC position statement.

138. Andrus, Richard E., E.F. Karlin, & S.S. Talbot, 1992. Rare and endangered *Sphagnum* species in North America. *Biological Conservation* 59: 247-254. [1] *A.T.:* mosses.

139. Angel, Martin V., 1991. Variations in time and space: is biogeography relevant to studies of long-time scale change? *J. of the Marine Biological Association* 71: 191-206. [1] *A.T.:* Northeast Atlantic; ostracods.

140. ———, 1993. Biodiversity of the pelagic ocean. *CB* 7: 760-772. [1] *A.T.:* species richness; latitudinal diversity gradients.

141. Angelsen, Arild, 1995. Shifting cultivation and deforestation: a study from Indonesia. *World Development* 23: 1713-1729. [1] *A.T.:* Sumatra.

142. Angelstam, P.K., V.M. Anufriev, et al., 1997. Biodiversity and sustainable forestry in European forests: how East and West can learn from each other. *Wildlife Society Bull.* 25: 38-48. [1] *A.T.:* forest management; Russia; Sweden.

143. Angermeier, Paul L., 1994. Does biodiversity include artificial diversity? *CB* 8: 600-602. [1].

144. ———, 1995. Ecological attributes of extinction-prone species: loss of freshwater fishes of Virginia. *CB* 9: 143-158. [1].

145. Angermeier, Paul L., & J.R. Karr, 1994. Biological integrity versus biological diversity as policy directives. *Bioscience* 44: 690-697. [2].

146. Angermeier, Paul L., & I.J. Schlosser, 1989. Species-area relation-

ships for stream fishes. *Ecology* 70: 1450-1462. [2] *A.T.:* Minnesota; Illinois; Panama.

147. Angermeier, Paul L., & J.E. Williams, Jan. 1994. Conservation of imperiled species and reauthorization of the Endangered Species Act of 1973. *Fisheries* 19(1): 26-29. [1] American Fisheries Society policy statement.

148. Angermeier, Paul L., & M.R. Winston, 1997. Assessing conservation value of stream communities: a comparison of approaches based on centres of density and species richness. *Freshwater Biology* 37: 699-710. *A.T.:* Index of Centers of Density; Virginia; Tennessee.

149. Angle, J.S., 1994. Release of transgenic plants: biodiversity and population-level considerations. *Molecular Ecology* 3: 45-50. [1] *A.T.:* soil fauna.

150. Annand, E.M., & F.R. Thompson, 1997. Forest bird response to regeneration practices in Central hardwood forests. *J. of Wildlife Management* 61: 159-171. *A.T.:* Missouri Ozarks; breeding birds.

151. Anonymous, 1984. Action plan for biosphere reserves. *Nature and Resources* 20(4): 11-22. [1] *A.T.:* MAB; nongovernmental organizations.

152. ———, July-Aug. 1991. Endangered marine finfish: a useful concept? [report] *Fisheries* 16(4): 23-26. [1] Discusses the criteria for vulnerability applied to threatened finfish.

153. ———, 1991. AIBS mission, goals, and objectives—1991. *Bioscience* 41: 791-792. [1].

154. ———, 1991. What biodiversity? There is a danger that the concept of biodiversity will become an aspect of political correctness [editorial]. *Nature* 352: 2. [1] Concerning the arguments for bdc.

155. ———, Feb. 1992. Biological diversity in forest ecosystems: a position of the Society of American Foresters. *J. of Forestry* 90(2): 42-44. [1].

156. ———, 13 June 1992. Biodivisive: the Earth Conference. *Economist* 323(7763): 93-94. [1] Short overview of UNCED and CBD.

157. ———, Sept.-Oct. 1992. Last stand for many species. *Futurist* 26(5): 53-54. [1] *A.T.:* deforestation; human population growth.

158. ———, July 1993. Halting Hawaiian extinctions. *American Horticulturist* 72(7): 12. [1] *A.T.:* endangered plants; Center for Plant Conservation; environmental degradation.

159. ———, 2 Oct. 1993. Buying diversity. *Economist* 329(7831): 18. [1] Concerning operations of the GEF. *A.T.:* World Bank.

160. ———, 19 March 1994. Fish: the tragedy of the oceans. *Economist* 330(7855): 21-24. [1] *A.T.:* marine fisheries; FAO.

161. ———, March 1994. An Appalachian deathwatch? *Environment* 36(2): 22-23. [1] Considers ozone pollution and excess nitrogen deposition and their effects on Appalachian forests.

162. ———, 5 Sept. 1994. China's marine nature reserves. *Beijing Re-*

view 37(36): 29-31. [1].

163. ———, Dec. 1995. Death and taxa. *Environment* 37(10): 26-27. [1] Discusses reauthorization of the Endangered Species Act.

164. ———, Aug. 1996. Keeping track of nature. *New York State Conservationist* 51(1): 12-13. [1] Describes the Return a Gift to Wildlife program.

165. ———, 1 Sept. 1996. A recipe for dispossession [interview]. *UNESCO Courier* 49(8): 20-23. [1] The views of Smitu Kothari on bd and cultural pluralism.

166. ———, 1996. Biodiversity. *J. of Soil and Water Conservation* 51: 204-205. [1] Dwells on the causes and effects of environmental deterioration.

167. ———, 1998. Mapping out the path to biodiversity. *Nature* 392: 643. Describes 1998 Mexico conference that considered means of implementing the CBD.

168. Anton, Danilo J., 1995. *Diversity, Globalization, and the Ways of Nature*. Ottawa: International Development Research Centre. 223p. [1/1/1] *A.T.:* bdc.

169. Anton, Donald K., 1997. Law for the sea's biological diversity. *Columbia J. of Transnational Law* 36: 341-371. *A.T.:* commons.

170. Anyinam, Charles, 1995. Ecology and ethnomedicine: exploring links between current environmental crisis and indigenous medical practices. *Social Science & Medicine* 40: 321-329. [1] *A.T.:* environmental degradation.

171. Aplet, Gregory H., R.D. Laven, & P.L. Fiedler, 1992. The relevance of conservation biology to natural resource management. *CB* 6: 298-300. [1] *A.T.:* ecosystem management.

172. Aradanas, J.S., 1998. Aboriginal whaling—biological diversity meets cultural diversity. *Northwest Science* 72: 142-145. *A.T.:* Washington; gray whales.

173. Arai, Ryoichi, Masahiro Kato, & Yoshimichi Doi, eds., 1995. *Biodiversity and Evolution* [symposium proceedings]. Tokyo: National Science Museum Foundation. 336p. [1/1/1].

174. Archer, Michael, & Georgina Clayton, eds., 1984. *Vertebrate Zoogeography and Evolution in Australasia (Animals in Space and Time)*. Carlisle, Western Australia: Hesperian Press. 1203p. [1/1/1].

175. Archibald, J. David, 1993. The importance of phylogenetic analysis for the assessment of species turnover: a case history of Paleocene mammals in North America. *Paleobiology* 19: 1-27. [1] *A.T.:* cladistics.

176. ———, 1996. *Dinosaur Extinction and the End of an Era: What the Fossils Say*. New York: Columbia University Press. 237p. [2/2/1] *A.T.:* Cretaceous-Tertiary boundary.

177. Ardern, S.L., & D.M. Lambert, 1997. Is the black robin in genetic peril? *Molecular Ecology* 6: 21-28. *A.T.:* population bottlenecks; conservation genetics.

178. Arends, Amy, Jan Cerovsky, & Gabriela Pickova, eds., 1995. *Transboundary Biodiversity Conservation: Selected Case Studies From Central Europe*. Praha: Ecopoint. 48p. [1/1/1].

179. Argus, George W., 1992. The phytogeography of rare vascular plants in Ontario and its bearing on plant conservation. *Canadian J. of Botany* 70: 469-490. [1].

180. Ariansen, Per, 1997. *The Non-utility Value of Nature: A Contribution to Understanding the Value of Biological Diversity*. Ås, Norway: Meddelelser fra Skogforsk 47.19. 45p. [1/1/1].

181. Arita, Héctor T., 1993. Conservation biology of the cave bats of Mexico. *J. of Mammalogy* 74: 693-702. [1].

182. Arita, Héctor T., J.G. Robinson, & K.H. Redford, 1990. Rarity in Neotropical forest mammals and its ecological correlates. *CB* 4: 181-192. [2].

183. Armesto, J.J., & S.T.A. Pickett, 1985. Experiments on disturbance in old-field plant communities: impact on species richness and abundance. *Ecology* 66: 230-240. [2] *A.T.:* New Jersey.

184. Armstrong, A.J., & H.J. van Hensbergen, 1997. Evaluation of afforestable montane grasslands for wildlife conservation in the north-eastern Cape, South Africa. *Biological Conservation* 81: 179-190.

185. Arnold, R.J., & S.J. Midgley, 1995. Conserving biodiversity—the work of the Australian Tree Seed Centre. *Commonwealth Forestry Review* 74: 121-128. [1].

186. Arrandale, Tom, 1988. Saving America's biological diversity. *Editorial Research Reports* 1: 70-79. [1] *A.T.:* Endangered Species Act.

187. ———, 1991. Endangered species. *CQ Researcher* 1: 395-415. [1] Considers the tradeoff between bdc and economic sacrifices. *A.T.:* Endangered Species Act.

188. Arrhenius, O., 1921. Species and area. *J. of Ecology* 9: 95-99. [2] On the relation between size of area and number of species.

189. Arrow, Kenneth, Bert Bolin, et al., 1995. Economic growth, carrying capacity, and the environment. *Science* 268: 520-521. [2] *A.T.:* ecological economics.

190. Arthington, Angela H., 1991. Ecological and genetic impacts of introduced and translocated freshwater fishes in Australia. *Canadian J. of Fisheries and Aquatic Sciences* 48, Suppl. 1: 33-43. [1].

191. Arthur, Jeffrey L., & Mark Hachey, 1997. Finding all optimal solutions to the reserve site selection problem: formulation and computational analysis. *Environmental and Ecological Statistics* 4: 153-165. *A.T.:* maximal covering problem (MCP); Oregon.

192. Arts, Bas, 1998. *The Political Influence of Global NGOs: Case Studies on the Climate and Biodiversity Conventions*. Utrecht: International Books. 352p. *A.T.:* CBD; nongovernmental organizations.

193. Artson, Bradley S., Sept.-Oct. 1997. Each after their own kind. *Tikkun* 12(5): 43-45. Examines relation of bdc to Judaism.

194. Artuso, Anthony, 1997. *Drugs of Natural Origin: Economic and Policy Aspects of Discovery, Development, and Marketing.* New York: Pharmaceutical Products Press. 201p. [1/1/-].

195. Asebey, Edgar J., & J.D. Kempenaar, 1995. Biodiversity prospecting: fulfilling the mandate of the Biodiversity Convention. *Vanderbilt J. of Transnational Law* 28: 703-754. [1].

196. Ashmore, M.R., & J.N.B. Bell, 1991. The role of ozone in global change. *Annals of Botany* 67, Suppl. 1: 39-48. [1] *A.T.:* climatic change.

197. Asian Development Bank, 1995. *Biodiversity Conservation in the Asia and Pacific Region: Constraints and Opportunities* [conference proceedings]. Manila. 508p. [1/1/1].

198. Askins, Robert A., 1993. Population trends in grassland, shrubland, and forest birds in eastern North America. *Current Ornithology* 11: 1-34. [2].

199. ———, 1994. Open corridors in a heavily forested landscape: impact on shrubland and forest-interior birds. *Wildlife Society Bull.* 22: 339-347. [2] *A.T.:* habitat fragmentation.

200. ———, 1995. Hostile landscapes and the decline of migratory songbirds. *Science* 267: 1956-1957. [1] *A.T.:* environmental indicators.

201. Askins, Robert A., J.F. Lynch, & Russell Greenberg, 1990. Population declines in migratory birds in eastern North America. *Current Ornithology* 7: 1-57. [2].

202. Askins, Robert A., M.J. Philbrick, & D.S. Sugeno, 1987. Relationship between the regional abundance of forest and the composition of forest bird communities. *Biological Conservation* 39: 129-152. [2] *A.T.:* forest remnants; Connecticut.

203. Aspinall, R., & K. Matthews, 1994. Climate change impact on distribution and abundance of wildlife species: an analytical approach using GIS. *Environmental Pollution* 86: 217-223. [1] *A.T.:* Scotland.

204. Assis, Luiz Fernando Soares de, 1991. *A Regional View of Negotiations on Biodiversity.* United Nations Economic Commission for Latin America and the Caribbean. 35p. [1/1/1] *A.T.:* Latin America.

205. Athanasiou, Tom, Oct.-Dec. 1992. After the summit. *Socialist Review* 22(4): 56-92. [1] Reviews major political forces shaping implementation of UNCED ideals.

206. Atkinson, I.A.E., 1988. Presidential address: opportunities for ecological restoration [editorial]. *New Zealand J. of Ecology* 11: 1-12. [1] *A.T.:* New Zealand.

207. Atran, Scott, 1993. Itza Maya tropical agro-forestry. *Current Anthropology* 34: 633-700. [1] *A.T.:* traditional agriculture.

208. Attfield, Robin, 1991. *The Ethics of Environmental Concern* (2nd ed.). Athens, GA: University of Georgia Press. 249p. [2/2/2] *A.T.:* environ-

mental ethics.

209. Attridge, Ian, ed., 1996. *Biodiversity Law and Policy in Canada: Review and Recommendations*. Toronto: Canadian Institute for Environmental Law and Policy. 506, 24p. [1/1/1].

210. Attrill, M.J., P.M. Ramsay, et al., 1996. An estuarine biodiversity hot-spot. *J. of the Marine Biological Association* 76: 161-175. [1].

211. Ault, T.R., & C.R. Johnson, 1998. Spatial variation in fish species richness on coral reefs: habitat fragmentation and stochastic structuring processes. *Oikos* 82: 354-364. *A.T.:* patch reefs.

212. Austin, Daniel F., 1978. Exotic plants and their effects in southeastern Florida. *Environmental Conservation* 5: 25-34. [1] *A.T.:* introduced plants.

213. Austin, Michael P., 1998. An ecological perspective on biodiversity investigations: examples from Australian eucalypt forests. *Annals of the Missouri Botanical Garden* 85: 2-17. *A.T.:* New South Wales; BioRap.

214. Austin, Michael P., & J.A. Meyers, 1996. Current approaches to modelling the environmental niche of eucalypts: implication for management of forest biodiversity. *Forest Ecology and Management* 85: 95-106. *A.T.:* Generalised Linear Modelling; Generalised Additive Modelling.

215. Australian and New Zealand Environment and Conservation Council, 1996. *The National Stategy for the Conservation of Australia's Biological Diversity* [conference papers]. Canberra: Commonwealth Dept. of the Environment, Sport, and Territories. 54p. [1/1/1] *A.T.:* environmental policy.

216. Australian National Parks and Wildlife Service, 1991. *Plant Invasions: The Incidence of Environmental Weeds in Australia*. Canberra. 188p. [1/1/1].

217. Ausubel, Ken, 1994. *Seeds of Change: The Living Treasure: The Passionate Story of the Growing Movement to Restore Biodiversity and Revolutionize the Way We Think About Food*. San Francisco: HarperSanFrancisco. 232p. [2/2/1] *A.T.:* food crops.

218. Avery, Dennis T., 1993. *Biodiversity: Saving Species With Biotechnology*. Indianapolis: Hudson Institute. 44p. [1/1/1] *A.T.:* food; agricultural technology.

219. ———, fall 1997. Saving nature's legacy through better farming. *Issues in Science and Technology* 14(1): 59-64. *A.T.:* agricultural ecology; bdc.

220. Avise, John C., 1995. Mitochondrial DNA polymorphism and a connection between genetics and demography of relevance to conservation. *CB* 9: 686-690. [2].

221. Avise, John C., & J.L. Hamrick, eds., 1996. *Conservation Genetics: Case Histories From Nature* [case studies]. New York: Chapman & Hall. 512p. [1/1/2] *A.T.:* animal germplasm resources.

222. Aylward, Bruce A., 1993. *The Economic Value of Pharmaceutical*

Prospecting and Its Role in Biodiversity Conservation. London: Discussion Paper DP 93-05, London Environmental Economics Centre. 76p. [1/1/1] *A.T.:* bioprospecting.

223. Aylward, Bruce A., & E.B. Barbier, 1992. Valuing environmental functions in developing countries. *Biodiversity and Conservation* 1: 34-50. [1].

224. Ayres, J.M., R.E. Bodmer, & R.A. Mittermeier, 1991. Financial considerations of reserve design in countries with high primate diversity. *CB* 5: 109-114. [1] *A.T.:* island biogeography; Brazil; Madagascar; Indonesia.

225. Ayres, R.U., 1993. Cowboys, cornucopians and long-run sustainability. *Ecological Economics* 8: 189-207. [1] Concerning the debate between neo-Malthusians and technological solutions advocates.

226. Azeta, Masanori, et al., eds., 1997. *Biodiversity and Aquaculture for Sustainable Development* [symposium proceedings]. Nansei, Mie, Japan: National Research Institute of Aquaculture and Fisheries Agency. 190p. [1/-/-].

227. Babbitt, Bruce, Jan.-Feb. 1994. Protecting biodiversity. *Nature Conservancy* 44(1): 16-21. [1] Speaks out against "us-versus-them" thinking in dealing with environmental issues.

228. Bachmann, Peter, Michael Köhl, & Risto Päivinen, eds., 1998. *Assessment of Biodiversity for Improved Forest Planning* [conference proceedings]. Dordrecht and Boston: Kluwer Academic. 421p.

229. Baer, Karen W., 1995. A theory of intellectual property and the biodiversity treaty. *Syracuse J. of International Law and Commerce* 21: 259-281. [1] *A.T.:* international law.

230. Bagarinao, T., 1998. Nature parks, museums, gardens, and zoos for biodiversity conservation and environment education: the Philippines. *Ambio* 27: 230-237.

231. Bahls, Peter, 1992. The status of fish populations and management of high mountain lakes in the western United States. *Northwest Science* 66: 183-193. [1] *A.T.:* stocking.

232. Bailey, Robert G., 1995. *Description of the Ecoregions of the United States* (2nd ed.). Washington, DC: Forest Service, USDA. 108p. [2/1/1].

233. ———, 1998. *Ecoregions: The Ecosystem Geography of the Oceans and Continents.* New York: Springer. 176p.

234. Bailey, Ronald, ed., 1995. *The True State of the Planet.* New York: Free Press. 472p. [2/2/1].

235. Bailey, R.C., 1988. Correlations between species richness and exposure: freshwater molluscs and macrophytes. *Hydrobiologia* 162: 183-191. [1].

236. Baillie, Jonathan, & Brian Groombridge, eds., 1996. *1996 IUCN Red List of Threatened Animals.* Gland, Switzerland: IUCN. 448p. [2/1/2].

237. Baker, C.S., A. Perry, et al., 1993. Abundant mitochondrial DNA variation and world-wide population structure in humpback whales. *PNAS*

90: 8239-8243. [2] *A.T.:* population bottlenecks; hunting.

238. Baker, Douglas S., L.M. Ferreira, & P.W. Saile, eds., 1997. *Proceedings and Papers of the International Workshop on Biodiversity Monitoring in Federal Protected Areas: Defining the Methodology.* Brasilia: Instituto Brasileiro do Meio Ambiente e dos Recursos Naturais Renováveis (IBAMA). 246p. [1/1/-].

239. Baker, J.T., J.D. Bell, & P.T. Murphy, 1996. Australian deliberations on access to its terrestrial and marine biodiversity. *J. of Ethnopharmacology* 51: 229-237. [1] *A.T.:* intellectual property.

240. Baker, J.T., R.P. Borris, et al., 1995. Natural product drug discovery and development—new perspectives on international collaboration. *J. of Natural Products* 58: 1325-1357. [1].

241. Baker, Robert J., 1994. Some thoughts on conservation, biodiversity, museums, molecular characters, systematics, and basic research. *J. of Mammalogy* 75: 277-287. [1] *A.T.:* mammals; phylogenetics; molecular genetics.

242. Baker, R., 1991. Diversity in biological control. *Crop Protection* 10: 85-94. [1].

243. Baker, William L., 1989. Landscape ecology and the nature reserve design in the Boundary Waters Canoe Area, Minnesota. *Ecology* 70: 23-35. [2] *A.T.:* patch dynamics; fire ecology.

244. ———, 1992. Effects of settlement and fire suppression on landscape structure. *Ecology* 73: 1879-1887. [1] *A.T.:* Minnesota; landscape ecology.

245. Balakrishnan, Mundanthra, Reidar Borgström, & S.W. Bie, eds., 1994. *Tropical Ecosystems: A Synthesis of Tropical Ecology and Conservation.* Lebanon, NH: International Science Publishers. 441p. [1/1/1] *A.T.:* rainforest conservation; rainforest management.

246. Balandrin, Manuel F., J.A. Klocke, et al., 1985. Natural plant chemicals: sources of industrial and medicinal materials. *Science* 228: 1154-1160. [2] *A.T.:* biotechnology; natural products.

247. Balchand, Alungal N., 1994. A case-study on biodiversity: pesticide engendering of ecological imbalance and resultant environmental consequences. *Environmental Conservation* 21: 164-166. [1] *A.T.:* environmental disturbance.

248. Baldi, A., & T. Kisbenedek, 1997. Orthopteran assemblages as indicators of grassland naturalness in Hungary. *Agriculture, Ecosystems & Environment* 66: 121-129. *A.T.:* ecological disturbance.

249. Baldwin, A. Dwight, Jr., Judith De Luce, & Carl Pletsch, eds., 1994. *Beyond Preservation: Restoring and Inventing Landscapes* [conference papers]. Minneapolis: University of Minnesota Press. 280p. [1/1/1] *A.T.:* restoration ecology; landscape restoration.

250. Balick, Michael J., 1990. Ethnobotany and the identification of therapeutic agents from the rainforest. *Ciba Foundation Symposium*, no. 154: 22-

39. [1] *A.T.:* medicinal plants.

251. ———, 1994. Ethnobotany, drug development and biodiversity conservation—exploring the linkages. *Ciba Foundation Symposium,* no. 185: 4-18. [1].

252. Balick, Michael J., & Robert Mendelsohn, 1992. Assessing the economic value of traditional medicines from tropical rain forests. *CB* 6: 128-130. [2] *A.T.:* medicinal plants; Belize.

253. Balick, Michael J., Rosita Arvigo, & Leopoldo Romero, 1994. The development of an ethnobiomedical forest reserve in Belize: its role in the preservation of biological and cultural diversity. *CB* 8: 316-317. [1].

254. Balick, Michael J., Elaine Elisabetsky, & S.A. Laird, eds., 1996. *Medicinal Resources of the Tropical Forest: Biodiversity and Its Importance to Human Health.* New York: Columbia University Press. 440p. [2/2/1] *A.T.:* traditional medicine; ethnobotany.

255. Ball, George E., & H.V. Danks, eds., 1993. *Systematics and Entomology: Diversity, Distribution, Adaptation, and Application* [symposium papers]. Ottawa: Memoirs of the Entomological Society of Canada No. 165. 272p. [1/1/1].

256. Ballou, Jonathan D., M.E. Gilpin, & T.J. Foose, eds., 1995. *Population Management for Survival and Recovery: Analytical Methods and Strategies in Small Population Conservation.* New York: Columbia University Press. 375p. [1/1/1] *A.T.:* wildlife conservation.

257. Balmford, Andrew, 1996. Extinction filters and current resilience: the significance of past selection pressures for conservation biology. *TREE* 11: 193-196. [1].

258. Balmford, Andrew, & Adrian Long, 1994. Avian endemism and forest loss. *Nature* 372: 623-624. [1] *A.T.:* deforestation; bd loss.

259. ———, 1995. Across-country analyses of biodiversity congruence and current conservation effort in the tropics. *CB* 9: 1539-1547. [1] *A.T.:* birds.

260. Balmford, Andrew, M.J.B. Green, & M.G. Murray, 1996. Using higher-taxon richness as a surrogate for species richness. I. Regional tests. *PRSL* B 263: 1267-1274. [2] *A.T.:* taxonomic inventories.

261. Balmford, Andrew, A.H.M. Jayasuriya, & M.J.B. Green, 1996. Using higher-taxon richness as a surrogate for species richness. II. Local applications. *PRSL* B 263: 1571-1575. [1] *A.T.:* woody plants; Sri Lanka.

262. Balon, Eugene K., 1993. Dynamics of biodiversity and mechanisms of change: a plea for balanced attention to form creation and extinction. *Biological Conservation* 66: 5-16. [1] *A.T.:* speciation.

263. Baltanás, Angel, 1992. On the use of some methods for the estimation of species richness. *Oikos* 65: 484-492. [2].

264. Baltz, Donald M., 1991. Introduced fishes in marine systems and inland seas. *Biological Conservation* 56: 151-177. [1] *A.T.:* biological in-

vasions.

265. Bambach, Richard K., 1977. Species richness in marine benthic environments through the Phanerozoic. *Paleobiology* 3: 152-167. [2] *A.T.:* invertebrates.

266. Bamforth, Stuart S., 1995. Interpreting soil ciliate biodiversity. *Plant and Soil* 170: 159-164. [1].

267. Banerjee, Amit, & G.E. Boyajian, 1996. Changing biologic selectivity of extinction in the foraminifera over the past 150 m.y. *Geology* 24: 607-610. [1] *A.T.:* extinction risk.

268. Bangs, Richard, May-June 1992. Lemurs in the mist. *Wildlife Conservation* 95(3): 22-33. [1] Describes the plight of nature in Madagascar. *A.T.:* habitat degradation; deforestation.

269. Bannan, Jan, May-June 1993. The cornucopia tree. *Wildlife Conservation* 96(3): 10. [1] Notes the many uses of the neem tree of Africa and South Asia. *A.T.:* natural products.

270. Baptista, Luis F., & P.W. Trail, 1992. The role of song in the evolution of passerine diversity. *Systematic Biology* 41: 242-247. [1].

271. Baradas, Faustina C., & E.H. Belen, eds., 1993. *Conservation of Biological Diversity in the Philippines* [workshop papers]. Los Banos, Laguna, Philippines: Philippine Council for Agriculture. 73p. [1/1/1].

272. Baram, Michael, 1996. LMOs: treasure chest or Pandora's box? *Environmental Health Perspectives* 104: 704-707. [1] *A.T.:* living modified organisms.

273. Barbault, Robert, Sept. 1995. Biodiversity: stakes and opportunities. *Nature and Resources* 31(3): 18-26. [1] *A.T.:* Diversitas; CBD.

274. ———, 1995. Biodiversity dynamics: from population and community ecology approaches to a landscape ecology point of view. *Landscape and Urban Planning* 31: 89-98. [1].

275. Barbault, Robert, & M.E. Hochberg, 1992. Population and community level approaches to studying biodiversity in international research programs. *Acta Oecologica* 13: 137-146. [1].

276. Barber, Charles V., & N.C. Johnson, 1994. Getting to the roots of forest loss: lessons from Indonesia. *Environmental Science & Technology* 28: 32A-34A. [1] *A.T.:* environmental policy.

277. Barber, Charles V., Suraya Afiff, & Agus Purnomo, 1995. *Tiger by the Tail? Reorienting Biodiversity Conservation and Development in Indonesia*. Washington, DC: WRI. 61p. [1/1/1] *A.T.:* environmental policy.

278. Barber, Charles V., K.R. Miller, & W.V. Reid, Sept.-Oct. 1992. The new conservation challenge. *Defenders* 67(5): 16-23. [1] *A.T.:* CBD.

279. Barbier, Edward B., 1989. *Economics, Natural Resource Scarcity and Development: Conventional and Alternative Views*. London: Earthscan. 223p. [1/1/2] *A.T.:* environmental economics.

280. ———, 1995. The economics of forestry and conservation: eco-

nomic values and policies. *Commonwealth Forestry Review* 74: 26-34. [1] *A.T.:* valuation; forest management.

281. Barbier, Edward B., & B.A. Aylward, 1996. Capturing the pharmaceutical value of biodiversity in a developing country. *Environmental & Resource Economics* 8: 157-181. [1] *A.T.:* natural products; Costa Rica; bioprospecting.

282. Barbier, Edward B., & Carl-Erik Schulz, 1997. Wildlife, biodiversity and trade. *Environment and Development Economics* 2: 145-172. *A.T.:* resource management; international trade.

283. Barbier, Edward B., J.C. Burgess, & Carl Folke, 1994. *Paradise Lost? The Ecological Economics of Biodiversity.* London: Earthscan. 267p. [1/2/2].

284. Barbier, Edward B., J.C. Burgess, & Anil Markandya, 1991. The economics of tropical deforestation. *Ambio* 20: 55-58. [1] *A.T.:* greenhouse effect; Brazil; Zaire; Indonesia.

285. Barbier, Edward B., J.C. Burgess, et al., 1990. *Elephants, Economics, and Ivory.* London: Earthscan. 154p. [1/1/1] *A.T.:* ivory trade; wild animal trade.

286. Barbour, Clyde D., & J.H. Brown, 1974. Fish species diversity in lakes. *American Naturalist* 108: 473-489. [1].

287. Barbour, Michael G., B. Pavlik, et al., 1993. *California's Changing Landscapes: Diversity and Conservation of California Vegetation.* Sacramento: California Native Plant Society. 244p. [1/1/1].

288. Barbour, Russell, & Rene Rabezandria, 1992. *The Broken Forest: Applying the Integrated Conservation and Development Paradigm to Madagascar's Protected Areas.* Bethesda, MD: DAI. 2 vols. [1/1/1] *A.T.:* forest management.

289. Barbujani, Guido, Paolo Vian, & Luigi Fabbris, 1992. Cultural barriers associated with large gene frequency differences among Italian populations. *Human Biology* 64: 479-495. [1] *A.T.:* human genetic diversity; gene flow.

290. Barde, Jean-Philippe, & David W. Pearce, eds., 1991. *Valuing the Environment: Six Case Studies.* London: Earthscan. 271p. [1/1/1] *A.T.:* environmental policy.

291. Bardman, Cynthia A., 1997. Applicability of biodiversity impact assessment methodologies to transportation projects. *Transportation Research Record,* no. 1601: 35-41.

292. Barel, C.D.N., R. Dorit, et al., 1985. Destruction of fisheries in Africa's lakes. *Nature* 315: 19-20. [2] *A.T.:* biological invasions; Lake Victoria; Lake Malawi; Nile perch.

293. Baringa, Marcia, 1990. Where have all the froggies gone? *Science* 247: 1033-1034. [2] Reports on the alarming worldwide decline in amphibian numbers.

294. Barisse, M., 1985. Action plan for biosphere reserves. *Environmental Conservation* 12: 17-27. [1] *A.T.:* MAB.

295. Barker, Jerry R., & D.T. Tingey, eds., 1992. *Air Pollution Effects on Biodiversity* [workshop papers]. New York: Van Nostrand Reinhold. 322p. [1/1/1] *A.T.:* environmental policy.

296. Barker, Rocky, 1993. *Saving All the Parts: Reconciling Economics and the Endangered Species Act.* Washington, DC: Island Press. 268p. [2/2/1] *A.T.:* Pacific Northwest; Rocky Mountains.

297. ———, 1995. *Mending Fences: Lessons in Island Biodiversity Protection From Hawai'i.* Honolulu: Program on Environment, East-West Center. 46p. [1/1/1].

298. Barkham, J.P., 1995. Environmental needs and social justice. *Biodiversity and Conservation* 4: 857-868. [1] *A.T.:* North-South relations.

299. Barlow, Connie, spring 1997. Re-storying biodiversity by way of science. *Wild Earth* 7(1): 14-18. [1] *A.T.:* environmental ethics; storytelling.

300. Barlow, N.D., 1994. Size distributions of butterfly species and the effect of latitude on species sizes. *Oikos* 71: 326-332. [1].

301. Barnes, Richard S.K., 1991. Dilemmas in the theory and practice of biological conservation as exemplified by British coastal lagoons. *Biological Conservation* 55: 315-328. [1] *A.T.:* sea-level rise.

302. ———, ed., 1998. *The Diversity of Living Organisms.* Oxford, U.K., and Malden, MA: Blackwell Science. 345p. Reviews all major groups.

303. Barnes, Robert D., 1989. Diversity of organisms: how much do we know? *American Zoologist* 29: 1075-1084. [1] *A.T.:* invertebrates; diversification; evolution.

304. Barnes, R.F.W., 1990. Deforestation trends in tropical Africa. *African J. of Ecology* 28: 161-173. [1].

305. Barns, Susan M., & S.A. Nierzwicki-Bauer, 1997. Microbial diversity in ocean, surface and subsurface environments. In Jillian F. Banfield & K.H. Nealson, eds., *Geomicrobiology: Interactions Between Microbes and Minerals* (Washington, DC: Mineralogical Society of America): 35-79. *A.T.:* archaebacteria; eubacteria.

306. Barns, Susan M., R.E. Fundyga, et al., 1994. Remarkable archaeal diversity detected in a Yellowstone National Park hot spring environment. *PNAS* 91: 1609-1613. [3] *A.T.:* archaebacteria.

307. Barr, T.C., Jr., & J.R. Holsinger, 1985. Speciation in cave faunas. *ARES* 16: 313-337. [1] *A.T.:* troglobites; North America.

308. Barraclough, T.G., A.P. Vogler, & P.H. Harvey, 1998. Revealing the factors that promote speciation. *PTRSL B* 353: 241-249. *A.T.:* diversification.

309. Barreto-de-Castro, L.A., 1996. Sustainable use of biodiversity—components of a model project for Brazil. *Brazilian J. of Medical and Biological Research* 29: 687-699. [1].

310. Barrett, Gary W., & J.D. Peles, 1994. Optimizing habitat frag-

mentation: an agrolandscape perspective. *Landscape and Urban Planning* 28: 99-105. [1] *A.T.:* landscape ecology; Ohio; Indiana.

311. Barrett, Gary W., H.A. Ford, & H.F. Recher, 1994. Conservation of woodland birds in a fragmented rural landscape. *Pacific Conservation Biology* 1: 245-256. [1].

312. Barrett, Gary W., J.D. Peles, & E.P. Odum, 1997. Transcending processes and the levels-of-organization concept. *Bioscience* 47: 531-535. *A.T.:* environmental education; interdisciplinary studies.

313. Barrett, Scott, 1994. The biodiversity supergame. *Environmental & Resource Economics* 4: 111-122. [1] *A.T.:* international cooperation; developing countries.

314. Barry, J.P., C.H. Baxter, et al., 1995. Climate-related, long-term faunal changes in a California rocky intertidal community. *Science* 267: 672-675. [2] *A.T.:* invertebrates; range change.

315. Barthlott, Wilhelm, & Matthias Winiger, eds., 1998. *Biodiversity: Challenge for Development Research and Policy*. Berlin and New York: Springer. 429p. *A.T.:* bdc.

316. Bartlein, Patrick J., Cathy Whitlock, & S.L. Shafter, 1997. Future climate in the Yellowstone National Park region and its potential impact on vegetation. *CB* 11: 782-792. *A.T.:* simulation studies.

317. Barton, Anthony, Dennis Brault, & Roger Wamboldt, 1982. Genetic control of global cybernetic feedback. *Nature and System* 4: 3-12. [1] Postulates the possible existence of "Gaian genes." *A.T.:* Gaia hypothesis.

318. Barton, John H., 1992. Biodiversity at Rio. *Bioscience* 42: 773-776. [1] Reviews UNCED.

319. Bascompte, Jordi, & R.V. Solé, 1996. Habitat fragmentation and extinction thresholds in spatially explicit models. *J. of Animal Ecology* 65: 465-473. [2] *A.T.:* metapopulation dynamics.

320. Basiago, A.D., 1994. Sustainable development in tropical forest ecosystems. *International J. of Sustainable Development and World Ecology* 1: 34-40. [1] *A.T.:* developing countries.

321. ———, 1995. Methods of defining "sustainability." *Sustainable Development* 3: 109-119. [1].

322. ———, 1995. Sustainable development in Indonesia: a case study of an indigenous regime of environmental law and policy. *International J. of Sustainable Development and World Ecology* 2: 199-211. [1] *A.T.:* land use practices.

323. Baskin, Yvonne, 1994. Ecosystem function of biodiversity. *Bioscience* 44: 657-660. [1] Describes activities of SCOPE.

324. ———, 1996. Curbing undesirable invaders. *Bioscience* 46: 732-736. [1] *A.T.:* biological invasions.

325. ———, 1997. *The Work of Nature: How the Diversity of Life Sustains Us*. Washington, DC: Island Press. 263p. [2/2/-] *A.T.:* human ecology;

environmentalism.

326. Basset, Yves, 1991. The taxonomic composition of the arthropod fauna associated with an Australian rainforest tree. *Australian J. of Zoology* 39: 171-190. [1] *A.T.:* Queensland; canopy faunas.

327. Basset, Yves, G.A. Samuelson, et al., 1996. How many species of host-specific insects feed on a species of tropical tree? *Biological J. of the Linnean Society* 59: 201-216. [1] *A.T.:* canopy faunas; beetles.

328. Bastian, O., 1998. Landscape-ecological goals as guiding principles to maintain biodiversity at different planning scales. *Ecology Bratislava* 17: 49-61.

329. Batabyal, Amitrajeet A., 1998. An optimal stopping approach to the conservation of biodiversity. *Ecological Modelling* 105: 293-298. *A.T.:* optimization models.

330. Bataillon, T.M., J.L. David, & D.J. Schoen, 1996. Neutral genetic markers and conservation genetics: simulated germplasm collections. *Genetics* 144: 409-417. [1] *A.T.:* collection sampling.

331. Bates, Henry Walter, 1863. *The Naturalist on the River Amazons.* London: J. Murray. 2 vols. [2/-/1] A classic nineteenth-century scientific travel work by the originator of the concept of protective mimicry. *A.T.:* insects.

332. Batianoff, George N., & Roslyn Burgess, 1993. Problems in the documentation of rare plants—the Australian experience. *Biodiversity Letters* 1: 168-171. [1].

333. Batisse, Michel, 1986. Development of the biosphere reserve concept. *Nature and Resources* 22(3): 2-11. [1].

334. ———, 1993. The silver jubilee of MAB and its revival. *Environmental Conservation* 20: 107-112. [1] *A.T.:* international cooperation.

335. ———, June 1997. Biosphere reserves: a challenge for biodiversity conservation & regional development. *Environment* 39(5): 6-15, 31-33. *A.T.:* MAB; CBD.

336. Battaglia, Bruno, Jose Valencia, & D.W.H. Walton, eds., 1997. *Antarctic Communities: Species, Structure, and Survival* [symposium papers]. Cambridge, U.K., and New York: CUP. 464p. [1/1/-].

337. Batten, L.A., 1972. Breeding bird species diversity in relation to increasing urbanisation. *Bird Study* 19: 157-166. [1].

338. Bauder, Ellen, D.A. Kreager, & Scott McMillan, 1997. *Vernal Pools of California: Draft Recovery Plan.* Portland, OR: U.S. Fish and Wildlife Service, Region 1. 152p. [1/1/1] *A.T.:* Southern California; wetlands.

339. Baumann, Miges, Janet Bell, et al., eds., 1996. *The Life Industry: Biodiversity, People and Profits* [symposium papers]. London: Intermediate Technology Publications. 206p. [1/1/1] Looks at the economic and ethical implications of commercializing bioresources.

340. Baur, A., & B. Baur, 1992. Effect of corridor width on animal

dispersal: a simulation study. *Global Ecology and Biogeography Letters* 2: 52-56. [1] *A.T.:* snails.

341. Bawa, Kamaljit S., & Shaily Menon, 1997. Biodiversity monitoring: the missing ingredients. *TREE* 12: 42. [1].

342. Bawa, Kamaljit S., & Reinmar Seidler, 1998. Natural forest management and conservation of biodiversity in tropical forests. *CB* 12: 46-55. *A.T.:* logging; sustainable forestry.

343. Baydack, Richard K., & Henry Campa, eds., 1998. *Practical Approaches to the Conservation of Biodiversity* [conference papers]. Washington, DC: Island Press. 320p.

344. Bayley, Peter B., 1995. Understanding large river-floodplain ecosystems. *Bioscience* 45: 153-158. [2] *A.T.:* environmental restoration.

345. Bazzaz, Fakhri A., 1990. The response of natural ecosystems to the rising CO_2 levels. *ARES* 21: 167-196. [3] A heavily cited review. *A.T.:* climatic change.

346. ———, 1998. Tropical forests in a future climate: changes in biological diversity and impact on the global carbon cycle. *Climatic Change* 39: 317-336. *A.T.:* carbon dioxide; biogeochemical cycles.

347. Bazzaz, Fakhri A., & E.D. Fajer, Jan. 1992. Plant life in a CO_2-rich world. *Scientific American* 266(1): 68-74. [2] *A.T.:* global carbon cycle.

348. Beamish, Richard J., 1993. Climate and exceptional fish production off the west coast of North America. *Canadian J. of Fisheries and Aquatic Sciences* 50: 2270-2291. [2] *A.T.:* North Pacific; marine fisheries.

349. Bean, Michael J., Sarah Fitzgerald, & M.A. O'Connell, 1991. *Reconciling Conflicts Under the Endangered Species Act: The Habitat Conservation Planning Experience.* Washington, DC: WWF. 109p. [1/1/1] *A.T.:* environmental policy.

350. Bear, Dinah, 1994. Using the National Environmental Policy Act to protect biological diversity. *Tulane Environmental Law J.* 8: 77-96. [1] *A.T.:* environmental impact statements.

351. Beardall, J., S. Beer, & J.A. Raven, 1998. Biodiversity of marine plants in an era of climate change: some predictions based on physiological performance. *Botanica Marina* 41: 113-123. *A.T.:* carbon dioxide.

352. Beardmore, J.A., G.C. Mair, & R.I. Lewis, 1997. Biodiversity in aquatic systems in relation to aquaculture. *Aquaculture Research* 28: 829-839.

353. Beardsley, Karen, & David Stoms, 1993. Compiling a digital map of areas managed for biodiversity in California. *Natural Areas J.* 13: 177-190. [1] *A.T.:* land use management.

354. Beardsley, Tim, 1986. Tropical rain forests: ecologists unite for diversity. *Nature* 323: 193. [1] Reports on comments by E.O. Wilson and Paul Ehrlich prior to the 1986 BioDiversity symposium.

355. ———, Nov. 1994. Some like it hot—and cold. *Scientific American*

271(5): 30. [1] Treats the influence of temperature on plant species distribution patterns.

356. Beare, M.H., D.C. Coleman, et al., 1995. A hierarchical approach to evaluating the significance of soil biodiversity to biogeochemical cycling. *Plant and Soil* 170: 5-22. [2] *A.T.:* species richness; keystone species.

357. Beatley, Timothy, 1991. Protecting biodiversity in coastal environments: introduction and overview. *Coastal Management* 19: 1-19. [1] *A.T.:* coastal zone management; marine reserves.

358. ———, 1994. *Habitat Conservation Planning: Endangered Species and Urban Growth.* Austin: University of Texas Press. 234p. [2/1/2] *A.T.:* urban planning.

359. Beattie, Andrew J., 1992. Discovering new biological resources—chance or reason? *Bioscience* 42: 290-292. [1] Argues for a more deliberate effort at discovering new bioresources.

360. ———, ed., 1993. *Rapid Biodiversity Assessment: Proceedings of the Biodiversity Assessment Workshop, 1993.* Sydney: Research Unit for Biodiversity and Bioresources, Macquarie University. 92p. [1/1/1] *A.T.:* Australia.

361. ———, 1995. Biodiversity and bioresources. *Australasian Biotechnology* 5: 212. [1].

362. ———, 1995. Natural history at the cutting edge. *Ecological Economics* 13: 93-97. [1].

363. ———, 1996. Putting biodiversity to work: how evolutionary adaptation is guiding the work of pharmacologists, engineers, architects and robotics experts. *Search* 27: 111-113. [1].

364. Beattie, Jeffrey E., 1992. *Human Benefits From Protecting Biological Diversity: The Use of Pacific Yew in Cancer Treatment.* Washington, DC: Forest Policy Center, American Forestry Association. 45p. [1/1/1] *A.T.:* medicinal plants; Washington.

365. Beaumont, Peter B., G.H. Miller, & J.C. Vogel, 1992. Contemplating old clues to the impact of future greenhouse climates in South Africa. *South African J. of Science* 88: 490-498. [1] *A.T.:* climatic change.

366. Beccaloni, G.W., & K.J. Gaston, 1995. Predicting the species richness of Neotropical forest butterflies: Ithomiinae (Lepidoptera: Nymphalidae) as indicators. *Biological Conservation* 71: 77-86. [2] *A.T.:* indicator species.

367. Beck, Michael W., 1996. On discerning the cause of late Pleistocene megafaunal extinctions. *Paleobiology* 22: 91-103. [1] *A.T.:* blitzkreig model.

368. Becker, Hank, Dec. 1996. Searching for parasitic "roots": the past is key to understanding the present. *Agricultural Research* 44(12): 4-7. [1] *A.T.:* indicator species.

369. Becker, P., 1992. Colonization of islands by carnivorous and herbivorous Heteroptera and Coleoptera: effects of island area, plant species

richness, and "extinction" rates. *J. of Biogeography* 19: 163-171. [1].

370. Beckon, W.N., 1993. The effect of insularity on the diversity of land birds in the Fiji islands: implications for refuge design. *Oecologia* 94: 318-329. [1].

371. Bederman, David J., 1991. International control of marine "pollution" by exotic species. *Ecology Law Quarterly* 18: 677-717. [1] *A.T.:* biological invasions; international law; international cooperation.

372. Bedward, Michael, R.L. Pressey, & D.A. Keith, 1992. A new approach for selecting fully representative reserve networks: addressing efficiency, reserve design and land suitability with an iterative analysis. *Biological Conservation* 62: 115-125. [2] *A.T.:* CODA; New South Wales.

373. Beebee, Trevor J.C., 1996. *Ecology and Conservation of Amphibians.* London and New York: Chapman & Hall. 214p. [2/1/-].

374. ———, 1997. Changes in dewpond numbers and amphibian diversity over 20 years on chalk downland in Sussex, England. *Biological Conservation* 81: 215-219.

375. Begossi, Alpina, 1996. Use of ecological methods in ethnobotany: diversity indices. *Economic Botany* 50: 280-289. [1].

376. Begossi, Alpina, & P.J. Richerson, 1993. Biodiversity, family income and ecological niche: a study on the consumption of animal foods on Buzios Island (Brazil). *Ecology of Food and Nutrition* 30: 51-61. [1] *A.T.:* human food preferences.

377. Beier, P., 1993. Determining minimum habitat areas and habitat corridors for cougars. *CB* 7: 94-108. [2] *A.T.:* corridors; simulation studies; California.

378. Beisel, J.N., & J.C. Moreteau, 1997. A simple formula for calculating the lower limit of Shannon's diversity index. *Ecological Modelling* 99: 289-292. *A.T.:* Pielou's index; Hurlbert's index; diversity indices.

379. Bekele, Frances L., & Isaac Bekele, 1996. A sampling of the phenetic diversity of cacao in the International Cocoa Gene Bank of Trinidad. *Crop Science* 36: 57-64. [1] *A.T.:* plant germplasm banks.

380. Bekele, Tsegaye, 1997. The significance of biodiversity for sustaining agricultural production and role of women in the traditional sector: the Ethiopian experience. *Agriculture, Ecosystems & Environment* 62: 215-227. *A.T.:* agricultural practices.

381. Belasky, P., 1996. Biogeography of Indo-Pacific foraminifera and scleractinian corals: a probabilistic approach to estimating taxonomic diversity, faunal similarity, and sampling bias. *Palaeogeography, Palaeoclimatology, Palaeoecology* 122: 119-141. [1] *A.T.:* Indo-Pacific region; biogeography.

382. Belbin, L., 1995. A multivariate approach to the selection of biological reserves. *Biodiversity and Conservation* 4: 951-963. [1] *A.T.:* cluster analysis.

383. Bell, David E., 1993. The 1992 Convention on Biological Diversity: the continuing significance of U.S. objections at the Earth Summit. *George Washington J. of International Law and Economics* 26: 479-538. [1] *A.T.:* UNCED; technology transfer; intellectual property.

384. Bell, E.A., 1993. Mankind and plants: the need to conserve biodiversity. *Parasitology* 106, Suppl.: S47-S53. [1] Emphasizes the medicinal value of plants.

385. Bell, Graham, & Austin Burt, 1991. The comparative biology of parasite species diversity: internal helminths of freshwater fish. *J. of Animal Ecology* 60: 1047-1064. [1] *A.T.:* Canada.

386. Bellan-Santini, Denise, Gilles Bonin, & Christian Emig, eds., 1995. *Functioning and Dynamics of Natural and Perturbed Ecosystems* [conference papers]. Paris: Lavoisier. 819p. [1/1/1] *A.T.:* ecosystem disturbance.

387. Bellan-Santini, Denise, P.M. Arnaud, et al., 1996. The influence of the introduced tropical alga *Caulerpa taxifolia*, on the biodiversity of the Mediterranean marine biota. *J. of the Marine Biological Association* 76: 235-237. [1] *A.T.:* biological invasions.

388. Bellon, Mauricio R., & S.B. Brush, 1994. Keepers of maize in Chiapas, Mexico. *Economic Botany* 48: 196-209. [1] Describes how local farmers maintain maize varieties through seed selection.

389. Belsky, M.H., 1995. Implementing the ecosystem management approach: optimism or fantasy? *Ecosystem Health* 1: 214-221. [1] *A.T.:* UNCED; environmental law.

390. Belson, Neil A., spring 1997. Marketing biodiversity: the U.S. regulatory structure for natural products. *Diversity* 13(1): 18-21. [1] *A.T.:* government regulation.

391. Benckiser, Gero, ed., 1997. *Fauna in Soil Ecosystems: Recycling Processes, Nutrient Fluxes, and Agricultural Production.* New York: Marcel Dekker. 414p. [1/1/-].

392. Benford, G., 1992. Saving the "library of life." *PNAS* 89: 11098-11101. [1] Advocates cryopreservation as a major way of preserving bd germplasm for future use.

393. Bengtsson, Bo, & Carl-Gustaf Thornström, 1998. *Biodiversity and Future Genetic Policy: A Study of Sweden.* Washington, DC: World Bank. 79p. *A.T.:* plant germplasm resources.

394. Bengtsson, Jan, S.R. Baillie, & J. Lawton, 1997. Community variability increases with time. *Oikos* 78: 249-256. *A.T.:* U.K.; birds; community ecology.

395. Benirschke, Kurt, & A.T. Kumamoto, 1991. Mammalian cytogenetics and conservation of species. *J. of Heredity* 82: 187-191. [1] *A.T.:* captive animals.

396. Benkman, Craig W., 1993. Adaptation to single resources and the evolution of crossbill (*Loxia*) diversity. *Ecological Monographs* 63: 305-

325. [2] *A.T.:* diversification; Pacific Northwest.

397. Bennett, Andrew F., 1990. Habitat corridors and the conservation of small mammals in a fragmented forest environment. *Landscape Ecology* 4: 109-122. [2] *A.T.:* Victoria, Australia.

398. ———, 1990. *Habitat Corridors: Their Role in Wildlife Management and Conservation.* Melbourne: Department of Conservation and Environment, Victoria. 37p. [1/1/1] *A.T.:* Australia; landscape ecology.

399. Bennett, Andrew F., G.N. Backhouse, & T.W. Clark, eds., 1995. *People and Nature Conservation: Perspectives on Private Land Use and Endangered Species Recovery* [symposium papers]. Mosman, New South Wales: Royal Zoological Society of New South Wales. 228p. [1/1/1] *A.T.:* Australia; wildlife conservation.

400. Bennett, Bradley C., 1992. Plants and people of the Amazonian rainforests: the role of ethnobotany in sustainable development. *Bioscience* 42: 599-607. [1] *A.T.:* indigenous peoples.

401. Bennett, E.L., & C.J. Reynolds, 1993. The value of a mangrove area in Sarawak. *Biodiversity and Conservation* 2: 359-375. [1] *A.T.:* coastal zone management.

402. Bennett, Graham, ed., 1994. *Conserving Europe's Natural Heritage: Towards a European Ecological Network* [conference proceedings]. Boston: Martinus Nijhoff. 334p. [1/1/1] *A.T.:* environmental policy; international cooperation.

403. Bennett, K.D., P.C. Tzedakis, & K.J. Willis, 1991. Quaternary refugia of North European trees. *J. of Biogeography* 18: 103-115. [2].

404. Benton, Michael J., 1985. Mass extinction among non-marine tetrapods. *Nature* 316: 811-814. [1] Reviews patterns of family diversity among amphibians, reptiles, birds, and mammals since the Devonian.

405. ———, 1986. More than one event in the late Triassic mass extinction. *Nature* 321: 857-861. [1].

406. ———, 1987. The history of the biosphere: equilibrium and non-equilibrium models of global diversity. *TREE* 2: 153-156. [1] *A.T.:* mass extinctions; diversification.

407. ———, 1989. Mass extinctions among tetrapods and the quality of the fossil record. *PTRSL B* 325: 369-386. [1].

408. ———, 1991. What really happened in the Late Triassic? *Historical Biology* 5: 263-278. [1].

409. ———, 1995. Diversification and extinction in the history of life. *Science* 268: 52-58. [2] Provides an overview of marine and terrestrial diversification in Phanerozoic times.

410. Benton, T.G., 1995. Biodiversity and biogeography of Henderson Island's insects. *Biological J. of the Linnean Society* 56: 245-259. [1] *A.T.:* colonization.

411. Bequette, France, June 1995. Turkey, land of diversity. *UNESCO*

Courier 48(6): 43-45. [1] Reviews efforts to protect Turkey's bd.

412. Berg, Å., B. Ehnström, et al., 1994. Threatened plant, animal, and fungus species in Swedish forests: distribution and habitat associations. *CB* 8: 718-731. [2] *A.T.:* endangered species.

413. Berger, Joel, 1990. Persistence of different-sized populations: an empirical assessment of rapid extinctions in bighorn sheep. *CB* 4: 91-98. [2] *A.T.:* local extinction.

414. ———, 1991. Funding asymmetries for endangered species, feral animals, and livestock. *Bioscience* 41: 105-106. [1] Opines that relatively too much federal support is being applied to preserving feral animals and livestock.

415. ———, 1994. Science, conservation, and black rhinos. *J. of Mammalogy* 75: 298-308. [1] *A.T.:* endangered species; in situ conservation.

416. ———, 1997. Population constraints associated with the use of black rhinos as an umbrella species for desert herbivores. *CB* 11: 69-78. [1].

417. Berger, John J., ed., 1990. *Environmental Restoration: Science and Strategies for Restoring the Earth* [conference papers]. Washington, DC: Island Press. 398p. [2/1/1] *A.T.:* restoration ecology.

418. Bergh, Jeroen C.J.M. van den, 1996. *Ecological Economics and Sustainable Development: Theory, Methods, and Applications*. Cheltenham, U.K., and Brookfield, VT: Edward Elgar. 312p. [1/1/1].

419. Bergh, Øivind, K.Y. Børsheim, et al., 1989. High abundance of viruses found in aquatic environments. *Nature* 340: 467-468. [3] *A.T.:* bacteriophages.

420. Bergmann, C.G.L.C., 1847. Über die Verhältnisse der Wärmeökonomie der Thiere zu ihrer Grösse. *Göttinger Studien* 3(1): 595-708. [2] The work describing what came to be known as "Bergmann's rule."

421. Bergmans, Bernhard, 1990. Industrial property and biological diversity of plant and animal species. *J. of the Patent and Trademark Office Society* 72: 600-609. [1] *A.T.:* patents.

422. Berkes, Fikret, ed., 1989. *Common Property Resources: Ecology and Community-based Sustainable Development* [conference papers]. London and New York: Belhaven Press. 302p. [1/1/2].

423. Bernard, E.C., 1992. Soil nematode biodiversity. *Biology and Fertility of Soils* 14: 99-103. [1] *A.T.:* soil fauna.

424. Berner, R.A., & D.E. Canfield, 1989. A new model for atmospheric oxygen over Phanerozoic time. *American J. of Science* 298: 333-361. [2].

425. Bernes, Claes, ed., 1994. *Biological Diversity in Sweden: A Country Study*. Vaxjo, Sweden: Ingvar Bingman. 280p. [1/1/1].

426. Berrens, Robert P., D.S. Brookshire, et al., 1998. Implementing the safe minimum standard approach: two case studies from the U.S. Endangered Species Act. *Land Economics* 74: 147-161. *A.T.:* endangered fishes; Colorado River Basin; environmental law.

427. Berry, R.J., 1983. Diversity and differentiation: the importance of island biology for general theory. *Oikos* 41: 523-529. [1] *A.T.:* Founder effect.

428. Bertiller, M.B., & D.A. Aloia, 1997. Seed bank strategies in Patagonian semi-arid grasslands in relation to their management and conservation. *Biodiversity and Conservation* 6: 639-650. *A.T.:* Argentina.

429. Bertness, M.D., & G.H. Leonard, 1997. The role of positive interactions in communities: lessons from intertidal habitats. *Ecology* 78: 1976-1989. *A.T.:* community ecology; New England.

430. Bertollo, P., 1998. Assessing ecosystem health in governed landscapes: a framework for developing core indicators. *Ecosystem Health* 4: 33-51. *A.T.:* sea-level rise; coastal zone management; Italy; ecological indicators.

431. Best, Brinley J., & Michael Kessler, 1995. *Biodiversity and Conservation in Tumbesian Ecuador and Peru.* Cambridge, U.K.: BirdLife International. 218p. [1/1/1] *A.T.:* birds.

432. Beus, Curtis E., & R.E. Dunlap, 1990. Conventional versus alternative agriculture: the paradigmatic roots of the debate. *Rural Sociology* 55: 590-616. [2] *A.T.:* traditional farming; sustainable agriculture.

433. Beveridge, Malcolm C.M., Lindsay Ross, & L.A. Kelly, 1994. Aquaculture and biodiversity. *Ambio* 23: 497-502. [1].

434. Bevers, Michael, John Hof, et al., 1995. Sustainable forest management for optimizing multispecies wildlife habitat: a coastal Douglas-fir example. *Natural Resource Modeling* 9: 1-23. [1] *A.T.:* optimization models.

435. Bezdicek, David F., & David Granatstein, 1989. Crop rotation efficiencies and biological diversity in farming systems. *American J. of Alternative Agriculture* 4: 111-119. [1] *A.T.:* sustainable agriculture.

436. Bhandari, N., 1991. Collisions with Earth over geologic times and their consequences to the terrestrial environment. *Current Science* 61: 97-104. [1].

437. Bharathie, K.P. Sri, 1993. Consumption and conservation of forest resources: the conflict in Sri Lanka. *Commonwealth Forestry Review* 72: 29-30. [1] *A.T.:* buffer zones.

438. Bhat, J.L., & Desh Bandhu, 1994. *Biodiversity for Sustainable Development.* Delhi: Indian Environmental Society. 100p. [1/1/1].

439. Bibby, Colin J., N.J. Collar, et al., 1992. *Putting Biodiversity on the Map: Priority Areas for Global Conservation.* Girton, Cambridge, U.K.: International Council for Bird Preservation. 90p. [2/2/2] *A.T.:* birds.

440. Bierregaard, Richard O., Jr., T.E. Lovejoy, et al., 1992. The biological dynamics of tropical rainforest fragments. *Bioscience* 42: 859-866. [2] *A.T.:* Amazonia.

441. Bilderbeek, Simone, 1993. Biodiversity as political game. *Politics and the Life Sciences* 12: 265-272. [1].

442. Bilderbeek, Simone, Ankie Wijgerde, & Netty van Schaik, eds.,

1992. *Biodiversity and International Law: The Effectiveness of International Environmental Law* [conference report]. Amsterdam and Washington, DC: IOS Press. 213p. [1/1/1].

443. Bildstein, Keith L., G.T. Bancroft, et al., 1991. Approaches to the conservation of coastal wetlands in the Western Hemisphere. *Wilson Bull.* 103: 218-254. [1] *A.T.:* birds; global warming; environmental policy; coastal zone management.

444. Bills, Gerald F., 1995. Analyses of microfungal diversity from a user's perspective. *Canadian J. of Botany* 73, Suppl. 1: S33-S41. [1] *A.T.:* industrial microbiology.

445. Binggeli, P., 1996. A taxonomic, biogeographical and ecological overview of invasive woody plants. *J. of Vegetation Science* 7: 121-124. [1] *A.T.:* biological invasions; bd databases.

446. Binns, Tony, 1990. Is desertification a myth? *Geography* 75: 106-113. [1].

447. Biodiversity Convention Office, Environment Canada, 1995. *Canadian Biodiversity Strategy: Canada's Response to the Convention on Biological Diversity*. Hull, Quebec, Canada. 80p. [1/1/1] *A.T.:* sustainable development.

448. Biodiversity Support Program, 1993. *African Biodiversity: Foundation for the Future: A Framework for Integrating Biodiversity Conservation and Sustainable Development*. Washington, DC. 149p. [1/1/1].

449. ———, 1995. *A Regional Analysis of Geographic Priorities for Biodiversity Conservation in Latin America and the Caribbean*. Washington, DC. 116p. [1/1/1].

450. Birckhead, Jim, Terry De Lacy, & Laurajane Smith, eds., 1993. *Aboriginal Involvement in Parks and Protected Areas* [conference papers]. Canberra: Aboriginal Studies Press. 390p. [1/1/1] *A.T.:* Australia.

451. Birkeland, Charles, ed., 1997. *Life and Death of Coral Reefs*. New York: Chapman & Hall. 536p. [1/1/-].

452. Birnie, Patricia W., & A.E. Boyle, 1992. *International Law and the Environment*. New York: OUP. 563p. [1/1/2].

453. Bisby, F.A., G.F. Russell, & R.J. Pankhurst, eds., 1993. *Designs for a Global Plant Species Information System* [symposium proceedings]. New York: OUP. 350p. [1/1/1] *A.T.:* computer databases.

454. Bishop, Christine A., & K.E. Pettit, eds., 1992. *Declines in Canadian Amphibian Populations: Designing a National Monitoring Strategy* [workshop proceedings]. Ottawa: Occasional Paper No. 76, Canadian Wildlife Service. 120p. [1/1/1] *A.T.:* wildlife conservation.

455. Bishop, Richard C., 1978. Endangered species and uncertainty: the economics of a safe minimum standard. *American J. of Agricultural Economics* 60: 10-18. [2] *A.T.:* game theory; environmental economics.

456. ———, 1993. Economic efficiency, sustainability, and biodiversity.

Ambio 22: 69-73. [2] *A.T.:* safe minimum standard.

457. Bissonette, John A., ed., 1997. *Wildlife and Landscape Ecology: Effects of Pattern and Scale* [includes conference papers]. New York: Springer. 410p. [1/1/-].

458. Bjørndalen, J.E., 1992. Tanzania's vanishing rain forests—assessment of nature conservation values, biodiversity and importance for water catchment. *Agriculture, Ecosystems & Environment* 40: 313-334. [1] *A.T.:* environmental degradation; deforestation.

459. Bjørnstad, Anders, A. Demissie, et al., 1997. The distinctness and diversity of Ethiopian barleys. *Theoretical and Applied Genetics* 94: 514-521. *A.T.:* plant germplasm resources.

460. Bjørse, G., & R. Bradshaw, 1998. 2000 years of forest dynamics in southern Sweden: suggestions for forest management. *Forest Ecology and Management* 104: 15-26. *A.T.:* sustainable forestry.

461. Blackburn, G.A., & E.J. Milton, 1996. Filling the gaps: remote sensing meets woodland ecology. *Global Ecology and Biogeography Letters* 5: 175-191. [1] *A.T.:* GIS; canopy gaps; England.

462. Blackburn, Tim M., & K.J. Gaston, 1996. A sideways look at patterns in species richness, or why there are so few species outside the tropics. *Biodiversity Letters* 3: 44-53. [1] *A.T.:* latitudinal diversity gradients; extratropical species.

463. ———, 1996. Spatial patterns in the species richness of birds in the New World. *Ecography* 19: 369-376. [2] *A.T.:* latitudinal diversity gradients.

464. ———, 1996. Spatial patterns in the body sizes of bird species in the Old World. *Oikos* 77: 436-446. [1] *A.T.:* Bergmann's rule; latitudinal gradients.

465. ———, 1997. A critical assessment of the form of the interspecific relationship between abundance and body size in animals. *J. of Animal Ecology* 66: 233-249. *A.T.:* allometric relationships.

466. ———, 1997. The relationship between geographic area and the latitudinal gradient in species richness in New World birds. *Evolutionary Ecology* 11: 195-204. *A.T.:* latitudinal diversity gradients; extratropical species.

467. Blackmore, Stephen, 1996. Knowing the earth's biodiversity: challenges for the infrastructure of systematic biology. *Science* 274: 63-64. [1] *A.T.:* systematic biology collections.

468. Blackwell, M., & K. Jones, 1997. Taxonomic diversity and interactions of insect-associated ascomycetes. *Biodiversity and Conservation* 6: 689-699. *A.T.:* fungi.

469. Blaikie, Piers M., & Sally Jeanrenaud, 1996. *Biodiversity and Human Welfare.* Geneva: Discussion Paper 72, United Nations Research Institute for Social Development. 82p. [1/1/1].

470. Blair, Robert B., & A.E. Launer, 1997. Butterfly diversity and human land use: species assemblages along an urban gradient. *Biological Conser-*

vation 80: 113-125. *A.T.:* urbanization; gradient analysis; California.

471. Blake, John G., J.M. Hanowski, et al., 1994. Annual variation in bird populations of mixed conifer-northern hardwood forests. *Condor* 96: 381-399. [1] Relates temporal variation in abundance to varying spatial scale. *A.T.:* Wisconsin; Michigan.

472. Blake, John G., & J.R. Karr, 1984. Species composition of bird communities and the conservation benefit of large versus small forests. *Biological Conservation* 30: 173-187. [2] *A.T.:* habitat patches; Illinois.

473. Blaney, Carol, 1995. Crossing the lines to extinction. *Bioscience* 45: 744-745. [1] Discusses the high extinction risk to rare plants that hybridize easily.

474. Blaustein, Andrew R., 1994. Chicken Little or Nero's fiddle? A perspective on declining amphibian populations. *Herpetologica* 50: 85-97. [2].

475. Blaustein, Andrew R., & D.B. Wake, 1990. Declining amphibian populations: a global phenomenon? *TREE* 5: 203-204. [2].

476. ———, April 1995. The puzzle of declining amphibian populations. *Scientific American* 272(4): 52-57. [2] *A.T.:* Washington; Oregon; *Rana cascadae*; *Bufo boreas*.

477. Blaustein, Andrew R., D.B. Wake, & W.P. Sousa, 1994. Amphibian declines: judging stability, persistence, and susceptibility of population to local and global extinctions. *CB* 8: 60-71. [3] *A.T.:* recolonization.

478. Blaustein, Andrew R., P.D. Hoffman, et al., 1994. UV repair and resistance to solar UV-B in amphibian eggs: a link to population declines? *PNAS* 91: 1791-1795. [2] *A.T.:* amphibian declines.

479. Blaustein, Richard J., 1996. Biodiversity and the law [book review]. *Environmental Law* 26: 1313-1318. [1] *A.T.:* environmental law.

480. Blay, Sam K.N., & R.W. Piotrowicz, 1993. Biodiversity and conservation in the twenty-first century: a critique of the Earth Summit 1992. *Environmental and Planning Law J.* 10: 450-469. [1] *A.T.:* UNCED.

481. Bloch, Julie B., 1992. Preserving biological diversity in the United States: the case for moving to an ecosystems approach to protect the nation's biological wealth. *Pace Environmental Law Review* 10: 175-225. [1] An overview of the reasons for protecting bd, and of potential approaches to bdc.

482. Blockhus, Jill M., M. Dillenbeck, et al., eds., 1992. *Conserving Biological Diversity in Managed Tropical Forests* [workshop proceedings]. Gland, Switzerland: IUCN/ITTO. 244p. [1/1/1] *A.T.:* tropical forest conservation.

483. Blockstein, David E., summer 1989. Toward a federal plan for biological diversity. *Issues in Science and Technology* 5(4): 63-67. [1] *A.T.:* environmental policy; endangered species.

484. Blockstein, David E., May-June 1992. An aquatic perspective on U.S. biodiversity policy. *Fisheries* 17(3): 26-30. [1].

485. Blockstein, David E., 1995. A strategic approach for biodiversity

conservation. *Wildlife Society Bull.* 23: 365-369. [1] *A.T.:* environmental policy; national strategies.

486. Bloemers, G.F., M. Hodda, et al., 1997. The effects of forest disturbance on diversity of tropical soil nematodes. *Oecologia* 111: 575-582. *A.T.:* Cameroon.

487. Blum, Elissa, May 1993. Making biodiversity conservation profitable: a case study of the Merck/INBio agreement. *Environment* 35(4): 16-20, 38-45. [1] *A.T.:* Costa Rica; private companies.

488. Bobek, B., K. Perzanowski, et al., 1994. The systems of managing wildlife and forest in Central Europe. *Forestry Chronicle* 70: 550-554. [1].

489. Bock, Carl E., & R.E. Ricklefs, 1983. Range size and local abundance of some North American songbirds: a positive correlation. *American Naturalist* 122: 295-299. [2].

490. Bodansky, Daniel M., 1995. International law and the protection of biological diversity. *Vanderbilt J. of Transnational Law* 28: 623-634. [1].

491. Bodmer, Richard E., J.F. Eisenberg, & K.H. Redford, 1997. Hunting and the likelihood of extinction of Amazonian mammals. *CB* 11: 460-466. *A.T.:* local extinction.

492. Bodmer, Richard E., T.G. Fang, et al., 1994. Managing wildlife to conserve Amazonian forests: population biology and economic considerations of game hunting. *Biological Conservation* 67: 29-35. [2] *A.T.:* Peru.

493. Bogan, Arthur E., 1993. Freshwater bivalve extinctions (Mollusca: Unionoida): a search for causes. *American Zoologist* 33: 599-609. [2].

494. Bohlen, P.J., W.M. Edwards, & C.A. Edwards, 1995. Earthworm community structure and diversity in experimental agricultural watersheds in northeastern Ohio. *Plant and Soil* 170: 233-239. [1].

495. Böhning-Gaese, Katrin, 1997. Determinants of avian species richness at different spatial scales. *J. of Biogeography* 24: 49-60. *A.T.:* Lake Constance region.

496. Bohnsack, James A., fall 1993. Marine reserves: they enhance fisheries, reduce conflicts, and protect resources. *Oceanus* 36(3): 63-71. [2].

497. Boice, L. Peter, Jan.-Feb. 1997. Defending our nation and its biodiversity. *Endangered Species Bull.* 22(1): 4-5. [1] Describes conservation management of military-administered public lands.

498. Bojórquez Tapia, Luis A., Iván Azuara, & Exequiel Ezcurra, 1995. Identifying conservation priorities in Mexico through geographic information systems and modeling. *Ecological Applications* 5: 215-231. [1] *A.T.:* gap analysis; species-rich areas.

499. Bojórquez Tapia, Luis A., Patricia Balvanera, & A.D. Cuarón, 1994. Biological inventories and computer data bases: their role in environmental assessments. *Environmental Management* 18: 775-785. [1] *A.T.:* Mexico.

500. Bolen, Eric G., & W.L. Robinson, 1999. *Wildlife Ecology and Management* (4th ed.). Englewood Cliffs, NJ: Prentice Hall. 605p. [2/1/1].

501. Bolger, Douglas T., A.C. Alberts, & M.E. Soulé, 1991. Occurrence patterns of bird species in habitat fragments: sampling, extinction, and nested species subsets. *American Naturalist* 137: 155-166. [2] *A.T.:* urban ecology; chaparral; California.

502. Bolgiano, Chris, July-Aug. 1993. The aliens. *American Forests* 99(7-8): 17-19, 53-54. [1] *A.T.:* biological invasions; Forest Service; forest diversity.

503. Bolin, Bert, B.R. Doos, et al., eds., 1986. *The Greenhouse Effect, Climatic Change, and Ecosystems*. New York: Wiley. 541p. [2/1/3] *A.T.:* carbon cycle.

504. Bolker, B.M., S.W. Pacala, et al., 1995. Species diversity and ecosystem response to carbon dioxide fertilization: conclusions reached from a temperate forest model. *Global Change Biology* 1: 373-381. [2] *A.T.:* climatic change.

505. Bonan, Gordon B., D. Pollard, & S.L. Thompson, 1992. Effects of boreal forest vegetation on global climate. *Nature* 359: 716-718. [2] *A.T.:* climatic change; global warming.

506. Bonan, Gordon B., H.H. Shugart, & D.L. Urban, 1990. The sensitivity of high-latitude forests to climatic parameters. *Climatic Change* 16: 9-29. [2].

507. Bonavita, P., C. Chemini, et al., 1998. Biodiversity and stress level in four forests of the Italian Alps. *Chemosphere* 36: 1055-1060. *A.T.:* fungi; arthropods.

508. Bond, W.J., & D.M. Richardson, 1990. What can we learn from extinctions and invasions about the effects of climate change? *South African J. of Science* 86: 429-433. [1].

509. Bonner, Raymond, 1993. *At the Hand of Man: Peril and Hope for Africa's Wildlife*. New York: Knopf. 322p. [2/3/1] *A.T.:* wildlife conservation; wildlife management.

510. Bonta, Marcia, Feb. 1990. A Noah's ark for endangered plants. *American Horticulturist* 69(2): 20-24. [1] *A.T.:* California; botanical gardens.

511. Boo, Elizabeth, 1990. *Ecotourism: The Potentials and Pitfalls* [case studies]. Washington, DC: WWF. 2 vols. [2/1/2] *A.T.:* Latin America.

512. Boone, D. Daniel, & G.H. Aplet, 1994. *Sustaining Biodiversity in the Southern Appalachians* [report]. Washington, DC: Wilderness Society. 76p. [1/1/1] *A.T.:* environmental policy.

513. Boot, R.G.A., 1997. Extraction of non-timber forest products from tropical rain forests: does diversity come at a price? *Netherlands J. of Agricultural Science* 45: 439-450. *A.T.:* forest management; domestication.

514. Booth, W., 1989. Monitoring the fate of the forests from space [editorial]. *Science* 243: 1428-1429. [1] Asks why remote sensing is not being more widely used in bd studies.

515. Borchsenius, E., 1997. Patterns of plant species endemism in Ec-

uador. *Biodiversity and Conservation* 6: 379-399. *A.T.:* biogeography.

516. Borkenhagen, Lea M., & J.N. Abramovitz, eds., 1992. *Proceedings of the International Conference on Women and Biodiversity.* Cambridge, MA: Committee on Women and Biodiversity, WRI USA. 98p. [1/1/1].

517. Bormann, F. Herbert, & S.R. Kellert, eds., 1991. *Ecology, Economics, Ethics: The Broken Circle.* New Haven: Yale University Press. 233p. [2/2/1] *A.T.:* environmental ethics.

518. Borneman, James, & E.W. Triplett, 1997. Molecular microbial diversity in soils from eastern Amazonia: evidence for unusual microorganisms and microbial population shifts associated with deforestation. *Applied and Environmental Microbiology* 63: 2647-2653. [2].

519. Borneman, James, P.W. Skroch, et al., 1996. Molecular microbial diversity of an agricultural soil in Wisconsin. *Applied and Environmental Microbiology* 62: 1935-1943. [3].

520. Bornette, Gudrun, Christophe Henry, et al., 1994. Theoretical habitat templets, species traits, and species richness: aquatic macrophytes in the Upper Rhône River and its floodplain. *Freshwater Biology* 31: 487-505. [1] *A.T.:* river ecology.

521. Bosselmann, Klaus, 1996. Plants and politics: the international legal regime concerning biotechnology and biodiversity. *Colorado J. of International Environmental Law and Policy* 7: 111-148. [1] *A.T.:* international cooperation; CBD.

522. Bosselmann, Klaus, & P. Taylor, 1995. The New Zealand law and conservation. *Pacific Conservation Biology* 2: 113-121. [1] *A.T.:* Resource Management Act; sustainable development.

523. Botkin, Daniel B., 1990. *Discordant Harmonies: A New Ecology for the Twenty-first Century.* New York: OUP. 241p. [3/2/2] *A.T.:* environmental policy; environmental protection.

524. Boucher, Douglas H., 1990. Growing back after hurricanes: catastrophes may be critical to rain forest dynamics. *Bioscience* 40: 163-166. [1] *A.T.:* rainforest ecology; ecological climax.

525. Boucher, G., 1997. Structure and biodiversity of nematode assemblages in the SW lagoon of New Caledonia. *Coral Reefs* 16: 177-186. *A.T.:* coral reefs.

526. Boucher, G., & P.J.D. Lambshead, 1995. Ecological biodiversity of marine nematodes in samples from temperate, tropical, and deep-sea regions. *CB* 9: 1594-1604. [2].

527. Boucher, Norman, Aug.-Sept. 1995. Back to the Everglades. *Technology Review* 98(6): 24-35. [1] Describes the Everglades recovery project.

528. Bouchet, P., T. Jaffre, & J.-M. Veillon, 1995. Plant extinction in New Caledonia: protection of sclerophyll forests urgently needed. *Biodiversity and Conservation* 4: 415-428. [1] *A.T.:* endangered plants.

529. Boucot, Arthur J., 1975. *Evolution and Extinction Rate Controls.*

Amsterdam and New York: Elsevier Scientific. 427p. [2/1/2] *A.T.:* fossil brachiopods.

530. ———, 1983. Area-dependent-richness hypotheses and rates of parasite/pest evolution. *American Naturalist* 121: 294-300. [1].

531. Boulinier, T., J.D. Nichols, et al., 1998. Higher temporal variability of forest breeding bird communities in fragmented landscapes. *PNAS* 95: 7497-7501. *A.T.:* habitat fragmentation; migratory birds; eastern U.S.

532. ———, 1998. Estimating species richness: the importance of heterogeneity in species detectability. *Ecology* 79: 1018-1028. *A.T.:* birds.

533. Boulter, Michael C., 1997. Plant macroevolution through the Phanerozoic. *Geology Today* 13: 102-106.

534. Bournaud, Michel, 1994. Theoretical habitat templets, species traits, and species richness: birds in the Upper Rhône River and its floodplain. *Freshwater Biology* 31: 469-485. [1].

535. Bourne, Joel, Sept.-Oct. 1992. All creatures, great and small. *Defenders* 67(5): 8-15. [1] Describes E.O. Wilson's advice that conservationists should shift to an approach emphasizing ecosystem-level studies.

536. Bourriau, Janine, ed., 1992. *Understanding Catastrophe* [lectures]. Cambridge, U.K., and New York: CUP. 213p. [2/2/1].

537. Bowler, P.A., 1992. Biodiversity conservation in Europe and North America. II. Shrublands in defense of disturbed land. *Restoration & Management Notes* 10: 144-149. [1] *A.T.:* disturbance ecology.

538. Bowles, Ian A., & G.T. Prickett, 1994. *Reframing the Green Window: An Analysis of the GEF Pilot Phase Approach to Biodiversity and Global Warming and Recommendations for the Operational Phase.* Washington, DC: Conservation International. 133p. [1/1/1].

539. Bowles, Ian A., R.E. Rice, & R.A. Mittermeier, 1998. Logging and tropical forest conservation. *Science* 280: 1899-1900. *A.T.:* World Bank; environmental policy.

540. Bowles, Ian A., et al., 1996. *Encouraging Private Sector Support for Biodiversity Conservation: The Use of Economic Incentives and Legal Tools.* Washington, DC: Conservation International. 22p. [1/1/1].

541. Bowles, Martin L., & C.J. Whelan, eds., 1994. *Restoration of Endangered Species: Conceptual Issues, Planning, and Implementation* [symposium papers]. Cambridge, U.K., and New York: CUP. 394p. [2/1/1] *A.T.:* restoration ecology.

542. Bowman, D.M.J.S., 1993. Tropical rain forests [research review]. *Progress in Physical Geography* 17: 484-492. Surveys 1991 and 1992 literature. [1].

543. Bowman, J.P., M.V. Brown, & D.S. Nichols, 1997. Biodiversity and ecophysiology of bacteria associated with Antarctic sea ice. *Antarctic Science* 9: 134-142.

544. Bowman, Michael, & Catherine Redgwell, eds., 1996. *International*

Law and the Conservation of Biological Diversity. London and Boston: Kluwer Law International. 334p. [1/1/1] *A.T.:* CBD.

545. Bowman, Robert I., Margaret Berson, & A.E. Leviton, eds., 1983. *Patterns of Evolution in Galápagos Organisms* [symposium papers]. San Francisco: Pacific Division, AAAS. 568p. [1/2/1].

546. Boyajian, George E., 1986. Phanerozoic trends in background extinction: consequence of an aging fauna. *Geology* 14: 955-958. [1] *A.T.:* paleobiogeography.

547. ———, 1992. Taxon age, origination, and extinction through geological time. *Historical Biology* 6: 281-291. [1].

548. Boyce, A.J., & C.G.N. Mascie-Taylor, eds., 1996. *Molecular Biology and Human Diversity*. Cambridge, U.K., and New York: CUP. 305p. [1/1/1] *A.T.:* human genetics; molecular evolution.

549. Boyce, M.S., 1992. Population viability analysis. *ARES* 23: 481-506. [2] *A.T.:* local extinction; extinction risk.

550. Boyd, R.S., J.D. Freeman, et al., 1995. Forest herbicide influences on floristic diversity seven years after broadcast pine release treatments in central Georgia, USA. *New Forests* 10: 17-34. [1].

551. Boyden, Stephen V., 1992. *Biohistory: The Interplay Between Human Society and the Biosphere*. Park Ridge, NJ: Parthenon. 265p. [2/1/1] *A.T.:* human ecology; human evolution.

552. Boyle, K.J., & R.C. Bishop, 1987. Valuing wildlife in benefit-cost analyses: a case study involving endangered species. *Water Resources Research* 23: 943-950. [1] *A.T.:* wildlife valuation; Wisconsin.

553. Boyle, T.J.B., 1992. Biodiversity of Canadian forests: current status and future challenges. *Forestry Chronicle* 68: 444-453. [1] *A.T.:* forest management.

554. Boyle, T.J.B., & B. Boontawee, eds., 1995. *Measuring and Monitoring Biodiversity in Tropical and Temperate Forests* [symposium proceedings]. Bogor, Indonesia: Center for International Forestry Research. 395p. [1/1/1] *A.T.:* forest genetics.

555. Boyle, T.J.B., & C.E.B. Boyle, eds., 1994. *Biodiversity, Temperate Ecosystems, and Global Change* [workshop proceedings]. Berlin and New York: Springer. 456p. [1/1/1] *A.T.:* climatic change.

556. Boyle, T.J.B., & J.A. Sayer, 1995. Measuring, monitoring and conserving biodiversity in managed tropical forests. *Commonwealth Forestry Review* 74: 20-25. [1].

557. Boza, M.A., 1993. Conservation in action: past, present, and future of the national park system of Costa Rica. *CB* 7: 239-247. [1].

558. Braatz, Susan M., G. Davis, et al., 1992. *Conserving Biological Diversity: A Strategy for Protected Areas in the Asia-Pacific Region*. Washington, DC: World Bank Technical Paper No. 193. 66p. [1/1/1] *A.T.:* environmental policy.

559. Bradford, David F., 1989. Allotopic distribution of native frogs and introduced fishes in high Sierra Nevada lakes of California: implication of the negative effect of fish introductions. *Copeia* (3): 775-778. [2].

560. Bradsen, John, 1992. Biodiversity legislation: species, vegetation, habitat. *Environmental and Planning Law J.* 9: 175-180. [1] *A.T.:* endangered species; Australia.

561. Bradshaw, A.D., 1983. The reconstruction of ecosystems [address]. *J. of Applied Ecology* 20: 1-17. [2] *A.T.:* reintroduction.

562. Bradstock, Ross A., ed., 1995. *Conserving Biodiversity: Threats and Solutions* [conference proceedings]. Chipping Norton, NSW: Surrey Beatty. 428p. [1/1/1] *A.T.:* New South Wales; wildlife conservation; environmental degradation.

563. Bragdon, Susan H., 1992. National sovereignty and global environmental responsibility: can the tension be reconciled for the conservation of biological diversity? *Harvard International Law J.* 33: 381-392. [1].

564. ———, 1996. The evolution and future of the law of sustainable development: lessons from the Convention on Biological Diversity. *Georgetown International Environmental Law Review* 8: 423-436. [1].

565. ———, 1996. The Convention on Biological Diversity [editorial]. *Global Environmental Change: Human and Policy Dimensions* 6: 177-179. [1].

566. Bramwell, David, 1990. Conserving biodiversity in the Canary Islands. *Annals of the Missouri Botanical Garden* 77: 28-37. [1] *A.T.:* plant conservation.

567. Bramwell, David, O. Hamann, et al., eds., 1987. *Botanic Gardens and the World Conservation Strategy* [conference proceedings]. London and Orlando: Academic Press. 350p. [1/1/1] *A.T.:* plant conservation.

568. Brandani, A., G.S. Hartshorn, & G.H. Orians, 1988. Internal heterogeneity of gaps and species richness in Costa Rican tropical wet forests. *J. of Tropical Ecology* 4: 99-119. [2] *A.T.:* treefall gaps.

569. Brandon, Katrina E., 1995. People, parks, forests or fields: a realistic view of tropical forest conservation. *Land Use Policy* 12: 137-144. [1] Supports the park protection method of bdc.

570. Brandon, Katrina E., & Michael Wells, 1992. Planning for people and parks: design dilemmas. *World Development* 20: 557-570. [1] *A.T.:* ICDPs; protected areas.

571. Brandon, Katrina E., K.H. Redford, & S.E. Sanderson, eds., 1998. *Parks in Peril: People, Politics, and Protected Areas* [case studies]. Washington, DC: Nature Conservancy and Island Press. 519p. *A.T.:* Latin America; bdc.

572. Brash, Alexander R., 1987. The history of avian extinction and forest conversion on Puerto Rico. *Biological Conservation* 39: 97-111. [1] *A.T.:* extinct birds.

573. Braus, Judy, 1995. Environmental education. *Bioscience* 45, Suppl.: 45-51. [1] Discusses environmental education in the U.S., and the WWF's bd-oriented programs.

574. Bravard, J.P., C. Amoros, et al., 1997. River incision in south-east France: morphological phenomena and ecological effects. *Regulated Rivers* 13: 75-90. *A.T.:* river basin management; alluvial plains; stream communities.

575. Brazner, John C., 1997. Regional, habitat, and human development influences on coastal wetland and beach fish assemblages in Green Bay, Lake Michigan. *J. of Great Lakes Research* 23: 36-51. *A.T.:* environmental degradation; lake ecology.

576. Breceda, Aurora, A. Castellanos, et al., 1995. Nature conservation in Baja California Sur, Mexico: protected areas. *Natural Areas J.* 15: 267-273. [1].

577. Brechin, Steven R., & Willett Kempton, 1994. Global environmentalism: a challenge to the postmaterialism thesis? *Social Science Quarterly* 75: 245-269. [2] Disputes the often-held belief that people in developing countries lack environmental values.

578. Breckenridge, Lee P., 1992. Protection of biological and cultural diversity: emerging recognition of local community rights and ecosystems under international environmental law. *Tennessee Law Review* 59: 735-785. [1].

579. Breece, Gary Allen, & B.J. Ward, 1996. Utility terrestrial biodiversity issues. *Environmental Management* 20: 799-803. [1].

580. Breininger, D.R., M.J. Barkaszi, et al., 1998. Prioritizing wildlife taxa for biological diversity conservation at the local scale. *Environmental Management* 22: 315-321. *A.T.:* endangered species; Florida.

581. Brendler, Thomas, & Henry Carey, March 1998. Community forestry, defined. *J. of Forestry* 96(3): 21-23. *A.T.:* sustainable forestry.

582. Brener, A.G.F., & Adriana Ruggiero, 1994. Leaf-cutting ants (*Atta* and *Acromyrmex*) inhabiting Argentina: patterns in species richness and geographical range sizes. *J. of Biogeography* 21: 391-399. [1] *A.T.:* dispersal barriers.

583. Brett, C.E., L.C. Ivany, & K.M. Schopf, 1996. Coordinated stasis: an overview. *Palaeogeography, Palaeoclimatology, Palaeoecology* 127: 1-20. [2] *A.T.:* evolution; speciation; extinction.

584. Brey, Thomas, Michael Klages, et al., 1994. Antarctic benthic diversity [letter]. *Nature* 368: 297. [1] *A.T.:* Weddell Sea; invertebrates.

585. Breymeyer, Alicja I., R.D. Noble, et al., eds., 1996. *Biodiversity Conservation in Transboundary Protected Areas* [workshop proceedings]. Washington, DC: National Academy Press. 279p. [1/1/1] *A.T.:* Eastern Europe.

586. Bridge, Paul D., Peter Jeffries, et al., eds., 1998. *Information Tech-*

nology, Plant Pathology, and Biodiversity [conference proceedings]. Wallingford, Oxon, U.K., and New York: CAB International. 478p. *A.T.:* plant diseases.

587. Bridgewater, P.B., 1988. Biodiversity and landscape. *Earth-Science Reviews* 25: 486-491. [1] *A.T.:* landscape heterogeneity.

588. Briggs, John C., 1974. *Marine Zoogeography.* New York: McGraw-Hill. 475p. [2/1/3].

589. ———, 1987. *Biogeography and Plate Tectonics.* Amsterdam and New York: Elsevier. 204p. [1/1/2] *A.T.:* paleobiogeography.

590. ———, 1990. *Global Extinctions, Recoveries, and Evolutionary Consequences.* Chicago: University of Chicago. 47p. [1/1/1] *A.T.:* sea-level change; paleobiogeography.

591. ———, 1991. A Cretaceous-Tertiary mass extinction? Were most of earth's species killed off? *Bioscience* 41: 619-624. [1] *A.T.:* sea-level change.

592. ———, 1991. Global species diversity [editorial]. *J. of Natural History* 25: 1403-1406. [1].

593. ———, 1992. The marine East Indies: centre of origin? *Global Ecology and Biogeography Letters* 2: 149-156. [1] *A.T.:* paleobiogeography.

594. ———, 1994. Species diversity: land and sea compared. *Systematic Biology* 43: 130-135. [1].

595. ———, 1995. *Global Biogeography.* Amsterdam and New York: Elsevier. 452p. [1/1/1] *A.T.:* paleobiogeography; extinction.

596. ———, 1996. Tropical diversity and conservation. *CB* 10: 713-718. [1] *A.T.:* wildlife conservation.

597. Bright, Chris, May-June 1992. Who's running America's Forest? A wake up call for all citizens. *Wildlife Conservation* 95(3): 62-67. [1] *A.T.:* old-growth forests; logging; endangered species.

598. Brisbin, I.L., Jr., 1995. Conservation of the wild ancestors of domestic animals. *CB* 9: 1327-1328. [1].

599. Briscoe, D.A., J.M. Malpica, et al., 1992. Rapid loss of genetic variation in large captive populations of *Drosophila* flies: implications for the genetic management of captive populations. *CB* 6: 416-425. [2] *A.T.:* heterozygosity; conservation genetics.

600. British Crop Protection Council, 1997. *Biodiversity and Conservation in Agriculture: Proceedings of an International Symposium.* Farnham, Surrey, U.K. 88p. [1/1/-] *A.T.:* agricultural ecology.

601. Brittingham, M.C., & S.A. Temple, 1983. Have cowbirds caused forest songbirds to decline? *Bioscience* 33: 31-35. [3].

602. Broadus, James M., fall 1992. Biodiversity, Riodiversity. *Oceanus* 35(3): 6-9. [1] *A.T.:* CBD; UNCED; marine conservation.

603. Brock, M.K., & B.N. White, 1992. Application of DNA fingerprinting to the recovery program of the endangered Puerto Rican parrot.

PNAS 89: 11121-11125. [1] *A.T.:* captive breeding; inbreeding.

604. Brockway, D.G., & C.E. Lewis, 1997. Long-term effects of dormant-season prescribed fire on plant community diversity, structure and productivity in a longleaf pine wiregrass ecosystem. *Forest Ecology and Management* 96: 167-183. *A.T.:* Georgia.

605. Brodribb, John, 13 Dec. 1997. Diversity shrinks: fuelling the future. *New Scientist* 156(2112): S4. [1] A general discussion of bdc-related issues.

606. Brokaw, Nicholas V.L., & L.R. Walker, 1991. Summary of the effects of Caribbean hurricanes on vegetation. *Biotropica* 23: 442-447. [2].

607. Bromley, Daniel W., ed., 1992. *Making the Commons Work: Theory, Practice, and Policy.* San Francisco: ICS Press. 339p. [1/1/2].

608. Bronstad, K., K. Dronen, et al., 1996. Phenotypic diversity and antibiotic resistance in soil bacterial communities. *J. of Industrial Microbiology & Biotechnology* 17: 253-259. [1].

609. Brookfield, Harold, & Christine Padoch, June 1994. Appreciating agrodiversity: a look at the dynamism and diversity of indigenous farming practices. *Environment* 36(5): 6-11, 37-45. [1] *A.T.:* traditional agriculture; developing countries.

610. Brooks, Cheri, March-April 1996. War and wildlife in Georgia. *Wildlife Conservation* 99(2): 12. [1] *A.T.:* poaching; Noah's Ark Center; Georgia (former USSR).

611. Brooks, Daniel R., 1990. The unified theory, macroevolution, and historical ecology. In P. Baas, K. Kalkman, & R. Geesink, eds., *The Plant Diversity of Malesia* (Dordrecht and Boston: Kluwer Academic): 379-386. [1] *A.T.:* phylogenetic analysis.

612. Brooks, Daniel R., R.L. Mayden, & D.A. McLennan, 1992. Phylogeny and biodiversity: conserving our evolutionary legacy. *TREE* 7: 55-59. [2] *A.T.:* evolution.

613. Brooks, Daniel R., John Collier, et al., 1989. Entropy and information in evolving biological systems. *Biology & Philosophy* 4: 407-432. [1].

614. Brooks, John Langdon, 1950. Speciation in ancient lakes. *Quarterly Review of Biology* 25: 30-60, 131-176. [1]. *A.T.:* Lake Baikal; Lake Tanganyika; Lake Nyasa.

615. Brooks, Thomas M., & Andrew Balmford, 1996. Atlantic forest extinctions. *Nature* 380: 115. [1] *A.T.:* Brazil; deforestation.

616. Brooks, Thomas M., S.L. Pimm, & N.J. Collar, 1997. Deforestation predicts the number of threatened birds in insular Southeast Asia. *CB* 11: 382-394. *A.T.:* extinction.

617. Brosnan, D.M., & L.L. Crumrine, 1994. Effects of human trampling on marine rocky shore communities. *J. of Experimental Marine Biology and Ecology* 177: 79-97. [1] *A.T.:* ecosystem disturbance.

618. Brothers, Timothy S., & Arthur Spingarn, 1992. Forest fragmentation and alien plant invasion of central Indiana old-growth forests. *CB* 6: 91-

100. [2].

619. Brouha, Paul, April 1995. Beyond fear and loathing: educating the landowner about biodiversity [editorial]. *Fisheries* 20(4): 4. [1] *A.T.:* private landowners; bd education.

620. Browder, John O., 1992. The limits of extractivism: tropical forest strategies beyond extractive reserves. *Bioscience* 42: 174-182. [1] *A.T.:* Bolivia; Brazil; sustainable forestry.

621. Brower, Kenneth, March 1989. State of the reef. *Audubon* 91(2): 56-81. [1] Overview of the conservation status of coral reefs.

622. Brower, Kenneth, & Deanne Klopfer, winter 1989. Losing paradise. *Wilderness* 53(187): 20-31. [1] Concerning the loss of native Hawaiian species.

623. Brower, Lincoln P., & S.B. Malcolm, 1991. Animal migrations: endangered phenomena. *American Zoologist* 31: 265-276. [1] Concerning the vulnerability of spectacular life history phenomena such as mass migrations, even when the entire population is not at risk. *A.T.:* monarch butterflies.

624. Brown, A.H.D., 1989. Core collections: a practical approach to genetic resources management. *Genome* 31: 818-824. [2].

625. Brown, A.H.D., O.H. Frankel, et al., eds., 1989. *The Use of Plant Genetic Resources*. Cambridge, U.K., and New York: CUP. 382p. [1/1/1] *A.T.:* plant germplasm resources.

626. Brown, Barbara E., & J.C. Ogden, Jan. 1993. Coral bleaching. *Scientific American* 268(1): 64-70. [1] *A.T.:* environmental stress.

627. Brown, David E., Frank Reichenbacher, & Susan E. Franson, 1998. *A Classification of North American Biotic Communities*. Salt Lake City: University of Utah Press. 141p.

628. Brown, Donald A., 1996. Thinking globally and acting locally: the emergence of global environmental problems and the critical need to develop sustainable development programs at state and local levels in the United States. *Dickinson J. of Environmental Law & Policy* 5: 175-214. [1] *A.T.:* environmental policy; environmental law.

629. Brown, George G., 1995. How do earthworms affect microfloral and faunal community diversity? *Plant and Soil* 170: 209-231. [1] *A.T.:* soil fauna; soil microflora.

630. Brown, G., & S. Porembski, 1997. The maintenance of species diversity by miniature dunes in a sand-depleted *Haloxylon salicornicum* community in Kuwait. *J. of Arid Environments* 37: 461-473.

631. Brown, James H., 1971. Mammals on mountaintops: nonequilibrium insular biogeography. *American Naturalist* 105: 467-478. [2] *A.T.:* Great Basin.

632. ———, 1978. The theory of insular biogeography and the distribution of boreal birds and mammals. In K.T. Harper & J.L. Reveal, eds., *Intermountain Biogeography: A Symposium* (Provo, UT: Great Basin Naturalist

Memoirs No. 2): 209-227. [2].

633. ———, 1981. Two decades of homage to Santa Rosalia: toward a general theory of diversity. *American Zoologist* 21: 877-888. [2] *A.T.:* community ecology; allocation rules; capacity rules.

634. ———, 1984. On the relationship between abundance and distribution of species. *American Naturalist* 124: 255-279. [3] Discusses the generality that population density tends to increase toward the center of a species' geographical range.

635. ———, 1986. Two decades of interaction between the MacArthur-Wilson model and the complexities of mammalian distributions. *Biological J. of the Linnean Society* 28: 231-251. [1] *A.T.:* island biogeography.

636. ———, 1995. *Macroecology.* Chicago: University of Chicago Press. 269p. [2/2/3] *A.T.:* biogeography; macroevolution.

637. Brown, James H., & Astrid Kodric-Brown, 1977. Turnover rates in insular biogeography: effect of immigration on extinction. *Ecology* 58: 445-449. [3] *A.T.:* Arizona; arthropods.

638. ———, April 1996. Biodiversity on the borderlands. *Natural History* 105(4): 58-61. [1] Describes diversity trends in the Mexican-American borderland region.

639. Brown, James H., & M.V. Lomolino, 1998. *Biogeography* (2nd ed.). Sunderland, MA: Sinauer Associates. 691p. [2/2/3].

640. Brown, James H., & B.A. Maurer, 1987. Evolution of species assemblages: effects of energetic constraints and species dynamics on the diversification of the American avifauna. *American Naturalist* 130: 1-17. [2]. *A.T.:* North America.

641. Brown, James H., & P.F. Nicoletto, 1991. Spatial scaling of species composition: body masses of North American land mammals. *American Naturalist* 138: 1478-1512. [2].

642. Brown, James H., David Mehlman, & G.C. Stevens, 1995. Spatial variation in abundance. *Ecology* 76: 2028-2043. [2] *A.T.:* biogeography; birds; North America.

643. Brown, Katrina, 1992. *Medicinal Plants, Indigenous Medicine and Conservation of Biodiversity in Ghana.* Norwich, East Anglia, and London: CSERGE Working Paper GEC 92-36, Centre for Social and Economic Research on the Global Environment. 30p. [1/1/1] *A.T.:* traditional medicine.

644. ———, 1994. Approaches to valuing plant medicines: the economics of culture or the culture of economics? *Biodiversity and Conservation* 3: 734-750. [1] *A.T.:* bd valuation; conservation policy.

645. ———, 1997. Plain tales from the grasslands: extraction, value and utilization of biomass in Royal Bardia National Park, Nepal. *Biodiversity and Conservation* 6: 59-74. *A.T.:* buffer zones.

646. ———, 1998. The political ecology of biodiversity, conservation and development in Nepal's Terai: confused meanings, means and ends.

Ecological Economics 24: 73-87. *A.T.:* bd definitions; conservation policy.

647. Brown, Katrina, & Dominic Moran, 1993. *Valuing Biodiversity: The Scope and Limitations of Economic Analysis.* Norwich, East Anglia, and London: CSERGE Working Paper GEC 93-09, Centre for Social and Economic Research on the Global Environment. 49p. [1/1/1].

648. Brown, Katrina, David Pearce, et al., 1993. *Economics and the Conservation of Global Biological Diversity.* Washington, DC: Global Environment Facility Working Paper No. 2. 75p. [1/1/1].

649. Brown, Keith S., Jr., 1997. Diversity, disturbance, and sustainable use of Neotropical forests: insects as indicators for conservation monitoring. *J. of Insect Conservation* 1: 25-42. *A.T.:* Amazonia.

650. Brown, Lester R., 1981. *Building a Sustainable Society.* New York: Norton. 433p. [3/3/2].

651. Brown, Michael, & Barbara Wyckoff-Baird, 1992. *Designing Integrated Conservation and Development Projects.* Washington, DC: Biodiversity Support Program. 62p. [1/1/1] *A.T.:* developing countries; Nepal; environmental policy.

652. Brown, Nick, & Malcolm Press, 14 March 1992. Logging rainforests the natural way. *New Scientist* 133(1812): 25-29. [1] *A.T.:* deforestation; sustainable forestry.

653. Brown, Sandra, & L.R. Iverson, 1992. Biomass estimates for tropical forests. *World Resource Review* 4: 366-384. [1] *A.T.:* Asia.

654. Brown, Sandra, A.J.R. Gillespie, & A.E. Lugo, 1991. Biomass of tropical forests of South and Southeast Asia. *Canadian J. of Forest Research* 21: 111-117. [1] *A.T.:* deforestation.

655. Brownell, Robert L., Jr., Katherine Ralls, & W.F. Perrin, fall-winter 1995. Marine mammal biodiversity: three diverse orders encompass 119 species. *Oceanus* 38(2): 30-33. [1] *A.T.:* morphological diversity; endangered species.

656. Browning, Graeme, 1992. Biodiversity battle. *National Journal* 24: 1827-1830. [1] Discusses the reasons for the U.S. unwillingness to ratify the CBD. *A.T.:* biotechnology industry.

657. Bruemmer, Fred, Jan.-Feb. 1994. Northern oases. *Canadian Geographic* 114(1): 54-63. [1] Spotlights polynyas. *A.T.:* Greenland; Canadian Arctic.

658. Bruenig, Eberhard F., 1996. *Conservation and Management of Tropical Rainforests: An Integrated Approach to Sustainability.* Wallingford, Oxon, U.K.: CAB International. 339p. [1/1/1].

659. Bruggen, Adolf Cornelius van, S.M. Wells, & T.C.M. Kemperman, eds., 1995. *Biodiversity and Conservation of the Mollusca: Proceedings of the Alan Solem Memorial Symposium on the Biodiversity and Conservation of the Mollusca.* Oegstgeest-Leiden: Backhuys. 228p. [1/1/1].

660. Brumfield, Robb T., & A.P. Capparella, 1996. Historical diversi-

fication of birds in northwestern South America: a molecular perspective on the role of vicariant events. *Evolution* 50: 1607-1624. [2].

661. Brunello, Anthony, et al., 1997. *Biodiversity Conservation in the Russian Federation*. Washington, DC: Environment Dept., World Bank. 67p. [1/1/-].

662. Brush, Stephen B., 1986. Genetic diversity and conservation in traditional farming systems. *J. of Ethnobiology* 6: 151-167. [1] *A.T.:* Peru; genetic erosion.

663. ———, 1989. Rethinking crop genetic resource conservation. *CB* 3: 19-29. [1] *A.T.:* crop germplasm conservation; in situ conservation; rice.

664. ———, 1991. A farmer-based approach to conserving crop germplasm. *Economic Botany* 45: 153-165. [1] *A.T.:* in situ conservation; landraces.

665. ———, 1992. Ethnoecology, biodiversity, and modernization in Andean potato agriculture. *J. of Ethnobiology* 12: 161-185. [1].

666. ———, 1992. Farmers' rights and genetic conservation in traditional farming systems. *World Development* 20: 1617-1630. [1] *A.T.:* intellectual property rights; in situ conservation; crop germplasm conservation.

667. ———, 1992. Reconsidering the green revolution: diversity and stability in cradle areas of crop domestication. *Human Ecology* 20: 145-167. [1] *A.T.:* Andean agriculture; Asian agriculture.

668. ———, 1993. Indigenous knowledge of biological resources and intellectual property rights: the role of anthropology [essay]. *American Anthropologist* 95: 653-671. [1].

669. ———, 1995. *In situ* conservation of landraces in centers of crop diversity. *Crop Science* 35: 346-354. [1].

670. ———, 1996. Valuing crop genetic resources. *J. of Environment & Development* 5: 416-433. [1] *A.T.:* valuation.

671. Brush, Stephen B., & Doreen Stabinsky, eds., 1996. *Valuing Local Knowledge: Indigenous People and Intellectual Property Rights*. Washington, DC: Island Press. 337p. [1/1/1] *A.T.:* human ecology; sustainable development.

672. Brush, Stephen B., J.E. Taylor, & M.R. Bellon, 1992. Technology adoption and biological diversity in Andean potato agriculture. *J. of Development Economics* 39: 365-387. [1] *A.T.:* in situ conservation.

673. Brussaard, Lijbert, J.P. Bakker, & H. Olff, 1996. Biodiversity of soil biota and plants in abandoned arable fields and grasslands under restoration management. *Biodiversity and Conservation* 5: 211-221. [1] *A.T.:* Western Europe.

674. Brussaard, Lijbert, V.M. Behan-Pelletier, et al., 1997. Biodiversity and ecosystem functioning in soil. *Ambio* 26: 563-570. *A.T.:* soil ecology; soil biota.

675. Brussard, Peter F., 1982. The role of field stations in the preservation

of biological diversity. *Bioscience* 32: 327-330. [1] *A.T.:* environmental education.

676. ———, 1985. The current status of conservation biology. *Bull. of the Ecological Society of America* 66: 9-11. [1].

677. Brussard, Peter F., D.D. Murphy, & R.F. Noss, 1992. Strategy and tactics for conserving biological diversity in the United States [editorial]. *CB* 6: 157-159. [1].

678. Bruton, M.N., 1995. Have fishes had their chips? The dilemma of threatened fishes. *Environmental Biology of Fishes* 43: 1-27. [1].

679. Bryant, Coralie, 1995. Towards developing environmental institutional capacity in Brazil's Amazonian states. *International J. of Technical Cooperation* 1: 171-182. [1] *A.T.:* environmental policy; Natural Resource Policy Project.

680. Bryant, R.L., 1992. Political ecology: an emerging research agenda in Third-World studies. *Political Geography* 11: 12-36. [2] *A.T.:* developing countries.

681. Buchanan, Joseph B., R.J. Fredrickson, & D.E. Seaman, 1998. Mitigation of habitat "take" and the core area concept [comment]. *CB* 12: 238-240. *A.T.:* habitat conservation plans.

682. Buchanan, Rob, Michael Passoff, & Ira Silver, March-April 1993. Looking for rainforest heroes: between rhetoric and reality on an ecotour in Brazil. *Utne Reader,* no. 56: 77-87. [1] *A.T.:* ecotourism.

683. Bucher, E.H., & P.C. Huszar, 1995. Critical environmental costs of the Paraguay-Parana waterway project in South America. *Ecological Economics* 15: 3-9. [1] *A.T.:* environmental impact; environmental management; Pantanal.

684. Buchmann, Stephen L., & G.P. Nabhan, 1996. *The Forgotten Pollinators.* Washington, DC: Island Press/Shearwater Books. 292p. [2/3/1] *A.T.:* pollination; community ecology.

685. Buckley, G. Peter, ed., 1989. *Biological Habitat Reconstruction.* London and New York: Belhaven Press. 363p. [2/1/1] *A.T.:* restoration ecology.

686. Buckley, P.A., M.S. Foster, et al., eds., 1985. *Neotropical Ornithology.* Washington, DC: American Ornithologists' Union. 1041p. [1/1/1].

687. Budiansky, Stephen, 1994. Extinction or miscalculation? [letter] *Nature* 370: 105. [1] Concerning the species-area relation.

688. ———, 1995. *Nature's Keepers: The New Science of Nature Management.* New York: Free Press. 310p. [2/2/1] *A.T.:* nature conservation.

689. ———, 25 Nov. 1996. Killing with kindness. *U.S. News and World Report* 121(21): 48-49. [1] Looks at the difficulties inherent in trying to interest Third World peoples in wildlife conservation.

690. Budiansky, Stephen, David Tilman, & Clarence Lehman, 1996. Species fragmentation or area loss? *Nature* 382: 215-216. [1].

691. Buffett, Howard G., 1996. The partnership of biodiversity and high yield agricultural production. In *Research in Domestic and International Agribusiness Management* (Greenwich, CT, and London: JAI Press), Vol. 12: 1-88. [1].

692. Buffington, John D., 1993. Sustaining the Diversity of Birds: Intercontinental Experiences. Nairobi: African Centre for Technology Studies. 32p. [1/1/1] *A.T.:* bird conservation.

693. Bull, Alan T., 1996. Biotechnology for environmental quality: closing the circles. *Biodiversity and Conservation* 5: 1-25. [1].

694. Bull, Alan T., Michael Goodfellow, & J.H. Slater, 1992. Biodiversity as a source of innovation in biotechnology. *Annual Review of Microbiology* 46: 219-252. [2] *A.T.:* microbial diversity; microbial taxonomy.

695. Bunce, Robert G.H., Lech Ryszkowski, & M.G. Paoletti, eds., 1993. *Landscape Ecology and Agroecosystems* [symposium papers]. Boca Raton, FL: Lewis Publishers. 241p. [1/1/1].

696. Bunge, J., & M. Fitzpatrick, 1993. Estimating the number of species: a review. *J. of the American Statistical Association* 88: 364-373. [2] *A.T.:* classification.

697. Bunkley-Williams, Lucy, & E.H. Williams, Jr., April 1990. Global assault on coral reefs: what's killing the great reefs of the world? *Natural History* 99(4): 46-54. [1] *A.T.:* coral bleaching; Puerto Rico.

698. Bunnell, Fred L., 1995. Forest-dwelling vertebrate faunas and natural fire regimes in British Columbia: patterns and implications for conservation. *CB* 9: 636-644. [1] *A.T.:* forest management.

699. ———, 1998. Overcoming paralysis by complexity when establishing operational goals for biodiversity. *J. of Sustainable Forestry* 7(3-4): 145-164. *A.T.:* forest management; bd definitions.

700. Bunnell, Fred L., & Jacklyn F. Johnson, eds., 1998. *Policy and Practices for Biodiversity in Managed Forests: The Living Dance*. Vancouver, BC: UBC Press. 162p. *A.T.:* forest genetics.

701. Bunyard, Peter, 1987. The significance of the Amazon Basin for global climatic equilibrium. *Ecologist* 17: 139-141. [1] *A.T.:* deforestation; climatic change; vegetation and climate.

702. Burbidge, A.A., & N.L. McKenzie, 1989. Patterns in the modern decline of Western Australia's vertebrate fauna: causes and conservation implications. *Biological Conservation* 50: 143-198. [2].

703. Burbidge, A.A., M.R. Williams, & Ian Abbott, 1997. Mammals of Australian islands: factors influencing species richness. *J. of Biogeography* 24: 703-715.

704. Burbrink, Frank T., C.A. Phillips, & E.J. Heske, 1995. *Factors Determining Richness of Reptile and Amphibian Species in a Riparian Wildlife Dispersal Corridor*. Champaign, IL: Center for Biodiversity, Natural History Survey (Illinois). 53p. [1/1/1] *A.T.:* Illinois.

705. Burel, Françoise, & Jacques Baudry, 1995. Species biodiversity in changing agricultural landscapes: a case study in the Pays d'Auge, France. *Agriculture, Ecosystems & Environment* 55: 193-200. [1] *A.T.:* landscape structure.

706. Burger, William C., 1981. Why are there so many kinds of flowering plants? *Bioscience* 31: 572, 577-581. [1].

707. Burgess, Joanne C., 1992. *The Impact of Wildlife Trade on Endangered Species.* London: Discussion Paper DP 92-02, London Environmental Economics Centre. 29, 17, 10p. [1/1/1] *A.T.:* wildlife conservation; wild animal trade.

708. Burgess, Robert L., 1988. Community organization: effects of landscape fragmentation. *Canadian J. of Botany* 66: 2687-2690. [1] *A.T.:* island biogeography.

709. Burgess, Robert L., & D.M. Sharpe, eds., 1981. *Forest Island Dynamics in Man-dominated Landscapes.* New York: Springer. 310p. [2/1/2] *A.T.:* island biogeography; forest ecology; habitat fragmentation.

710. Burgman, Mark A., H.R. Akçakaya, & S.S. Loew, 1988. The use of extinction models for species conservation. *Biological Conservation* 43: 9-25. [2] *A.T.:* population biology models.

711. Burgman, Mark A., S. Ferson, & H.R. Akçakaya, 1993. *Risk Assessment in Conservation Biology.* London and New York: Chapman & Hall. 314p. [1/1/3].

712. Burhenne, Wolfgang E., 1992. Biodiversity—the legal aspects. *Environmental Policy and Law* 22: 324-326. [1].

713. Burk, Dan L., Kenneth Barovsky, & G.H. Monroe, 1993. Biodiversity and biotechnology. *Science* 260: 1900-1901. [1] Why the U.S. refused to sign the CBD in Rio.

714. Burke, I.C., T.G.F. Kittel, et al., 1991. Regional analysis of the central Great Plains. *Bioscience* 41: 685-692. [2] *A.T.:* ecosystem modelling; carbon balance; grasslands.

715. Burke, V.J., & J.W. Gibbons, 1995. Terrestrial buffer zones and wetland conservation: a case study of freshwater turtles in a Carolina bay. *CB* 9: 1365-1369. [1] *A.T.:* GIS.

716. Burkey, Tormod V., 1989. Extinction in nature reserves: the effect of fragmentation and the importance of migration between reserve fragments. *Oikos* 55: 75-81. [2] *A.T.:* extinction modelling.

717. ———, 1994. Tropical tree species diversity: a test of the Janzen-Connell model. *Oecologia* 97: 533-540. [1].

718. ———, 1995. Faunal collapse in East African game reserves revisited. *Biological Conservation* 71: 107-110. [1] *A.T.:* large mammals; Malay Archipelago.

719. Burnett, Michael R., P.V. August, et al., 1998. The influence of geomorphological heterogeneity on biodiversity I. A patch-scale perspective.

CB 12: 363-370. *A.T.:* North America; woody plants.

720. Burney, David A., 1993. Recent animal extinctions: recipes for disaster. *American Scientist* 81: 530-541. [1] *A.T.:* Pleistocene extinctions; human population growth.

721. Burney, David A., & R.D.E. MacPhee, July 1988. Mysterious island: what killed Madagascar's large native animals? *Natural History* 97(7): 46-55. [1].

722. Burnham, Charles R., 1988. The restoration of the American chestnut. *American Scientist* 76: 478-487. [1].

723. Burrows, Beth, winter 1998. Biopiracy, patenting, and international trade agreements. *Synthesis/Regeneration* (winter): 31-34. *A.T.:* developing countries; intellectual property.

724. Burton, P.J., A.C. Balisky, et al., 1992. The value of managing for biodiversity. *Forestry Chronicle* 68: 225-237. [2] *A.T.:* bd management; bd value.

725. Busch, David E., & S.D. Smith, 1995. Mechanisms associated with decline of woody species in riparian ecosystems of the southwestern U.S. *Ecological Monographs* 65: 347-370. [1] *A.T.:* Colorado River; ecosystem disturbance.

726. Busch, Lawrence, W.B. Lacy, et al., 1995. *Making Nature, Shaping Culture: Plant Biodiversity in Global Context.* Lincoln: University of Nebraska Press. 261p. [1/2/1] *A.T.:* plant germplasm resources; food crops.

727. Bush, A.O., J.M. Aho, & C.R. Kennedy, 1990. Ecological versus phylogenetic determinants of helminth parasite community richness. *Evolutionary Ecology* 4: 1-20. [2].

728. Bush, Mark B., 1994. Amazonian speciation: a necessarily complex model. *J. of Biogeography* 21: 5-17. [2] *A.T.:* vicariance; refugia; climatic change.

729. ———, 1996. Amazonian conservation in a changing world. *Biological Conservation* 76: 219-228. [1] *A.T.:* reserve selection; refugia.

730. Bush, Mark B., & R.J. Whittaker, 1991. Krakatau: colonization patterns and hierarchies. *J. of Biogeography* 18: 341-356. [1] *A.T.:* succession; species turnover.

731. Busing, R.T., & P.S. White, 1997. Species diversity and small-scale disturbance in an old-growth temperate forest: a consideration of gap partitioning concepts. *Oikos* 78: 562-568. *A.T.:* trees; Appalachians.

732. Butcher, B., spring 1996. Local biodiversity planning—the lessons from Mendip. *ECOS* 17(2): 35-41. [1] *A.T.:* action plans; England.

733. Butler, Brett J., & R.L. Chazdon, 1998. Species richness, spatial variation, and abundance of the soil seed bank of a secondary tropical rain forest. *Biotropica* 30: 214-222. *A.T.:* Costa Rica.

734. Butler, Declan, 1995. Genetic diversity proposal fails to impress international ethics panel. *Nature* 377: 373. [1] Describes controversy over

the Human Genome Diversity Project.

735. Butler, James N., James Burnett-Herkes, et al., Jan.-Feb. 1993. The Bermuda fisheries: a tragedy of the commons averted? *Environment* 35(1): 6-15, 25-33. [1].

736. Butman, Cheryl Ann, & J.T. Carlton, 1995. Marine biological diversity: some important issues, opportunities and critical research needs. *Reviews of Geophysics* 33, Suppl. B: 1201-1209. [1].

737. Butman, Cheryl Ann, J.T. Carlton, & S.R. Palumbi, 1995. Whaling effects on deep-sea biodiversity. *CB* 9: 462-464. [1] *A.T.:* marine mammals.

738. Butterfield, R.P., 1995. Promoting biodiversity: advances in evaluating native species for reforestation. *Forest Ecology and Management* 75: 111-121. [1] *A.T.:* Costa Rica.

739. Buys, M.H., P.J. Vorster, & J.J.A. Van der Walt, 1995. Using the WORLDMAP PC program for measuring biodiversity in order to choose prioritized conservation areas in Southern Africa. *South African J. of Botany* 61: 80-84. [1].

740. Buzas, Martin A., 1972. Patterns of species diversity and their explanation. *Taxon* 21: 275-286. [1].

741. Buzas, Martin A., & S.J. Culver, 1991. Species diversity and dispersal of benthic foraminifera. *Bioscience* 41: 483-489. [1] *A.T.:* biogeography.

742. Bye, Robert, Edelmira Linares, & Eric Estrada, 1995. Biological diversity of medicinal plants in Mexico. In J.T. Arnason, Rachel Mata, & J.T. Romeo, eds., *Phytochemistry of Medicinal Plants* (New York: Plenum): 65-82. [1].

743. Byrd, John E., 1997. The analysis of diversity in archaeological faunal assemblages: complexity and subsistence strategies in the Southeast during the Middle Woodland Period. *J. of Anthropological Archaeology* 16: 49-72.

744. Byron, N., & M.R. Perez, 1996. What future for the tropical moist forests 25 years hence? *Commonwealth Forestry Review* 75: 124-129. [1] *A.T.:* deforestation; bd loss.

745. Cade, Tom J., & S.A. Temple, 1995. Management of threatened bird species: evaluation of the hands-on approach. *Ibis* 137, Suppl.: S161-S172. [1].

746. Cadwallader, Eva H., 1994. Ultimate reality and meaning in the conflict between globalism and anti-globalism. *Ultimate Reality and Meaning* 17: 232-245. [1] *A.T.:* environmental philosophy.

747. Caicco, S.L., J.M. Scott, et al., 1995. A gap analysis of the management status of the vegetation of Idaho (USA). *CB* 9: 498-511. [1] *A.T.:* remote sensing.

748. Cairns, John, Jr., 1986. The myth of the most sensitive species: multispecies testing can provide valuable evidence for protecting the environment. *Bioscience* 36: 670-672. [2] *A.T.:* environmental monitoring; indi-

749. Cairns, John, Jr., & J.R. Bidwell, 1996. The modification of human society by natural systems: discontinuities caused by the exploitation of endemic species and the introduction of exotics [editorial]. *Environmental Health Perspectives* 104: 1142-1145. [1] Argues that an indifference to the possible consequences of human activities may lead to a world dominated by pest species.

750. Cairns, John, Jr., & J.R. Heckman, 1996. Restoration ecology: the state of an emerging field. *Annual Review of Energy and the Environment* 21: 167-189. [1].

751. Cairns, Michael A., & R.T. Lackey, May-June 1992. Biodiversity and management of natural resources: the issues. *Fisheries* 17(3): 6-10. [1] *A.T.:* land use.

752. Cairns, Michael A., & R.A. Meganck, 1994. Carbon sequestration, biological diversity, and sustainable development: integrated forest management. *Environmental Management* 18: 13-22. [1].

753. Cairns, Michael A., Rodolfo Dirzo, & Frank Zadroga, July 1995. Forests of Mexico: a diminishing resource? *J. of Forestry* 93(7): 21-24. [1] *A.T.:* deforestation; sustainable forestry.

754. Caldecott, Julian O., 1994. *Terrestrial Biodiversity Management in Indonesia: Study and Recommendations.* Halifax, NS: School for Resource and Environmental Studies, Dalhousie University. 59, 89p. [1/1/1].

755. ———, 1996. *Designing Conservation Projects.* Cambridge, U.K., and New York: CUP. 312p. [1/1/1] *A.T.:* tropical forests.

756. Caldecott, Julian O., M.D. Jenkins, et al., 1996. Priorities for conserving global species richness and endemism. *Biodiversity and Conservation* 5: 699-727. [1] *A.T.:* environmental policy and planning.

757. Caldwell, Martyn M., A.H. Teramura, et al., 1995. Effects of increased solar ultraviolet radiation on terrestrial plants. *Ambio* 24: 166-173. [2] *A.T.:* ozone.

758. Cale, P.G., & R.J. Hobbs, 1994. Landscape heterogeneity indices: problems of scale and applicability, with particular reference to animal habitat description. *Pacific Conservation Biology* 1: 183-193. [1].

759. Caletrio, J., J.M. Fernández, et al., 1996. Spanish national inventory on road mortality of vertebrates. *Global Biodiversity* 5(4): 15-18. [1].

760. Caley, M. Julian, & Dolph Schluter, 1997. The relationship between local and regional diversity. *Ecology* 78: 70-80. [2]. *A.T.:* spatial scale; sampling methods, Australia; North America; species richness.

761. Callaghan, T.V., & S. Jonasson, 1995. Arctic terrestrial ecosystems and environmental change. *PTRSL A* 352: 259-276. [1].

762. Callahan, Joan R., 1996. Vanishing biodiversity. *Environmental Health Perspectives* 104: 386-388. [1] Discusses the 1995 *Global Biodiversity Assessment* report by UNEP.

763. Callicott, J. Baird, 1980. Animal liberation: a triangular affair. *Environmental Ethics* 2: 311-338. [2] Compares the ethics of the animal liberation movement with those of Aldo Leopold's "land ethic." *A.T.:* environmental ethics.

764. ———, 1989. *In Defense of the Land Ethic: Essays in Environmental Philosophy.* Albany: State University of New York Press. 325p. [2/2/2] *A.T.:* Aldo Leopold; environmental ethics.

765. ———, 1990. Whither conservation ethics? *CB* 4: 15-20. [1] *A.T.:* Aldo Leopold; environmental ethics.

766. ———, 1994. *Earth's Insights: A Survey of Ecological Ethics From the Mediterranean Basin to the Australian Outback.* Berkeley: University of California Press. 285p. [2/2/1].

767. ———, 1995. A review of some problems with the concept of ecosystem health. *Ecosystem Health* 1: 101-112. [2].

768. ———, 1995. The value of ecosystem health. *Environmental Values* 4: 345-361. [1] *A.T.:* environmental value; environmental ethics.

769. ———, 1996. Should wilderness areas become biodiversity reserves? *George Wright Forum* 13(2): 32-38. [1].

770. Callicott, J. Baird, & Karen Mumford, 1997. Ecological sustainability as a conservation concept. *CB* 11: 32-40.

771. Callicott, J. Baird, & M.P. Nelson, eds., 1998. *The Great New Wilderness Debate* [anthology]. Athens, GA: University of Georgia Press. 697p. *A.T.:* nature philosophy.

772. Callow, J.A., B.V. Ford-Lloyd, & H.J. Newbury, eds., 1997. *Biotechnology and Plant Genetic Resources: Conservation and Use.* Wallingford, Oxon, U.K., and New York: CAB International. 308p. [1/1/-] *A.T.:* plant germplasm resources.

773. Camargo, Julio A., 1992. New diversity index for assessing structural alterations in aquatic communities. *Bull. of Environmental Contamination and Toxicology* 48: 428-434. [1] *A.T.:* pollution monitoring.

774. ———, 1997. A thermodynamic perspective on natural selection. *Acta Biotheoretica* 46: 65-75. *A.T.:* competitive exclusion.

775. Cameron, D.S., D.J. Leopold, & D.J. Raynal, 1997. Effect of landscape position on plant diversity and richness on electric transmission rights-of-way in New York State. *Canadian J. of Botany* 75: 242-251. *A.T.:* introduced plants.

776. Cameron, Mary M., 1996. Biodiversity and medicinal plants in Nepal: involving Untouchables in conservation and development. *Human Organization* 55: 84-92. [1].

777. Campbell, David G., & H.D. Hammond, eds., 1989. *Floristic Inventory of Tropical Countries: The Status of Plant Systematics, Collections, and Vegetation, Plus Recommendations for the Future.* Bronx, NY: New York Botanical Garden. 545p. [1/1/1] *A.T.:* vegetation surveys.

778. Campbell, Jeffrey Y., 1987. *Tropical Forestry and Biological Diversity in India and the Role of USAID/New Delhi.* New Haven: Tropical Resources Institute, Yale School of Forestry and Environmental Studies. 73p. [1/1/1].

779. Canadian Forest Service (Science Branch), 1997. *Biodiversity in the Forest: The Canadian Forest Service Three-Year Action Plan.* Ottawa. 38p. [1/1/-] *A.T.:* Canada; forest conservation; environmental policy.

780. Canhos, V.P., D.A.L. Canhos, & S. de Souza, 1993. Establishment of a computerized biodiversity/biotechnology network: the Brazilian effort. *J. of Biotechnology* 31: 67-73. [1] *A.T.:* computer programs.

781. Canhos, V.P., D. Lange, et al., eds., 1992. *Needs and Specifications for a Biodiversity Information Network: Proceedings of an International Workshop.* Nairobi: UNEP. 265p. [1/1/1] *A.T.:* computer networks.

782. Cannon, John R., 1996. Whooping crane recovery: a case study in public and private cooperation in the conservation of endangered species. *CB* 10: 813-821. [1] *A.T.:* environmental policy.

783. Cannon, L.R.G., & K.B. Sewell, 1994. Symbionts and biodiversity. *Memoirs of the Queensland Museum* 36: 33-40. [1] *A.T.:* invertebrates; crayfishes; Australia.

784. Cannon, P.F., 1997. Strategies for rapid assessment of fungal diversity. *Biodiversity and Conservation* 6: 669-680. *A.T.:* sampling analysis.

785. Cao, K.-F., & R. Peters, 1997. Species diversity of Chinese beech forests in relation to warmth and climatic disturbances. *Ecological Research* 12: 175-189.

786. Cao, M., & J. Zhang, 1997. Tree species diversity of tropical forest vegetation in Xishuangbanna, SW China. *Biodiversity and Conservation* 6: 995-1006.

787. Cao, Y., W.P. Williams, & A.W. Bark, 1997. Similarity measure bias in river benthic aufwuchs community analysis. *Water Environment Research* 69: 95-106. Compares nine similarity measures. *A.T.:* invertebrates.

788. Caporale, L.H., 1995. Chemical ecology: a view from the pharmaceutical industry. *PNAS* 92: 75-82. [1] *A.T.:* natural products; medicinal chemistry; drug discovery.

789. Capparella, Angelo P., 1991. Neotropical avian diversity and riverine barriers. *Acta XX Congressus Internationalis Ornithologici*, Vol. 1: 307-316. [1].

790. Cardoso da Silva, J.M., 1995. Avian inventory of the cerrado region, South America: implications for biological conservation. *Bird Conservation International* 5: 291-304. [1] *A.T.:* Brazil; sampling.

791. Cardskadden, H., & D. Lober, 1998. Environmental stakeholder management as business strategy: the case of the corporate wildlife habitat enhancement programme. *J. of Environmental Management* 52: 183-202. Focuses on the benefits to corporations of taking part in bdc programs.

792. Carey, Andrew B., 1996. Interactions of Northwest forest canopies and arboreal mammals. *Northwest Science* 70, Special Issue: 72-78. [1] *A.T.:* rodents.

793. Carey, Andrew B., & R.O. Curtis, 1996. Conservation of biodiversity: a useful paradigm for forest ecosystem management. *Wildlife Society Bull.* 24: 610-620. [1] *A.T.:* Oregon; Washington.

794. Carlowicz, Michael, 1997. Was Cambrian explosion the result of wandering continents? *Eos* 78: 381-382. *A.T.:* paleobiogeography; continental drift; diversification.

795. Carlquist, Sherwin, 1974. *Island Biology*. New York: Columbia University Press. 660p. [2/2/2] *A.T.:* island biota.

796. Carlson, Cynthia, 1988. NEPA and the conservation of biological diversity. *Environmental Law* 19: 15-36. [1].

797. Carlson, T.J., M.M. Iwu, et al., spring 1997. Medicinal plant research in Nigeria: an approach for compliance with the Convention on Biological Diversity. *Diversity* 13(1): 29-33. [1].

798. Carlton, James T., 1985. Transoceanic and interoceanic dispersal of coastal marine organisms: the biology of ballast water. *Oceanography and Marine Biology* 23: 313-371. [2].

799. ———, 1989. Man's role in changing the face of the ocean: biological invasions and implications for conservation of near-shore environments. *CB* 3: 265-273. [2].

800. ———, 1993. Neoextinctions of marine invertebrates. *American Zoologist* 33: 499-509. [1] Gives four case studies.

801. ———, 1996. Biological invasions and cryptogenic species. *Ecology* 77: 1653-1655. [2].

802. Carlton, James T., & Cheryl Ann Butman, fall 1995. Understanding marine biodiversity. *Oceanus* 38(2): 4-8. [1] An overview of environmental issues affecting marine bd. *A.T.:* National Research Council.

803. Carlton, James T., & J.B. Geller, 1993. Ecological roulette: the global transport of nonindigenous marine organisms. *Science* 261: 78-82. [3] *A.T.:* ballast water; biological invasions; Oregon; Japan.

804. Carlton, James T., & J. Hodder, 1995. Biogeography and dispersal of coastal marine organisms: experimental studies on a replica of a 16th-century sailing vessel. *Marine Biology* 121: 721-730. [1] *A.T.:* Oregon; California; fouling species.

805. Carney, R.S., 1997. Basing conservation policies for the deep-sea floor on current-diversity concepts: a consideration of rarity. *Biodiversity and Conservation* 6: 1463-1485. *A.T.:* Gulf of Mexico; Western Atlantic; waste dumping.

806. Caro, T.M., & M.K. Laurenson, 1994. Ecological and genetic factors in conservation: a cautionary tale. *Science* 263: 485-486. [2] *A.T.:* cheetahs.

807. Caro, T.M., N. Pelkey, & M. Grigione, 1994. Effects of conservation

biology education on attitudes toward nature. *CB* 8: 846-852. [1] *A.T.:* human ecology.

808. Caron, David A., & R.J. Gast, fall 1995. Probing biodiversity. *Oceanus* 38(2): 11-15. [1] Examines use of molecular biology techniques to investigate bd.

809. Carpaneto, Giuseppe M., & F.P. Germi, 1992. Diversity of mammals and traditional hunting in Central African rain forests. *Agriculture, Ecosystems & Environment* 40: 335-354. [1] *A.T.:* Zaire; Uganda; food preferences.

810. Carpenter, J.F., 1998. Internally motivated development projects: a potential tool for biodiversity conservation outside of protected areas. *Ambio* 27: 211-216.

811. Carpenter, S.R., S.G. Fisher, et al., 1992. Global change and freshwater ecosystems. *ARES* 23: 119-139. [2] *A.T.:* climatic change; fishes.

812. Carr, M.H., & D.C. Reed, 1993. Conceptual issues relevant to marine harvest refuges: examples from temperate reef fishes. *Canadian J. of Fisheries and Aquatic Science* 50: 2019-2028. [1] *A.T.:* recruitment rates.

813. Carr, Thomas A., H.L. Pedersen, & Sunder Ramaswamy, Sept. 1993. Rain forest entrepreneurs: cashing in on conservation. *Environment* 35(7): 12-15, 33-38. [1] Describes how certain companies have become involved in rainforest preservation ventures. *A.T.:* Latin America.

814. Carr, Timothy R., & J.A. Kitchell, 1980. Dynamics of taxonomic diversity. *Paleobiology* 6: 427-433. [1] *A.T.:* evolution; mathematical models.

815. Carroll, C. Ronald, J.H. Vandermeer, & Peter Rosset, eds., 1990. *Agroecology.* New York: McGraw-Hill. 641p. [1/1/1].

816. Carroll, Sean B., 1994. Developmental regulatory mechanisms in the evolution of insect diversity. *Development,* 1994 Supplement: 217-223. [1] *A.T.:* homeobox genes.

817. Carroll, S.S., & D.L. Pearson, 1998. Spatial modeling of butterfly species richness using tiger beetles (Cicindelidae) as a bioindicator taxon. *Ecological Applications* 8: 531-543. *A.T.:* spatial pattern; North America.

818. ———, 1998. The effects of scale and sample size on the accuracy of spatial predictions of tiger beetle (Cicindelidae) species richness. *Ecography* 21: 401-414. *A.T.:* spatial scale; western North America.

819. Carson, Hampton L., J.P. Lockwood, & E.M. Craddock, 1990. Extinction and recolonization of local populations on a growing shield volcano. *PNAS* 87: 7055-7057. [1] *A.T.: Drosophila*; Hawaii.

820. Carson, Rachel, 1962. *Silent Spring.* Boston: Houghton Mifflin. 368p. [3/-/3] This famous study of the ecological impact of DDT and other pesticides helped launch the environmental movement. *A.T.:* environmental conservation; pollution.

821. Carté, Brad K., 1996. Biomedical potential of marine natural products: marine organisms are yielding novel molecules for use in basic research and medical applications. *Bioscience* 46: 271-286. [2].

822. Carter, Brandon, 1983. The anthropic principle and its implications for biological evolution. *PTRSL A* 310: 347-363. [1].
823. Carter, J., J. Gibson, et al., 1994. Creation of the Hol Chan Marine Reserve in Belize: a grass-roots approach to barrier reef conservation. *Environmental Professional* 16: 220-231. [1] *A.T.:* marine parks.
824. Carter, L.E., 1992. Wilderness and its role in the preservation of biodiversity: the need for a shift in emphasis. *Australian Zoologist* 28: 28-36. [1] *A.T.:* environmental legislation; Australia.
825. Case, Ted J., 1978. A general explanation for insular body size trends in terrestrial vertebrates. *Ecology* 59: 1-18. [1] *A.T.:* Gulf of California; biogeography; mammals; reptiles.
826. Case, Ted J., & D.T. Bolger, 1991. The role of introduced species in shaping the distribution and abundance of island reptiles. *Evolutionary Ecology* 5: 272-290. [1] *A.T.:* Pacific islands.
827. Cassidy, Kelly M., M.R. Smith, et al., 1997. *Gap Analysis of Washington State: An Evaluation of the Protection of Biodiversity* [report]. Seattle: Washington Cooperative Fish and Wildlife Research Unit, University of Washington. 137p. [1/1/-] *A.T.:* GIS.
828. Castañeda, Porfirio G., 1993. Management planning for the Palawan Biosphere Reserve. *Nature and Resources* 29: 35-38. [1] *A.T.:* Philippines; protected areas management.
829. Casteels, Peter, Jill Romagnolo, et al., 1994. Biodiversity of apidaecin-type peptide antibiotics: prospects of manipulating the antibacterial spectrum and combating acquired resistance. *J. of Biological Chemistry* 269: 26107-26115. [1] *A.T.:* insects.
830. Castello, John D., D.J. Leopold, & P.J. Smallidge, 1995. Pathogens, patterns, and processes in forest ecosystems. *Bioscience* 45: 16-24. [2] *A.T.:* plant pathogens; forest ecology.
831. Castilleja, Guillermo, P.J. Poole, & C.C. Geisler (ed. S.H. Davis), 1993. *The Social Challenge of Biodiversity Conservation*. Washington, DC: Working Paper No. 1, Global Environment Facility. 43p. [1/1/1].
832. Castley, J.G., & G.I.H. Kerley, 1996. The paradox of forest conservation in South Africa. *Forest Ecology and Management* 85: 35-46. [1] *A.T.:* wildlife conservation.
833. Castro Parga, Isabel, & J.C. Moreno Saiz, 1996. Strengthening the natural and national park system of Iberia to conserve vascular plants. *Botanical J. of the Linnean Society* 121: 189-205. [1] *A.T.:* rarity; species richness; character richness.
834. Catling, Peter C., 1994. Bushfires and prescribed burning: protecting native fauna. *Search* 25: 37-40. [1].
835. Caughley, Graeme, 1994. Directions in conservation biology. *J. of Animal Ecology* 63: 215-244. [3] Discusses two major themes in conservation biology: the "small population" paradigm and the "declining popula-

tion" paradigm.

836. Caughley, Graeme, & Anne Gunn, 1996. *Conservation Biology in Theory and Practice*. Cambridge, MA: Blackwell Science. 459p. [2/2/2].

837. Cavalier-Smith, T., 1987. The origin of eukaryote and archaebacterial cells. *Annals of the New York Academy of Sciences* 503: 17-54. [2].

838. ———, 1993. Kingdom Protozoa and its 18 phyla. *Microbiological Reviews* 57: 953-994. [3] A major study on classification.

839. Cavanaugh, Colleen M., 1994. Microbial symbiosis: patterns of diversity in the marine environment. *American Zoologist* 34: 79-89. [2] *A.T.:* evolutionary potential; habitat loss.

840. Cawley, R. McGreggor, 1988. Biodiversity: lessons from the U.S. Wilderness Act. *Society & Natural Resources* 1: 205-214. [1] *A.T.:* environmental policy.

841. Ceballos, Gerardo, & J.H. Brown, 1995. Global patterns of mammalian diversity, endemism, and endangerment. *CB* 9: 559-568. [2].

842. Ceballos, Gerardo, & Andrés Garcia, 1995. Conserving Neotropical biodiversity: the role of dry forests in western Mexico. *CB* 9: 1349-1353. [1].

843. Ceballos, Gerardo, P. Rodriguez, & R.A. Medellín, 1998. Assessing conservation priorities in megadiverse Mexico: mammalian diversity, endemicity, and endangerment. *Ecological Applications* 8: 8-17.

844. Ceballos-Lascuráin, Héctor, ed., 1996. *Tourism, Ecotourism, and Protected Areas: The State of Nature-based Tourism Around the World and Guidelines for Its Development* [papers from workshops]. Gland, Switzerland, and Cambridge, U.K.: IUCN. 301p. [1/1/1].

845. Center, T.D., J.H. Frank, & F.A. Dray, Jr., 1995. Biological invasions: stemming the tide in Florida. *Florida Entomologist* 78: 45-55. [1] *A.T.:* introduced species; control programs; insects.

846. Center for Wildlife Law (University of New Mexico), & Defenders of Wildlife, 1996. *Saving Biodiversity: A Status Report on State Laws, Policies and Programs*. Albuquerque. 218p. [1/1/1] *A.T.:* environmental law; environmental policy.

847. Centre for Ecological Sciences, Indian Institute of Science, 1991. *Proceedings of the Workshop on Linkages Between Wastelands and Biodiversity*. Bangalore. 1076p. [1/1/1] *A.T.:* India.

848. Cervigni, Raffaello, 1995. *North-South Transfers, Incremental Cost and the Rio Environment Conventions*. Norwich, U.K.: CSERGE Working Paper GEC 95-33, Centre for Social and Economic Research on the Global Environment. 37p. [1/1/1] *A.T.:* CBD.

849. ———, 1998. Incremental cost in the Convention on Biological Diversity. *Environmental & Resource Economics* 11: 217-241. *A.T.:* environmental economics.

850. Chabanet, P., H. Ralambondrainy, et al., 1997. Relationships between coral reef substrata and fish. *Coral Reefs* 16: 93-102. *A.T.:* community

ecology; Réunion.

851. Chadwick, Derek J., & Joan Marsh, eds., 1994. *Ethnobotany and the Search for New Drugs* [symposium papers]. Chichester, U.K., and New York: Wiley. 280p. [1/1/1] *A.T.:* medicinal plants; ethnopharmacology.

852. Chadwick, Douglas H., May-June 1990. The biodiversity challenge. *Defenders* 65(3): 19-30. [1].

853. ———, Sept.-Oct. 1992. U.S. imperative: networking habitats. *Defenders* 67(5): 26-33. [1] *A.T.:* gap analysis; GIS; Idaho.

854. Chaloner, William G., & Anthony Hallam, eds., 1994. *Evolution and Extinction: Proceedings of a Joint Symposium of the Royal Society and the Linnean Society.* Cambridge, U.K., and New York: CUP. 248p. [1/1/1] *A.T.:* paleobiology.

855. Chaloupka, M.Y., & S.B. Domm, 1986. Role of anthropochory in the invasion of coral cays by alien flora. *Ecology* 67: 1536-1547. [1] *A.T.:* plant invasions; Great Barrier Reef.

856. Chambers, J.R., 1991. Coastal degradation and fish population losses. In Richard H. Stroud, ed., *Stemming the Tide of Coastal Fish Habitat Loss* (Savannah, GA: National Coalition for Marine Conservation): 45-51. [1].

857. Chandel, K.P.S., 1996. Biodiversity and strategies for plant germplasm conservation in India. *Tropical Ecology* 37: 21-29. [1].

858. Chandran, M.D.S., & J.D. Hughes, 1997. The sacred groves of South India: ecology, traditional communities and religious change. *Social Compass* 44: 413-427.

859. Chang, T.T., 1984. Conservation of rice genetic resources: luxury or necessity? *Science* 224: 251-256. [1] *A.T.:* rice germplasm.

860. Chapin, F. Stuart, III, & Christian Korner, eds., 1995. *Arctic and Alpine Biodiversity: Patterns, Causes, and Ecosystem Consequences* [workshop papers]. Berlin and New York: Springer. 332p. [1/1/2].

861. Chapin, F. Stuart, III, E.-D. Schulze, & H.A. Mooney, 1992. Biodiversity and ecosystem processes. *TREE* 7: 107-108. [1].

862. Chapin, F. Stuart, III, O.E. Sala, et al., 1998. Ecosystem consequences of changing biodiversity. *Bioscience* 48: 45-52. *A.T.:* bd loss; extinction; human disturbance.

863. Chapin, F. Stuart, III, G.R. Shaver, et al., 1995. Responses of arctic tundra to experimental and observed changes in climate. *Ecology* 76: 694-711. [2] *A.T.:* Alaska.

864. Chapin, F. Stuart, III, B.H. Walker, et al., 1997. Biotic control over the functioning of ecosystems. *Science* 277: 500-504. *A.T.:* human impact; keystone species.

865. Chapin, Mac, 1990. Introduction: the value of biological and cultural diversity. *Cultural Survival Quarterly* 14(4): 2-3. [1].

866. Chapleau, F., C.S. Findlay, & E. Szenasy, 1997. Impact of pisciv-

orous fish introductions on fish species richness of small lakes in Gatineau Park, Quebec. *Écoscience* 4: 259-268.

867. Chapman, Duane, Sept.-Oct. 1995. Managing nature in South Africa. *American Enterprise* 6(5): 67-69. [1] *A.T.:* Kruger National Park; wildlife management.

868. Chapman, Lauren J., C.A. Chapman, & Mark Chandler, 1996. Wetland ecotones as refugia for endangered fishes. *Biological Conservation* 78: 263-270. [1] *A.T.:* introduced fishes; Nile perch; Uganda.

869. Charles-Dominique, P., P. Blanc, et al., 1998. Forest perturbations and biodiversity during the last 10 thousand years in French Guiana. *Acta Oecologica* 19: 295-302. *A.T.:* disturbance ecology.

870. Chatelain, C., L. Gautier, & R. Spichiger, 1996. A recent history of forest fragmentation in southwestern Ivory Coast. *Biodiversity and Conservation* 5: 37-53. [1] *A.T.:* remote sensing.

871. Chatterjee, Pratap, & Matthias Finger, 1994. *The Earth Brokers: Power, Politics, and World Development*. London and New York: Routledge. 191p. [1/2/1] Concerning corporate sector influences on the UNCED process. *A.T.:* environmental policy.

872. Chatterjee, Sudipto, 1995. Global "hotspots" of biodiversity. *Current Science* 68: 1178-1180. [1].

873. Chayax, Reginaldo, & D.F. Whitacre, 1997. A plea from the last of the Maya-Itza [editorial]. *CB* 11: 1457-1458. Says conservation biologists should follow agenda with a stronger political and social emphasis.

874. Chazeau, Jean, 1993. Research on New Caledonian terrestrial fauna: achievements and prospects. *Biodiversity Letters* 1: 123-129. [1].

875. Cheever, Federico, 1997. Human population and the loss of biological diversity: two aspects of the same problem. *International J. of Environment and Pollution* 7: 62-71.

876. Chen, Xu, & Jia Yu Rong, 1991. Concepts and analysis of mass extinction with the Late Ordovician event as an example. *Historical Biology* 5: 107-121. [1] *A.T.:* invertebrates; sea-level change; global cooling.

877. Cheney, Jim, 1987. Ecofeminism and deep ecology. *Environmental Ethics* 9: 115-145. [1].

878. Cherfas, Jeremy, 1991. Disappearing mushrooms: another mass extinction? *Science* 254: 1458. [1] *A.T.:* fungi; Europe.

879. ———, 9 May 1992. Farming goes back to its roots. *New Scientist* 134(1820): 12-13. [1] Describes how farmers in developing countries are being asked to return to the cultivation of original landraces.

880. ———, 6 Aug. 1994. How many species do we need? *New Scientist* 143(1937): 36-40. [1] Describes the facility known as Ecotron. *A.T.:* microcosms.

881. Cherrill, A.J., C. McClean, et al., 1995. Predicting the distributions of plant species at the regional scale: a hierarchical matrix model. *Landscape*

Ecology 10: 197-207. [1].

882. Chesser, Ronald K., 1991. Influence of gene flow and breeding tactics on gene diversity within populations. *Genetics* 129: 573-583. [2] *A.T.:* inbreeding.

883. Chesson, Peter, & Marissa Pantastico-Caldas, 1994. The forest architecture hypothesis for diversity maintenance. *TREE* 9: 79-80. [1].

884. Chester, Charles C., Oct. 1996. Controversy over Yellowstone's biological resources. *Environment* 38(8): 10-15, 34-36. [1] *A.T.:* thermophiles; natural resource management.

885. Chiappe, L.M., 1995. The first 85 million years of avian evolution. *Nature* 378: 349-355. [2] *A.T.:* Mesozoic; phylogeny.

886. Chipeniuk, Raymond, 1995. Childhood foraging as a means of acquiring competent human cognition about biodiversity. *Environment and Behavior* 27: 490-512. [1] *A.T.:* bd awareness; Canada; environmental perception.

887. Chiu, S.W., A.M. Ma, et al., 1996. Genetic homogeneity of cultivated strains of shiitake (*Lentinula edodes*) used in China as revealed by the polymerase chain reaction. *Mycological Research* 100: 1393-1399. [1] *A.T.:* mushrooms; genetic diversity.

888. Choi, Jae S., 1998. Lake ecosystem responses to rapid climate change. *Environmental Monitoring and Assessment* 49: 281-290. *A.T.:* life history studies; Canada.

889. Chomitz, Kenneth M., & Kanta Kumari, 1998. The domestic benefits of tropical forests: a critical review. *World Bank Research Observer* 13: 13-35. *A.T.:* conservation value.

890. Choudhary, S.K., 1990. Species diversity of phytoplankton as a potent tool for monitoring River Ganga pollution. *Environment & Ecology* 8(1A): 115-118. [1] *A.T.:* biomonitoring; India.

891. Chown, S.L., K.J. Gaston, & P.H. Williams, 1998. Global patterns in species richness of pelagic seabirds: the Procellariiformes. *Ecography* 21: 342-350. *A.T.:* latitudinal diversity gradients.

892. Chrietiennot-Dinet, M.J., & C. Courties, 1997. Biodiversity of unicellular algae: example of pico- and ultraplanktonic eucaryotes of the Thau Lagoon. *Vie et Milieu* 47: 317-324. *A.T.:* France; phytoplankton.

893. Christensen, Morten, & Jens Emborg, 1996. Biodiversity in natural versus managed forest in Denmark. *Forest Ecology and Management* 85: 47-51. [1] *A.T.:* forest dynamics.

894. Christian, D.P., P.T. Collins, et al., 1997. Bird and small mammal use of short-rotation hybrid poplar plantations. *J. of Wildlife Management* 61: 171-182. [1] *A.T.:* north central U.S.; biomass plantations.

895. Christiansen, Mette Bohn, & Elin Pitter, 1997. Species loss in a forest bird community near Lagoa Santa in southeastern Brazil. *Biological Conservation* 80: 23-32. *A.T.:* habitat fragmentation.

896. Christiansen, Scott, & C.M. Anthea Vaughan, 1997. *Conservation, Management and Sustainable Use of Dryland Biodiversity Within Priority Agro-ecosystems of the Near East* [workshop highlights]. Aleppo, Syria: International Center for Agricultural Research in the Dry Areas. 54p. [1/1/-].

897. Christie, Edward, 1993. The eternal triangle: the Biodiversity Convention, endangered species legislation and the precautionary principle. *Environmental and Planning Law J.* 10: 470-485. [1] *A.T.:* CBD; Australia.

898. Christie, J., 1995. Indigenous peoples, biodiversity and intellectual property rights. *Australasian Biotechnology* 5: 241-243. [1].

899. Christie, W.J., 1974. Changes in the fish species composition of the Great Lakes. *J. of the Fisheries Research Board of Canada* 31: 827-854. [2] *A.T.:* sea lamprey.

900. Churchill, Steven P., Henrik Balslev, et al., eds., 1995. *Biodiversity and Conservation of Neotropical Montane Forests* [symposium proceedings]. Bronx, NY: New York Botanical Garden. 702p. [1/1/1] *A.T.:* Andes region.

901. Chyba, C.F., P.J. Thomas, et al., 1990. Cometary delivery of organic molecules to the early Earth. *Science* 249: 366-373. [2] *A.T.:* origin of life; prebiotic organic molecules.

902. Cicin-Sain, Biliana, & R.W. Knecht, 1995. Measuring progress on UNCED implementation. *Ocean & Coastal Management* 29: 1-11. [1] *A.T.:* Commission on Sustainable Development; marine and coastal zone management.

903. Ciesla, William M., 1993. Recent introductions of forest insects and their effects: a global overview. *FAO Plant Protection Bull.* 41: 3-13. [1].

904. Ciriacy-Wantrup, Siegfried V., & R.C. Bishop, 1975. "Common property" as a concept in natural resources policy. *Natural Resources J.* 15: 713-727. [2] *A.T.:* commons; environmental law.

905. Cislaghi, Cesare, & P.L. Nimis, 1997. Lichens, air pollution and lung cancer. *Nature* 387: 463-464. *A.T.:* Italy; biological indicators.

906. Claridge, Michael F., H.A. Dawah, & M.R. Wilson, eds., 1997. *Species: The Units of Biodiversity* [conference papers]. London and New York: Chapman & Hall. 439p. [1/1/-] *A.T.:* species concept.

907. Clark, Colin W., 1973. The economics of overexploitation. *Science* 181: 630-634. [2] *A.T.:* renewable resources depletion.

908. ———, 1992. Empirical evidence for the effect of tropical deforestation on climatic change. *Environmental Conservation* 19: 39-47. [1].

909. Clark, Dana, & David Downes, 1996. What price biodiversity? Economic incentives and biodiversity conservation in the United States. *J. of Environmental Law and Litigation* 11: 9-89. [1] *A.T.:* environmental policy; bd legislation.

910. Clark, Deborah A., 1998. Deciphering landscape mosaics of neotropical trees: GIS and systematic sampling provide new views of tropical

rain forest diversity. *Annals of the Missouri Botanical Garden* 85: 18-33. *A.T.:* landscape patterns.

911. Clark, Deborah A., & D.B. Clark, 1992. Life history diversity of canopy and emergent trees in a Neotropical rain forest. *Ecological Monographs* 62: 315-344. [2] *A.T.:* Costa Rica; rainforest ecology.

912. Clark, Jim, Jan.-Feb. 1995. Rediscovering the land ethic. *Endangered Species Bull.* 20(1): 10-11. [1] *A.T.:* Aldo Leopold.

913. Clark, Jo, 1996. Biodiversity laws: state experiences. *Environmental Management* 20: 919-923. [1].

914. Clark, Karen L., & Ian Attridge, 1997. *Protecting the Biodiversity of the Americas: Legal and Policy Mechanisms Concerning Genetic Resources in Canada.* Toronto: Canadian Institute for Environmental Law and Policy. 90p. [1/1/-] *A.T.:* bd legislation.

915. Clark, Tanza E., & M.J. Samways, 1996. Dragonflies (Odonata) as indicators of biotope quality in the Kruger National Park, South Africa. *J. of Applied Ecology* 33: 1001-1012. [1] *A.T.:* bioindicators.

916. Clark, Tim W., 1993. Creating and using knowledge for species and ecosystem conservation: science, organizations, and policy. *Perspectives in Biology and Medicine* 36: 497-525. [2] *A.T.:* environmental policy; conservation biology.

917. ———, 1997. *Averting Extinction: Reconstructing Endangered Species Recovery.* New Haven: Yale University Press. 270p. [2/1/1] *A.T.:* environmental policy; black-footed ferret.

918. Clark, Tim W., R.P. Reading, & A.L. Clarke, eds., 1994. *Endangered Species Recovery: Finding the Lessons, Improving the Process* [conference papers; case studies]. Washington, DC: Island Press. 450p. [2/1/1] *A.T.:* wildlife conservation.

919. Clark, William C., & R.E. Munn, eds., 1986. *Sustainable Development of the Biosphere.* Cambridge, U.K., and New York: CUP. 491p. [2/2/2] *A.T.:* environmental policy.

920. Clarke, Andrew, 1992. Is there a latitudinal diversity cline in the sea? [editorial] *TREE* 7: 286-287. [1].

921. ———, 1993. Temperature and extinction in the sea: a physiologist's view. *Paleobiology* 19: 499-518. [1] *A.T.:* climatic change; invertebrate physiology.

922. Clarke, A.R., & G.H. Walter, 1995. "Strains" and the classical biological control of insect pests. *Canadian J. of Zoology* 73: 1777-1790. [1] *A.T.:* cryptic species; genetic diversity.

923. Clarke, G.M., 1993. Fluctuating asymmetry of invertebrate populations as a biological indicator of environmental quality. *Environmental Pollution* 82: 207-211. [2] *A.T.:* shrimp; blood worms.

924. Clarke, K.R., & R.M. Warwick, 1998. Quantifying structural redundancy in ecological communities. *Oecologia* 113: 278-289. *A.T.:* North

Atlantic; France; England; marine macrobenthos.

925. Clay, Jason, Jan.-Feb. 1991. The knowledge of nations. *Ceres* (FAO) 23(1): 29-33. [1] Concerning biopiracy and the rights of indigenous peoples.

926. ———, Jan.-Feb. 1996. The diversity debate: counterpoint. *Utne Reader*, no. 73: 36-37. [1] Decries the impact of the global culture on the environment and biodiversity.

927. Clay, Keith, 1995. Correlates of pathogen species richness in the grass family. *Canadian J. of Botany* 73, Suppl. 1: S42-S49. [1] *A.T.:* fungal pathogens; U.S.

928. Clayton, Creed, & Robert Mendelsohn, 1993. The value of watchable wildlife: a case study of McNeil River. *J. of Environmental Management* 39: 101-106. [1] *A.T.:* Alaska; contingent valuation; ecotourism; bears; bd value.

929. Clayton, Margaret N., 1994. Evolution of the Antarctic marine benthic algal flora. *J. of Phycology* 30: 897-904. [1] *A.T.:* biogeography; diversification.

930. Cleaver, Kevin M., M. Munasinghe, et al., eds., 1993. *Conservation of West and Central African Rainforests* [conference papers]. Washington, DC: World Bank Environment Paper No. 1. 353p. [1/1/1].

931. Cleland, David T., & R.E. Scott, 1990. The U.S. Forest Service perspective on managing for biodiversity on federal lands. *Michigan Academician* 22: 355-363. [1].

932. Clemens, William A., J.D. Archibald, & P.M. Sheehan et al., 1992. Dinosaur diversity and extinction [three letters]. *Science* 256: 159-161. [1].

933. Clement, Charles R., 1989. A center of crop genetic diversity in western Amazonia. *Bioscience* 39: 624-631. [1].

934. Clements, F.E., 1936. Nature and structure of the climax. *J. of Ecology* 24: 252-284. [2] *A.T.:* biotic communities.

935. Clemmons, Janine R., & Richard Buchholz, eds., 1997. *Behavioral Approaches to Conservation in the Wild.* New York: CUP. 382p. [1/1/-].

936. Clergeau, Philippe, J.P.L. Savard, et al., 1998. Bird abundance and diversity along an urban-rural gradient: a comparative study between two cities on different continents. *Condor* 100: 413-425. *A.T.:* Canada; France; urban ecology.

937. Cleveland, David A., Daniela Soleri, & S.E. Smith, 1994. Do folk crop varieties have a role in sustainable agriculture? *Bioscience* 44: 740-751. [1].

938. Clinebell, Richard R., II, O.L. Phillips, et al., 1995. Prediction of Neotropical tree and liana species richness from soil and climatic data. *Biodiversity and Conservation* 4: 56-90. [1] *A.T.:* nutrient cycling; rainfall.

939. Cloud, Joseph G., & G.H. Thorgaard, eds., 1993. *Genetic Conservation of Salmonid Fishes.* New York: Plenum Press. 314p. [1/1/1] *A.T.:* germplasm resources.

940. Cloudsley-Thompson, John L., 1993. The adaptational diversity of desert biota. *Environmental Conservation* 20: 227-231. [1] *A.T.:* morphological diversity.

941. Clout, M.N., & A.J. Saunders, 1995. Conservation and ecological restoration in New Zealand. *Pacific Conservation Biology* 2: 91-98. [1] *A.T.:* biological invasions.

942. Clube, S.V.M., ed., 1989. *Catastrophes and Evolution: Astronomical Foundations* [conference papers]. Cambridge, U.K., and New York: CUP. 239p. [1/1/1].

943. Clunies-Ross, Tracey, 1995. Marigolds, manure and mixtures: the importance of crop diversity on British farms. *Ecologist* 25: 181-187. [1] *A.T.:* organic farming.

944. Coblentz, Bruce E., 1978. The effects of feral goats (*Capra hircus*) on island ecosystems. *Biological Conservation* 13: 279-286. [1].

945. ———, 1990. Exotic organisms: a dilemma for conservation biology. *CB* 4: 261-265. [1] *A.T.:* biological invasions; eradication programs.

946. Coccioni, Rodolfo, & Simone Galeotti, 1994. K-T boundary extinction: geologically instantaneous or gradual event? Evidence from deep-sea benthic foraminifera. *Geology* 22: 779-782. [2] *A.T.:* Spain.

947. Cockburn, Andrew, 1991. *An Introduction to Evolutionary Ecology.* Oxford, U.K., and Boston: Blackwell Scientific. 370p. [2/2/2].

948. Cody, M.L., & H.A. Mooney, 1978. Convergence versus nonconvergence in Mediterranean-climate ecosystems. *ARES* 9: 265-321. [2] *A.T.:* convergent evolution; California; Chile.

949. Cognetti, G., & M. Curini-Galletti, 1993. Biodiversity conservation problems in the marine environment [editorial]. *Marine Pollution Bull.* 26: 179-183. [1] *A.T.:* environmental degradation; bd loss.

950. Cohen, Andrew S., ed., 1991. *Report on the First International Conference on the Conservation and Biodiversity of Lake Tanganyika.* Washington, DC: Biodiversity Support Program. 128p. [1/1/1] *A.T.:* fishery conservation.

951. ———, 1995. Paleoecological approaches to the conservation biology of benthos in ancient lakes: a case study from Lake Tanganyika. *Marine & Freshwater Research* 46: 654-668. [1] *A.T.:* zoobenthos.

952. Cohen, Andrew S., Roger Bills, et al., 1993. The impact of sediment pollution on biodiversity in Lake Tanganyika. *CB* 7: 667-677. [1] *A.T.:* ostracods; fish.

953. Cohen, Gary, spring 1996. Toward a spirituality based on justice and ecology. *Social Policy* 26(3): 6-18. [1] *A.T.:* deep ecology.

954. Cohen, Joel E., 1995. Population growth and Earth's human carrying capacity. *Science* 269: 341-346. [2].

955. Cohen, Joel E., & David Tilman, 1996. Biosphere 2 and biodiversity: the lessons so far. *Science* 274: 1150-1151. [1].

956. Cohen, Joel I., J.B. Alcorn, & C.S. Potter, 1991. Utilization and conservation of genetic resources: international projects for sustainable agriculture. *Economic Botany* 45: 190-199. [1] *A.T.:* germplasm resources; gene banks.

957. Cohen, Joel I., J.T. Williams, et al., 1991. *Ex situ* conservation of plant genetic resources: global development and environmental concerns. *Science* 253: 866-872. [1] *A.T.:* gene banks; germplasm resources.

958. Cohen, Margot, 11 Jan. 1996. Forest fire. *Far Eastern Economic Review* 159(2): 66-69. [1] Describes international concern over the regulation of bioprospecting.

959. Cohen, Tracey, April 1993. Pharmaceuticals from the sea [editorial]. *Technology Review* 96(3): 15-16. [1] Concerning the importance of marine bd as a potential source of new pharmaceuticals.

960. Cohn, Jeffrey P., 1989. Gauging the biological impacts of the greenhouse effect. *Bioscience* 39: 142-146. [1].

961. ———, 1992. Decisions at the zoo: ethics, politics, profit and animal-rights concerns affect the progress of balancing conservation goals and the public interests. *Bioscience* 42: 654-659. [1].

962. ———, April 1993. Defenders of biodiversity. *Government Executive* 25(4): 18-22. [1] How the Interior Department is trying to improve its endangered species protection program.

963. ———, 1994. Salamanders slip-sliding away or too surreptitious to count? *Bioscience* 44: 219-223. [1] Concerning the apparent decline in salamander numbers.

964. ———, Feb. 1998. Negotiating nature. *Government Executive* 30(2): 50-53. A look at the Fish and Wildlife Service's efforts to protect endangered species on private lands.

965. Colaninno, Andrew, June 1996. What's the big deal? Putting extinction in perspective [editorial]. *J. of Forestry* 94(6): 60. [1].

966. Colchester, Marcus, 1994. *Salvaging Nature: Indigenous Peoples, Protected Areas and Biodiversity Conservation.* Geneva: Discussion Paper 55, United Nations Research Institute for Social Development. 76p. [1/1/1] Argues that indigenous peoples should have more input into protected areas management.

967. Colchester, Marcus, & Larry Lohmann, eds., 1993. *The Struggle for Land and the Fate of the Forests.* London and Atlantic Highlands, NJ: Zed Books. 389p. [1/2/1] *A.T.:* deforestation; rainforest conservation.

968. Cole, F. Russell, DeeAnn Reeder, & D.E. Wilson, 1994. A synopsis of distribution patterns and the conservation of mammal species. *J. of Mammalogy* 75: 266-276. [1] *A.T.:* species diversity; geographic distribution.

969. Cole, F. Russell, A.C. Medeiros, et al., 1992. Effects of the Argentine ant on arthropod fauna of Hawaiian high-elevation shrubland. *Ecology* 73: 1313-1322. [2] *A.T.:* biological invasions; endemic species.

970. Cole, Monica M., 1986. *The Savannas: Biogeography and Geobotany*. London and Orlando, FL: Academic Press. 438p. [1/1/2].
971. Coleman, Bernard D., M.A. Mares, et al., 1982. Randomness, area and species richness. *Ecology* 63: 1121-1133. [2] *A.T.:* Pennsylvania; breeding birds; island biogeography.
972. Coleman, William, 1979. Bergmann's rule: animal heat as a biological phenomenon. *Studies in History of Biology* 3: 67-88. [1].
973. Coleman, William G., 1996. Biodiversity and industry ecosystem management. *Environmental Management* 20: 815-825. [1].
974. Colgan, M.W., 1987. Coral reef recovery on Guam Micronesia after catastrophic predation by *Acanthaster planci*. *Ecology* 68: 1592-1605. [1].
975. Collar, Nigel J., M.J. Crosby, & A.J. Stattersfield, 1994. *Birds to Watch 2: The World List of Threatened Birds*. Cambridge, U.K.: BirdLife International. 407p. [1/1/2].
976. Collar, Nigel J., L.P. Gonzaga, et al., 1992. *Threatened Birds of the Americas: The ICBP/IUCN Red Data Book, Third Edition, Part 2*. Washington, DC, and London: Smithsonian Institution Press. 1150p. [1/1/2].
977. Collier, Kevin, 1993. Review of the status, distribution and conservation of freshwater invertebrates in New Zealand. *New Zealand J. of Marine and Freshwater Research* 27: 339-356. [1].
978. Collier, Michael P., R.H. Webb, & E.D. Andrews, Jan. 1997. Experimental flooding in Grand Canyon. *Scientific American* 276(1): 82-89. [1].
979. Collins, Harold P., G.P. Robertson, & M.J. Klug, eds., 1995. *The Significance and Regulation of Soil Biodiversity: Selected Papers From the Proceedings of the International Symposium on Soil Biodiversity*. Dordrecht and Boston: Kluwer Academic. 239p. [1/1/1] *A.T.:* soil ecology.
980. Collins, N. Mark, ed., 1990. *The Last Rain Forests: A World Conservation Atlas*. New York: OUP. 200p. [3/2/1].
981. Collins, N. Mark, & J.A. Thomas, eds., 1991. *The Conservation of Insects and Their Habitats* [symposium papers]. London and San Diego: Academic Press. 450p. [1/1/1].
982. Collins, N. Mark, & Susan Wells, 1983. Invertebrates—who needs them? *New Scientist* 98: 441-444. [1].
983. Collins, N. Mark, J.A. Sayer, & T.C. Whitmore, eds., 1991. *The Conservation Atlas of Tropical Forests: Asia and the Pacific*. London: Macmillan. 256p. [2/2/2].
984. Collins, Scott L., & S.M. Glenn, 1991. Importance of spatial and temporal dynamics in species regional abundance and distribution. *Ecology* 72: 654-664. [2] *A.T.:* immigration-extinction model; core-satellite hypothesis; Kansas; tallgrass prairie; plants.
985. ———, 1997. Effects of organismal and distance scaling on analysis of species distribution and abundance. *Ecological Applications* 7: 543-551. *A.T.:* Kansas; core-satellite hypothesis; tallgrass prairie.

986. Collins, Scott L., A.K. Knapp, et al., 1998. Modulation of diversity by grazing and mowing in native tallgrass prairie. *Science* 280: 745-747. *A.T.:* Kansas; plant species diversity.

987. Collins, Wanda W., & C.O. Qualset, eds., 1998. *Biodiversity in Agroecosystems.* Boca Raton, FL: CRC Press. 334p.

988. Colwell, Mark A., A.V. Dubynin, et al., 1997. Russian nature reserves and conservation of biological diversity. *Natural Areas J.* 17: 56-68. [1].

989. Colwell, Rita R., 1994. Biodiversity and utilization of microorganisms. *Annals of the New York Academy of Sciences* 745: 395-398. [1].

990. Colwell, Robert K., & J.A. Coddington, 1994. Estimating terrestrial biodiversity through extrapolation. *PTRSL B* 345: 101-118. [2] *A.T.:* species richness; complementarity measures.

991. Colwell, Robert K., & G.C. Hurtt, 1994. Nonbiological gradients in species richness and a spurious Rapoport effect. *American Naturalist* 144: 570-595. [2] A.T. sampling bias.

992. Combes, C., 1996. Parasites, biodiversity and ecosystem stability. *Biodiversity and Conservation* 5: 953-962. [1].

993. Comiskey, J.A., G.E. Ayzanoa, & F. Dallmeier, 1995. A data management system for monitoring forest dynamics. *J. of Tropical Forest Science* 7: 419-427. [1] *A.T.:* MAB; bd monitoring.

994. Commission of the European Communities, 1992. *Biological Diversity: A Challenge to Science, the Economy and Society* [conference proceedings]. Brussels. 601p. [1/1/1] *A.T.:* farming; industry.

995. Common, Michael S., & T.W. Norton, 1992. Biodiversity: its conservation in Australia. *Ambio* 21: 258-265. [1] *A.T.:* bd loss; reserve systems.

996. ———, 1994. Biodiversity, natural resource accounting and ecological monitoring. *Environmental & Resource Economics* 4: 29-53. [1].

997. Common, Michael S., & Charles Perrings, 1992. Towards an ecological economics of sustainability. *Ecological Economics* 6: 7-34. [2].

998. Conant, Sheila, 1988. Saving endangered species by translocation. *Bioscience* 38: 254-257. [1] *A.T.:* Hawaii.

999. Conant, Sheila, R.C. Fleischer, et al., 1992. When endangered species are aliens: some thoughts on the conservation of rare species. *Pacific Science* 46: 401-402. [1].

1000. Condit, Richard, S.P. Hubbell, et al., 1996. Species-area and species-individual relationships for tropical trees: a comparison of three 50-ha plots. *J. of Ecology* 84: 549-562. [2] *A.T.:* Malaysia; India; Panama; quadrat methods.

1001. Condon, Marty A., 1994. Tom Sawyer meets insects: how biodiversity opens science to the public. *Biodiversity Letters* 2: 159-162. [1] *A.T.:* bd education.

1002. ———, 1995. Biodiversity, systematics, and Tom Sawyer science.

CB 9: 711-714. [1] *A.T.:* bd education.

1003. Connell, Joseph H., 1978. Diversity in tropical rain forests and coral reefs. *Science* 199: 1302-1310. [3] A classic and perennially heavily-cited article in which Connell argues that high diversity is related to frequency of natural disturbance.

1004. ———, 1989. Some processes affecting the species composition in forest gaps. *Ecology* 70: 560-562. [2] *A.T.:* treefall gaps; patch dynamics.

1005. Connell, Joseph H., & M.D. Lowman, 1989. Low-diversity tropical rain forests: some possible mechanisms for their existence. *American Naturalist* 134: 88-119. [2] *A.T.:* fungi.

1006. Connell, Joseph H., & Eduardo Orias, 1964. The ecological regulation of species diversity. *American Naturalist* 98: 399-414. [2] *A.T.:* stability; productivity.

1007. Connell, Joseph H., & W.P. Sousa, 1983. On the evidence needed to judge ecological stability or persistence. *American Naturalist* 121: 789-824. [3].

1008. Connell, Joseph H., J.G. Tracey, & L.J. Webb, 1984. Compensatory recruitment, growth, and mortality as factors maintaining rain forest tree diversity. *Ecological Monographs* 54: 141-164. [2] *A.T.:* population dynamics.

1009. Connor, Edward F., & E.D. McCoy, 1979. The statistics and biology of the species-area relationship. *American Naturalist* 113: 791-833. [3].

1010. Conrad, Jon M., 1991. *Economic Strategies for Coevolution: Parks, Buffer Zones and Biodiversity*. Ithaca, NY: Working Papers in Agricultural Economics 91-8, Dept. of Agricultural Economics, New York College of Agriculture and Life Sciences. 24p. [1/1/1].

1011. Conroy, Michael J., 1996. Designing surveys of forest diversity using statistical sampling procedures. In Michael Köhl & George Z. Gertner, eds., *Statistics, Mathematics and Computers* (Birmensdorf, Switzerland: Swiss Federal Institute for Forest, Snow and Landscape Research): 117-143. [1] *A.T.:* bd monitoring.

1012. Conroy, Michael J., & B.R. Noon, 1996. Mapping of species richness for conservation of biological diversity: conceptual and methodological issues. *Ecological Applications* 6: 763-773. [2] *A.T.:* gap analysis; reserve selection.

1013. Conservation International, 1997. *The First Decade: 1987-1997*. Washington, DC. 64p. [1/1/-] Describes the organization's work.

1014. Consultative Group on International Agricultural Research, 1997. *Ethics and Equity in Conservation and Use of Genetic Resources for Sustainable Food Security* [workshop proceedings]. Rome: IPGRI. 197p. [1/1/-] *A.T.:* food crops; developing countries; plant germplasm resources.

1015. Contreras, Juventino, D.F. Austin, et al., 1995. Biodiversity of sweetpotato (*Ipomoea batatas*, Convolvulaceae) in southern Mexico. *Eco-*

nomic Botany 49: 286-296. [1].
  1016. Contreras-B., Salvador, & M.L. Lozano-V., 1994. Water, endangered fishes, and development perspectives in arid lands of Mexico. *CB* 8: 379-387. [1].
  1017. Conway, William G., 1995. Wild and zoo animal interactive management and habitat conservation. *Biodiversity and Conservation* 4: 573-594. [1] *A.T.:* wildlife conservation.
  1018. Conway Morris, Simon, 1998. The evolution of diversity in ancient ecosystems: a review. *PTRSL B* 353: 327-345.
  1019. Cook, A.G., A.C. Janetos, & W.T. Hinds, 1990. Global effects of tropical deforestation: towards an integrated perspective. *Environmental Conservation* 17: 201-212. [1] *A.T.:* research programs.
  1020. Cook, James H., Jan Beyea, & K.H. Keeler, 1991. Potential impacts of biomass production in the United States on biological diversity. *Annual Review of Energy and the Environment* 16: 401-431. [1] *A.T.:* energy resources.
  1021. Cook, Laurence M., 1991. *Genetic and Ecological Diversity: The Sport of Nature*. London and New York: Chapman & Hall. 192p. [1/1/1].
  1022. Cook, M.A., T. Jones, & D. Pittis, 1995. Project to identify developing world's creatures and CARINET. *International J. of Environmental Studies A* 48: 49-54. [1] *A.T.:* BioNET International; biosystematic services.
  1023. Cook, Robert E., 1969. Variation in species density of North American birds. *Systematic Zoology* 18: 63-84. [2] *A.T.:* areography.
  1024. Cook, R. James, 1993. Making greater use of introduced microorganisms for biological control of plant pathogens. *Annual Review of Phytopathology* 31: 53-80. [2].
  1025. Cook, R.R., & J.F. Quinn, 1995. The influence of colonization in nested species subsets. *Oecologia* 102: 413-424. [2] *A.T.:* island biogeography.
  1026. Coomes, Oliver T., 1995. A century of rain forest use in Western Amazonia: lessons for extraction-based conservation of tropical forest resources. *Forest & Conservation History* 39: 108-120. [1] *A.T.:* Peru.
  1027. Coope, G.R., 1994. The response of insect faunas to glacial-interglacial climatic fluctuations. *PTRSL B* 344: 19-26. [1] *A.T.:* Quaternary.
  1028. Cooper, Alan, & David Penny, 1997. Mass survival of birds across the Cretaceous-Tertiary boundary: molecular evidence. *Science* 275: 1109-1113. [2] *A.T.:* mass extinctions.
  1029. Cooper, David, Renée Vellvé, & Henk Hobbelin, eds., 1992. *Growing Diversity: Genetic Resources and Local Food Security*. London: Intermediate Technology Publications. 166p. [1/1/1] *A.T.:* germplasm resources; food crops; developing countries.
  1030. Cooper, Gregory, 1998. Teleology and environmental ethics. *American Philosophical Quarterly* 35: 195-207.

1031. Cooperrider, Allen Y., summer 1994. Saving nature's legacy. *Defenders* 69(3): 17-24. [1] *A.T.:* land management.

1032. Cooperrider, Allen Y., & D.S. Wilcove, 1995. *Defending the Desert: Conserving Biodiversity on BLM Lands in the Southwest.* New York: Environmental Defense Fund. 148p. [1/1/1] *A.T.:* public lands; desert environments.

1033. Cooperrider, Allen Y., R.J. Boyd, & H.R. Stuart, compilers/eds., 1986. *Inventory and Monitoring of Wildlife Habitat.* Denver: Bureau of Land Management, U.S. Dept. of the Interior. 858p. [2/1/1].

1034. Corbley, Kevin P., Feb. 1996. Utah finds biodiversity gaps using GIS. *Geo Info Systems* 6(2): 32-36. [1] *A.T.:* gap analysis; tortoises.

1035. ———, Aug. 1996. National GIS fills "gaps" in biological diversity. *GIS World* 9(8): 50-53. [1] *A.T.:* gap analysis; U.S. National Biological Survey.

1036. Cork, S.J., & P.C. Catling, 1996. Modelling distributions of arboreal and ground-dwelling mammals in relation to climate, nutrients, plant chemical defences and vegetation structure in the eucalypt forests of southeastern Australia. *Forest Ecology and Management* 85: 163-175. [1] *A.T.:* marsupials; New South Wales.

1037. Cornell, Howard V., & B.A. Hawkins, 1993. Accumulation of native parasitoid species on introduced herbivores: a comparison of hosts as natives and hosts as invaders. *American Naturalist* 141: 847-865. [2] *A.T.:* insects.

1038. Cornell, Howard V., & R.H. Karlson, 1996. Species richness of reef-building corals determined by local and regional processes. *J. of Animal Ecology* 65: 233-241. [2] *A.T.:* community ecology.

1039. Cornell, Howard V., & J.H. Lawton, 1992. Species interactions, local and regional processes, and limits to the richness of ecological communities: a theoretical perspective. *J. of Animal Ecology* 61: 1-12. [3] *A.T.:* species richness; local and regional richness; saturation.

1040. Cornell International Institute for Food, Agriculture, and Development, 1992. *Sustainable Development and Biodiversity: Conflicts and Complementarities* [symposium proceedings]. Ithaca, NY. 77p. [1/1/1].

1041. Corona, Piermaria, 1993. Applying biodiversity concepts to plantation forestry in northern Mediterranean landscapes. *Landscape and Urban Planning* 24: 23-31. [1] *A.T.:* forest management; afforestation.

1042. Corpe, W.A., & T.E. Jensen, 1996. The diversity of bacteria, eukaryotic cells and viruses in an oligotrophic lake. *Applied Microbiology and Biotechnology* 46: 622-630. [1] *A.T.:* New York.

1043. Costanza, Robert, ed., 1991. *Ecological Economics: The Science and Management of Sustainability* [workshop papers]. New York: Columbia University Press. 525p. [2/2/2] *A.T.:* sustainable development.

1044. Costanza, Robert, & H.E. Daly, 1992. Natural capital and sustainable development. *CB* 6: 37-46. [2].

1045. Costanza, Robert, W.M. Kemp, & W.R. Boynton, 1993. Predictability, scale, and biodiversity in coastal and estuarine ecosystems: implications for management. *Ambio* 22: 88-96. [1] *A.T.:* coastal zone management; ecosystem stability.

1046. Costanza, Robert, B.G. Norton, & B.D. Haskell, eds., 1992. *Ecosystem Health: New Goals for Environmental Management.* Washington, DC: Island Press. 269p. [2/1/2].

1047. Costanza, Robert, Olman Segura, & Juan Martínez-Alier, eds., 1996. *Getting Down to Earth: Practical Applications of Ecological Economics* [workshop papers]. Washington, DC: Island Press. 472p. [2/1/1] *A.T.:* sustainable development.

1048. Costanza, Robert, Ralph d'Arge, et al., 1997. The value of the world's ecosystem services and natural capital. *Nature* 387: 253-260. [3].

1049. Cotgreave, Peter, & P.H. Harvey, 1994. Associations among biogeography, phylogeny and bird species diversity. *Biodiversity Letters* 2: 46-55. [1] *A.T.:* latitudinal diversity gradients.

1050. ———, 1994. Evenness of abundance in bird communities. *J. of Animal Ecology* 63: 365-374. [1] *A.T.:* community ecology.

1051. Cotterill, F.P.D., 1995. Systematics, biological knowledge and environmental conservation. *Biodiversity and Conservation* 4: 183-205. [1] *A.T.:* natural history collections.

1052. Cottingham, Kathryn L., & S.R. Carpenter, 1998. Population, community, and ecosystem variates as ecological indicators: phytoplankton responses to whole-lake enrichment. *Ecological Applications* 8: 508-530.

1053. Coufal, James E., 1997. Biodiversity and environmental ethics: a personal reflection. *J. of the American Water Resources Association* 33: 13-19. [1] *A.T.:* aesthetics.

1054. Coughlin, Michael D., Jr., 1993. Using the Merck-INBio agreement to clarify the Convention on Biological Diversity. *Columbia J. of Transnational Law* 31: 337-376. [1].

1055. Coulter, John, 1993. Protecting our biodiversity from corporate exploitation. *Search* 24: 280-282. [1].

1056. Council of Europe, 1989. *A Political Action Plan for the Preservation of the World's Tropical Forests and Their Biodiversity: Summary Report on the Elaboration of the Action Plan.* Strasbourg. 108p. [1/1/1].

1057. ———, 1995. *Symposium on the United Nations Conference on Environment and Development (UNCED), the Convention on Biological Diversity, and the Bern Convention: The Next Steps.* Strasbourg. 149p. [1/1/1].

1058. ———, 1996. *Pan-European Biological and Landscape Diversity Strategy.* Strasbourg. 68p. [1/1/1] *A.T.:* landscape conservation.

1059. ———, 1997. *Colloquy on Conservation, Management and Restoration of Habitats for Invertebrates: Enhancing Biological Diversity.* Strasbourg. 161p. [1/1/-] *A.T.:* Europe.

1060. Council on Environmental Quality, 1981. *Biological Diversity.* Washington, DC. 50p. [1/1/1] *A.T.:* human ecology.

1061. ———, 1993. *Incorporating Biodiversity Considerations Into Environmental Impact Analysis Under the National Environmental Policy Act.* Washington, DC. 29p. [2/1/1].

1062. Courtenay, Walter R., Jr., & J.R. Stauffer, Jr., 1990. The introduced fish problem and the aquarium fish industry. *J. of the World Aquaculture Society* 21: 145-159. [1] *A.T.:* U.S.

1063. Courtillot, Vincent E., Oct. 1990. What caused the mass extinctions: a volcanic eruption. *Scientific American* 263(4): 85-92. [1] *A.T.:* Cretaceous-Tertiary boundary.

1064. Courtillot, Vincent E., & Y. Gaudemer, 1996. Effects of mass extinctions on biodiversity. *Nature* 381: 146-148. [1] *A.T.:* diversification; Phanerozoic.

1065. Cousins, Steven H., 1989. Species richness and the energy theory [letter]. *Nature* 340: 350-351. [1] *A.T.:* birds; Great Britain; body size.

1066. ———, 1991. Species diversity measurement: choosing the right index. *TREE* 6: 190-192. [2] *A.T.:* diversity indices.

1067. Cowan, D.A., 1997. The marine biosphere: a global resource for biotechnology. *Trends in Biotechnology* 15: 129-131. [1].

1068. Cowie, Robert H., 1995. Variation in species diversity and shell shape in Hawaiian land snails: *in situ* speciation and ecological relationships. *Evolution* 49: 1191-1202. [1] *A.T.:* adaptive radiation; island biogeography.

1069. Cowling, Richard M., ed., 1992. *The Ecology of Fynbos: Nutrients, Fire, and Diversity.* Cape Town: OUP. 411p. [1/1/2] *A.T.:* South Africa; fire ecology.

1070. Cowling, Richard M., & M.J. Samways, 1994. Predicting global patterns of endemic plant species richness. *Biodiversity Letters* 2: 127-131. [1] *A.T.:* latitudinal diversity gradients.

1071. Cowling, Richard M., I.A.W. Macdonald, & M.T. Simmons, 1996. The Cape Peninsula, South Africa: physiographical, biological and historical background to an extraordinary hot-spot of biodiversity. *Biodiversity and Conservation* 5: 527-550. [1].

1072. Cowling, Richard M., K.J. Esler, et al., 1994. Plant functional diversity, species diversity, and climate in arid and semi-arid Southern Africa. *J. of Arid Environments* 27: 141-158. [1] *A.T.:* morphological diversity.

1073. Cowling, Richard M., P.W. Rundel, et al., 1996. Plant diversity in Mediterranean-climate regions. *TREE* 11: 362-366. [1] *A.T.:* fire; climatic change.

1074. Cowling, Richard M., P.W. Rundel, et al., 1998. Extraordinary high regional-scale plant diversity in Southern African arid lands: subcontinental and global comparisons. *Diversity & Distributions* 4: 27-36. *A.T.:* Karoo-

Namib region.

1075. Cox, C. Barry, & P.D. Moore, 1993. *Biogeography: An Ecological and Evolutionary Approach* (5th ed.) [textbook]. Oxford, U.K., and Boston: Blackwell Scientific. 326p. [3/2/2].

1076. Cox, Paul A., 1997. *Nafanua: Saving the Samoan Rain Forest.* New York: W.H. Freeman. 238p. [2/-/-] *A.T.:* ethnobotany.

1077. Cox, Paul A., & M.J. Balick, June 1994. The ethnobotanical approach to drug discovery. *Scientific American* 270(6): 82-87. [1] *A.T.:* medicinal plants.

1078. Cox, Paul A., & T. Elmqvist, 1997. Ecocolonialism and indigenous-controlled rainforest preserves in Samoa. *Ambio* 26: 84-89. [1] *A.T.:* traditional knowledge.

1079. Cox, Paul A., T. Elmqvist, et al., 1991. Flying foxes as strong interactors in South Pacific island ecosystems: a conservation hypothesis. *CB* 5: 448-454. [1] *A.T.:* keystone species; pollinators; Guam; Samoa.

1080. Coyle, K.J., 1993. The new advocacy for aquatic species conservation. *J. of the North American Benthological Society* 12: 185-188. [1].

1081. Cracraft, Joel, 1983. Species concepts and speciation analysis. *Current Ornithology* 1: 159-187. [2] *A.T.:* phylogenetics; endemism.

1082. ———, 1985. Biological diversification and its causes. *Annals of the Missouri Botanical Garden* 72: 794-822. [2] *A.T.:* macroevolution; geomorphological complexity.

1083. ———, 1991. Patterns of diversification within continental biotas: hierarchical congruence among the areas of endemism of Australian vertebrates. *Australian Systematic Botany* 4: 211-227. [1] *A.T.:* vicariance; cladistic analysis.

1084. ———, 1994. Species diversity, biogeography, and the evolution of biotas. *American Zoologist* 34: 33-47. [1] *A.T.:* Australia; birds; dispersal; vicariance.

1085. ———, 1995. The urgency of building global capacity for biodiversity science. *Biodiversity and Conservation* 4: 463-475. [1] *A.T.:* research programs.

1086. Cracraft, Joel, & R.O. Prum, 1988. Patterns and processes of diversification: speciation and historical congruence in some Neotropical birds. *Evolution* 42: 603-620. [2] *A.T.:* vicariance.

1087. Crain, Ian, Gareth Lloyd, & Gwynneth Martin, 1995. *Guidelines for Information Management: In the Context of the Convention on Biological Diversity.* Nairobi: UNEP. 57p. [1/1/1] *A.T.:* bd databases.

1088. Crame, J.A., 1997. An evolutionary framework for the polar regions. *J. of Biogeography* 24: 1-9. *A.T.:* latitudinal diversity gradients; biogeography; evolutionary rates.

1089. Cranbrook, Gathorne Gathorne-Hardy, Earl of, & D.S. Edwards, 1994. *A Tropical Rainforest: The Nature of Biodiversity in Borneo at Bela-*

long, Brunei. London: Royal Geographical Society. 389p. [1/1/1].

1090. Crandall, K.A., 1998. Conservation phylogenetics of Ozark crayfishes: assigning priorities for aquatic habitat protection. *Biological Conservation* 84: 107-117. *A.T.:* Missouri; Arkansas; biogeography.

1091. Crane, Peter R., & Scott Lidgard, 1989. Angiosperm diversification and paleolatitudinal gradients in Cretaceous floristic diversity. *Science* 246: 675-678. [1] *A.T.:* latitudinal diversity gradients.

1092. Cranston, P.S., 1990. Biomonitoring and invertebrate taxonomy. *Environmental Monitoring and Assessment* 14: 265-273. [1].

1093. Creech, David, 15 Dec. 1996. In support of biodiversity. *American Nurseryman* 184(12): 38-43. [1] Describes work at a Texas arboretum contributing to endangered species preservation.

1094. Crews, D., 1997. Species diversity and the evolution of behavioral controlling mechanisms. *Annals of the New York Academy of Sciences* 807: 1-21.

1095. Crisp, Michael D., H.P. Linder, & P.H. Weston, 1995. Cladistic biogeography of plants in Australia and New Guinea: congruent pattern reveals two endemic tropical tracks. *Systematic Biology* 44: 457-473. [1] *A.T.:* angiosperms.

1096. Crisp, P.N., K.J.M. Dickinson, & G.W. Gibbs, 1998. Does native invertebrate diversity reflect native plant diversity: a case study from New Zealand and implications for conservation. *Biological Conservation* 83: 209-220. *A.T.:* beetles; rapid assessment.

1097. Crist, Patrick, July 1996. Helping habitats. *Planning* 62(7): 16-17. [1] Describes the National Biological Service's GAP program. *A.T.:* gap analysis; GIS.

1098. Croizat, Leon, 1958. *Panbiogeography, or, An Introductory Synthesis of Zoogeography, Phytogeography and Geology*. Caracas: L. Croizat. 2 vols. [1/-/2].

1099. Croizat, Leon, Gareth Nelson, & D.E. Rosen, 1974. Centers of origin and related concepts. *Systematic Zoology* 23: 265-287. [2] *A.T.:* generalized tracks; biogeography; evolution.

1100. Cronin, T.M., & M.E. Raymo, 1997. Orbital forcing of deep-sea benthic species diversity. *Nature* 385: 624-627. *A.T.:* Pliocene; ostracods; stability-time hypothesis.

1101. Cronin, T.M., & C.E. Schneider, 1990. Climatic influences on species: evidence from the fossil record. *TREE* 5: 275-279. [1] *A.T.:* climatic change; evolution.

1102. Cronk, Q.C.B., 1992. Relict floras of Atlantic islands: patterns assessed. *Biological J. of the Linnean Society* 46: 91-103. [1].

1103. ———, 1997. Islands: stability, diversity, conservation. *Biodiversity and Conservation* 6: 477-493. *A.T.:* endemism; St. Helena; island biogeography.

1104. Cronk, Q.C.B., & J.L. Fuller, 1995. *Plant Invaders: The Threat to Natural Ecosystems*. London and New York: Chapman & Hall. 241p. [1/1/2] *A.T.:* plant conservation.

1105. Crosby, Alfred W., 1972. *The Columbian Exchange: Biological and Cultural Consequences of 1492*. Westport, CT: Greenwood. 268p. [3/2/2] *A.T.:* diseases; cultural biogeography.

1106. ———, 1986. *Ecological Imperialism: The Biological Expansion of Europe, 900-1900*. Cambridge, U.K., and New York: CUP. 368p. [3/2/2] *A.T.:* cultural biogeography.

1107. Cross, Andrew, June 1990. Species and habitat: the analysis and impoverishment of variety. *Geographical Magazine* 62(6): 42-47. [1] A brief but effective survey written at an introductory level. *A.T.:* hydrobiosphere; geobiosphere; Ecuador.

1108. Crossley, D.A., Jr., B.R. Mueller, & J.C. Perdue, 1992. Biodiversity of microarthropods in agricultural soils: relations to processes. *Agriculture, Ecosystems & Environment* 40: 37-46. [1] *A.T.:* mites; Collembola; fungi.

1109. Crossman, E.J., 1991. Introduced freshwater fishes: a review of the North American perspective with emphasis on Canada. *Canadian J. of Fisheries and Aquatic Sciences* 48, Suppl. 1: 46-57. [1] *A.T.:* biological invasions.

1110. Crow, G.E., 1993. Species diversity in aquatic angiosperms: latitudinal patterns. *Aquatic Botany* 44: 229-258. [1] *A.T.:* latitudinal diversity gradients.

1111. Crowe, Tim M., 1991. Genetics, biotechnology and the conservation of biodiversity? [editorial] *South African J. of Science* 87: 81-84. [1].

1112. ———, 1996. Developing a national strategy for the protection and sustainable use of South Africa's biodiversity [editorial]. *South African J. of Science* 92: 218-219. [1] *A.T.:* CBD; environmental policy; bd education.

1113. Crowe, Tim M., & A.A. Crowe, 1982. Patterns of distribution, diversity and endemism in Afrotropical birds. *J. of Zoology* 198: 417-442. [2] *A.T.:* refugia.

1114. Crowe, Tim M., & W.R. Siegfried, 1993. Conserving Africa's biodiversity: stagnation versus innovation [editorial]. *South African J. of Science* 89: 208-210. [1].

1115. Crowley, Thomas J., & G.R. North, 1988. Abrupt climate change and extinction events in earth history. *Science* 240: 996-1002. [1] *A.T.:* Phanerozoic.

1116. Crozier, R.H., 1992. Genetic diversity and the agony of choice. *Biological Conservation* 61: 11-15. [2] Concerning the assessment of evolutionary distinctiveness in assigning conservation priorities. *A.T.:* phylogenetic analysis.

1117. ———, 1997. Preserving the information content of species: genetic diversity, phylogeny, and conservation worth. *ARES* 28: 243-268. *A.T.:*

phylogenetic analysis; higher-taxon richness.

1118. Crucible Group, 1994. *People, Plants and Patents: The Impact of Intellectual Property on Biodiversity, Conservation, Trade, and Rural Society.* Ottawa: International Development Research Centre. 116p. [1/1/1] *A.T.:* plant germplasm resources.

1119. Crump, Martha L., F.R. Hensley, & K.L. Clark, 1992. Apparent decline of the golden toad: underground or extinct? *Copeia* (2): 413-420. [2] *A.T.:* Costa Rica; endangered species; amphibian declines.

1120. Cruz, Maria C.J., & S.H. Davis, 1997. *Social Assessment in World Bank and GEF-funded Biodiversity Conservation Projects: Case Studies From India, Ecuador, and Ghana.* Washington, DC: World Bank Environment Dept. Paper No. 43. 38p. [1/1/-].

1121. Csuti, Blair, S. Polasky, et al., 1997. A comparison of reserve selection algorithms using data on terrestrial vertebrates in Oregon. *Biological Conservation* 80: 83-97. [2] *A.T.:* heuristic algorithms.

1122. Cuccia, Christine M., 1995. Protecting animals in the name of biodiversity: effects of the Uruguay round of measures regulating methods of harvesting. *Boston University International Law J.* 13: 481-502. [1] *A.T.:* GATT; dolphins.

1123. Cullen, Vicky, fall-winter 1995. Diversity—nature's insurance policy against catastrophe. *Oceanus* 38(2): 2-3. [1] *A.T.:* bd value.

1124. Culotta, Elizabeth, 1991. Biological immigrants under fire. *Science* 254: 1444-1447. [1] *A.T.:* biological invasions.

1125. ———, 1994. Is marine biodiversity at risk? *Science* 263: 918-920. [1] *A.T.:* marine extinctions; bd loss; coastal environments.

1126. ———, 1995. Many suspects to blame in Madagascar extinctions. *Science* 268: 1568-1569. [1] *A.T.:* megafauna.

1127. ———, 1996. Exploring biodiversity's benefits. *Science* 273: 1045-1046. [1] *A.T.:* ecosystem health; ecosystem diversity; David Tilman.

1128. Culver, David C., 1982. *Cave Life: Evolution and Ecology.* Cambridge, MA: Harvard University Press. 189p. [2/1/2].

1129. Culver, David C., & J.R. Holsinger, 1992. How many species of troglobites are there? *NSS Bull.* 54: 79-80. [1] *A.T.:* cave biota.

1130. Cumming, D.H.M., M.B. Fenton, et al., 1997. Elephants, woodlands and biodiversity in Southern Africa. *South African J. of Science* 93: 231-236.

1131. Cummings, Ralph W., Jr., 1991. Reverse technology flow as a key to future world and U.S. agriculture. *Bioscience* 41: 775-778. [1] *A.T.:* North-South relations; plant breeding.

1132. Cummings, Ronald G., D.S. Brookshire, et al., 1986. *Valuing Environmental Goods: An Assessment of the Contingent Valuation Method.* Totowa, NJ: Rowman & Allanheld. 270p. [2/2/3].

1133. Cunningham, A.B., summer 1991. Indigenous knowledge and biodiversity. *Cultural Survival Quarterly* 15(3): 4-8. [1] *A.T.:* traditional knowledge; indigenous peoples.

1134. ———, 1994. Integrating local plant resources and habitat management. *Biodiversity and Conservation* 3: 104-115. [1] *A.T.:* Africa; botanical gardens.

1135. ———, 1996. *Ethics, Biodiversity, and New Natural Products Development* (new ed.). Gland, Switzerland: WWF. 44p. [1/1/1] *A.T.:* medicinal plants.

1136. Cunningham, Richard L., 1990. *The Biological Diversity of Food Plants: Some Interpretive Thoughts.* San Francisco: Western Regional Office, National Park Service. 51p. [1/1/1].

1137. Curnutt, John L., Julie Lockwood, et al., 1994. Hotspots and species diversity [letter]. *Nature* 367: 326-327. [2] *A.T.:* birds; Australia.

1138. Currie, David J., 1991. Energy and large-scale patterns of animal- and plant-species richness. *American Naturalist* 137: 27-49. [3] *A.T.:* North America; vertebrates; trees; species-energy theory; spatial scale.

1139. Currie, David J., & J.T. Fritz, 1993. Global patterns of animal abundance and species energy use. *Oikos* 67: 56-68. [2] *A.T.:* allometry; body size.

1140. Currie, David J., & Viviane Paquin, 1987. Large-scale biogeographical patterns of species richness of trees. *Nature* 329: 326-327. [2] *A.T.:* North America; species-energy theory.

1141. Curtis, Allan, & Terry De Lacy, winter 1996. Landcare in Australia: beyond the expert farmer. *Agriculture and Human Values* 13(1): 20-31. [1] *A.T.:* land management; sustainable agriculture.

1142. Cury, Philippe, 1994. Obstinate nature: an ecology of individuals: thoughts on reproductive behavior and biodiversity. *Canadian J. of Fisheries and Aquatic Sciences* 51: 1664-1673. [1] *A.T.:* marine turtles; fish; natal homing.

1143. Cushing, D.H., & R.R. Dickson, 1976. The biological response in the sea to climatic changes. *Advances in Marine Biology* 14: 1-122. [2].

1144. Cushman, J.H., J.H. Lawton, & B.F.J. Manly, 1993. Latitudinal patterns in European ant assemblages: variation in species richness and body size. *Oecologia* 95: 30-37. [2] *A.T.:* latitudinal diversity gradients.

1145. Cutler, Alan H., 1991. Nested faunas and extinction in fragmented habitats. *CB* 5: 496-505. [2] *A.T.:* habitat fragmentation; Great Basin; mammals; birds.

1146. ———, 1994. Nested biotas and biological conservation: metrics, mechanisms, and meaning of nestedness. *Landscape and Urban Planning* 28: 73-82. [1] *A.T.:* habitat fragmentation; reserve design.

1147. Cutler, M. Rupert, 1992. Biodiversity—the political/legal context. *Proc. of the Society of American Foresters National Convention 1992*: 221-226. [1] *A.T.:* forest management; bd legislation.

1148. ———, Feb. 1993. A land exchange program to protect biodiversity. *J. of Forestry* 91(2): 25-26, 28-29. [1] *A.T.:* Wilderness Act; environmental policy.

1149. Cutright, Noel J., 1996. Joint implementation: biodiversity and greenhouse gas offsets. *Environmental Management* 20: 913-918. [1].

1150. Czarnetzy, John M., 1988. Altering nature's blueprints for profits: patenting multicellular animals. *Virginia Law Review* 74: 1327-1362. [1] *A.T.:* biotechnology.

1151. Czech, Brian, & P.R. Krausman, 1997. Distribution and causation of species endangerment in the United States [letter]. *Science* 277: 1116-1117.

1152. Da Silva, J.M.C., 1997. Endemic bird species and conservation in the cerrado region, South America. *Biodiversity and Conservation* 6: 435-450. *A.T.:* biogeography.

1153. Dabrowska-Prot, Eliza, 1995. The effect of forest-field ecotones on biodiversity of entomofauna and its functioning in agricultural landscape. *Ekologia Polska* 43: 51-78. [1] *A.T.:* Poland.

1154. Dahlberg, Anders, Lena Jonsson, & Jan-Erik Nylund, 1997. Species diversity and distribution of biomass above and below ground among ectomycorrhizal fungi in an old-growth Norway spruce forest in south Sweden. *Canadian J. of Botany* 75: 1323-1335.

1155. Dahlberg, Kenneth A., 1992. The conservation of biological diversity and U.S. agriculture: goals, institutions, and policies. *Agriculture, Ecosystems & Environment* 42: 177-193. [1] *A.T.:* environmental policy; agricultural practices.

1156. Daily, Gretchen C., 1995. Restoring value to the world's degraded lands. *Science* 269: 350-354. [1] *A.T.:* environmental degradation.

1157. ———, ed., 1997. *Nature's Services: Societal Dependence on Natural Ecosystems.* Washington, DC: Island Press. 392p. [2/2/2] *A.T.:* human ecology.

1158. Daily, Gretchen C., & P.R. Ehrlich, 1992. Population, sustainability, and Earth's carrying capacity: a framework for estimating population sizes and lifestyles that could be sustained without undermining future generations. *Bioscience* 42: 761-771. [2] *A.T.:* human population growth.

1159. ———, 1995. Preservation of biodiversity in small rainforest patches: rapid evaluations using butterfly trapping. *Biodiversity and Conservation* 4: 35-55. [2] *A.T.:* corridors; moths; butterflies; Costa Rica.

1160. ———, 1996. Nocturnality and species survival. *PNAS* 93: 11709-11712. [1] *A.T.:* moths; butterflies; dispersal.

1161. Dale, Virginia H., & H.M. Rauscher, 1994. Assessing impacts of climate change on forests: the state of biological modeling. *Climatic Change* 28: 65-90. [1].

1162. Dale, Virginia H., S.M. Pearson, et al., 1994. Relating patterns of land-use change to faunal biodiversity in the central Amazon. *CB* 8: 1027-

1036. [1] *A.T.:* habitat fragmentation.

1163. Dallmeier, Francisco, ed., 1992. *Long-term Monitoring of Biological Diversity in Tropical Forest Areas: Methods for Establishment and Inventory of Permanent Plots.* Paris: MAB Digest No. 11, UNESCO. 72p. [1/1/1].

1164. Dallmeier, Francisco, & J.A. Comiskey, eds., 1998. *Forest Biodiversity in Europe, Africa and Australasia: Research and Monitoring.* Paris: UNESCO, and New York: Parthenon. 696p.

1165. ———, eds., 1998. *Forest Biodiversity in North, Central and South America, and the Caribbean: Research and Monitoring* [symposium papers]. Paris: UNESCO, and New York: Parthenon. 768p.

1166. ———, eds., 1998. *Forest Biodiversity Research, Monitoring and Modeling: Conceptual Background and Old World Case Studies* [conference papers]. Paris: UNESCO, and New York: Parthenon. 671p.

1167. Daly, Herman E., & K.N. Townsend, eds., 1993. *Valuing the Earth: Economics, Ecology, Ethics.* Cambridge, MA: MIT Press. 387p. [2/2/1].

1168. Damania, Ardeshir B., ed., 1993. *Biodiversity and Wheat Improvement* [workshop papers]. Chichester, U.K., and New York: Wiley. 434p. [1/1/1] *A.T.:* plant germplasm resources.

1169. ———, 1994. In situ conservation of biodiversity of wild progenitors of cereal crops in the Near East. *Biodiversity Letters* 2: 56-60. [1] *A.T.:* gene banks.

1170. Damodaran, Aswath, 1992. Local self-governments and geometry of biodiversity conservation: roots of the incompatibility. *Economic and Political Weekly* 27(8): 419-424. [1]

1171. Danell, Kjell, P. Lundberg, & P. Niemelä, 1996. Species richness in mammalian herbivores: patterns in the boreal zone. *Ecography* 19:404-409. [1].

1172. Daniels, R.C., T.W. White, & K.K. Chapman, 1993. Sea-level rise: destruction of threatened and endangered species habitat in South Carolina. *Environmental Management* 17: 373-385. [1] *A.T.:* spatial scale; coastal environments.

1173. Daniels, R.J. Ranjit, 1996. Landscape ecology and conservation of birds in the Western Ghats, South India. *Ibis* 138: 64-69. [1].

1174. ———, 1997. Taxonomic uncertainties and conservation assessment of the Western Ghats. *Current Science* 73: 169-170. Concerning the Western Ghats hot-spot. *A.T.:* endemic species.

1175. Daniels, R.J. Ranjit, M.D.S. Chandran, & Madhav Gadgil, 1993. A strategy for conserving the biodiversity of the Uttara Kannada district in South India. *Environmental Conservation* 20: 131-138. [1] *A.T.:* nature reserves.

1176. Daniels, R.J. Ranjit, Madhav Gadgil, & N.V. Joshi, 1995. Impact of human extraction on tropical humid forests in the Western Ghats in Uttara

Kannada, South India. *J. of Applied Ecology* 32: 866-874. [1] *A.T.:* logging; disturbance.

1177. Danielsen, F., 1997. Stable environments and fragile communities: does history determine the resilience of avian rain forest communities to habitat degradation? *Biodiversity and Conservation* 6: 423-433. *A.T.:* habitat fragmentation; logging.

1178. Danish, K.W., 1995. The promise of national environmental funds in developing countries. *International Environmental Affairs* 7: 150-175. [1].

1179. Danks, H.V., 1994. Regional diversity of insects in North America. *American Entomologist* 40: 50-55. [1] *A.T.:* geographical distribution.

1180. D'Antonio, C.M., & P.M. Vitousek, 1992. Biological invasions by exotic grasses, the grass/fire cycle, and global change. *ARES* 23: 63-87. [2].

1181. Darlington, Philip J., 1957. *Zoogeography: The Geographical Distribution of Animals.* New York: Wiley. 675p. [3/-/2].

1182. Darwin, Charles, 1859. *On the Origin of Species by Means of Natural Selection, Or, The Preservation of Favoured Races in the Struggle for Life.* London: John Murray. 502p. [3/-/3] A logical starting point for just about any study of the origins of biodiversity. *A.T.:* evolution.

1183. Das, Indraneil, 1994. Evaluating biodiversity: the Batu Apoi experience. *Ambio* 23: 238-242. [1] *A.T.:* Brunei; forest conservation.

1184. Dasmann, Raymond F., 1988. Biosphere reserves, buffers, and boundaries [commentary]. *Bioscience* 38: 487-489. [1] *A.T.:* wilderness areas.

1185. Daugherty, C.H., A. Cree, et al., 1990. Neglected taxonomy and continuing extinctions of tuatara (*Sphenodon*). *Nature* 347: 177-179. [2] *A.T.:* endangered species; New Zealand.

1186. Dauvin, J.-C., M. Kendall, et al., 1994. An initial assessment of polychaete diversity in the northeastern Atlantic Ocean. *Biodiversity Letters* 2: 171-181. [1] *A.T.:* vertical distribution; latitudinal diversity gradients.

1187. Davey, S.M., D.R.B. Stockwell, & D.G. Peters, 1995. Managing biological diversity with intelligent systems. *AI Applications* 9(2): 69-89. [1] *A.T.:* artificial intelligence; Australia.

1188. Davidson, Art, 1990. *In the Wake of the Exxon Valdez: The Devastating Impact of the Alaska Oil Spill.* San Francisco: Sierra Club Books. 333p. [3/3/1].

1189. Davidson, D.W., W.D. Newmark, et al., 1996. Selecting wilderness areas to conserve Utah's biological diversity. *Great Basin Naturalist* 56: 95-118. [1] *A.T.:* Bureau of Land Management; reserve selection; environmental policy.

1190. Davidson, Osha Gray, 1998. *The Enchanted Braid: Coming to Terms With Nature on the Coral Reef.* New York: Wiley. 269p. [2/-/-].

1191. Davies, R.G., 1997. Termite species richness in fire-prone and fire-protected dry deciduous dipterocarp forest in Doi Suthep-Pui National Park, northern Thailand. *J. of Tropical Ecology* 13: 153-160.

1192. Davis, Frank W., 1995. Information systems for conservation research, policy, and planning. *Bioscience* 45, Suppl.: S36-S42. [1] *A.T.:* bd information; online databases; environmental policy.

1193. Davis, Frank W., D.M. Stoms, et al., 1990. An information systems approach to the preservation of biological diversity. *International J. of Geographical Information Systems* 4: 55-78. [2] *A.T.:* GIS; California; gap analysis.

1194. Davis, George W., 1995. International concern with functional aspects of biodiversity. *South African J. of Science* 91: 61-62. [1] *A.T.:* international cooperation; SCOPE.

1195. Davis, George W., & D.M. Richardson, eds., 1995. *Mediterranean-type Ecosystems: The Function of Biodiversity*. Berlin and New York: Springer. 366p. [1/1/1].

1196. Davis, George W., G.F. Midgley, & M.T. Hoffmann, 1994. Linking biodiversity to ecosystem function: a challenge to fynbos ecology. *South African J. of Science* 90: 319-321. [1] *A.T.:* South Africa; water catchment.

1197. Davis, Mary B., ed., 1996. *Eastern Old-growth Forests: Prospects for Rediscovery and Recovery*. Washington, DC: Island Press. 383p. [2/1/1].

1198. Davis, Peter, 1996. *Museums and the Natural Environment: The Role of Natural History Museums in Biological Conservation*. London and New York: Leicester University Press. 286p. [1/1/1].

1199. Davis, Stephen D., S.J.M. Droop, et al., 1986. *Plants in Danger: What Do We Know?* Gland, Switzerland: IUCN. 461p. [2/1/1] *A.T.:* endangered plants; plant conservation.

1200. Davis, Steven M., & J.C. Ogden, eds., 1994. *Everglades: The Ecosystem and Its Restoration*. Delray Beach, FL: St. Lucie Press. 826p. [2/2/2] *A.T.:* restoration ecology.

1201. Davis, William J., 1996. Focal species offer management tool. *Science* 271: 1362-1363. [1] *A.T.:* Australia; wildlife management.

1202. Davison, Angus, & J.R. Bridle, 1996. Exploding bird diversity brings biological species into question [letter]. *TREE* 11: 509. [1] *A.T.:* systematics.

1203. Davison, A.W., & J.D. Barnes, 1998. Effects of ozone on wild plants. *New Phytologist* 139: 135-151.

1204. Dawson, I.K., R. Waugh, et al., 1997. Simple sequence repeats provide a direct estimate of pollen-mediated gene dispersal in the tropical tree *Gliricidia sepium*. *Molecular Ecology* 6: 179-183. *A.T.:* Guatemala.

1205. Day, Kelly, 1996. Agriculture's links to biodiversity. *Agricultural Outlook*, no. 236: 32-37. [1].

1206. Dayton, Leigh, 15 Feb. 1992. Third World countries should control their genetic resources [editorial]. *New Scientist* 133(1808): 17. [1] *A.T.:* Global Biodiversity Strategy.

1207. Dayton, Paul K., B.J. Mordida, & F. Bacon, 1994. Polar marine communities. *American Zoologist* 34: 90-99. [1] A general review.

1208. Dayton, Paul K., M.J. Tegner, et al., 1998. Sliding base lines, ghosts, and reduced expectations in kelp forest communities. *Ecological Applications* 8: 309-322. *A.T.:* California; community ecology.

1209. De Fontaubert, A. Charlotte, et al., 1996. *Biodiversity in the Seas: Implementing the Convention on Biological Diversity in Marine and Coastal Habitats.* Gland, Switzerland: IUCN Environmental Policy and Law Paper No. 32. 82p. [1/1/1] *A.T.:* international cooperation.

1210. De Freese, Duane E., 1991. Threats to biological diversity in marine and estuarine ecosystems of Florida. *Coastal Management* 19: 73-101. [1] *A.T.:* coastal zone management.

1211. ———, 1995. Land acquisition: a tool for biological diversity protection in the Indian River Lagoon, Florida. *Bull. of Marine Science* 57: 14-27. [1] *A.T.:* coastal zone management.

1212. De Selincourt, Kate, 15 Feb. 1992. South Africa's other bush war. *New Scientist* 133(1808): 46-49. [1] Describes threats to the fynbos.

1213. De Vecchi, S., & Z. Coppes, 1996. Marine fish digestive proteases: relevance to food industry and the South West Atlantic region: a review. *J. of Food Biochemistry* 20: 193-214. [1] *A.T.:* enzymes.

1214. Deacon, Robert T., 1995. Assessing the relationship between government policy and deforestation. *J. of Environmental Economics and Management* 28: 1-18. [1].

1215. Dean, Denis J., K.R. Wilson, & C.H. Flather, 1997. Spatial error analysis of species richness for a gap analysis map. *Photogrammetric Engineering and Remote Sensing* 63: 1211-1217. *A.T.:* Oregon; GIS.

1216. Deane, G.C., 1994. The role of GIS in the management of natural resources. *ASLIB Proc.* 46: 157-161. [1] *A.T.:* bd databases.

1217. Dearden, Philip, 1995. Biodiversity and development in northern Thailand. *J. of Business Administration* (Vancouver) 22-23: 255-269. [1] *A.T.:* sustainable development.

1218. Dearden, Philip, S. Chettamart, et al., 1996. National parks and hill tribes in northern Thailand: a case study of Doi Inthanon. *Society & Natural Resources* 9: 125-141. [1] *A.T.:* indigenous peoples; protected areas management.

1219. Death, Russell G., & M.J. Winterbourn, 1995. Diversity patterns in stream benthic invertebrate communities: the influence of habitat stability. *Ecology* 76: 1446-1460. [2] *A.T.:* New Zealand.

1220. DeBano, Leonard F., G.J. Gottfried, et al., coordinators, 1995. *Biodiversity and Management of the Madrean Archipelago: The Sky Islands of Southwestern United States and Northwestern Mexico* [conference proceedings]. Fort Collins, CO: General Technical Report RM GTR-264, Rocky Mountain Forest and Range Experiment Station. 669p. [1/1/1].

1221. Debinski, Diane M., & P.F. Brussard, 1994. Using biodiversity data to assess species-habitat relationships in Glacier National Park, Montana. *Ecological Applications* 4: 833-843. [1] *A.T.:* birds; butterflies.

1222. Debinski, Diane M., & P.S. Humphrey, 1997. An integrated approach to biological diversity management. *Natural Areas J.* 17: 355-365. *A.T.:* bd assessment.

1223. Deblinger, R.D., & R.E. Jenkins, Jr., 1991. Preserving coastal biodiversity: the private, nonprofit approach. *Coastal Management* 19: 103-112. [1] *A.T.:* Natural Heritage Program; institutional cooperation.

1224. DeBuhr, Larry E., 1995. Public understanding of biodiversity. *Bioscience* 45, Suppl.: S43-S44. [1] On the need for the scientific community to help educators with bd education.

1225. Decher, Jan, 1997. Conservation, small mammals, and the future of sacred groves in West Africa. *Biodiversity and Conservation* 6: 1007-1026. *A.T.:* Ghana; traditional knowledge.

1226. Decker, Daniel J., & G.R. Goff, eds., 1987. *Valuing Wildlife: Economic and Social Perspectives* [symposium papers]. Boulder, CO: Westview Press. 424p. [2/1/1].

1227. DeGraaf, Richard M., & R.I. Miller, eds., 1996. *Conservation of Faunal Diversity in Forested Landscapes*. London: Chapman & Hall. 633p. [1/1/1] *A.T.:* vertebrates; wildlife conservation.

1228. DeGraaf, Richard M., & J.H. Rappole, 1995. *Neotropical Migratory Birds: Natural History, Distribution, and Population Change*. Ithaca, NY: Comstock. 676p. [2/2/1].

1229. DeGraaf, Richard M., & D.D. Rudis, 1990. Herpetofaunal species composition and relative abundance among three New England forest types. *Forest Ecology and Management* 32: 155-165. [1] *A.T.:* New Hampshire.

1230. Deharveng, Louis, 1996. Soil Collembola diversity, endemism, and reforestation: a case study in the Pyrenees (France). *CB* 10: 74-84. [1].

1231. Dehgan, Bijan, 15 Jan. 1996. Permian permanence. *American Nurseryman* 183(2): 66-81. [1] Concerning the conservation of cycads. *A.T.:* endangered species; nursery industry.

1232. Dejean, Alain, Jean Luc Durand, & Barry Bolton, 1996. Ants inhabiting *Cubitermes* termitaries in African rain forests. *Biotropica* 28: 701-713. [1] *A.T.:* Cameroon; nesting sites.

1233. Delcourt, H.R., P.A. Delcourt, & T. Webb III, 1983. Dynamic plant ecology: the spectrum of vegetation change in space and time. *Quater-nary Science Reviews* 1: 153-175. [2] *A.T.:* climatic change; Quaternary; spatial-temporal scale.

1234. Delettre, Y.R., N. Morvan, et al., 1998. Local biodiversity and multi-habitat use in empidoid flies (Insecta: Diptera, Empidoidea). *Biodiversity and Conservation* 7: 9-25. *A.T.:* France; life cycle analysis; animal behavior.

1235. Delibes, Miguel, & M.C. Blázquez, 1998. Tameness of insular lizards and loss of biological diversity [comment]. *CB* 12: 1142-1143. *A.T.:* animal behavior.

1236. Delille, D., 1996. Biodiversity and function of bacteria in the Southern Ocean. *Biodiversity and Conservation* 5: 1505-1523. [1] *A.T.:* bacterioplankton.

1237. Della Sala, Dominick A., J.R. Strittholt, et al., 1996. A critical role for core reserves in managing Inland Northwest landscapes for natural resources and biodiversity. *Wildlife Society Bull.* 24: 209-221. [1] *A.T.:* protected areas; disturbance landscapes.

1238. DeLong, Don C., Jr., 1996. Defining biodiversity. *Wildlife Society Bull.* 24: 738-749. [1].

1239. DeLong, Edward F., 1997. Marine microbial diversity: the tip of the iceberg. *Trends in Biotechnology* 15: 203-207. [1] *A.T.:* microorganisms; bd assessment; bacteria.

1240. DeLong, S. Craig, & D. Tanner, 1996. Managing the pattern of forest harvest: lessons from wildfire. *Biodiversity and Conservation* 5: 1191-1205. [1] *A.T.:* forest management; fire ecology; British Columbia.

1241. Demain, Arnold L., 1998. Biodiversity: microbial natural products: alive and well in 1998. *Nature Biotechnology* 16: 3-4. *A.T.:* antibiotics.

1242. DeMaynadier, Phillip G., & M.L. Hunter, Jr., 1998. Effects of silvicultural edges on the distribution and abundance of amphibians in Maine. *CB* 12: 340-352. *A.T.:* edge effects; forest management; microhabitat.

1243. Demissie, A., & A. Björnstad, 1997. Geographical, altitude and agro-ecological differentiation of isozyme and hordein genotypes of landrace barleys from Ethiopia: implications to germplasm conservation. *Genetic Resources and Crop Evolution* 44: 43-55. *A.T.:* genetic diversity.

1244. Dennis, Brian, P.L. Munholland, & J.M. Scott, 1991. Estimation of growth and extinction parameters for endangered species. *Ecological Monographs* 61: 115-143. [2] *A.T.:* stochastic models.

1245. Dennis, John, & Michael Ruggiero, 1990. Conserving biological diversity. *Trends* (National Park Service) 27(4): 4-7, 47. [1] *A.T.:* national parks.

1246. Dennis, Roger L.H., 1997. An inflated conservation load for European butterflies: increases in rarity and endemism accompany increases in species richness. *J. of Insect Conservation* 1: 43-62.

1247. Dennis, Roger L.H., & T.G. Shreeve, 1997. Diversity of butterflies on British Islands: ecological influences underlying the roles of area, isolation and the size of the faunal source. *Biological J. of the Linnean Society* 60: 257-275. *A.T.:* island biogeography.

1248. Denniston, Derek, Nov.-Dec. 1993. Saving the Himalaya. *World Watch* 6(6): 10-21. [1] *A.T.:* environmental degradation.

1249. Denny, Patrick, 1994. Biodiversity and wetlands. *Wetlands Ecology and Management* 3: 55-61. [1] *A.T.:* CBD; habitat conservation.

1250. Denslow, Julie Sloan, 1980. Patterns of plant species diversity during succession under different disturbance regimes. *Oecologia* 46: 18-21. [2].
1251. ———, 1987. Tropical rainforest gaps and tree species diversity. *ARES* 18: 431-451. [3] *A.T.:* treefall gaps; forest turnover.
1252. Department of State (U.S.), 1982. *Proceedings of the U.S. Strategy Conference on Biological Diversity, November 16-18, 1981*. Washington, DC. 126p. [1/1/1] *A.T.:* germplasm resources.
1253. Department of the Environment, Sport and Territories (Australia), 1993. *Biological Diversity: Its Future Conservation in Australia* [conference proceedings]. Canberra: Dept. of the Environment, Sport and Territories. 212p. [1/1/1].
1254. Department of the Environment (Great Britain), 1994. *Biodiversity: The UK Action Plan*. London: HMSO. 188p. [1/1/1].
1255. Desai, M.C., R.N. Zuckermann, & W.H. Moos, 1994. Recent advances in the generation of chemical diversity libraries. *Drug Development Research* 33: 174-188. [2] *A.T.:* molecular diversity; combinatorial chemistry.
1256. Deshaye, Jean, & Pierre Morisset, 1989. Species-area relationships and the SLOSS effect in a subarctic archipelago. *Biological Conservation* 48: 265-276. [1] *A.T.:* Quebec; island biogeography; vascular plants.
1257. Detwiler, R.P., & C.A.S. Hall, 1988. Tropical forests and the global carbon cycle. *Science* 239: 42-47. [2] *A.T.:* carbon dioxide; deforestation.
1258. Deutsch, J.S., E. Mouchel-Vielh, & J.-M. Gibert, 1997. The developmental origin of biodiversity: the case of the barnacles (Cirripedes). *Vie et Milieu* 47: 325-331. *A.T.:* phylogenetics; homeotic genes.
1259. Devall, Bill, & George Sessions, 1985. *Deep Ecology*. Salt Lake City: Gibbs Smith. 266p. [2/2/3].
1260. Devine, Robert, Jan.-Feb. 1994. Botanical barbarians. *Sierra* 79(1): 50-57, 71. [1] Concerning the problem of invasive plant species in the U.S.
1261. DeVries, Philip J., Debra Murray, & Russell Lande, 1997. Species diversity in vertical, horizontal, and temporal dimensions of a fruit-feeding butterfly community in an Ecuadorian rainforest. *Biological J. of the Linnean Society* 62: 343-364. *A.T.:* vertical distribution.
1262. Dhar, Uppeandra, ed., 1993. *Himalayan Biodiversity: Conservation Strategies*. Nainital, India: Gyanodaya Prakashan. 553p. [1/1/1].
1263. ———, ed., 1997. *Himalayan Biodiversity: Action Plan* [workshop papers]. Nainital, India: Gyanodaya Prakashan. 136p. [1/1/-].
1264. Dhar, Uppeandra, R.S. Rawal, & S.S. Samant, 1997. Structural diversity and representativeness of forest vegetation in a protected area of Kumaun Himalaya, India: implications for conservation. *Biodiversity and Conservation* 6: 1045-1062. *A.T.:* altitudinal gradients.
1265. Di Castri, Francesco, & Talal Younès, eds., 1996. *Biodiversity, Science and Development: Towards a New Partnership* [forum papers].

Wallingford, Oxon, U.K.: CAB International. 646p. [1/1/1].

1266. Di Castri, Francesco, A.J. Hansen, & M. Debussche, eds., 1990. *Biological Invasions in Europe and the Mediterranean Basin*. Dordrecht and Boston: Kluwer Academic. 463p. [1/1/1].

1267. Di Castri, Francesco, J.R. Vernhes, & Talal Younès, eds., 1992. *Inventorying and Monitoring Biodiversity: A Proposal for an International Network* [meeting report]. Paris: Biology International Special Issue No. 27, IUBS. 28p. [1/1/1].

1268. Di Cello, F., A. Bevivino, et al., 1997. Biodiversity of a *Burkholderia cepacia* population isolated from the maize rhizosphere at different plant growth stages. *Applied and Environmental Microbiology* 63: 4485-4493. *A.T.:* genetic diversity; bacteria.

1269. Dial, Kenneth P., & J.M. Marzluff, 1988. Are the smallest organisms the most diverse? *Ecology* 69: 1620-1624. [2] Considers whether speciosity is related to body size.

1270. Diamond, Antony W., & F.L. Filion, eds., 1987. *The Value of Birds* [conference proceedings]. Cambridge, U.K.: International Council for Bird Preservation. 267p. [1/1/1] *A.T.:* bird conservation.

1271. Diamond, Antony W., & T.E. Lovejoy, eds., 1985. *Conservation of Tropical Forest Birds: Proceedings of a Workshop and Symposium*. Cambridge, U.K.: International Council for Bird Preservation. 318p. [1/1/1].

1272. Diamond, Antony W., R.L. Schreiber, et al., 1989. *Save the Birds*. Boston: Houghton Mifflin. 384p. [2/1/1] *A.T.:* bird conservation.

1273. Diamond, Irene, & G.F. Orenstein, eds., 1990. *Reweaving the World: The Emergence of Ecofeminism* [essays]. San Francisco: Sierra Club Books. 320p. [2/2/2].

1274. Diamond, Jared M., 1972. Biogeographic kinetics: estimation of relaxation times for avifaunas of southwest Pacific islands. *PNAS* 69: 3199-3203. [2] *A.T.:* island biogeography; equilibrium theory.

1275. ———, 1975. The island dilemma: lessons of modern biogeographic studies for the design of natural reserves. *Biological Conservation* 7: 129-146. [2].

1276. ———, 1987. Extant unless proven extinct? Or, extinct unless proven extant? *CB* 1: 77-79. [1].

1277. ———, 1988. Factors controlling species diversity: overview and synthesis. *Annals of the Missouri Botanical Garden* 75: 117-129. [1] *A.T.:* latitudinal diversity gradients; altitudinal diversity gradients.

1278. ———, 1989. The present, past and future of human-caused extinctions. *PTRSL B* 325: 469-477. [1] *A.T.:* extinction rates; Pleistocene extinctions.

1279. ———, 1989. Quaternary megafaunal extinctions: variations on a theme by Paganini. *J. of Archaeological Science* 16: 167-175. [1].

1280. Diamond, Jared M., & C.R. Veitch, 1981. Extinctions and introductions in the New Zealand avifauna: cause and effect? *Science* 211: 499-501. [1].

1281. Diamond, Peter A., & J.A. Hausman, fall 1994. Contingent valuation: is some number better than no number? *J. of Economic Perspectives* 8(4): 45-64. [3].

1282. Dice, Lee R., 1945. Measures of the amount of ecologic association between species. *Ecology* 26: 297-302. [3].

1283. Dickinson, K.J.M., & J.B. Kirkpatrick, 1987. The short-term effects of clearfelling and slash-burning on the richness, diversity and relative abundance of higher plant species in two types of eucalypt forest on dolerite in Tasmania. *Australian J. of Botany* 35: 601-616. [1] *A.T.:* fire.

1284. Dickman, C.R., 1987. Habitat fragmentation and vertebrate species richness in an urban environment. *J. of Applied Ecology* 24: 337-351. [1] *A.T.:* England.

1285. Didham, R.K., J. Ghazoul, et al., 1996. Insects in fragmented forests: a functional approach. *TREE* 11: 255-260. [2] *A.T.:* habitat fragmentation.

1286. Didham, R.K., P.M. Hammond, et al., 1998. Beetle species responses to tropical forest fragmentation. *Ecological Monographs* 68: 295-323. *A.T.:* Amazonia; body size; edge effects.

1287. Diego Gómez, Luis, 1993. Perspectives on Mexican biodiversity. *CB* 7: 961-971. [1].

1288. Diekman, Bruce, ed., 1997. *Sustainable Use of Wildlife: Utopian Dream or Unrealistic Nightmare?* [seminar proceedings] Sydney, New South Wales: Nature Conservation Council of NSW. 218p. [1/1/-] *A.T.:* wildlife conservation.

1289. Diekman, Bruce, et al., eds., 1997. *Ecology at the Cutting Edge: Information Techniques for Managing Biodiversity and Ecological Processes* [seminar proceedings]. Sydney, New South Wales: Nature Conservation Council of NSW. 199p. [1/1/-] *A.T.:* Australia; bd databases.

1290. Diner, David N., 1994. Army and the Endangered Species Act: who's endangering whom? *Military Law Review* 143: 161-223. [1] *A.T.:* military bases; endangered species.

1291. Dinerstein, Eric, & E.D. Wikramanayake, 1993. Beyond "hotspots": how to prioritize investments to conserve biodiversity in the Indo-Pacific region. *CB* 7: 53-65. [1] *A.T.:* protected areas.

1292. Dinerstein, Eric, D.M. Olson, et al., 1995. *A Conservation Assessment of the Terrestrial Ecoregions of Latin America and the Caribbean.* Washington, DC: World Bank. 129p. [1/1/1].

1293. Dirkse, G.M., & G.F.P. Martakis, 1998. Species density of phanerogams and bryophytes in Dutch forests. *Biodiversity and Conservation* 7: 147-157.

1294. Dirzo, Rodolfo, & A. Miranda, 1990. Contemporary Neotropical defaunation and forest structure, function, and diversity: a sequel to John Terborgh. *CB* 4: 444-447. [1].

1295. DiSilvestro, Roger L., 1989. *The Endangered Kingdom: The Struggle to Save America's Wildlife*. New York: Wiley. 241p. [3/2/1].

1296. ——, 1993. *Reclaiming the Last Wild Places: A New Agenda for Biodiversity*. New York: Wiley. 266p. [2/2/1] *A.T.:* wildlife conservation; environmental philosophy.

1297. Diwan, Noa, G.R. Bauchan, & M.S. McIntosh, 1994. A core collection for the United States annual Medicago germplasm collection. *Crop Science* 34: 279-285. [1].

1298. Dix, M.E., Ned Klopfenstein, et al., 1994. Biodiversity can enhance biological control of pests in Great Plains agroecosystems. *Proc. of the Great Plains Agricultural Council 1994*: 37-53. [1] *A.T.:* weeds; plant diseases.

1299. Dixon, Anthony F.G., Pavel Kindlmann, et al., 1987. Why there are so few species of aphids, especially in the tropics. *American Naturalist* 129: 580-592. [2] *A.T.:* host plants.

1300. Dixon, Bernard, 1990. Biotech's effects on biodiversity debated [editorial]. *Bio/technology* 8: 499. [1].

1301. Dixon, John A., fall 1993. Economic benefits of marine protected areas. *Oceanus* 36(3): 35-40. [1] *A.T.:* bdc; ecotourism.

1302. Dixon, John A., & P.B. Sherman, 1990. *Economics of Protected Areas: A New Look at Benefits and Costs*. Washington, DC: Island Press. 234p. [2/2/2] *A.T.:* developing countries; environmental policy.

1303. Dixon, John A., Louise Fallon Scura, & Tom Van't Hof, 1993. Meeting ecological and economic goals: marine parks in the Caribbean. *Ambio* 22: 117-125. [2] *A.T.:* bdc; developing countries.

1304. Doak, D.F., & L.S. Mills, 1994. A useful role for theory in conservation. *Ecology* 75: 615-626. [2] *A.T.:* conservation biology; ecological theory.

1305. Doak, D.F., D. Bigger, et al., 1998. The statistical inevitability of stability-diversity relationships in community ecology. *American Naturalist* 151: 264-276.

1306. Dobbeler, P., 1997. Biodiversity of bryophilous ascomycetes. *Biodiversity and Conservation* 6: 721-738. *A.T.:* bryophytes; ascomycetes; fungal parasites.

1307. Dobkin, David S., A.C. Rich, & W.H. Pyle, 1998. Habitat and avifaunal recovery from livestock grazing in a riparian meadow system of the northwestern Great Basin. *CB* 12: 209-221.

1308. Dobrovol'skii, G.V., 1996. The role of soils in preservation of biodiversity. *Eurasian Soil Science* 29: 626-629. [1] *A.T.:* soil fauna; soil diversity.

1309. Dobson, Andrew P., Sept. 1992. Withering heats: global warming will be the last straw for many threatened species. *Natural History* 101(9):

2-8. [1].

1310. ———, 1995. Biodiversity and human health [editorial]. *TREE* 10: 390-391. [1].

1311. ———, 1995. *Green Political Thought* (2nd ed.). London and New York: Routledge. 225p. [2/2/2] *A.T.:* Green movement; environmental politics.

1312. ———, 1996. *Conservation and Biodiversity.* New York: Scientific American Library. 264p. [2/2/1].

1313. Dobson, Andrew P., & Robin Absher, 1991. How to pay for tropical rain forests. *TREE* 6: 348-351. [1].

1314. Dobson, Andrew P., & Robin Carper, 1993. Biodiversity. *Lancet* 342: 1096-1099. [1] Examines the impact of climatic change on disease incidence and spread.

1315. Dobson, Andrew P., A.D. Bradshaw, & A.J.M. Baker, 1997. Hopes for the future: restoration ecology and conservation biology. *Science* 277: 515-522. *A.T.:* environmental degradation; land management.

1316. Dobson, Andrew P., A. Jolly, & D. Rubenstein, 1989. The greenhouse effect and biological diversity. *TREE* 4: 64-68. [2]

1317. Dobson, Andrew P., J.P. Rodriguez, et al., 1997. Geographic distribution of endangered species in the United States. *Science* 275: 550-553. [2] *A.T.:* hotspots.

1318. Dobson, Tracy A., 1992. Loss of biodiversity: an international environmental policy perspective. *North Carolina J. of International Law and Commercial Regulation* 17: 277-309. [1] *A.T.:* endangered species.

1319. Dobyns, John R., 1997. Effects of sampling intensity on the collection of spider (Araneae) species and the estimation of species richness. *Environmental Entomology* 26: 150-162. *A.T.:* Georgia.

1320. Dobzhansky, Theodosius, 1950. Evolution in the tropics. *American Scientist* 38: 209-221. [2].

1321. Dodd, C. Kenneth, Jr., 1992. Biological diversity of a temporary pond herpetofauna in north Florida sandhills. *Biodiversity and Conservation* 1: 125-142. [1].

1322. Dodd, C. Kenneth, Jr., & R.A. Seigel, 1991. Relocation, repatriation and translocation of amphibians and reptiles: are they conservation strategies that work? *Herpetologica* 47: 336-350. [1].

1323. Dodds, John H., ed., 1991. *In Vitro Methods for Conservation of Plant Genetic Resources.* London and New York: Chapman & Hall. 247p. [1/1/1] *A.T.:* plant germplasm resources.

1324. Dodson, Edward O., 1971. The kingdoms of organisms. *Systematic Zoology* 20: 265-281. [1].

1325. Dodson, Stanley I., 1991. Species richness of crustacean zooplankton in European lakes of different size and latitude. *Verhandlungen der Internationalen Vereinigung für Theoretische und Angewandte Limnologie*

24: 1223-1229. [1].

1326. ———, 1992. Predicting crustacean zooplankton species richness. *Limnology and Oceanography* 37: 848-856. [1] *A.T.:* North America; lakes.

1327. Dogsé, Peter, & Bernd von Droste, 1990. *Debt-for-Nature Exchanges and Biosphere Reserves: Experiences and Potential.* Paris: MAB Digest No. 6, UNESCO. 88p. [1/1/1].

1328. Doherty, P.J., & D.M. Williams, 1988. The replenishment of coral reef fish populations. *Oceanography and Marine Biology* 26: 487-551. [3] An important review.

1329. Doledec, Sylvain, & Bernhard Statzner, 1994. Theoretical habitat templets, species traits, and species richness: 548 plant and animal species in the Upper Rhône River and its floodplain. *Freshwater Biology* 31: 523-538. [1] *A.T.:* river ecology.

1330. Dombrowski, Daniel A., 1990. Nature as personal. *Philosophy, Theology* 5: 81-96. [1] *A.T.:* Gaia hypothesis; metaphysics; James Lovelock.

1331. Dompka, Victoria, ed., 1996. *Human Population, Biodiversity and Protected Areas: Science and Policy Issues* [workshop report]. Washington, DC: American Association for the Advancement of Science. 254p. [1/1/1] *A.T.:* environmental policy.

1332. Donaldson, J.S., & G. Scott, 1994. Aspects of human dependence on plant diversity in the Cape mediterranean-type ecosystem. *South African J. of Science* 90: 338-342. [1] *A.T.:* South Africa; fynbos; resource utilization.

1333. Done, T.J., 1995. Ecological criteria for evaluating coral reefs and their implications for managers and researchers. *Coral Reefs* 14: 183-192. [2] *A.T.:* ecological risk assessment.

1334. Donovan, Stephen K., ed., 1989. *Mass Extinctions: Processes and Evidence.* New York: Columbia University Press. 266p. [2/2/1].

1335. Doolan, Seán, ed., 1997. *African Rainforests and the Conservation of Biodiversity* [conference proceedings]. Oxford, U.K.: Earthwatch Europe. 170p. [1/1/-].

1336. Doolittle, Russell F., Da-fei Feng, et al., 1996. Determining divergence times of the major kingdoms of living organisms with a protein clock. *Science* 271: 470-477. [3] *A.T.:* amino acid sequence data.

1337. Dore, M.H.I., & J.M. Nogueira, 1994. The Amazon rain forest, sustainable development and the Biodiversity Convention: a political economy perspective. *Ambio* 23: 491-496. [1] *A.T.:* deforestation; CBD.

1338. Doremus, Holly, 1991. Patching the ark: improving legal protection of biological diversity. *Ecology Law Quarterly* 18: 265-333. [2] *A.T.:* Endangered Species Act; endangered species.

1339. Dottavio, F. Dominic, P.F. Brussard, & J.D. McCrone, eds., 1990. *Protecting Biological Diversity in the National Parks* [workshop report]. Washington, DC: National Park Service, U.S. Dept. of the Interior. 79p.

[1/1/1] *A.T.:* environmental management.

1340. Dougall, T.A.G., & J.C. Dodd, 1997. A study of species richness and diversity in seed banks and its use for the environmental mitigation of a proposed holiday village development in a coniferized woodland in southeast England. *Biodiversity and Conservation* 6: 1413-1428. *A.T.:* diversity indices.

1341. Douglas, M., & P.S. Lake, 1994. Species richness of stream stones: an investigation of the mechanisms generating the species-area relationship. *Oikos* 69: 387-396. [1] *A.T.:* habitat diversity; stream stone fauna; Victoria, Australia.

1342. Dourojeanni, Marc J., 1990. Entomology and biodiversity conservation in Latin America. *American Entomologist* 36: 88-93. [1].

1343. Dove, Michael R., 1993. A revisionist view of tropical deforestation and development. *Environmental Conservation* 20: 17-24, 56. [2] *A.T.:* forest resources; Indonesia; resource exploitation.

1344. Dove, Michael R., & P.E. Sajise, 1997. *The Conditions of Biodiversity Maintenance in Asia: The Policy Linkages Between Environmental Conservation and Sustainable Development* [research report]. Honolulu: Program on Environment, East-West Center. 346p. [1/1/-].

1345. Dovers, Stephen R., T.W. Norton, & J.W. Handmer, 1996. Uncertainty, ecology, sustainability and policy. *Biodiversity and Conservation* 5: 1143-1167. [1] *A.T.:* Australia; environmental policy.

1346. Downes, B.J., & J. Jordan, 1993. Effects of stone topography on abundance of net-building caddisfly larvae and arthropod diversity in an upland stream. *Hydrobiologia* 252: 163-174. [1] *A.T.:* Victoria, Australia.

1347. Downes, David R., 1994. The Convention on Biological Diversity: seeds of green trade? *Tulane Environmental Law J.* 8: 163-180. [1] *A.T.:* international trade.

1348. Downing, Theodore E., S.B. Hecht, et al., eds., 1992. *Development or Destruction: The Conversion of Tropical Forest to Pasture in Latin America.* Boulder, CO: Westview Press. 405p. [1/1/1] *A.T.:* deforestation.

1349. Doyle, Jamie K., & John Schelhas, eds., 1993. *Forest Remnants in the Tropical Landscape: Benefits and Policy Implications* [symposium proceedings]. Washington, DC: Smithsonian Migratory Bird Center, Smithsonian Institution. 103p. [1/1/1] *A.T.:* environmental policy; environmental management.

1350. Doyle, Rodger, Jan. 1997. Threatened mammals. *Scientific American* 276(1): 32. [1] A succinct summary of worldwide trends.

1351. Dragun, Andrew K., & K.M. Jakobsson, eds., 1997. *Sustainability and Global Environmental Policy: New Perspectives.* Cheltenham, U.K., and Lyme, NH: Edward Elgar. 319p. [1/1/-].

1352. Drake, James A., 1990. Communities as assembled structures: do rules govern pattern? *TREE* 5: 159-164. [2] *A.T.:* assembly rule; community assembly.

1353. Drake, James A., H.A. Mooney, et al., eds., 1989. *Biological Invasions: A Global Perspective* [workshop papers]. Chichester, U.K., and New York: Wiley. 525p. [1/2/2].

1354. Dransfield, John, M.J.E. Coode, & D.A. Simpson, eds., 1997. *Plant Diversity in Malesia III: Proceedings of the Third International Flora Malesiana Symposium 1995.* London: Royal Botanic Gardens, Kew. 449p. [1/1/-].

1355. Drinkrow, D.R., M.I. Cherry, & W.R. Siegfried, 1994. The role of natural history museums in preserving biodiversity in South Africa. *South African J. of Science* 90: 470-479. [1] *A.T.:* research programs.

1356. Droege, Sam, André Cyr, & Jacques Larivée, 1998. Checklists: an under-used tool for the inventory and monitoring of plants and animals. *CB* 12: 1134-1138. *A.T.:* Quebec; birds.

1357. Drohan, Joy, Aug.-Sept. 1996. Sustainably developing the DMZ [editorial]. *Technology Review* 99(6): 17-18. [1] *A.T.:* Korea; demilitarized zones; protected areas.

1358. Drost, Charles A., & G.M. Fellers, 1996. Collapse of a regional frog fauna in the Yosemite area of the California Sierra Nevada, USA. *CB* 10: 414-425. [2] *A.T.:* introduced fish; drought; amphibian declines.

1359. Droste, Bernd von, Nov. 1991. No room in the ark. *UNESCO Courier* 44(11): 36-39. [1] A general plan for greater worldwide government and private involvement in bdc. *A.T.:* MAB.

1360. Drozdowski, James, 1995. Saving an endangered act: the case for a biodiversity approach to ESA conservation efforts. *Case Western Reserve Law Review* 45: 553-602. [1] *A.T.:* Endangered Species Act.

1361. Drury, William H., 1974. Rare species. *Biological Conservation* 6: 162-169. [1].

1362. Dryzek, John S., 1990. Green reason: communicative ethics for the biosphere. *Environmental Ethics* 12: 195-210. [1].

1363. Du Toit, J.T., 1995. Determinants of the composition and distribution of wildlife communities in Southern Africa. *Ambio* 24: 2-6. [1] *A.T.:* large mammals; environmental factors.

1364. Duchesne, L.C., 1994. Defining Canada's old-growth forests—problems and solutions. *Forestry Chronicle* 70: 739-744. [1] *A.T.:* forest management.

1365. Duckworth, W. Donald, H.H. Genoways, & C.L. Rose, 1993. *Preserving Natural Science Collections: Chronicle of Our Environmental Heritage* [report]. Washington, DC: National Institute for the Conservation of Cultural Property. 140p. [1/1/1] *A.T.:* museum collections; plant specimens.

1366. Dudley, Joseph P., 1992. Rejoinder to Rohlf and O'Connell: biodiversity as a regulatory criterion [letter]. *CB* 6: 587-589. [1].

1367. Dudley, Nigel, 1992. *Forests in Trouble: A Review of the Status of*

Temperate Forests Worldwide. Gland, Switzerland: World Wide Fund for Nature. 260p. [1/1/1].

1368. ———, 1993. Conservation, biodiversity and European forestry. *European Environment* 3(1): 9-12. [1] *A.T.:* forest decline; environmental management.

1369. Duelli, Peter, 1997. Biodiversity evaluation in agricultural landscapes: an approach at two different scales. *Agriculture, Ecosystems & Environment* 62: 81-91. *A.T.:* inventories; rapid assessment; arthropods.

1370. Duelli, Peter, & M.K. Obrist, 1998. In search of the best correlates for local organismal biodiversity in cultivated areas. *Biodiversity and Conservation* 7: 297-309. *A.T.:* indicator species; flight and pitfall traps.

1371. Duellman, William E., ed., 1979. *The South American Herpetofauna: Its Origin, Evolution, and Dispersal* [symposium papers]. Lawrence, KS: Museum of Natural History, University of Kansas. 485p. [1/1/1].

1372. ———, 1988. Patterns of species diversity in anuran amphibians in the American tropics. *Annals of the Missouri Botanical Garden* 75: 79-104. [1].

1373. Duellman, William E., & E.R. Pianka, 1990. Biogeography of nocturnal insectivores: historical events and ecological filters. *ARES* 21: 57-68. [1] *A.T.:* lizards; frogs; diversification.

1374. Duffy, David C., 1992. Biodiversity and research on seabirds. *Colonial Waterbirds* 15: 155-158. [1] *A.T.:* research programs.

1375. Dugan, Jenifer E., & G.E. Davis, 1993. Applications of marine refugia to coastal fisheries management. *Canadian J. of Fisheries and Aquatic Sciences* 50: 2029-2042. [2].

1376. Dugan, Patrick J., ed., 1990. *Wetland Conservation: A Review of Current Issues and Required Action*. Gland, Switzerland: IUCN. 96p. [1/1/1].

1377. Duivenvoorden, Joost F., 1996. Patterns of tree species richness in rain forests of the middle Caqueta area, Colombia, NW Amazonia. *Biotropica* 28: 142-158. [1].

1378. Duke, N.C., M.C. Ball, & J.C. Ellison, 1998. Factors influencing biodiversity and distributional gradients in mangroves. *Global Ecology and Biogeography Letters* 7: 27-47. *A.T.:* Australia.

1379. Dumont, H.J., & H. Segers, 1996. Estimating lacustrine zooplankton species richness and complementarity. *Hydrobiologia* 341: 125-132. [1].

1380. Dunlap, Paul V., fall-winter 1995. New insights on marine bacterial diversity: molecular techniques complement culturing. *Oceanus* 38(2): 16-19. [1] *A.T.:* molecular biology.

1381. Dunlap, Thomas R., 1988. *Saving America's Wildlife*. Princeton, NJ: Princeton University Press. 222p. [2/3/1] *A.T.:* wildlife conservation history.

1382. Dunn, Christopher P., M.L. Bowles, & G.B. Rabb, 1997. Endangered species "hot spots" [discussion]. *Science* 276: 513-517. Discusses a report on the geographical distribution of endangered species in the U.S.

1383. Dunn, Christopher P., R. Stearns, et al., 1993. Ecological benefits of the Conservation Reserve Program. *CB* 7: 132-139. [1] *A.T.:* soil conservation.

1384. Dunning, J.B., Jr., D.J. Stewart, et al., 1995. Spatially explicit population models: current forms and future uses. *Ecological Applications* 5: 3-11. [2] *A.T.:* landscape modelling.

1385. Durbin, J.C., & J.A. Ralambo, 1994. The role of local people in the successful maintenance of protected areas in Madagascar. *Environmental Conservation* 21: 115-120. [1].

1386. Durbin, Kathie, winter 1995. Mapping Oregon's biodiversity. *Defenders* 71(1): 26-33. [1] *A.T.:* Defenders of Wildlife.

1387. Durning, Alan T., 1992. *How Much Is Enough? The Consumer Society and the Future of the Earth.* New York: Norton. 200p. [2/2/1] *A.T.:* consumer behavior.

1388. ———, 1993. *Saving the Forests: What Will It Take?* Washington, DC: Worldwatch Paper 117. 51p. [2/1/1] *A.T.:* forest conservation; deforestation.

1389. Durrell, Lee, 1986. *State of the Ark.* Garden City, NY: Doubleday. 224p. [2/2/1] *A.T.:* nature conservation.

1390. Dutfield, Graham, 1997. *Can the TRIPs Agreement Protect Biological and Cultural Diversity?* Nairobi: African Centre for Technology Studies. 49p. [1/1/-].

1391. Dutton, John, 1994. Introduced mammals in São Tomé and Principe: possible threats to biodiversity. *Biodiversity and Conservation* 3: 927-938. [1] *A.T.:* bd monitoring.

1392. Duvick, D.N., 1984. Genetic diversity in major farm crops on the farm and in reserve. *Economic Botany* 38: 161-178. [1] *A.T.:* U.S.

1393. Dvorak, W.S., 1990. CAMCORE: industry and governments' efforts to conserve threatened forest species in Guatemala, Honduras and Mexico. *Forest Ecology and Management* 35: 151-157. [1] *A.T.:* forest species preservation.

1394. Dwyer, L.E., & D.D. Murphy, 1995. Fulfilling the promise: reconsidering and reforming the California Endangered Species Act. *Natural Resources J.* 35: 735-770. [1] *A.T.:* environmental legislation; environmental policy.

1395. Dykhuizen, Daniel E., 1998. Santa Rosalia revisited: why are there so many species of bacteria? *Antonie van Leeuwenhoek* 73: 25-33. *A.T.:* speciation; diversification; DNA hybridization; species definitions.

1396. Dynesius, M., & C. Nilsson, 1994. Fragmentation and flow regulation of river systems in the northern third of the world. *Science* 266: 753-762. [2] *A.T.:* habitat fragmentation; riverine species.

1397. Earhart, John E., 1990. Biodiversity: a status report on the conservation of the world's tropical moist forests. *Proc. of the Society of American Foresters National Convention 1990*: 136-140. [1] *A.T.:* deforestation; environmental degradation.

1398. Earle, Sylvia A., 1991. Sharks, squids, and horseshoe crabs—the significance of marine biodiversity. *Bioscience* 41: 506-509. [1] Concerning the progress being made in marine bdc.

1399. Easteal, Simon, Chris Collet, & David Betty, 1995. *The Mammalian Molecular Clock*. New York: Springer. 170p. [1/1/2] *A.T.:* mammalian genetics; mammalian evolution.

1400. Easter-Pilcher, Andrea L., 1996. Implementing the Endangered Species Act: assessing the listing of species as endangered or threatened. *Bioscience* 46: 355-363. [1].

1401. Ebenhard, Torbjorn, 1995. Conservation breeding as a tool for saving animal species from extinction. *TREE* 10: 438-443. [1] *A.T.:* captive breeding.

1402. Eberhart, Steve A., et al., eds., 1998. *Intellectual Property Rights III Global Genetic Resources: Access and Property Rights* [workshop papers]. Madison, WI: Crop Science Society of America and American Society of Agronomy. 176p. *A.T.:* germplasm resources.

1403. Ebersole, J.L., W.J. Liss, & C.A. Frissell, 1997. Restoration of stream habitats in the western United States: restoration as re-expression of habitat capacity. *Environmental Management* 21: 1-14. *A.T.:* ecological restoration; stream biota; habitat classification.

1404. Eckstrom, Chris, March-April 1994. Up Borneo's tropical tower. *International Wildlife* 24(2): 52-59. [1] Description of the Mt. Kinabalu environs.

1405. Economic Commission for Europe (United Nations), 1992. *Code of Practice for the Conservation of Threatened Animals and Plants and Other Species of International Significance*. New York. 61p. [1/1/1].

1406. Edvenson, June C., 1994. Predator control and regulated killing: a biodiversity analysis. *UCLA J. of Environmental Law & Policy* 13: 31-86. [1].

1407. Edwards, Brian, 8 June 1994. Biodiversity and building design. *Architects' J.* 199(23): 36-37. [1] Treats the matter of bd-conscious design.

1408. Edwards, David S., W.E. Booth, & S.C. Choy, eds., 1996. *Tropical Rainforest Research—Current Issues* [conference proceedings]. Dordrecht and Boston: Kluwer Academic. 566p. [1/1/1] *A.T.:* rainforest ecology; rainforest conservation.

1409. Edwards, D.G.W., & Y.A. El-Kassaby, 1993. *Ex situ* conservation of forest biodiversity in British Columbia. *FRDA Report*, no. 210: 65-67. [1].

1410. Edwards, M., & D.R. Morse, 1995. The potential for computer-aided identification in biodiversity research. *TREE* 10: 153-158. [1] Focuses on computer-aided species identification.

1411. Edwards, Peter J., & C. Abivardi, 1998. The value of biodiversity: where ecology and economy blend. *Biological Conservation* 83: 239-246. *A.T.:* ecosystem services; ecological economics.

1412. Edwards, Peter J., R.M. May, & N.R. Webb, eds., 1994. *Large Scale Ecology and Conservation Biology* [symposium papers]. Oxford, U.K., and Boston: Blackwell Scientific. 375p. [1/1/1].

1413. Edwards, Thomas C., 1996. Data defensibility and GAP analysis [letter]. *Bioscience* 46: 75-76. [1] *A.T.:* wildlife census; vegetation maps.

1414. Edwards, Victoria M., 1995. *Dealing in Diversity: America's Market for Nature Conservation.* Cambridge, U.K., and New York: CUP. 182p. [2/1/1] *A.T.:* bd value; markets.

1415. Eggen, M., 1997. That sinking feeling: do "artificial reefs" in BC waters increase biodiversity or waste? *Alternatives* 23: 7. *A.T.:* recreation; ecotourism.

1416. Eggleton, Paul, P.H. Williams, & K.J. Gaston, 1994. Explaining global termite diversity: productivity or history? *Biodiversity and Conservation* 3: 318-330. [2] *A.T.:* energy-diversity theory.

1417. Ehrenfeld, David W., 12 June 1986. Thirty million cheers for diversity. *New Scientist* 110: 38-43. [1].

1418. Ehrlich, Paul R., 1982. Human carrying capacity, extinctions, and nature reserves. *Bioscience* 32: 331-333. [1].

1419. ———, spring 1987. Habitats in crisis: why we should care about the loss of species. *Wilderness* 50(176): 12-15. [1] *A.T.:* western U.S.; overgrazing; public lands management; habitat degradation.

1420. ———, 1992. Environmental deterioration, biodiversity and the preservation of civilisation [lecture]. *Environmentalist* 12: 9-14. [1].

1421. ———, 1992. Population biology of checkerspot butterflies and the preservation of global biodiversity. *Oikos* 63: 6-12. [2].

1422. ———, 1994. Energy use and biodiversity loss. *PTRSL B* 344: 99-104. [1] *A.T.:* extinction; habitat destruction.

1423. ———, 1996. Conservation in temperate forests: what do we need to know and do? *Forest Ecology and Management* 85: 9-19. [1] *A.T.:* species diversity; population diversity.

1424. Ehrlich, Paul R., & G.C. Daily, 1993. Population extinction and saving biodiversity. *Ambio* 22: 64-68. [2] Distinguishes between population- and species-level bdc. *A.T.:* environmental policy.

1425. Ehrlich, Paul R., & A.H. Ehrlich., 1981. *Extinction: The Causes and Consequences of the Disappearance of Species.* New York: Random House. 305p. [3/3/2] *A.T.:* endangered and threatened species.

1426. ———, 1990. *The Population Explosion.* New York: Simon and Schuster. 320p. [3/3/2] *A.T.:* human population growth.

1427. ———, 1992. The value of biodiversity. *Ambio* 21: 219-226. [2].

1428. ———, fall 1996. Biodiversity and the brownlash. *Defenders* 71(4): 6-17. [1] Attacks critics of the Green movement. *A.T.:* anti-Greens.

1429. ———, 1996. *Betrayal of Science and Reason: How Anti-environmental Rhetoric Threatens Our Future.* Washington, DC: Island Press. 335p. [3/3/2].

1430. Ehrlich, Paul R., & H.A. Mooney, 1983. Extinction, substitution, and ecosystem services. *Bioscience* 33: 248-254. [1].

1431. Ehrlich, Paul R., & Mark Penberthy, Sept.-Oct. 1993. Is the extinction crisis real? *Wildlife Conservation* 96(5): 66-67. [1] *A.T.:* habitat destruction; bd loss.

1432. Ehrlich, Paul R., & Brian Walker, 1998. Rivets and redundancy. *Bioscience* 48: 387. Concerning the nature of redundancy in ecosystem makeup. *A.T.:* rivet-popper hypothesis; redundancy hypothesis.

1433. Ehrlich, Paul R., & E.O. Wilson, 1991. Biodiversity studies: science and policy. *Science* 253: 758-762. [2] *A.T.:* bd research; environmental policy.

1434. Ehrlich, Paul R., D.S. Dobkin, & Darryl Wheye, 1992. *Birds in Jeopardy: The Imperiled and Extinct Birds of the United States and Canada Including Hawaii and Puerto Rico.* Stanford, CA: Stanford University Press. 259p. [2/2/1].

1435. Ehrlich, Paul R., A.H. Ehrlich, & G.C. Daily, 1993. Food security, population, and environment. *Population and Development Review* 19: 1-32. [2] *A.T.:* sustainable agriculture; human population growth.

1436. Eisner, Thomas, winter 1989. Prospecting for nature's chemical riches. *Issues in Science and Technology* 6(2): 31-34. [1] *A.T.:* extinction; chemical prospecting.

1437. Eisner, Thomas, & E.A. Beiring, 1994. Biotic Exploration Fund—protecting biodiversity through chemical prospecting. *Bioscience* 44: 95-98. [1] *A.T.:* natural products.

1438. Eisner, Thomas, Hans Eisner, et al., 1981. Conservation of tropical forests [letter]. *Science* 213: 1314. [1] Draws attention to the worldwide extinction crisis.

1439. Eiswerth, Mark E., ed., 1990. *Marine Biological Diversity* [meeting report]. Woods Hole, MA: Technical Report WHOI-90-13, Woods Hole Oceanographic Institution. 57p. [1/1/1].

1440. Eiswerth, Mark E., & J.C. Haney, 1992. Allocating conservation expenditures: accounting for inter-species genetic distinctiveness. *Ecological Economics* 5: 235-249. [1] *A.T.:* cranes; conservation planning.

1441. Ekim, Tuna, 1993. In Turkey. *Naturopa*, no. 73: 21. [1] Brief summary of bdc in Turkey.

1442. Elder, P.S., 1996. Biological diversity and Alberta law. *Alberta Law Review* 34: 293-351. [1].

1443. Eldredge, Niles, 1991. *The Miner's Canary: Unraveling the Mys-*

teries of Extinction. New York: Prentice Hall. 246p. [3/3/1].

1444. ———, ed., 1992. *Systematics, Ecology, and the Biodiversity Crisis* [symposium papers]. New York: Columbia University Press. 220p. [2/2/1].

1445. ———, June 1998. Evolution and environment: the two faces of biodiversity. *Natural History* 107(5): 54-55.

1446. ———, 1998. *Life in the Balance: Humanity and the Biodiversity Crisis.* Princeton, NJ: Princeton University Press. 224p. *A.T.:* human ecology.

1447. Elfring, Chris, 1984. Can technology save tropical forests? [editorial] *Bioscience* 34: 350-352. [1] *A.T.:* technology transfer; Office of Technology Assessment.

1448. Elias, Thomas S., ed., 1987. *Conservation and Management of Rare and Endangered Plants* [conference proceedings]. Sacramento: California Native Plant Society. 630p. [1/1/1] *A.T.:* California.

1449. Elisabetsky, Elaine, & L. Wannmacher, 1993. The status of ethnopharmacology in Brazil. *J. of Ethnopharmacology* 38: 137-143. [1] *A.T.:* medicinal plants.

1450. Ellegren, Hans, G. Hartman, et al., 1993. Major histocompatibility complex monomorphism and low levels of DNA fingerprinting variability in a reintroduced and rapidly expanding population of beavers. *PNAS* 90: 8150-8153. [2] *A.T.:* population bottlenecks; Scandinavia; Russia.

1451. Elliot, Robert, 1994. Extinction, restoration, naturalness. *Environmental Ethics* 16: 135-144. [1] *A.T.:* environmental ethics; habitat restoration.

1452. Elliott, David K., ed., 1986. *Dynamics of Extinction.* New York: Wiley. 294p. [2/1/1].

1453. Elliott, Katherine J., & Deidre Hewitt, March 1997. Forest species diversity in upper elevation hardwood forests in the Southern Appalachian Mountains. *Castanea* 62(1): 32-42. [1].

1454. Ellis, John, & D.N. Schramm, 1995. Could a nearby supernova explosion have caused a mass extinction? *PNAS* 92: 235-238. [1] *A.T.:* Cretaceous-Tertiary boundary.

1455. Ellis, Susie, & U.S. Seal, 1995. Tools of the trade to aid decision-making for species survival. *Biodiversity and Conservation* 4: 553-572. [1] *A.T.:* endangered species management; IUCN.

1456. Ellis, William S., 1988. Rondônia: Brazil's imperiled rain forest. *National Geographic* 174: 772-799. [1].

1457. Ellison, Aaron M., & E.J. Farnsworth, 1996. Anthropogenic disturbance of Caribbean mangrove ecosystems: past impacts, present trends, and future predictions. *Biotropica* 28: 549-565. [1].

1458. Ellstrand, Norman C., & D.R. Elam, 1993. Population genetic consequences of small population size: implications for plant conservation.

ARES 24: 217-242. [2] *A.T.:* genetic drift; inbreeding; gene flow; rare plants.

1459. Elphick, Chris S., 1997. Correcting avian richness estimates for unequal sample effort in atlas studies. *Ibis* 139: 189-190. *A.T.:* species richness.

1460. Elton, Charles S., 1958. *The Ecology of Invasions by Animals and Plants.* London: Methuen. 181p. [3/-/3] The classic early investigation of this subject.

1461. Eltringham, S.K., 1994. Can wildlife pay its way? *Oryx* 28: 163-168. [1] *A.T.:* protected areas; wildlife conservation.

1462. Embley, T.M., R.P. Hirt, & D.M. Williams, 1994. Biodiversity at the molecular level: the domains, kingdoms and phyla of life. *PTRSL B* 345: 21-33. [1] *A.T.:* microbial taxonomy.

1463. Emmer, Igino M., Josef Fanta, et al., 1998. Reversing borealization as a means to restore biodiversity in Central-European mountainforests: an example from the Krkonose Mountains, Czech Republic. *Biodiversity and Conservation* 7: 229-247. *A.T.:* forest decline; forest restoration.

1464. Emmons, Louise H., 1984. Geographic variation in densities and diversities of non-flying mammals in Amazonia. *Biotropica* 16: 210-222. [1] *A.T.:* hunting.

1465. Endler, John A., 1982. Problems in distinguishing historical from ecological factors in biogeography. *American Zoologist* 22: 441-452. [2] *A.T.:* vicariance biogeography; forest refuge hypothesis.

1466. Endress, Peter K., 1994. *Diversity and Evolutionary Biology of Tropical Flowers.* Cambridge, U.K., and New York: CUP. 511p. [2/2/3].

1467. Engels, Jan M.M., ed., 1995. *In Situ Conservation and Sustainable Use of Plant Genetic Resources for Food and Agriculture in Developing Countries* [workshop report]. Rome: IPGRI, and Feldafing, Germany: DSE/ZEL. 116p. [1/1/1] *A.T.:* plant germplasm resources.

1468. English, A.W., 1997. Veterinarians and the conservation of biodiversity. *Australian Veterinary J.* 75: 567. [1] *A.T.:* Australia.

1469. Enquist, B.J., M.A. Jordan, & J.H. Brown, 1995. Connections between ecology, biogeography, and paleobiology: relationship between local abundance and geographic distribution in fossil and recent molluscs. *Evolutionary Ecology* 9: 586-604. [1] *A.T.:* Gulf of California; geographical range; Mexico.

1470. Environment Canada, 1994. *Biodiversity in Canada: A Science Assessment for Environment Canada.* Ottawa. 245p. [1/1/1].

1471. Epperson, Brian K., 1993. Recent advances in correlation studies of spatial patterns of genetic variation. *Evolutionary Biology* 27: 95-155. [2].

1472. Epstein, P.R., 1995. Emerging diseases and ecosystem instability: new threats to public health. *American J. of Public Health* 85: 168-172. [2].

1473. Epstein, P.R., H.F. Diaz, et al., 1998. Biological and physical signs of climate change: focus on mosquito-borne diseases. *Bull. of the American*

Meteorological Society 79: 409-417. *A.T.:* public health; high altitude environmental change.

1474. Eriksson, Ove, 1993. The species-pool hypothesis and plant community diversity. *Oikos* 68: 371-374. [1] *A.T.:* vascular plants.

1475. Erman, Nancy A., & D.C. Erman, 1995. Spring permanence, Trichoptera species richness, and the role of drought. *J. of the Kansas Entomological Society* 68(2), Suppl.: 50-64. [1] *A.T.:* caddisflies.

1476. Erwin, Douglas H., 1990. The end-Permian mass extinction. *ARES* 21: 69-91. [1].

1477. ———, 1993. *The Great Paleozoic Crisis: Life and Death in the Permian.* New York: Columbia University Press. 327p. [2/2/2] *A.T.:* mass extinctions.

1478. ———, 1994. The Permo-Triassic extinction. *Nature* 367: 231-236. [2] *A.T.:* environmental change.

1479. ———, July 1996. The mother of mass extinctions. *Scientific American* 275(1): 72-78. [1] *A.T.:* Permo-Triassic extinctions.

1480. ———, 1998. After the end: recovery from extinction. *Science* 279: 1324-1325. *A.T.:* mass extinctions.

1481. Erwin, Douglas H., & R.L. Anstey, eds., 1995. *New Approaches to Speciation in the Fossil Record.* New York: Columbia University Press. 342p. [1/1/1].

1482. Erwin, Douglas H., & T.A. Vogel, 1992. Testing for causal relationships between large pyroclastic volcanic eruptions and mass extinctions. *Geophysical Research Letters* 19: 893-896. [1].

1483. Erwin, Douglas H., J.W. Valentine, & J.J. Sepkoski, Jr., 1987. A comparative study of diversification events: the early Paleozoic versus the Mesozoic. *Evolution* 41: 1177-1186. [2].

1484. Erwin, Terry L., 1982. Tropical forests: their richness in Coleoptera and other arthropod species. *Coleopterists' Bull.* 36: 74-75. [2].

1485. ———, 1983. Tropical forest canopies: the last biotic frontier. *Bull. of the Entomological Society of America* 29: 14-19. [1].

1486. ———, 1991. An evolutionary basis for conservation strategies. *Science* 253: 750-752. [2] *A.T.:* species richness.

1487. ———, 1991. How many species are there?: revisited. *CB* 5: 330-333. [1].

1488. Esch, Gerald W., & J.C. Fernandez, 1993. *A Functional Biology of Parasitism: Ecological and Evolutionary Implications.* London: Chapman & Hall. 337p. [1/1/2].

1489. Eshbaugh, W.H., 1995. Systematics Agenda 2000: an historical perspective. *Biodiversity and Conservation* 4: 455-462. [1].

1490. Estabrook, G.F., 1991. The size of nature reserves and the number of long lived plant species they contain. *Coenoses* 6: 39-45. [1] *A.T.:* species-area relation.

1491. Estep, Kenneth W., R. Sluys, & E.E. Syvertsen, 1993. "Linnaeus" and beyond: workshop report on multimedia tools for the identification and database storage of biodiversity. *Hydrobiologia* 269-270: 519-525. [1] *A.T.:* Expert Center for Taxonomic Identification; classification systems.

1492. Estes, J.A., D.O. Duggins, & G.B. Rathbun, 1989. The ecology of extinctions in kelp forest communities. *CB* 3: 252-264. [1] *A.T.:* sea otter; spiny lobster; Steller's sea-cow; local extinction.

1493. Estrada, Alejandro, Rosamond Coates-Estrada, & D.A. Meritt, Jr., 1997. Anthropogenic landscape changes and avian diversity at Los Tuxtlas, Mexico. *Biodiversity and Conservation* 6: 19-43. *A.T.:* forest fragments; agricultural habitats.

1494. Etter, Ron J., & J.F. Grassle, 1992. Patterns of species diversity in the deep sea as a function of sediment particle size diversity. *Nature* 360: 576-578. [2] *A.T.:* benthic diversity; vertical distribution.

1495. European Commission (Directorate-General for Environment, Nuclear Safety, and Civil Protection), 1998. *First Report on the Implementation of the Convention on Biological Diversity by the European Community.* Luxembourg: Office for Official Publications of the European Communities. 95p.

1496. Evans, K.E., & R.C. Szaro, 1990. Biodiversity: challenges for hardwood utilization. *Diversity* 6(2): 27-29. [1].

1497. Evans, Tom, Jan.-Feb. 1995. Spotlight on Laos. *Wildlife Conservation* 98(1): 52-57. [1] *A.T.:* bdc; wildlife reserves.

1498. Everett, Richard L., & J.F. Lehmkuhl, 1996. An emphasis-use approach to conserving biodiversity. *Wildlife Society Bull.* 24: 192-199. [1] *A.T.:* reserve networks; ecosystem management.

1499. Everett, Richard L., P.F. Hessburg, & T.R. Lillybridge, 1994. Emphasis areas as an alternative to buffer zones and reserved areas in the conservation of biodiversity and ecosystem processes. *J. of Sustainable Forestry* 2: 283-292. [1].

1500. Ewald, P.W., 1983. Host-parasite relations, vectors, and the evolution of disease severity. *ARES* 14: 465-485. [2].

1501. Eyre, M.D., & S.P. Rushton, 1989. Quantification of conservation criteria using invertebrates. *J. of Applied Ecology* 26: 159-171. [1] *A.T.:* England; beetles.

1502. Fa, John E., 1989. Conservation-motivated analysis of mammalian biogeography in the trans-Mexican neovolcanic belt. *National Geographic Research* 5: 296-316. [1] *A.T.:* endemic species.

1503. Fa, John E., Javier Juste, et al., 1995. Impact of market hunting on mammal species in Equatorial Guinea. *CB* 9: 1107-1115. [2] *A.T.:* hunting.

1504. Fagan, W.F., & P.M. Kareiva, 1997. Using compiled species lists to make biodiversity comparisons among regions: a test case using Oregon butterflies. *Biological Conservation* 80: 249-259. *A.T.:* sampling effort;

asymptotic diversity.

1505. Fahrig, Lenore, & G. Merriam, 1994. Conservation of fragmented populations. *CB* 8: 50-59. [3] *A.T.:* local extinction; recolonization; habitat patches.

1506. Fahrig, Lenore, & Jyri Paloheimo, 1988. Determinants of local population size in patchy habitats. *Theoretical Population Biology* 34: 194-212. [2].

1507. Faith, Daniel P., 1992. Conservation evaluation and phylogenetic diversity. *Biological Conservation* 61: 1-10. [2] *A.T.:* crested newts; bumble bees; taxonomic diversity.

1508. ———, 1992. Systematics and conservation: on predicting the feature diversity of subsets of taxa. *Cladistics* 8: 361-373. [1].

1509. ———, 1993. Biodiversity and systematics: the use and misuse of divergence information in assessing taxonomic diversity. *Pacific Conservation Biology* 1: 53-57. [1].

1510. ———, 1994. Genetic diversity and taxonomic priorities for conservation. *Biological Conservation* 68: 69-74. [1] *A.T.:* phylogenetic diversity; taxon weighting.

1511. ———, 1994. Phylogenetic pattern and the quantification of organismal biodiversity. *PTRSL B* 345: 45-58. [2] *A.T.:* phylogenetic diversity.

1512. ———, 1995. *Biodiversity and Regional Sustainability Analysis.* Lyneham, ACT, Australia: Division of Wildlife and Ecology, CSIRO. 30p. [1/1/1].

1513. ———, 1997. Biodiversity, biospecifics, and ecological services [letter]. *TREE* 12: 66. [1] *A.T.:* bd value.

1514. Faith, Daniel P., & R.H. Norris, 1989. Correlation of environmental variables with patterns of distribution and abundance of common and rare freshwater macroinvertebrates. *Biological Conservation* 50: 77-98. [2] *A.T.:* multidimensional scaling; Victoria, Australia.

1515. Faith, Daniel P., & P.A. Walker, 1996. How do indicator groups provide information about the relative biodiversity of different sets of areas?: on hotspots, complementarity and pattern-based approaches. *Biodiversity Letters* 3: 18-25. [1].

1516. ———, 1996. Environmental diversity: on the best-possible use of surrogate data for assessing the relative biodiversity of sets of areas. *Biodiversity and Conservation* 5: 399-415. [1] *A.T.:* indicator species.

1517. ———, 1996. Integrating conservation and development: incorporating vulnerability into biodiversity assessment of areas. *Biodiversity and Conservation* 5: 417-429. [1] *A.T.:* bd management.

1518. ———, 1996. Integrating conservation and development: effective trade-offs between biodiversity and cost in the selection of protected areas. *Biodiversity and Conservation* 5: 431-446. [1] *A.T.:* conservation weighting.

1519. Faith, Daniel P., P.A. Walker, et al., 1996. Integrating conservation and forestry production: exploring trade-offs between biodiversity and production in regional land-use assessment. *Forest Ecology and Management* 85: 251-260. [1] *A.T.:* New South Wales; bd surrogates.

1520. Fakir, Mohamed S., 1994. *Biodiversity and Biotechnology in South Africa: Some Issues for the Development of Future Policy.* Johannesburg: Working Paper 3, Land and Agriculture Policy Centre. 21p. [1/1/1].

1521. Falconer, Allan, ed., 1993. *Mapping Tomorrow's Resources: A Symposium on the Uses of Remote Sensing, Geographic Information Systems (GIS), and Global Positioning Systems (GPS) for Natural Resources Management.* Logan, UT: College of Natural Resources, Utah State University. 87p. [1/1/1].

1522. Falk, Bryce W., & George Bruening, 1994. Will transgenic crops generate new viruses and new diseases? [editorial] *Science* 263: 1395-1396. [2] *A.T.:* plant viruses.

1523. Falk, Donald A., 1990. Integrated strategies for conserving plant genetic diversity. *Annals of the Missouri Botanical Garden* 77: 38-47. [1].

1524. ———, 1990. Endangered forest resources in the U.S.: integrated strategies for conservation of rare species and genetic diversity. *Forest Ecology and Management* 35: 91-107. [1].

1525. Falk, Donald A., & K.E. Holsinger, eds., 1991. *Genetics and Conservation of Rare Plants* [conference papers]. New York: OUP. 283p. [2/1/2].

1526. Falk, Donald A., & P. Olwell, 1992. Scientific and policy considerations in restoration and reintroduction of endangered species. *Rhodora* 94(879): 287-315. [1].

1527. Falk, Donald A., C.I. Millar, & Margaret Olwell, eds., 1996. *Restoring Diversity: Strategies for Reintroduction of Endangered Plants.* Washington, DC: Island Press. 505p. [2/1/1] *A.T.:* restoration ecology.

1528. Falkner, Maurya B., & T.J. Stohlgren, 1997. Evaluating the contribution of small national park areas to regional biodiversity. *Natural Areas J.* 17: 324-330. *A.T.:* Rocky Mountains.

1529. Fallas, G.A., M.L. Hahn, et al., 1996. Molecular and biochemical diversity among islolates of *Radopholus* spp. from different areas of the world. *J. of Nematology* 28: 422-430. [1] *A.T.:* biochemical systematics.

1530. Farjon, Aljos, 1996. Biodiversity of *Pinus* (Pinaceae) in Mexico: speciation and palaeo-endemism. *Botanical J. of the Linnean Society* 121: 365-384. [1].

1531. Farmar-Bowers, Quentin, 1998. *The National Protocol System in 1997: A Cooperative Management System to Maintain Biodiversity* [research report]. Vermont South, Victoria, Australia: Research Report ARR No. 317, ARRB Transport Research. 44p. A.T: Australia; roads.

1532. Farnsworth, Norman R., & D.D. Soejarto, 1985. Potential consequence of plant extinction in the United States on the current and future availability of prescription drugs. *Economic Botany* 39: 231-240. [1] *A.T.:* medicinal plants.

1533. Farrell, Brian D., & Charles Mitter, 1994. Adaptive radiation in insects and plants: time and opportunity. *American Zoologist* 34: 57-69. [1] *A.T.:* coevolution.

1534. ———, 1998. The timing of insect/plant diversification: might *Tetraopes* (Coleoptera, Cerambycidae) and *Asclepias* (Asclepiadaceae) have co-evolved? *Biological J. of the Linnean Society* 63: 553-577.

1535. Farrell, Brian D., Charles Mitter, & D.J. Futuyma, 1992. Diversification at the insect-plant interface. *Bioscience* 42: 34-42. [1] *A.T.:* coevolution.

1536. Farrier, David, 1995. Conserving biodiversity on private land: incentives for management or compensation for lost expectations? *Harvard Environmental Law Review* 19: 303-408. [2] *A.T.:* federal programs; environmental policy.

1537. Farrow, S., 1995. Extinction and market forces: two case studies. *Ecological Economics* 13: 115-123. [1] *A.T.:* environmental economics; privatization.

1538. Fastovsky, David E., & D.B. Weishampel, 1996. *The Evolution and Extinction of the Dinosaurs.* Cambridge, U.K., and New York: CUP. 460p. [2/2/1].

1539. Fauth, J.E., B.I. Crother, & J.B. Slowinski, 1989. Elevational patterns of species richness, evenness, and abundance of the Costa Rican leaf-litter herpetofauna. *Biotropica* 21: 178-185. [1].

1540. Favre, David S., 1989. *International Trade in Endangered Species: A Guide to CITES.* Dordrecht and Boston: M. Nijhoff, 1989. 415p. [1/1/1].

1541. Fawcett, Dick, April 1991. Make room for diversity on the farm [opinion]. *Farm J.* 115(7): C4. [1].

1542. Fearnside, Philip M., 1990. The rate and extent of deforestation in Brazilian Amazonia. *Environmental Conservation* 17: 213-226. [1].

1543. ———, 1997. Environmental services as a strategy for sustainable development in rural Amazonia. *Ecological Economics* 20: 53-70. [1].

1544. Fearnside, Philip M., & J. Ferraz, 1995. A conservation gap analysis of Brazil's Amazonian vegetation. *CB* 9: 1134-1147. [1] *A.T.:* GIS.

1545. Feinsilver, Julie M., 1996. Biodiversity prospecting: a new panacea for development? *CEPAL Review*, no. 60: 115-132. [1] *A.T.:* Latin America; Merck; INBio.

1546. Feinsinger, P., 1996. Biodiversity knowledge in Chile: diagnosis and the first prescription [editorial]. *TREE* 11: 451-452. [1].

1547. Feistner, A.T.C., & J.B. Carroll, 1993. Breeding aye-ayes: an aid to preserving biodiversity. *Biodiversity and Conservation* 2: 283-289. [1] *A.T.:* captive breeding; Madagascar.

1548. Fekete, G., 1994. Foundations for developing a national strategy of biodiversity conservation. *Acta Zoologica Academiae Scientarum Hungaricae* 40: 289-327. [1] *A.T.:* Hungary.

1549. Feldmann, Rodney M., & R.B. Manning, 1992. Crisis in systematic biology in the "Age of Biodiversity" [editorial]. *J. of Paleontology* 66: 157-158. [1] *A.T.:* basic research; institutional support.

1550. Fellows, L., 1992. What are the forests worth? *Lancet* 339: 1330-1333. [1] General commentary on the global bd crisis.

1551. Felsenstein, Joseph, 1981. Skepticism towards Santa Rosalia, or why are there so few kinds of animals? *Evolution* 35: 124-138. [2] *A.T.:* species diversity.

1552. Fenchel, Tom, 1993. There are more small than large species? [editorial] *Oikos* 68: 375-378. [2] *A.T.:* body size; size distribution.

1553. Fenchel, Tom, & B.J. Finlay, 1994. The evolution of life without oxygen. *American Scientist* 82: 22-29. [1] *A.T.:* anaerobiosis; bacteria; protozoa; symbiosis.

1554. Fenger, Mike A., 1996. Implementing biodiversity conservation through the British Columbia Forest Practices code. *Forest Ecology and Management* 85: 67-77. [1] *A.T.:* forestry policy.

1555. Fenger, Mike A., E.H. Miller, et al., eds., 1993. *Our Living Legacy: Proceedings of a Symposium on Biological Diversity*. Victoria, BC: Royal British Columbia Museum. 392p. [1/1/1] *A.T.:* British Columbia.

1556. Fenner, M., W.G. Lee, & J.B. Wilson, 1997. A comparative study of the distribution of genus size in twenty angiosperm floras. *Biological J. of the Linnean Society* 62: 225-237. *A.T.:* phylogenetic diversity.

1557. Fenton, M. Brock, D.H.M. Cumming, et al., 1998. Bats and the loss of tree canopy in African woodlands. *CB* 12: 399-407. *A.T.:* Zimbabwe.

1558. Ferguson, David A., spring 1993. Biodiversity conservation in Asia and the Near East. *Fish and Wildlife News* (spring): 17-18. [1] *A.T.:* Oman; Philippines; India.

1559. Ferguson, Moira M., 1990. The genetic impact of introduced fishes on native species. *Canadian J. of Zoology* 68: 1053-1057. [1] *A.T.:* transgenics.

1560. Fernandes, G.W., & P.W. Price, 1988. Biogeographical gradients in galling species richness: tests of hypotheses. *Oecologia* 76: 161-167. [1].

1561. Fernandez-Duque, Eduardo, & Claudia Valeggia, 1994. Meta-analysis: a valuable tool in conservation research. *CB* 8: 555-561. [1] *A.T.:* logging; birds; statistical analysis.

1562. Fernando, C.H., ed., 1984. *Ecology and Biogeography in Sri Lanka*. The Hague: W. Junk. 505p. [1/1/1].

1563. Fernholm, Bo, Kåre Bremer, & Hans Jörnvall, eds., 1989. *The Hierarchy of Life: Molecules and Morphology in Phylogenetic Analysis* [symposium papers]. Amsterdam: Excerpta Medica. 499p. [1/1/1].

1564. Feron, E.M., 1995. New food sources, conservation of biodiversity and sustainable development: can unconventional animal species contribute to feeding the world? *Biodiversity and Conservation* 4: 233-240. [1] *A.T.:* livestock; wildlife utilization.

1565. Ferreira, Leandro V., 1997. Effects of the duration of flooding on species richness and floristic composition in three hectares in the Jau National Park in floodplain forests in central Amazonia. *Biodiversity and Conservation* 6: 1353-1363.

1566. Ferrier, Simon, & Graham Watson, 1997. *An Evaluation of the Effectiveness of Environmental Surrogates and Modelling Techniques in Predicting the Distribution of Biological Diversity.* Canberra: Environment Australia. 193p. [1/1/-] *A.T.:* mathematical models.

1567. Ferris-Kaan, Richard, ed., 1995. *Managing Forests for Biodiversity* [symposium papers]. Edinburgh: Forestry Commission Technical Paper No. 8. 51p. [1/1/1].

1568. Fet, Victor, & K.I. Atamuradov, eds., 1994. *Biogeography and Ecology of Turkmenistan.* Dordrecht and Boston: Kluwer, 1994. 650p. [1/1/1].

1569. Fiedler, Konrad, 1998. Diet breadth and host-plant diversity of tropical zone vs. temperate zone herbivores: South-East Asian and West Palearctic butterflies as a case study. *Ecological Entomology* 23: 285-297.

1570. Fiedler, Peggy L., & S.K. Jain, eds., 1992. *Conservation Biology: The Theory and Practice of Nature Conservation, Preservation, and Management.* New York: Chapman & Hall. 507p. [2/2/1].

1571. Fiedler, Peggy L., R.A. Leidy, et al., Jan. 1993. The contemporary paradigm in ecology and its implications for endangered species conservation. *Endangered Species Update* 10(3-4): 7-12. [1] *A.T.:* Endangered Species Act; single-species approach; wildlife conservation.

1572. Field, Christopher B., J.G. Osborn, et al., 1998. Mangrove biodiversity and ecosystem function. *Global Ecology and Biogeography Letters* 7: 3-14. *A.T.:* remote sensing.

1573. Field, K.G., G.J. Olsen, et al., 1988. Molecular phylogeny of the animal kingdom. *Science* 239: 748-753. [3] *A.T.:* ribosomal RNA; phylogenetic relationships.

1574. Fields, P.A., J.B. Graham, et al., 1993. Effects of expected global climate change on marine faunas. *TREE* 8: 361-367. [1] *A.T.:* global warming; marine ecology.

1575. Filser, Juliane, Henning Fromm, et al., 1995. Effects of previous intensive agricultural management on microorganisms and the biodiversity of soil fauna. *Plant and Soil* 170: 123-129. [1].

1576. Findlay, C. Scott, & Jeff Houlahan, 1997. Anthropogenic correlates of species richness in southeastern Ontario wetlands. *CB* 11: 1000-1009. *A.T.:* roads.

1577. Finizio, A., A. Di Guardo, & L. Cartmale, 1998. Hazardous air pollutants (HAPS) and their effects on biodiversity: an overview of the atmospheric pathways of persistent organic pollutants (POPS) and suggestions for future studies. *Environmental Monitoring and Assessment* 49: 327-336. *A.T.:* atmospheric change; pesticides.

1578. Finlay, Bland J., 1998. The global diversity of protozoa and other small species. *International J. for Parasitology* 28: 29-48.

1579. Finlay, Bland J., & T.M. Embley, Aug. 1992. On protozoan consortia and the meaning of biodiversity. *Society for General Microbiology Quarterly* 19(3): 67-69. [1].

1580. Finlay, Bland J., & G.F. Esteban, 1998. Planktonic ciliate species diversity as an integral component of ecosystem function in a fresh water pond. *Protist* 149: 155-165. *A.T.:* protozoa; niche filling.

1581. Finlay, Bland J., G.F. Esteban, & Tom Fenchel, 1996. Global diversity and body size [letter]. *Nature* 383: 132-133. [1].

1582. Finlay, Bland J., S.C. Maberly, & J.I. Cooper, 1997. Microbial diversity and ecosystem function. *Oikos* 80: 209-213. Points out that the niches of microorganisms may not be "conservable" in the same sense those of larger forms are.

1583. Finlay, Bland J., J.O. Corliss, et al., 1996. Biodiversity at the microbial level: the number of free-living ciliates in the biosphere. *Quarterly Review of Biology* 71: 221-237. [2] *A.T.:* microorganisms.

1584. Finsen, Lawrence, & Susan Finsen, 1994. *The Animal Rights Movement in America: From Compassion to Respect.* New York: Twayne Publishers. 309p. [2/1/1].

1585. Fischer, Alfred G., 1960. Latitudinal variations in organic diversity. *Evolution* 14: 64-81. [2] *A.T.:* latitudinal diversity gradients; climatic change.

1586. [not used]

1587. Fischman, Robert L., 1992. Biological diversity and environmental protection: authorities to reduce risk. *Environmental Law* 22: 435-502. [1] *A.T.:* environmental assessment.

1588. ———, 1997. The role of riparian water law in protecting biodiversity: an Indiana (USA) case study. *Natural Areas J.* 17: 30-37. *A.T.:* environmental policy.

1589. Fishelson, Lev, 1997. Ecology at the crossroads: a need for adaptation. *Israel J. of Zoology* 43: 89-92. *A.T.:* ecosystem analysis.

1590. Fisher, Brian L., 1998. Insect behavior and ecology in conservation: preserving functional species interactions. *Annals of the Entomological Society of America* 91: 155-158.

1591. Fisher, Brian L., & Robert Fisher, April 1994. Insect appreciation. *Science Teacher* 61(4): 22-23. [1] *A.T.:* bd education; urban bd.

1592. Fisher, Robert N., & H.B. Shaffer, 1996. The decline of amphibians in California's Great Central Valley. *CB* 10: 1387-1397. [2].

1593. Fisher, Ronald A., A.S. Corbet, & C.B. Williams, 1943. The relation between the number of species and the number of individuals in a random sample of an animal population. *J. of Animal Ecology* 12: 42-58. [2].

1594. Fitter, Richard S.R., 1986. *Wildlife for Man: How and Why We Should Conserve Our Species*. London: Collins. 223p. [1/1/1] *A.T.:* germplasm resources; wildlife conservation.

1595. Fitter, Richard S.R., & Maisie Fitter, eds., 1987. *The Road to Extinction: Problems of Categorizing the Status of Taxa Threatened With Extinction* [symposium proceedings]. Gland, Switzerland: IUCN. 121p. [1/1/1].

1596. Fitzgerald, John M., July-Aug. 1992. The biological diversity treaty of the United Nations Conference on Environment and Development. *Endangered Species Update* 9(9-10): 1-7, 12. [1] *A.T.:* CBD.

1597. Fitzgerald, Sarah, 1989. *International Wildlife Trade: Whose Business Is It?* Washington, DC: WWF. 459p. [2/1/1] *A.T.:* wild animal trade; endangered species.

1598. Fjeldså, J., 1995. Have ornithologists "slept during class"? On the response of ornithology to the "Biodiversity Crisis" and "Biodiversity Convention." *J. of Avian Biology* 26: 89-93. [1].

1599. Fjeldså, J., & J.C. Lovett, 1997. Biodiversity and environmental stability. *Biodiversity and Conservation* 6: 315-323.

1600. ———, 1997. Geographical patterns of old and young species in African forest biota: the significance of specific montane areas as evolutionary centres. *Biodiversity and Conservation* 6: 325-346. *A.T.:* refuge hypothesis; habitat stability.

1601. Fjeldså, J., D. Ehrlich, et al., 1997. Are biodiversity "hotspots" correlated with ecoclimatic stability? A pilot study using the NOAA-AVHRR remote sensing data. *Biodiversity and Conservation* 6: 401-422. *A.T.:* tropical Africa; endemic species.

1602. Flack, Stephanie, & Elaine Furlow, Nov.-Dec. 1996. America's least wanted. *Nature Conservancy* 46(6): 17-23. [1] Profiles some particularly invasive introduced species.

1603. Flannery, Tim F., 1995. *The Future Eaters: An Ecological History of the Australasian Lands and People*. New York: Braziller. 423p. [2/2/2] *A.T.:* human ecology; natural history.

1604. Flaster, T., 1990. Ethnobiologists emphasize germplasm conservation and sustainable agriculture. *Diversity* 6(2): 34-35. [1] *A.T.:* ethnobotany.

1605. Flather, Curtis H., 1996. Fitting species-accumulation functions and assessing regional land use impacts on avian diversity. *J. of Biogeography* 23: 155-168. [1].

1606. Flather, Curtis H., & J.R. Sauer, 1996. Using landscape ecology to test hypotheses about large-scale abundance patterns in migratory birds. *Ecology* 77: 28-35. [1] *A.T.:* Latin America; Neotropical migrant birds decline.

1607. Flather, Curtis H., M.S. Knowles, & I.A. Kendall, 1998. Threatened and endangered species geography. *Bioscience* 48: 365-376. *A.T.:* hot spots; U.S.; bdc.

1608. Flather, Curtis H., K.R. Wilson, et al., 1997. Identifying gaps in conservation networks: of indicators and uncertainty in geographic-based analyses. *Ecological Applications* 7: 531-542. [2] *A.T.:* gap analysis; indicator species.

1609. Fleischmann, K., 1997. Invasion of alien woody plants on the islands of Mahé and Silhouette, Seychelles. *J. of Vegetation Science* 8: 5-12. [1].

1610. Fleischner, T.L., 1994. Ecological costs of livestock grazing in western North America. *CB* 8: 629-644. [2] *A.T.:* environmental degradation; riparian habitats.

1611. Fleming, Ian A., & Kaare Ragaard, 1993. *Documentation and Measurement of Biodiversity*. Oslo: Norsk Institutt for Naturforskning. 23p. [1/1/1].

1612. Fleming, L. Vincent, ed., 1997. *Biodiversity in Scotland: Status, Trends and Initiatives* [conference papers]. Edinburgh: Stationery Office. 309p. [1/1/-].

1613. Fleming, R.A., & J.N. Candau, 1998. Influences of climatic change on some ecological processes of an insect outbreak system in Canada's boreal forests and the implications for biodiversity. *Environmental Monitoring and Assessment* 49: 235-249.

1614. Fleming, Theodore H., 1973. Numbers of mammal species in North and Central American forest communities. *Ecology* 54: 555-563. [1] *A.T.:* latitudinal diversity gradients.

1615. Fleming, Theodore H., Randall Breitwisch, & G.H. Whitesides, 1987. Patterns of tropical frugivore diversity. *ARES* 18: 91-109. [2].

1616. Flessa, Karl W., 1975. Area, continental drift and mammalian diversity. *Paleobiology* 1: 189-194. [1] *A.T.:* paleobiogeography.

1617. Flessa, Karl W., & David Jablonski, 1995. Biogeography of recent marine bivalve molluscs and its implications for paleobiogeography and the geography of extinction: a progress report. *Historical Biology* 10: 25-47. [1] *A.T.:* latitudinal diversity gradients.

1618. Flessa, Karl W., & J.J. Sepkoski, Jr., 1978. On the relationship between Phanerozoic diversity and changes in habitable area. *Paleobiology* 4: 359-366. [1].

1619. Flicker, John, Sept.-Oct. 1995. Fostering a culture of conservation [opinion]. *Audubon* 97(5): 6. [1] Summarizes role of the National Audubon Society in 20th-century conservation.

1620. Flint, Michael, 1991. *Biological Diversity and Developing Countries: Issues and Options (A Synthesis Paper)*. London: Overseas Development Administration. 50p. [1/1/1].

1621. Florman, Samuel C., July 1993. Progress for the birds [comment]. *Technology Review* 96(5): 63. [1] An optimistic view of the progress of bird conservation. *A.T.:* bird population declines.

1622. Fluegeman, Richard H., & M.L. McKinney, 1990. Causation and nonrandomness in biological and geological time series: temperature as a proximal control of extinction and diversity: discussion and reply [editorial]. *Palaios* 5: 486-488. [1] *A.T.:* Cenozoic.

1623. Flux, J.E.C., & P.J. Fullagar, 1992. World distribution of the rabbit *Oryctolagus cuniculus* on islands. *Mammal Review* 22: 151-205. [1] *A.T.:* introduced species.

1624. Foissner, W., 1997. Protozoa as bioindicators in agroecosystems, with emphasis on farming practices, biocides, and biodiversity. *Agriculture, Ecosystems & Environment* 62: 93-103.

1625. Folke, Carl, C.S. Holling, & Charles Perrings, 1996. Biological diversity, ecosystems, and the human scale. *Ecological Applications* 6: 1018-1024. [1] *A.T.:* functional diversity; ecosystem services; environmental policy.

1626. Folke, Carl, Charles Perrings, et al., 1993. Biodiversity conservation with a human face: ecology, economics and policy. *Ambio* 22: 62-63. [1] *A.T.:* environmental policy; ecosystem services.

1627. Fontenelle Bizerril, C.R.S., 1996. Identification of priority areas for management of the fish fauna's biological diversity: a case study in Baixada de Jacarepaguá, Rio de Janeiro, RJ, Brazil. *Arquivos de Biologia e Tecnologia* 39: 295-306. [1].

1628. Foote, Mike, 1991. Morphologic patterns of diversification: examples from trilobites. *Palaeontology* 34: 461-485. [2] *A.T.:* morphological diversity; taxonomic diversity; Ordovician.

1629. ———, 1992. Rarefaction analysis of morphological and taxonomic diversity. *Paleobiology* 18: 1-16. [2] *A.T.:* Paleozoic; marine invertebrates.

1630. ———, 1993. Discordance and concordance between morphological and taxonomic diversity. *Paleobiology* 19: 185-204. [2].

1631. ———, 1996. Perspective: evolutionary patterns in the fossil record. *Evolution* 50: 1-11. [1] *A.T.:* macroevolution; phylogenetic analysis.

1632. ———, 1997. Sampling, taxonomic description, and our evolving knowledge of morphological diversity. *Paleobiology* 23: 181-206. [1] *A.T.:* fossil marine invertebrates.

1633. ———, 1997. The evolution of morphological diversity. *ARES* 28: 129-152. *A.T.:* macroevolution.

1634. Ford, Michael J., 1982. *The Changing Climate: Responses of the Natural Fauna and Flora*. London and Boston: G. Allen & Unwin. 190p. [2/2/1].

1635. Ford, Tim, Dec. 1994. Pollutant effects on the microbial ecosystem. *Environmental Health Perspectives Supplements* 102(12): 45-48. [1] *A.T.:*

microorganisms; genetic exchange; ecosystem health; pathogens.

1636. Ford-Lloyd, Brian V., & Michael Jackson, 1986. *Plant Genetic Resources: An Introduction to Their Conservation and Use*. London: Edward Arnold. 146p. [1/1/1] *A.T.:* plant germplasm resources; genetic variation.

1637. Foresta, Ronald A., 1991. *Amazon Conservation in the Age of Development: The Limits of Providence*. Gainesville: University of Florida Press. 366p. [2/2/1].

1638. Forester, Deborah J., & G.E. Machlis, 1996. Modeling human factors that affect the loss of biodiversity. *CB* 10: 1253-1263. [1].

1639. Forey, Peter L., C.J. Humphries, & R.I. Vane-Wright, eds., 1994. *Systematics and Conservation Evaluation* [symposium papers]. Oxford, U.K.: Clarendon Press. 438p. [1/1/1].

1640. Forey, Peter L., C.J. Humphries, et al., 1992. *Cladistics: A Practical Course in Systematics* [workshop papers]. Oxford, U.K.: Clarendon Press. 191p. [2/1/2].

1641. Forman, Richard T.T., 1995. *Land Mosaics: The Ecology of Landscapes and Regions*. Cambridge, U.K., and New York: CUP. 632p. [1/1/2] *A.T.:* mosaic patterns.

1642. Forman, Richard T.T., & Michel Godron, 1986. *Landscape Ecology*. New York: Wiley. 619p. [2/1/3].

1643. Forstenzer, M., March-April 1997. What's wrong in the Sierra? *Audubon* 99(2): 14-18. [1] Laments the increasing environmental degradation and bd loss in the Sierra Nevada.

1644. Forsyth, Adrian, & Kenneth Miyata, 1984. *Tropical Nature: Life and Death in the Rain Forests of Central and South America*. New York: Scribner. 248p. [2/3/1] *A.T.:* rainforest ecology.

1645. Fortey, Richard A., 1989. There are extinctions and extinctions: examples from the lower Palaeozoic. *PTRSL B* 325: 327-355. [1] *A.T.:* mass extinctions; invertebrates.

1646. Foster, S.A., R.J. Scott, & W.A. Cresko, 1998. Nested biological variation and speciation. *PTRSL B* 353: 207-218. *A.T.:* sibling species; population variation.

1647. Fowler, Cary, & P.R. Mooney, 1990. *Shattering: Food, Politics, and the Loss of Genetic Diversity*. Tucson: University of Arizona Press. 278p. [2/2/1] *A.T.:* food crops; plant germplasm resources.

1648. Fowler, C. Mary R., & Verena Tunnicliffe, 1997. Hydrothermal vent communities and plate tectonics. *Endeavour* N.S. 21: 164-168. *A.T.:* biogeography; deep-sea environment.

1649. Fox, A.D., & J. Madsen, 1997. Behavioural and distributional effects of hunting disturbance on waterbirds in Europe: implications for refuge design. *J. of Applied Ecology* 34: 1-13. *A.T.:* wildlife management.

1650. Fox, B.J., J.E. Taylor, et al., 1997. Vegetation changes across edges of rainforest remnants. *Biological Conservation* 82: 1-13. *A.T.:* edge zones;

New South Wales.

1651. Fox, Jefferson, Pralad Yonzon, & Nancy Podger, 1996. Mapping conflicts between biodiversity and human needs in Langtang National Park, Nepal. *CB* 10: 562-569. [1] *A.T.*: red pandas; park-people relations; grazing.

1652. Fox, Laurel R., 1977. Species richness in streams—an alternative mechanism. *American Naturalist* 111: 1017-1021. [1].

1653. Fox, Warwick, 1989. The deep ecology-ecofeminism debate and its parallels. *Environmental Ethics* 11: 5-25. [1] *A.T.*: anthropocentrism.

1654. Frame, Bob, Joe Victor, & Yateendra Joshi, eds., 1993. *Biodiversity Conservation, Forests, Wetlands, and Deserts* [workshop papers]. New Delhi: Tata Energy Research Institute. 153p. [1/1/1] *A.T.*: India.

1655. France, Robert, 1992. The North American latitudinal gradient in species richness and geographical range of freshwater crayfish and amphipods. *American Naturalist* 139: 342-354. [2] *A.T.*: Rapoport's rule.

1656. France, Robert, & M. Sharp, 1992. Polynyas as centers of organization for structuring the integrity of arctic marine communities. *CB* 6: 442-446. [1].

1657. Francione, Gary L., 1995. *Animals, Property, and the Law*. Philadelphia: Temple University Press. 349p. [2/1/1] *A.T.*: animal rights and welfare.

1658. Francis, A.P., & D.J. Currie, 1998. Global patterns of tree species richness in moist forests: another look. *Oikos* 81: 598-602. *A.T.*: energy-richness hypothesis; evapotranspiration.

1659. Franco-Vizcaino, Ernesto, R.C. Graham, & E.B. Alexander, 1993. Plant species diversity and chemical properties of soils in the Central Desert of Baja California, Mexico. *Soil Science* 155: 406-416. [1] *A.T.*: soil chemistry; soil nutrients.

1660. Frankel, Otto H., 1974. Genetic conservation: our evolutionary responsibility. *Genetics* 78: 53-65. [1] *A.T.*: crop varieties; genetic resources.

1661. Frankel, Otto H., & Erna Bennett, eds., 1970. *Genetic Resources in Plants: Their Exploration and Conservation* [conference papers]. Philadelphia: F.A. Davis. 554p. [2/1/1] *A.T.*: plant gene banks.

1662. Frankel, Otto H., & J.G. Hawkes, eds., 1975. *Crop Genetic Resources for Today and Tomorrow*. Cambridge, U.K., and New York: CUP. 492p. [2/1/1] *A.T.*: plant germplasm resources; plant gene banks.

1663. Frankel, Otto H., & M.E. Soulé, 1981. *Conservation and Evolution*. Cambridge, U.K., and New York: CUP. 327p. [2/1/3].

1664. Frankel, Otto H., A.H.D. Brown, & J.J. Burdon, 1995. *The Conservation of Plant Biodiversity*. Cambridge, U.K., and New York: CUP. 299p. [2/1/2] *A.T.*: plant germplasm resources.

1665. Frankham, Richard, 1995. Conservation genetics. *Annual Review of Genetics* 29: 305-327. [2] *A.T.*: population genetics; inbreeding; genetic variation.

1666. ———, 1996. Relationship of genetic variation to population size in wildlife. *CB* 10: 1500-1508. [2] *A.T.:* genetic diversity; population genetics.

1667. ———, 1997. Do island populations have less genetic variation than mainland populations? *Heredity* 78: 311-327. [2] *A.T.:* inbreeding; population genetics.

1668. ———, 1998. Inbreeding and extinction: island populations. *CB* 12: 665-675.

1669. Franklin, Janet, & D.W. Steadman, 1991. The potential for conservation of Polynesian birds through habitat mapping and species translocation. *CB* 5: 506-521. [1] *A.T.:* GIS.

1670. Franklin, Jerry F., 1993. Preserving biodiversity: species, ecosystems, or landscapes? *Ecological Applications* 3: 202-205. [2] *A.T.:* Pacific Northwest; forest conservation.

1671. ———, 1994. Preserving biodiversity: species in landscapes. Response to Tracy and Brussard [letter]. *Ecological Applications* 4: 208-209. [1].

1672. Franklin, Jerry F., F.J. Swanson, et al., 1991. Effects of global climatic change on forests in northwestern North America. *Northwest Environmental J.* 7: 233-254. [1].

1673. Fränzle, Otto, 1994. Thermodynamic aspects of species diversity in tropical and ectropical plant communities. *Ecological Modelling* 75-76: 63-70. [1] *A.T.:* soil nutrients.

1674. Fraser, L.H., & P. Keddy, 1997. The role of experimental microcosms in ecological research. *TREE* 12: 478-481. *A.T.:* research methodology.

1675. Fraser, Robert H., & D.J. Currie, 1996. The species richness-energy hypothesis in a system where historical factors are thought to prevail: coral reefs. *American Naturalist* 148: 138-159. [1] *A.T.:* species diversity gradients.

1676. Frawley, B.J., & S. Walters, 1996. Reuse of annual set-aside lands: implications for wildlife. *Wildlife Society Bull.* 24: 655-659. [1] *A.T.:* Indiana; wildlife management; farm management.

1677. Freckman, Diana W., & R.A. Virginia, 1997. Low diversity Antarctic soil nematode communities: distribution and response to disturbance. *Ecology* 78: 363-369.

1678. Freckman, Diana W., T.H. Blackburn, et al., 1997. Linking biodiversity and ecosystem functioning of soils and sediments. *Ambio* 26: 556-562.

1679. Frederiksen, N.O., 1994. Paleocene floral diversities and turnover events in eastern North America and their relation to diversity models. *Review of Palaeobotany and Palynology* 82: 225-238. [1] *A.T.:* angiosperms.

1680. Fredrickson, James K., & T.C. Onstott, Oct. 1996. Microbes deep inside the earth. *Scientific American* 275(4): 68-73. [2] *A.T.:* microorganisms.

1681. Freedman, B., & S. Beauchamp, 1998. Implications of atmospheric change for biodiversity of aquatic ecosystems in Canada. *Environmental Monitoring and Assessment* 49: 271-280.

1682. Freedman, B., S. Woodley, & J. Loo, 1994. Forestry practices and biodiversity, with particular reference to the Maritime Provinces of eastern Canada. *Environmental Reviews* 2: 33-77. [1] *A.T.:* forest management; old-growth forests.

1683. Freedman, B., V. Zelazny, et al., 1996. Biodiversity implications of changes in the quantity of dead organic matter in managed forests. *Environmental Reviews* 4: 238-265. [1] *A.T.:* old-growth forests.

1684. Freeland, W.J., 1990. Large herbivorous mammals: exotic species in northern Australia. *J. of Biogeography* 17: 445-449. [1] *A.T.:* pathogen introduction.

1685. Freemark, K.E., 1995. Assessing effects of agriculture on terrestrial wildlife: developing a hierarchical approach for the U.S. EPA. *Landscape and Urban Planning* 31: 99-115. [1].

1686. Freese, Curtis H., ed., 1997. *Harvesting Wild Species: Implications for Biodiversity Conservation* [case studies]. Baltimore: Johns Hopkins University Press. 703p. [1/1/-].

1687. ———, 1998. *Wild Species as Commodities: Managing Markets and Ecosystems for Sustainability.* Washington, DC: Island Press. 319p.

1688. Freitag, S., & A.S. Van Jaarsveld, 1997. Relative occupancy, endemism, taxonomic distinctiveness and vulnerability: prioritizing regional conservation actions. *Biodiversity and Conservation* 6: 211-232. *A.T.:* South Africa; rarity; regional surveys.

1689. ———, 1997. Ranking priority biodiversity areas: an iterative conservation value-based approach. *Biological Conservation* 82: 263-272. *A.T.:* mammals; South Africa; reserve selection.

1690. Freitag, S., A.O. Nicholls, & A.S. Van Jaarsveld, 1998. Dealing with established reserve networks and incomplete distribution data sets in conservation planning. *South African J. of Science* 94: 79-86. *A.T.:* reserve selection.

1691. Frelich, Lee E., 1995. Old forest in the Lake States today and before European settlement. *Natural Areas J.* 15: 157-167. [1] *A.T.:* old-growth forest.

1692. Frenay, R., & M.S. Kelly, Jan.-Feb. 1997. Going to seed. *Audubon* 99(1): 22-25. [1] *A.T.:* seed banks; Center for Plant Conservation; germplasm resources.

1693. French, D.D., 1994. Hierarchical Richness Index (HRI): a simple procedure for scoring "richness," for use with grouped data. *Biological Conservation* 69: 207-212. [1] *A.T.:* diversity vs. richness.

1694. Friday, Laurie E., 1987. The diversity of macroinvertebrate and macrophyte communities in ponds. *Freshwater Biology* 18: 87-104. [1] *A.T.:* pH.

1695. Friend, Tim, Oct.-Nov. 1995. Power tool. *National Wildlife* 33(6): 16-23. [1] Concerning the use of DNA analysis in bd studies. *A.T.:* wildlife conservation.

1696. Frison, E.A., & Jerzy Serwinski, 1994. *Report of a Joint FAO/ IPGRI Mission to Survey Plant Genetic Resources Programmes in Central Asia and the Caucasus.* Rome: FAO. 49p. [1/1/1] *A.T.:* plant germplasm resources.

1697. Frissell, Christopher A., 1993. Topology of extinction and endangerment of native fishes in the Pacific Northwest and California (U.S.A.). *CB* 7: 342-354. [1] *A.T.:* extinction risk.

1698. Frissell, Christopher A., & David Bayles, 1996. Ecosystem management and the conservation of aquatic biodiversity and ecological integrity. *Water Resources Bull.* 32: 229-240. [1].

1699. Frisvold, George B., & P.T. Condon, July 1994. Biodiversity conservation and biotechnology development agreements. *Contemporary Economic Policy* 12(3): 1-9. [1] *A.T.:* environmental policy.

1700. ———, 1995. The Convention on Biological Diversity: implications for agriculture. *Technological Forecasting and Social Change* 50: 41-54. [1] *A.T.:* crop genetic resources; intellectual property rights.

1701. ———, 1998. The Convention on Biological Diversity and agriculture: implications and unresolved debates. *World Development* 26: 551-570. *A.T.:* plant genetic resources; intellectual property rights.

1702. Fritts, S.H., E.E. Bangs, & J.F. Gore, 1994. The relationship of wolf recovery to habitat conservation and biodiversity in the northwestern United States. *Landscape and Urban Planning* 28: 23-32. [1] *A.T.:* reintroduction.

1703. Frumhoff, Peter C., 1995. Conserving wildlife in tropical forests managed for timber. *Bioscience* 45: 456-464. [2].

1704. Fry, M.E., & N.R. Money, 1994. Biodiversity conservation in the management of utility rights-of-way. *Proc. of the 15th Annual Forest Vegetation Management Conference*: 94-103. [1].

1705. Fuccillo, Domenic A., Linda Sears, & Paul Stapleton, eds., 1997. *Biodiversity in Trust: Conservation and Use of Plant Genetic Resources in CGIAR Centres.* Cambridge, U.K., and New York: CUP. 371p. [1/1/1] *A.T.:* plant germplasm resources.

1706. Fuhrman, Jed A., & A.A. Davis, 1997. Widespread archaea and novel bacteria from the deep sea as shown by 16S rRNA gene sequences. *Marine Ecology Progress Series* 150: 275-285. [2].

1707. Fuhrman, Jed A, Kirk McCallum, & A.A. Davis, 1993. Phylogenetic diversity of subsurface marine microbial communities from the Atlantic and Pacific Oceans. *Applied and Environmental Microbiology* 59: 1294-

1302. [3].

1708. Fujisaka, Sam, German Escobar, & Erik Veneklaas, 1998. Plant community diversity relative to human land uses in an Amazon forest colony. *Biodiversity and Conservation* 7: 41-58. *A.T.:* pasture; slash and burn.

1709. Fujisaka, Sam, C. Castilla, et al., 1998. The effects of forest conversion on annual crops and pastures: estimates of carbon emissions and plant species loss in a Brazilian Amazon colony. *Agriculture, Ecosystems & Environment* 69: 17-26. *A.T.:* slash and burn; deforestation.

1710. Fujita, Marty S., & M.D. Tuttle, 1991. Flying foxes (Chiroptera: Pteropodidae): threatened animals of key ecological and economic importance. *CB* 5: 455-463. [1].

1711. Fuller, R.J., R.D. Gregory, et al., 1995. Population declines and range contractions among lowland farmland birds in Britain. *CB* 9: 1425-1441. [2].

1712. Funk, V.A., 1993. Uses and misuses of floras. *Taxon* 42: 761-772. [1] *A.T.:* biogeography.

1713. Furze, Brian, Terry De Lacy, & Jim Birckhead, 1996. *Culture, Conservation, and Biodiversity: The Social Dimension of Linking Local Level Development and Conservation Through Protected Areas* [case studies]. Chichester, U.K., and New York: Wiley. 269p. [1/1/1] *A.T.:* rural development.

1714. Fussey, George D., 1995. Biodiversity and species discovery curves: a teaching simulation. *J. of Biological Education* 29: 41-45. [1] *A.T.:* species richness estimation.

1715. Futter, Ellen V., Nov.-Dec. 1997. Toward a natural history museum for the 21st Century: biodiversity. *Museum News* 76(6): 40-42.

1716. Gaard, Greta C., ed., 1993. *Ecofeminism: Women, Animals, Nature*. Philadelphia: Temple University Press. 331p. [2/2/1].

1717. Gadgil, Madhav, 1987. Diversity: cultural and biological. *TREE* 2: 369-373. [1].

1718. ———, 1992. Conserving biodiversity as if people matter: a case study from India. *Ambio* 21: 266-270. [1] *A.T.:* protected areas management; human ecology.

1719. ———, 1993. Biodiversity and India's degraded lands. *Ambio* 22: 167-172. [1] *A.T.:* human ecology; developing countries.

1720. ———, 1994. Inventorying, monitoring and conserving India's biological diversity. *Current Science* 66: 401-406. [1].

1721. ———, 1994. A system of positive incentives to conserve biodiversity. *Economic and Political Weekly* 29: 2103-2107. [1].

1722. Gadgil, Madhav, & Preston Devasia, 1995. Intellectual property rights and biological resources: specifying geographical origins and prior knowledge of uses. *Current Science* 69: 637-639. [1].

1723. Gadgil, Madhav, Fikret Berkes, & Carl Folke, 1993. Indigenous knowledge for biodiversity conservation. *Ambio* 22: 151-156. [2] *A.T.:* community participation.

1724. Gage, John D., & R.M. May, 1993. Biodiversity: a dip into the deep seas [editorial]. *Nature* 365: 609-610. [1] *A.T.:* latitudinal diversity gradients; invertebrates; benthic fauna.

1725. Gage, John D., & P.A. Tyler, 1991. *Deep-sea Biology: A Natural History of Organisms at the Deep-sea Floor*. Cambridge, U.K., and New York: CUP. 504p. [1/2/3] *A.T.:* benthic animals.

1726. Gakhova, E.N., March 1998. Genetic cryobanks for conservation of biodiversity. The development and current status of this problem in Russia. *Cryo-Letters*, Suppl. No. 1: 57-64.

1727. Galatowitsch, Susan M., & A.G. van der Valk, 1994. *Restoring Prairie Wetlands: An Ecological Approach*. Ames, IA: Iowa State University Press. 246p. [2/1/1] *A.T.:* restoration ecology; wetlands ecology.

1728. Galbraith, C.A., P.V. Grice, et al., 1995. The role of the statutory bodies in ornithological conservation within the U.K. *Ibis* 137, Suppl. 1: S224-S231. [1] *A.T.:* environmental policy; wildlife conservation.

1729. Galeffi, C., 1993. The contribution of American plants to pharmaceutical sciences. *Il Farmaco* 48: 1175-1195. [1] *A.T.:* medicinal plants; natural products.

1730. Galil, B.S., winter 1997. Biodiversity and invasion: how resilient is the Levant Sea? *Israel Environment Bull.* 20(1): 20-21. [1] *A.T.:* biological invasions; Eastern Mediterranean.

1731. Galindo Leal, Carlos, & F.L. Bunnell, 1995. Ecosystem management: implications and opportunities of a new paradigm. *Forestry Chronicle* 71: 601-606. [1] *A.T.:* forest management; landscape ecology; Canada.

1732. Gall, G.A.E., & M. Staton, 1992. Integrating conservation biology and agricultural production: conclusions. *Agriculture, Ecosystems & Environment* 42: 217-230. [1].

1733. Galleni, Lodovico, 1995. How does the Teilhardian vision of evolution compare with contemporary theories? *Zygon* 30: 25-45. [1] *A.T.:* Gaia hypothesis.

1734. Gallina, S., S. Mandujano, & A. González Romero, 1996. Conservation of mammalian biodiversity in coffee plantations of central Veracruz, Mexico. *Agroforestry Systems* 33: 13-27. [1].

1735. Galzin, R., S. Planes, et al., 1994. Variation in diversity of coral reef fish between French Polynesian atolls. *Coral Reefs* 13: 175-180. [1] *A.T.:* biogeography.

1736. Gámez, Rodrigo, 1991. Biodiversity conservation through facilitation of its sustainable use: Costa Rica's National Biodiversity Institute [editorial]. *TREE* 6: 377-378. [1].

1737. Ganeshaiah, K.N., K. Chandrashekara, & A.R.V. Kumar, 1997. Avalanche Index: a new measure of biodiversity based on biological heterogeneity of the communities. *Current Science* 73: 128-133. *A.T.:* diversity indices; dung beetles; community structure.

1738. Garcia, A., 1992. Conserving the species-rich meadows of Europe. *Agriculture, Ecosystems & Environment* 40: 219-232. [1] *A.T.:* Spain; hay meadows.

1739. Garcia, Serafin, Oct. 1996. Biodiversity is a guarantee of evolution [interview]. *UNESCO Courier* 49(9): 4-7. [1] A chat with Swiss microbiologist Werner Arber.

1740. Gardner, Scott L., & M.L. Campbell, 1992. Parasites as probes for biodiversity. *J. of Parasitology* 78: 596-600. [1] *A.T.:* host-parasite relations; Bolivia; mammals; coevolution.

1741. Garnier, E., M.L. Navas, et al., 1997. A problem for biodiversity-productivity studies: how to compare the productivity of multispecific plant mixtures to that of monocultures? *Acta Oecologica* 18: 657-670.

1742. Garrabou, Joaquim, Enric Sala, et al., 1998. The impact of diving on rocky sublittoral communities: a case study of a bryozoan population. *CB* 12: 302-312. *A.T.:* Spain.

1743. Garrett, Laurie, 1994. *The Coming Plague: Newly Emerging Diseases in a World Out of Balance*. New York: Farrar, Straus & Giroux. 750p. [3/3/3] *A.T.:* epidemiology.

1744. Garrod, G.D., & K.G. Willis, 1994. Valuing biodiversity and nature conservation at a local level. *Biodiversity and Conservation* 3: 555-565. [1] *A.T.:* contingent valuation; U.K.

1745. ———, 1997. The non-use benefits of enhancing forest biodiversity: a contingent ranking study. *Ecological Economics* 21: 45-61. *A.T.:* U.K.; forest management.

1746. Garrott, Robert A., P.J. White, & C.A. Vanderbilt White, 1993. Overabundance: an issue for conservation biologists? *CB* 7: 946-949. [1].

1747. Garson, M., 1996. Is marine bioprospecting sustainable? *Search* 27: 114-117. [1] *A.T.:* bd value; Australia.

1748. Gascuel, D., & F. Menard, 1997. Assessment of a multispecies fishery in Senegal, using production models and diversity indices. *Aquatic Living Resources* 10: 281-288.

1749. Gasnier, N., J. Cabaret, et al., 1997. Species diversity in gastrointestinal nematode communities of dairy goats: species-area and species-climate relationships. *Veterinary Research* 28: 55-64. *A.T.:* France.

1750. Gasser, Charles S., & R.T. Fraley, 1989. Genetically engineering plants for crop improvement. *Science* 244: 1293-1299. [3] *A.T.:* transgenic plants.

1751. ———, June 1992. Transgenic crops. *Scientific American* 266(6): 62-69. [1] *A.T.:* biotechnology; genetic engineering.

1752. Gaston, Kevin J., 1990. Patterns in the geographical ranges of species. *Biological Reviews* 65: 105-129. [2] *A.T.:* areography.

1753. ———, 1991. Estimates of the near-imponderable: a reply to Erwin. *CB* 5: 564-566. [1].
1754. ———, 1991. The magnitude of global insect species richness. *CB* 5: 283-296. [2] *A.T.:* range size.
1755. ———, 1992. Regional numbers of insect and plant species. *Functional Ecology* 6: 243-247. [1].
1756. ———, 1994. *Rarity*. London and New York: Chapman & Hall. 205p. [1/1/3] *A.T.:* rare and threatened plants and animals; endangered species.
1757. ———, 1994. Biodiversity—measurement. *Progress in Physical Geography* 18: 565-574. [1] A review.
1758. ———, 1994. Spatial patterns of species description: how is our knowledge of the global insect fauna growing? *Biological Conservation* 67: 37-40. [1] *A.T.:* Coleoptera; Diptera; Hymenoptera.
1759. ———, 1995. Biodiversity: loss. *Progress in Physical Geography* 19: 255-264. [1] A review.
1760. ———, ed., 1996. *Biodiversity: A Biology of Numbers and Difference*. Oxford, U.K., and Cambridge, MA: Blackwell Science. 396p. [1/1/2].
1761. ———, 1996. Biodiversity—congruence. *Progress in Physical Geography* 20: 105-112. [2] A review.
1762. ———, 1996. Biodiversity—latitudinal gradients. *Progress in Physical Geography* 20: 466-476. [1] A review.
1763. ———, 1998. Species-range size distributions: products of speciation, extinction and transformation. *PTRSL B* 353: 219-230. *A.T.:* macroecology.
1764. Gaston, Kevin J., & T.M. Blackburn, 1995. Mapping biodiversity using surrogates for species richness: macro-scales and New World birds. *PRSL B* 262: 335-341. [2] *A.T.:* higher taxa.
1765. ———, 1996. The tropics as a museum of biological diversity: an analysis of the New World avifauna. *PRSL B* 263: 63-68. [1] *A.T.:* latitudinal clines; Latin America.
1766. ———, 1996. The spatial distribution of threatened species: macro-scales and New World birds. *PRSL B* 263: 235-240. [1] *A.T.:* geographical distribution; species richness.
1767. ———, 1996. Global scale macroecology: interactions between population size, geographic range size and body size in the Anseriformes. *J. of Animal Ecology* 65: 701-714. [2].
1768. ———, 1997. How many birds are there? *Biodiversity and Conservation* 6: 615-625.
1769. Gaston, Kevin J., & Rhian David, 1994. Hotspots across Europe. *Biodiversity Letters* 2: 108-116. [1] *A.T.:* higher taxa; species richness.
1770. Gaston, Kevin J., & E. Hudson, 1994. Regional patterns of diversity and estimates of global insect species richness. *Biodiversity and Conserva-*

tion 3: 493-500. [1] *A.T.:* biogeography.

1771. Gaston, Kevin J., & J.H. Lawton, 1990. Effects of scale and habitat on the relationship between regional distribution and local abundance. *Oikos* 58: 329-335. [2] *A.T.:* reference habitats.

1772. Gaston, Kevin J., & R.M. May, 1992. Taxonomy of taxonomists. *Nature* 356: 281-282. [1] Comments on the mismatch between numbers of species and numbers of taxonomists.

1773. Gaston, Kevin J., & L.A. Mound, 1993. Taxonomy, hypothesis testing and the biodiversity crisis. *PRSL B* 251: 139-142. [1] *A.T.:* synonymy; Thysanoptera; taxonomic description.

1774. Gaston, Kevin J., & A.O. Nicholls, 1995. Probable time to extinction of some rare breeding bird species in the United Kingdom. *PRSL B* 259: 119-123. [1].

1775. Gaston, Kevin J., & J.I. Spicer, 1998. *Biodiversity: An Introduction.* Oxford, U.K., and Malden, MA: Blackwell Science. 113p. *A.T.:* bdc.

1776. Gaston, Kevin J., & P.H. Williams, 1993. Mapping the world's species—the higher taxon approach. *Biodiversity Letters* 1: 2-8. [2] *A.T.:* species richness.

1777. Gaston, Kevin J., T.R. New, & M.J. Samways, eds., 1993. *Perspectives on Insect Conservation.* Andover, Hampshire, U.K.: Intercept. 250p. [1/1/1].

1778. Gaston, Kevin J., Malcolm Scoble, & Anne Crook, 1995. Patterns in species description: a case study using the Geometridae (Lepidoptera). *Biological J. of the Linnean Society* 55: 225-237. [1] *A.T.:* species richness; synonyms.

1779. Gaston, Kevin J., P.H. Williams, et al., 1995. Large scale patterns of biodiversity: spatial variation in family richness. *PRSL B* 260: 149-154. [2] *A.T.:* hotspots; latitudinal diversity gradients.

1780. Gates, David M., 1993. *Climate Change and Its Biological Consequences.* Sunderland, MA: Sinauer Associates. 280p. [2/2/2].

1781. Gautam, P.L., et al., eds., 1998. *Plant Germplasm Collecting: Principles and Procedures* [training program]. New Delhi: National Bureau of Plant Genetic Resources. 218p. *A.T.:* plant genetic engineering.

1782. Gee, Henry, 1992. The objective case for conservation [opinion]. *Nature* 357: 639. [1] Seeks to define bd in terms broader than endangered species conservation alone.

1783. Gee, J.M., & R.M. Warwick, 1996. A study of global biodiversity patterns in the marine motile fauna of hard substrata. *J. of the Marine Biological Association* 76: 177-184. [1].

1784. Geevan, C.P., 1995. Biodiversity conservation information network: a concept plan. *Current Science* 69: 906-913. [1].

1785. Gel'tser, Yu G., 1996. The significance of biodiversity for soil diagnostics. *Eurasian Soil Science* 29: 663-669. [1].

1786. Gentry, Alwyn H., 1982. Neotropical floristic diversity: phytogeographical connections between Central and South America, Pleistocene climatic fluctuations, or an accident of the Andean orogeny? *Annals of the Missouri Botanical Garden* 69: 557-593. [2].

1787. ———, 1982. Patterns of Neotropical plant species diversity. *Evolutionary Biology* 15: 1-84. [2].

1788. ———, 1988. Changes in plant community diversity and floristic composition on environmental and geographical gradients. *Annals of the Missouri Botanical Garden* 75: 1-34. [2].

1789. ———, 1988. Tree species richness of Upper Amazonian forests. *PNAS* 85: 156-159. [2].

1790. ———, ed., 1990. *Four Neotropical Rainforests* [symposium proceedings]. New Haven: Yale University Press. 627p. [2/2/2].

1791. ———, 1992. Tropical forest biodiversity: distributional patterns and their conservational significance. *Oikos* 63: 19-28. [2]. *A.T.:* endemism.

1792. ———, 1993. Tropical forest biodiversity and the potential for new medicinal plants. In A.D. Kinghorn & M.F. Balandrin, eds., *Human Medicinal Agents From Plants* (Washington, DC: ACS Symposium Series No. 534): 13-24. [1].

1793. Gentry, Alwyn H., & C. Dobson, 1987. Contribution of nontrees to species richness of a tropical rain forest. *Biotropica* 19: 149-156. [1].

1794. George, Susan, W.J. Snape III, & Rina Rodriguez, 1997. The public in action: using state citizen suit statutes to protect biodiversity. *University of Baltimore J. of Environmental Law* 6: 1-44.

1795. Georgiadis, Nicholas, & Andrew Balmford, 1992. The calculus of conserving biological diversity [editorial]. *TREE* 7: 321-322. [1].

1796. Gerard, Philip W., 1995. *Agricultural Practices, Farm Policy, and the Conservation of Biological Diversity*. Washington, DC: National Biological Service, U.S. Dept. of the Interior. 28p. [1/1/1] *A.T.:* environmental policy.

1797. German Federal Agency for Nature Conservation, 1997. *Biodiversity and Tourism: Conflicts on the World's Seacoasts and Strategies for Their Solution*. Berlin and New York: Springer. 343p. [1/1/-].

1798. Germida, James J., S.D. Siciliano, et al., 1998. Diversity of root-associated bacteria associated with field-grown canola (*Brassica napus L.*) and wheat (*Triticum aestivum L.*). *FEMS Microbiology Ecology* 26: 43-50. *A.T.:* Saskatchewan; endophytes.

1799. Gershon, Diane, 1992. If biological diversity has a price, who sets it and who should benefit? *Nature* 359: 565. [1] *A.T.:* Merck; Costa Rica.

1800. Ghimire, Krishna B., & M.P. Pimbert, eds., 1997. *Social Change and Conservation: Environmental Politics and Impacts of National Parks and Protected Areas*. London: Earthscan. 342p. [1/1/-].

1801. Ghosh, Asish K., Q.H. Baquri, & Ishwar Prakash, eds., 1996. *Fau-*

nal Diversity in the Thar Desert: Gaps in Research [conference proceedings]. Jodhpur, India: Scientific Publishers. 410p. [1/1/1] *A.T.:* India; Pakistan.

1802. Giampietro, M., 1997. Socioeconomic constraints to farming with biodiversity. *Agriculture, Ecosystems & Environment* 62: 145-167. *A.T.:* agricultural practices; technological progress.

1803. Gibbons, Edward F., Jr., B.S. Durrant, & Jack Demarest, eds., 1995. *Conservation of Endangered Species in Captivity: An Interdisciplinary Approach.* Albany: State University of New York Press. 810p. [1/1/1].

1804. Gibbons, J.W., V.J. Burke, et al., 1997. Perceptions of species abundance, distribution, and diversity: lessons from four decades of sampling on a government-managed reserve. Environmental Management 21: 259-268. *A.T.:* South Carolina; reptiles; amphibians.

1805. Gibbs, James P., 1993. Importance of small wetlands for the persistence of local populations of wetland-associated animals. *Wetlands* 13: 25-31. [1] *A.T.:* Maine; vertebrates.

1806. Gibbs, J.N., & D. Wainhouse, 1986. Spread of forest pests and pathogens in the Northern Hemisphere. *Forestry* 59: 141-153. [1].

1807. Gibert, Janine, D.L. Danielopol, & J.A. Stanford, eds., 1994. *Groundwater Ecology.* San Diego: Academic Press. 571p. [1/1/1].

1808. Gibson, Arthur C., ed., 1996. *Neotropical Biodiversity and Conservation.* Los Angeles: Mildred E. Mathias Botanical Garden, University of California Los Angeles. 202p. [1/1/1].

1809. Gibson, C.C., & S.A. Marks, 1995. Transforming rural hunters into conservationists: an assessment of community-based wildlife management programs in Africa. *World Development* 23: 941-957. [1] *A.T.:* hunting; Zambia.

1810. Gilbert, F., A. Gonzalez, & I. Evans-Freke, 1998. Corridors maintain species richness in the fragmented landscapes of a microecosystem. *PRSL B* 265: 577-582. *A.T.:* immigration; patch dynamics.

1811. Gilbert, G.S., & S.P. Hubbell, 1996. Plant diseases and the conservation of tropical forests. *Bioscience* 46: 98-106. [1].

1812. Gilbertson, D.D., M. Kent, & F.B. Pyatt, 1995 (reprint of 1985 ed.). *Practical Ecology for Geography and Biology: Survey, Mapping and Data Analysis.* London and Dover, NH: Hutchinson Education. 320p. [1/1/1] *A.T.:* ecological surveys.

1813. Gilinsky, N.L., & I.J. Good, 1991. Probabilities of origination, persistence, and extinction of families of marine invertebrate life. *Paleobiology* 17: 145-166. [1].

1814. Gill, A. Malcolm, & J.E. Williams, 1996. Fire regimes and biodiversity: the effects of fragmentation of southeastern Australian eucalypt forests by urbanisation, agriculture and pine plantations. *Forest Ecology and Management* 85: 261-278. [1].

1815. Gill, A. Malcolm, R.H. Groves, & I.R. Noble, eds., 1981. *Fire and the Australian Biota* [conference papers]. Canberra: Australian Academy of Science. 582p. [1/1/2] *A.T.:* fire ecology.

1816. Giller, P.S., 1996. The diversity of soil communities, the "poor man's tropical rainforest." *Biodiversity and Conservation* 5: 135-168. [1] *A.T.:* soil fauna; species richness; species abundance; community ecology.

1817. Gillis, Anna Maria, 1990. The new forestry: an ecosystem approach to land management. *Bioscience* 40: 558-562. [1].

1818. Gilpin, Michael E., & Ilkka Hanski, eds., 1991. *Metapopulation Dynamics: Empirical and Theoretical Investigations*. London and San Diego: Academic Press. 336p. [1/1/2].

1819. Gippoliti, Spartaco, & G.M. Carpaneto, 1997. Captive breeding, zoos, and good sense [editorial]. *CB* 11: 806-807. Defends captive breeding as a conservation strategy.

1820. Gipps, J.H.W., ed., 1991. *Beyond Captive Breeding: Re-introducing Endangered Mammals to the Wild* [symposium proceedings]. Oxford, U.K.: Clarendon Press. 284p. [1/1/1].

1821. Girel, J., & O. Manneville, 1998. Present species richness of plant communities in alpine stream corridors in relation to historical river management. *Biological Conservation* 85: 21-33. *A.T.:* habitat disturbance.

1822. Gitay, Habiba, J.B. Wilson, & W.G. Lee, 1996. Species redundancy: a redundant concept? *J. of Ecology* 84: 121-124. [1] Evaluates the concept.

1823. Gittleman, J.L., & A. Purvis, 1998. Body size and species-richness in carnivores and primates. *PRSL B* 265: 113-119. *A.T.:* size-diversity relation.

1824. Giudice, John H., & J.T. Ratti, 1995. *Interior Wetlands of the United States: A Review of Wetland Status, General Ecology, Biodiversity, and Management*. Vicksburg, MS: Technical Report WRP-SM-9, U.S. Army Engineer Waterways Experiment Station. 156p. [1/1/1].

1825. Given, David R., 1990. Conserving botanical diversity on a global scale. *Annals of the Missouri Botanical Garden* 77: 48-62. [1] *A.T.:* public awareness.

1826. ———, 1993. Changing aspects of endemism and endangerment in Pteridophyta. *J. of Biogeography* 20: 293-302. [1] *A.T.:* habitat fragmentation; biogeography.

1827. ———, 1994. *Principles and Practice of Plant Conservation*. Portland, OR: Timber Press. 292p. [2/1/2].

1828. ———, 1995. Forging a biodiversity ethic in a multicultural context. *Biodiversity and Conservation* 4: 877-891. [1] *A.T.:* environmental ethics; Maori people; New Zealand; resource management.

1829. Given, David R., & Warwick Harris, 1994. *Techniques and Methods of Ethnobotany: As an Aid to the Study, Evaluation, Conservation and Sustainable Use of Biodiversity*. London: Commonwealth Secretariat. 148p. [1/1/1].

1830. Given, David R., & D.A. Norton, 1993. A multivariate approach to assessing threat and for priority setting in threatened species conservation. *Biological Conservation* 64: 57-66. [1] *A.T.:* complex threat; plant conservation.

1831. Givnish, Thomas J., 1994. Does diversity beget stability? [letter] *Nature* 371: 113-114. [1] *A.T.:* David Tilman.

1832. Glanznig, Andreas, 1995. *Native Vegetation Clearance, Habitat Loss, and Biodiversity Decline: An Overview of Recent Native Vegetation Clearance in Australia and Its Implications for Biodiversity*. Canberra: Biodiversity Unit, Dept. of the Environment, Sport, and Territories. 46p. [1/1/1].

1833. Glaser, Paul H., 1992. Raised bogs in eastern North America—regional controls for species richness and floristic assemblages. *J. of Ecology* 80: 535-554. [1] *A.T.:* floristic regions; environmental factors; aquatic plants.

1834. Glausiusz, Josie, March 1997. Where insects fear to tread. *Discover* 18(3): 26. [1] Considers the absence of insects in the oceans and the evolutionary reasons therefor.

1835. Glazier, Douglas S., 1987. Energetics and taxonomic patterns of species diversity. *Systematic Zoology* 36: 62-71. [1] *A.T.:* bioenergetics.

1836. Gleason, H.A., 1922. On the relation between species and area. *Ecology* 3: 158-162. [2] Concerning Arrhenius's discussion of the relation between size of area and number of species.

1837. ———, 1926. The individualistic concept of the plant association. *Bull. of the Torrey Botanical Club* 53: 7-26. [2].

1838. Glen, William, 1990. What killed the dinosaurs? *American Scientist* 78: 354-370. [1] Defends catastrophic volcanism as the extinction agent. *A.T.:* Cretaceous-Tertiary boundary; mass extinctions.

1839. ———, ed., 1994. *The Mass-extinction Debates: How Science Works in a Crisis*. Stanford, CA: Stanford University Press. 370p. [2/2/1].

1840. Glenn, Susan M., 1990. Regional analysis of mammal distributions among Canadian parks: implications for park planning. *Canadian J. of Zoology* 68: 2457-2464. [1] *A.T.:* reserve design.

1841. Glenn, Susan M., & S.L. Collins, 1992. Effects of scale and disturbance on rates of immigration and extinction of species in prairies. *Oikos* 63: 273-280. [1] *A.T.:* local extinction; Kansas; spatial scale.

1842. Gliessman, Stephen R., ed., 1990. *Agroecology: Researching the Ecological Basis for Sustainable Agriculture*. New York: Springer. 380p. [1/1/1].

1843. ———, ed., 1998. *Agroecology: Ecological Processes in Sustainable Agriculture*. Chelsea, MI: Ann Arbor Press. 357p.

1844. Glitzenstein, Jeff S., P.A. Harcombe, & D.R. Streng, 1986. Disturbance, succession, and maintenance of species diversity in an East Texas forest. *Ecological Monographs* 56: 243-258. [1] *A.T.:* forest regeneration.

1845. Glover, F., Kuo Ching-Chung, & K.S. Dhir, 1995. A discrete optimization model for preserving biological diversity. *Applied Mathematical Modelling* 19: 696-701. [1] *A.T.:* mathematical models; cranes.

1846. Glowka, Lyle, & Cyrille de Klemm, 1996. International instruments, processes and non-indigenous species introductions: is a protocol necessary? *Environmental Policy and Law* 26: 247-255. [1] *A.T.:* CBD; introduced species; environmental policy.

1847. Glowka, Lyle, Françoise Burhenne-Guilmin, & Hugh Synge, 1994. *A Guide to the Convention on Biological Diversity*. Gland, Switzerland: Environmental Policy and Law Paper No. 30, IUCN. 161p. [1/1/1].

1848. Glynn, P.W., 1993. Coral reef bleaching: ecological perspectives. *Coral Reefs* 12: 1-17. [2] *A.T.:* coral reef ecology.

1849. Gobster, Paul H., Feb. 1995. Aldo Leopold's ecological esthetic: integrating esthetic and biodiversity values. *J. of Forestry* 93(2): 6-10. [1] *A.T.:* forest management.

1850. Godoy, Ricardo A., & Ruben Lubowski, 1992. Guidelines for the economic valuation of nontimber tropical-forest products. *Current Anthropology* 33: 423-433. [1] *A.T.:* bd value.

1851. Goehl, Thomas J., 1996. Pura vida [editorial]. *Environmental Health Perspectives* 104: 582-583. [1] Concerning bdc in Costa Rica. *A.T.:* INBio.

1852. Goerck, J.M., 1997. Patterns of rarity in the birds of the Atlantic forest of Brazil. *CB* 11: 112-118.

1853. Goh, Teik-Khiang, & K.D. Hyde, 1996. Biodiversity of freshwater fungi. *J. of Industrial Microbiology & Biotechnology* 17: 328-345. [1].

1854. Goklany, Indur M., & M.W. Sprague, Nov. 1992. Limits on technology would hurt biodiversity. *USA Today* 121(2570): 28-31. [1] *A.T.:* UNCED; sustainable development.

1855. ———, 1992. *Sustaining Development and Biodiversity: Productivity, Efficiency, and Conservation*. Washington, DC: Cato Institute. 21p. [1/1/1] *A.T.:* agriculture.

1856. Goldammer, Johann G., ed., 1990. *Fire in the Tropical Biota: Ecosystem Processes and Global Challenges* [symposium papers]. Berlin and New York: Springer. 497p. [1/1/1].

1857. Goldberg, Deborah E., & T.E. Miller, 1990. Effects of different resource additions on species diversity in an annual plant community. *Ecology* 71: 213-225. [2] *A.T.:* environmental resources; productivity.

1858. Goldblatt, Peter, ed., 1993. *Biological Relationships Between Africa and South America* [symposium proceedings]. New Haven: Yale University Press. 630p. [1/2/1] *A.T.:* biogeography; paleobiogeography.

1859. ———, 1997. Floristic diversity in the Cape flora of South Africa.

Biodiversity and Conservation 6: 359-377. *A.T.:* endemism; soil diversity.

1860. Goldman, Karen Anne, 1994. Compensation for use of biological resources under the Convention on Biological Diversity: compatibility of conservation measures and competitiveness of the biotechnology industry. *Law and Policy in International Business* 25: 695-726. [1] *A.T.:* technology transfer; intellectual property rights.

1861. Goldschmidt, Tijs, 1996. *Darwin's Dreampond: Drama in Lake Victoria.* Cambridge, MA: MIT Press. 274p. [2/3/1] *A.T.:* cichlids; diversification; evolution.

1862. Goldsmith, Frank Barrie, ed., 1991. *Monitoring for Conservation and Ecology.* London and New York: Chapman & Hall. 275p. [1/1/1].

1863. ———, 1993-1994. The formulation of biodiversity strategies. *Science Progress* 77: 221-232. [1] Describes UNCED and the national action plan strategies of the U.K., Jersey, and India.

1864. Goldstein, B.E., 1992. Can ecosystem management turn an administrative patchwork into a Greater Yellowstone Ecosystem? *Northwest Environmental J.* 8: 285-324. [1] *A.T.:* planning models.

1865. Goldstein, Jon H., 1991. The prospects for using market incentives to conserve biological diversity. *Environmental Law* 21: 985-1014. [1].

1866. Golley, Frank B., 1987. Deep ecology from the perspective of environmental science. *Environmental Ethics* 9: 45-55. [1] *A.T.:* environmental ethics; environmentalism.

1867. ———, 1993. *A History of the Ecosystem Concept in Ecology: More Than the Sum of the Parts.* New Haven: Yale University Press. 254p. [2/2/2].

1868. Gómez-Limón, F. Javier, & J.V. de Lucio, 1995. Recreational activities and loss of diversity in grasslands in Alta Manzanares Natural Park, Spain. *Biological Conservation* 74: 99-105. [1] *A.T.:* trampling.

1869. Gómez-Pompa, Arturo, & Andrea Kaus, 1992. Taming the wilderness myth. *Bioscience* 42: 271-279. [1] *A.T.:* public awareness; wilderness management.

1870. Gomiero, T., M. Giampietro, et al., 1997. Biodiversity use and technical performance of freshwater fish aquaculture in different socioeconomic contexts: China and Italy. *Agriculture, Ecosystems & Environment* 62: 169-185. *A.T.:* labor use.

1871. Gooday, Andrew J., B.J. Bett, et al., 1998. Deep-sea benthic foraminiferal species diversity in the NE Atlantic and NW Arabian Sea: a synthesis. *Deep-Sea Research, Part II* 45: 165-201.

1872. Goodland, Robert J.A., ed., 1990. *Race to Save the Tropics: Ecology and Economics for a Sustainable Future.* Washington, DC: Island Press. 219p. [2/1/1] *A.T.:* developing countries.

1873. Goodland, Robert J.A., & G. Ledec, Nov. 1989. Wildlands: balancing conversion with conservation in World Bank projects. *Environment*

31(9): 6-11, 27-35. [1] *A.T.:* wilderness value.

1874. Goodman, Billy, 1993. Drugs and people threaten diversity in Andean forests. *Science* 261: 293. [1] *A.T.:* deforestation; illegal drug cultivation; montane forests.

1875. Goodman, Daniel, 1975. The theory of diversity-stability relationships in ecology. *Quarterly Review of Biology* 50: 237-266. [2].

1876. Goodman, Steven M., & Bruce D. Patterson, eds., 1997. *Natural Change and Human Impact in Madagascar* [symposium papers]. Washington, DC: Smithsonian Institution Press. 432p. [1/1/-].

1877. Goodwin, Brian, 1995-1996. Emergent form: evolving beyond Darwinism. *Complexity* 1(5): 11-15. [1] *A.T.:* adaptation; diversification; behavioral diversity; morphological diversity.

1878. Goodwin, H., 1996. In pursuit of ecotourism. *Biodiversity and Conservation* 5: 277-291. [1] *A.T.:* bdc; nature reserves.

1879. Goote, Maas M., 1997. Current legal developments. Convention on Biological Diversity. The Jakarta Mandate on Marine and Coastal Biological Diversity. *International J. of Marine and Coastal Law* 12: 377-395. *A.T.:* coastal zone management.

1880. Gophen, M., P.B.O. Ochumba, & L.S. Kaufman, 1995. Some aspects of perturbation in the structure and biodiversity of the ecosystem of Lake Victoria (East Africa). *Aquatic Living Resources* 8: 27-41. [1] *A.T.:* Nile perch; biological invasions.

1881. Gordon, Donald, & Chris Magin, 1995. *Guidelines for a National Institutional Survey: In the Context of the Convention on Biological Diversity.* Nairobi: UNEP. 75p. [1/1/1] *A.T.:* bd databases.

1882. Gore, Albert, Jr., 1992. *Earth in the Balance: Ecology and the Human Spirit.* Boston: Houghton Mifflin. 407p. [3/3/3] *A.T.:* environmental protection; environmental policy.

1883. Goreau, T.J., & R.L. Hayes, 1994. Coral bleaching and ocean "hot spots." *Ambio* 23: 176-180. [2].

1884. Gotsch, N., & P. Rieder, 1995. Biodiversity, biotechnology, and institutions among crops: situation and outlook. *J. of Sustainable Agriculture* 5: 5-40. [1] *A.T.:* intellectual property rights; international cooperation.

1885. Gottlieb, Otto R., 1989. The role of oxygen in phytochemical evolution towards diversity. *Phytochemistry* 28: 2545-2558. [1] *A.T.:* molecular evolution.

1886. ———, 1992. Plant phenolics as expressions of biological diversity. *Basic Life Sciences* 59: 523-538. [1].

1887. Gould, Stephen Jay, Sept. 1990. The Golden Rule—a proper scale for our environmental crisis. *Natural History* 99(9): 24-30. [1].

1888. ———, Oct. 1994. The evolution of life on earth. *Scientific American* 271(4): 84-91. [1].

1889. Gould, Stephen Jay, & Niles Eldredge, 1977. Punctuated equilibria:

the tempo and mode of evolution reconsidered. *Paleobiology* 3: 115-151. [3] *A.T.:* phyletic gradualism.

1890. Gould, Stephen Jay, & R.C. Lewontin, 1979. The spandrels of San Marco and the Panglossian paradigm: a critique of the adaptationist programme. *PRSL B* 205: 581-598. [3] Gould & Lewontin's celebrated criticism of the logic of the concept of adaptation.

1891. Gould, William A., & M.D. Walker, 1997. Landscape-scale patterns in plant species richness along an Arctic river. *Canadian J. of Botany* 75: 1748-1765. *A.T.:* Northwest Territories; landscape heterogeneity; riparian habitats.

1892. Goulding, Michael, March 1993. Flooded forests of the Amazon. *Scientific American* 266(3): 114-120. [1] *A.T.:* river chemistry; flooding.

1893. Gouy, Manolo, & W.-H. Li, 1989. Molecular phylogeny of the kingdoms Animal, Plantae, and Fungi. *Molecular Biology and Evolution* 6: 109-122. [2] *A.T.:* RNA analysis.

1894. Goward, Trevor, summer 1994. Living antiquities. *Nature Canada* 23(3): 14-21. [1] Concerning lichens and forest health. *A.T.:* old-growth forest.

1895. Gowdy, John M., 1993. Economic and biological aspects of genetic diversity. *Society & Natural Resources* 6: 1-16. [1] *A.T.:* environmental policy; environmental economics.

1896. ———, 1994. The social context of natural capital: the social limits to sustainable development. *International J. of Social Economics* 21(8): 43-55. [1] Concerning human impact on natural systems.

1897. ———, 1996. Discounting, hierarchies, and the social aspects of biodiversity protection. *International J. of Social Economics* 23(4-6): 49-63. [1] *A.T.:* market value; land use policy.

1898. ———, 1997. The value of biodiversity. *Land Economics* 73: 25-41. *A.T.:* market value; exchange value.

1899. Gowdy, John M., & C.N. McDaniel, 1995. One world, one experiment: addressing the biodiversity-economics conflict. *Ecological Economics* 15: 181-192. [1] *A.T.:* market theory; environmental economics.

1900. Grace, James B., & B.H. Pugesek, 1997. A structural equation model of plant species richness and its application to a coastal wetland. *American Naturalist* 149: 436-460. *A.T.:* Louisiana; Mississippi; floodplains.

1901. Gradwohl, Judith, & Russell Greenberg, 1988. *Saving the Tropical Forests*. Washington, DC: Island Press. 214p. [3/2/2] *A.T.:* forest conservation; deforestation.

1902. Graham, Jeffrey B., Robert Dudley, et al., 1995. Implications of the late Palaeozoic oxygen pulse for physiology and evolution. *Nature* 375: 117-120. [2] *A.T.:* atmospheric oxygen; hyperoxia.

1903. Graham, Robin L., M.G. Turner, & V.H. Dale, 1990. How increasing CO_2 and climate change affect forests. *Bioscience* 40: 575-587. [2] *A.T.:*

global warming.

1904. Graham, R.W., 1985. Diversity and community structure of the late Pleistocene mammal fauna of North America. *Acta Zoologica Fennica* 170: 181-192. [1] *A.T.:* Pleistocene extinctions.

1905. Grainger, Alan, 1993. *Controlling Tropical Deforestation.* London: Earthscan. 310p. [1/1/1] *A.T.:* tropical agriculture; logging.

1906. Grandcolas, Philippe, 1993. The origin of biological diversity in a tropical cockroach lineage: a phylogenetic analysis of habitat choice and biome occupancy. *Acta Oecologica* 14: 259-270. [1] *A.T.:* diversification.

1907. ———, ed., 1997. *The Origin of Biodiversity in Insects: Phylogenetic Tests of Evolutionary Scenarios* [symposium papers]. Paris: Mémoires du Muséum National D'Histoire Naturelle Tome 173. 354p. [1/1/-] *A.T.:* diversification.

1908. Granjon, L., J.F. Cosson, et al., 1996. Influence of tropical rain forest fragmentation on mammal communities in French Guiana: short term effects. *Acta Oecologica* 17: 673-684. [1] *A.T.:* population density.

1909. Grant, Peter R., 1981. Speciation and the adaptive radiation of Darwin's finches. *American Scientist* 69: 653-663. [1] *A.T.:* intraspecific variation; Galápagos Islands.

1910. ———, ed., 1998. *Evolution on Islands* [discussion meeting papers]. Oxford, U.K., and New York: OUP. 334p. *A.T.:* island ecology.

1911. Grassle, J. Frederick, 1985. Hydrothermal vent animals: distribution and biology. *Science* 229: 713-717. [2] *A.T.:* invertebrates; dispersal.

1912. ———, 1989. Species diversity in deep-sea communities. *TREE* 4: 12-15. [2] *A.T.:* sampling.

1913. ———, 1991. Deep-sea benthic biodiversity. *Bioscience* 41: 464-469. [1].

1914. Grassle, J. Frederick, & N.J. Maciolek, 1992. Deep-sea species richness: regional and local diversity estimates from quantitative bottom samples. *American Naturalist* 139: 313-341. [2] *A.T.:* North Atlantic; continental slope; continental rise.

1915. Gray, Andrew, 1991. *Between the Spice of Life and the Melting Pot: Biodiversity Conservation and Its Impact on Indigenous Peoples.* Copenhagen: International Work Group for Indigenous Affairs. 74p. [1/1/1].

1916. Gray, Gary G., 1993. *Wildlife and People: The Human Dimensions of Wildlife Ecology.* Urbana, IL: University of Illinois Press. 260p. [2/1/1].

1917. Gray, J.S., 1997. Marine biodiversity: patterns, threats and conservation needs. *Biodiversity and Conservation* 6: 153-175. *A.T.:* species diversity; latitudinal diversity gradients; coastal zone management.

1918. Grayson, Arnold J., 1995. *The World's Forests: International Initiatives Since Rio.* Oxford, U.K.: Commonwealth Forestry Association, and Oxford Forestry Institute. 72p. [1/1/1].

1919. Grayson, Donald K., 1991. Late Pleistocene mammalian extinctions in North America: taxonomy, chronology, and explanations. *J. of World Prehistory* 5: 193-231. [1] *A.T.:* overkill hypothesis; climatic change.

1920. Grayson, Donald K., & S.D. Livingston, 1993. Missing mammals on Great Basin mountains: Holocene extinctions and inadequate knowledge. *CB* 7: 527-532. [1] *A.T.:* montane faunas; boreal mammals; Nevada.

1921. Graziano, Angela V., Sept.-Oct. 1994. Endangered species and wetlands conservation. *Endangered Species Technical Bull.* 19(5): 14-16. [1] *A.T.:* North American Wetlands Conservation Act.

1922. Greaves, Tom, ed., 1994. *Intellectual Property Rights for Indigenous Peoples: A Sourcebook.* Oklahoma City: Society for Applied Anthropology. 274p. [1/1/1] *A.T.:* developing countries.

1923. Greenberg, Cathryn H., S.H. Crownover, & D.R. Gordon, 1997. Roadside soils: a corridor for invasion of xeric scrub by nonindigenous plants. *Natural Areas J.* 17: 99-109. *A.T.:* Florida.

1924. Greenberg, Russell, Peter Bichier, et al., 1997. Bird populations in shade and sun coffee plantations in central Guatemala. *CB* 11: 448-459.

1925. Greene, Harry W., 1994. Systematics and natural history, foundations for understanding and conserving biodiversity. *American Zoologist* 34: 48-56. [1].

1926. Greene, Patricia, & Dean Apostol, April 1995. Design for biodiversity. *Landscape Architecture* 85(4): 62-65. [1] *A.T.:* Pacific Northwest; forest design; landscape architecture.

1927. Greenfield, Joyce, & Nicole Richer, 1997. *Biodiversity Initiatives: Canadian Agricultural Producers.* Ottawa: Environment Bureau, Agriculture and Agri-Food Canada. 66p. [1/1/-] *A.T.:* environmental policy.

1928. Greenslade, P.J.M., 1997. Are Collembola useful as indicators of the conservation value of native grasslands? *Pedobiologia* 41: 215-220. *A.T.:* Australia.

1929. Greenstreet, Simon P.R., & S.J. Hall, 1996. Fishing and the groundfish assemblage structure in the northwestern North Sea: an analysis of long-term and spatial trends. *J. of Animal Ecology* 65: 577-598. [2] *A.T.:* community structure.

1930. Greenway, James C., Jr., 1967. *Extinct and Vanishing Birds of the World* (2nd ed.). New York: Dover. 520p. [3/-/1].

1931. Greenwood, P.H., 1992. Are the major fish faunas well-known? *Netherlands J. of Zoology* 42: 131-138. [1].

1932. Gregg, William P., Jr., S.L. Krugman, & J.D. Wood, Jr., eds., 1989. *Worldwide Conservation: Proceedings of the Symposium on Biosphere Reserves.* Atlanta: National Park Service, U.S. Dept. of the Interior. 291p. [2/1/1].

1933. Gregory, R.D., 1994. Species abundance patterns of British birds. *PRSL B* 257: 299-301. [1] *A.T.:* frequency distributions.

1934. ———, 1998. Biodiversity and body size: patterns among British birds. *Ecography* 21: 87-91. *A.T.:* species diversity.

1935. Gregory, R.D., A.E. Keymer, & P.H. Harvey, 1996. Helminth parasite richness among vertebrates. *Biodiversity and Conservation* 5: 985-997. [1] *A.T.:* literature reviews; sampling effort.

1936. Grehan, J.R., 1991. Panbiogeography 1981-91: development of an earth/life synthesis. *Progress in Physical Geography* 15: 331-363. [1] *A.T.:* generalized tracks; biogeography.

1937. Greig, J.C., 1979. Principles of genetic conservation in relation to wildlife management in Southern Africa. *South African J. of Wildlife Research* 9: 57-78. [1].

1938. Gressitt, J. Linsley, ed., 1982. *Biogeography and Ecology of New Guinea.* The Hague: W. Junk. 2 vols. [1/1/1].

1939. Greuter, W., 1994. Extinctions in Mediterranean areas. *PTRSL B* 344: 41-46. [1] *A.T.:* vascular plants.

1940. Grey, J., J. Laybourn-Parry, et al., 1997. Temporal patterns of protozooplankton abundance and their food in Ellis Fjord, Princess Elizabeth Land, eastern Antarctica. *Estuarine, Coastal and Shelf Science* 45: 17-25. *A.T.:* ciliates; dinoflagellates; nanoflagellates.

1941. Gribbin, John R., 1990. *Hothouse Earth: The Greenhouse Effect and Gaia.* New York: Grove Weidenfeld. 272p. [3/2/1].

1942. Griffith, Brad, J.M. Scott, et al., 1989. Translocation as a species conservation tool: status and strategy. *Science* 245: 477-480. [2] *A.T.:* literature reviews.

1943. Griffiths, David, 1997. Local and regional species richness in North American lacustrine fish. *J. of Animal Ecology* 66: 49-56.

1944. Grifo, Francesca, & Joshua Rosenthal, eds., 1997. *Biodiversity and Human Health* [conference papers]. Washington, DC: Island Press. 379p. [1/-/-] *A.T.:* pharmacognosy.

1945. Grigg, Gordon C., P.T. Hale, & Daniel Lunney, eds., 1995. *Conservation Through the Sustainable Use of Wildlife* [conference papers]. Brisbane: Centre for Conservation Biology, University of Queensland. 362p. [1/1/1] *A.T.:* Australia; New Zealand; wildlife conservation.

1946. Grime, J.P., 1997. Biodiversity and ecosystem function: the debate deepens. *Science* 277: 1260-1261. [2].

1947. Grisolia, James S., 1991. The Human Genome Project and our sense of self. *Impact of Science on Society* 41: 45-48. [1]. *A.T.:* human genetic diversity.

1948. Gromov, B.V., A.V. Pinevich, & A.A. Vepritskii, 1993. Biodiversity of cyanobacteria. *Microbiology* (Russia) 62: 253-261. [1].

1949. Groombridge, Brian, ed., 1992. *Global Biodiversity: Status of the Earth's Living Resources: A Report.* London and New York: Chapman & Hall. 585p. [2/2/2] *A.T.:* sustainable development.

1950. Groombridge, Brian, & Martin Jenkins, eds., 1996. *Assessing*

Biodiversity Status and Sustainability. Cambridge, U.K.: World Conservation Press. 104p. [1/1/1].

1951. Groot, Rudolf S. de, 1992. *Functions of Nature: Evaluation of Nature in Environmental Planning, Management and Decision Making.* Groningen: Wolters-Noordhoff. 315p. [1/1/2] *A.T.:* environmental policy.

1952. Groot, Rudolf S. de, Pieter Ketner, & A.H. Ovaa, 1995. Selection and use of bio-indicators to assess the possible effects of climate change in Europe. *J. of Biogeography* 22: 935-943. [1] *A.T.:* Netherlands.

1953. Grootjans, A.P., W.H.O. Ernst, & P.J. Stuyfzand, 1998. European dune slacks: strong interactions of biology, pedogenesis, and hydrology. *TREE* 13: 96, 100. *A.T.:* wetlands; coastal ecosystems.

1954. Grosholz, Edwin D., 1996. Contrasting rates of spread for introduced species in terrestrial and marine systems. *Ecology* 77: 1680-1686. [1] *A.T.:* biological invasions; range expansion.

1955. Grossman, Dennis H., K.L. Goodin, & C.L. Reuss, eds., 1994. *Rare Plant Communities of the Conterminous United States: An Initial Survey.* Arlington, VA: Nature Conservancy. 620p. [1/1/1].

1956. Grove, Noel, Nov.-Dec. 1992. The species you save may be your own. *American Forests* 98(11-12): 26-29. [1] *A.T.:* human impact; grizzly bears.

1957. ———, 1992. *Preserving Eden: The Nature Conservancy.* New York: Harry N. Abrams. 176p. [2/2/1].

1958. Groves, Craig R., M.L. Klein, & T.F. Breden, 1995. Natural Heritage Programs: public-private partnerships for biodiversity conservation. *Wildlife Society Bull.* 23: 784-790. [1] *A.T.:* The Nature Conservancy.

1959. Groves, R.H., & J.J. Burdon, eds., 1986. *Ecology of Biological Invasions* [symposium papers]. Cambridge, U.K., and New York: CUP. 166p. [2/1/2].

1960. Groves, R.H., & Franceso Di Castri, eds., 1991. *Biogeography of Mediterranean Invasions.* Cambridge, U.K., and New York: CUP. 485p. [1/2/1] *A.T.:* biological invasions.

1961. Grubb, Michael J., Matthias Koch, et al., 1993. *The "Earth Summit" Agreements: A Guide and Assessment.* London: Earthscan. 180p. [1/1/2] *A.T.:* UNCED; CBD.

1962. Grubb, P.J., 1977. The maintenance of species-richness in plant communities: the importance of the regeneration niche. *Biological Reviews* 52: 107-145. [3].

1963. Grueso, Libia, June 1998. Territory, culture and biodiversity: the Black communities of the Pacific Coast of Colombia qualify globalization. *Development* (Rome) 41(2): 65-69.

1964. Grumbine, R. Edward, 1 Dec. 1988. Ecosystem management for native diversity: how to save our national parks and forests. *Forest Watch* 9(6): 21-27. [1].

1965. ———, 1990. Protecting biological diversity through the greater ecosystem concept. *Natural Areas J.* 10: 114-120. [1].

1966. ———, 1990. Viable populations, reserve size, and federal lands management: a critique. *CB* 4: 127-134. [1] *A.T.:* environmental policy; ecosystem management.

1967. ———, 1991. Cooperation or conflict? Interagency relationships and the future of biodiversity for US parks and forests. *Environmental Management* 15: 27-37. [1] *A.T.:* National Park Service; Forest Service.

1968. ———, 1992. *Ghost Bears: Exploring the Biodiversity Crisis.* Washington, DC: Island Press. 294p. [1/1/2]. *A.T.:* environmental ethics.

1969. ———, ed., 1994. *Environmental Policy and Biodiversity.* Washington, DC: Island Press. 416p. [1/1/1].

1970. ———, 1994. What is ecosystem management? *CB* 8: 27-38. [3] A general review featuring a history of the subject.

1971. ———, 1994. Wildness, wise use, and sustainable development. *Environmental Ethics* 16: 227-249. [1] Concerning ideas of wilderness. *A.T.:* environmental perception.

1972. ———, 1995. Using biodiversity as a justification for nature protection in the U.S. *Humboldt J. of Social Relations* 21(1): 35-59. [1] *A.T.:* environmentalism; environmental movement.

1973. ———, 1997. Reflections on "What is ecosystem management?" *CB* 11: 41-47.

1974. Guarino, Luigi, V. Ramanatha Rao, & Robert Reid, eds., 1995. *Collecting Plant Genetic Diversity: Technical Guidelines.* Wallingford, Oxon, U.K.: CAB International. 748p. [1/1/1] *A.T.:* plant germplasm resources.

1975. Guby, N.A.B., & M. Dobbertin, 1996. Quantitative estimates of coarse woody debris and standing dead trees in selected Swiss forests. *Global Ecology and Biogeography Letters* 5: 327-341. [1].

1976. Guégan, J.-F., A. Lambert, et al., 1992. Can host body size explain the parasite species richness in tropical freshwater fishes? *Oecologia* 90: 197-204. [2] *A.T.:* West Africa; host-parasite relations.

1977. Guertin, D.S., W.E. Easterling, & J.R. Brandle, 1997. Climate change and forests in the Great Plains: issues in modeling fragmented woodlands in intensively managed landscapes. *Bioscience* 47: 287-295. *A.T.:* carbon dioxide.

1978. Gujral, G.S., & Virinder Sharma, eds., 1996. *Changing Perspectives of Biodiversity Status in the Himalaya.* New Delhi: British Council Division, British High Commission. 186p. [1/1/1] *A.T.:* bdc.

1979. Gunn, Alastair S., 1980. Why should we care about rare species? *Environmental Ethics* 2: 17-38. [1] *A.T.:* rarity; environmental ethics.

1980. Gunningham, Neil, & M.D. Young, 1997. Toward optimal environmental policy: the case of biodiversity conservation. *Ecology Law Quarterly* 24: 243-298. *A.T.:* Australia.

1981. Gupta, Anil K., 1991. *Why Does Poverty Persist in Regions of High Biodiversity? A Case for Indigenous Property Right System.* Ahmedabad, India: Indian Institute of Management. 21p. [1/1/1] *A.T.:* developing countries; indigenous peoples.

1982. ———, 1994. Environmental policy analysis for maintaining diversity. *Indian J. of Social Science* 7: 1-31. [1] *A.T.:* India; natural resource management.

1983. Guruswamy, Lakshman D., & J.A. McNeely, eds., 1998. *Protection of Global Biodiversity: Converging Strategies* [conference papers]. Durham, NC: Duke University Press. 425p.

1984. Gustafson, J. Perry, R. Appels, & P.H. Raven, eds., 1993. *Gene Conservation and Exploitation* [symposium proceedings]. New York: Plenum Press. 224p. [1/1/1] *A.T.:* plant germplasm resources; genetic engineering.

1985. Gutierrez, D., 1997. Importance of historical factors on species richness and composition of butterfly assemblages (Lepidoptera, Rhopalocera) in a northern Iberian mountain range. *J. of Biogeography* 24: 77-88. [1] *A.T.:* Spain.

1986. Guzmán, Gastón, 1998. Inventorying the fungi of Mexico. *Biodiversity and Conservation* 7: 369-384. *A.T.:* deforestation.

1987. Gyasi, Edwin A., & J.I. Uitto, eds., 1997. *Environment, Biodiversity and Agricultural Change in West Africa: Perspectives From Ghana* [workshop proceedings]. Tokyo and New York: United Nations University Press. 141p. [1/1/-].

1988. Haas, Peter M., R.O. Keohane, & M.A. Levy, eds., 1993. *Institutions for the Earth: Sources of Effective International Environmental Protection.* Cambridge, MA: MIT Press. 448p. [2/2/2] *A.T.:* international organizations; international cooperation; environmental policy.

1989. Haas, Peter M., M.A. Levy, & E.A. Parson, Oct. 1992. Appraising the Earth Summit: how should we judge UNCED's success? *Environment* 34(8): 6-11, 26-33. [1].

1990. Hacker, S.D., & S.D. Gaines, 1997. Some implications of direct positive interactions for community species diversity. *Ecology* 78: 1990-2003. *A.T.:* symbiosis; commensalism; mutualism.

1991. Hackney, Courtney T., S.M. Adams, & W.H. Martin, eds., 1992. *Biodiversity of the Southeastern United States: Aquatic Communities.* New York: Wiley. 779p. [2/1/1].

1992. Hackstein, J.H.P., 1997. Eukaryotic molecular biodiversity: systematic approaches for the assessment of symbiotic associations. *Antonie van Leeuwenhoek* 72: 63-76. *A.T.:* parasites.

1993. Hadfield, Michael G., S.E. Miller, & A.H. Carwile, 1993. The decimation of endemic Hawai'ian tree snails by alien predators. *American Zoologist* 33: 610-622. [1].

1994. Hadley, M., 1994. Diversity and the management of tropical forests. *Ecodecision*, no. 13: 33-38. [1].

1995. Haeder, Paul K., Sept.-Oct. 1995. Vietnam's new war. *E: The Environmental Magazine* 6(5): 18-19. [1] A description of Vietnam's efforts at bdc.

1996. Haffer, Jürgen, 1969. Speciation in Amazonian forest birds. *Science* 165: 131-137. [2] *A.T.:* refuge theory.

1997. ———, 1997. Alternative models of vertebrate speciation in Amazonia: an overview. *Biodiversity and Conservation* 6: 451-476. *A.T.:* refuge theory; barriers; vicariance.

1998. Hagan, John M., III, & D.W. Johnston, eds., 1992. *Ecology and Conservation of Neotropical Migrant Landbirds* [symposium papers]. Washington, DC: Smithsonian Institution Press. 609p. [2/2/2] *A.T.:* Latin America.

1999. Hagan, John M., III, W.M. Vanderhaegen, & P.S. McKinley, 1996. The early development of forest fragmentation effects on birds. *CB* 10: 188-202. [2] *A.T.:* edge effects.

2000. Hagvar, S., 1994. Preserving the natural heritage: the process of developing attitudes. *Ambio* 23: 515-518. [1] *A.T.:* environmental perception; bd value.

2001. Haig, Susan M., J.D. Ballou, & S.R. Derrickson, 1990. Management options for preserving genetic diversity: reintroduction of Guam rails to the wild. *CB* 4: 290-300. [1].

2002. Haig, Susan M., D.W. Mehlman, & L.W. Oring, 1998. Avian movements and wetland connectivity in landscape conservation. *CB* 12: 749-758.

2003. Haigh, Nigel, 1991. The European Community and international environmental policy. *International Environmental Affairs* 3: 163-180. [1] *A.T.:* international cooperation; global warming; bd.

2004. Haila, Yrjö, 1994. Preserving ecological diversity in boreal forests: ecological background, research, and management. *Annales Zoologici Fennici* 31: 203-217. [1] *A.T.:* taiga; forest management.

2005. ———, 1996. Quantitative surveys in biodiversity research. *Ecography* 19: 321-322. [1].

2006. Haila, Yrjö, & J. Kouki, 1994. The phenomenon of biodiversity in conservation biology. *Annales Zoologici Fennici* 31: 5-18. [1] Considers the complexity of the notion of bd, and the unlikelihood there could be any single measure of it.

2007. Hails, A.J., ed., 1996. *Wetlands, Biodiversity and the Ramsar Convention: The Role of the Convention on Wetlands in the Conservation and Wise Use of Biodiversity*. Gland, Switzerland: Ramsar Convention Bureau. 196p. [1/1/1] *A.T.:* wetlands conservation.

2008. Hails, Christopher, & P.S. Sochaczewski, 1989. *The Importance of Biological Diversity*. Gland, Switzerland: WWF. 32p. [1/1/1] *A.T.:* WWF.

2009. Haines-Young, Roy H., D.R. Green, & Steven Cousins, eds., 1993. *Landscape Ecology and Geographic Information Systems.* London and New York: Taylor & Francis. 288p. [1/1/1] *A.T.:* remote sensing.

2010. Hairston, Nelson G., F.E. Smith, & L.B. Slobodkin, 1960. Community structure, population control, and competition. *American Naturalist* 94: 421-425. [3].

2011. Hajra, P.K., & V. Mudgal, eds., 1997. *Plant Diversity Hotspots in India: An Overview.* Calcutta: Botanical Survey of India. 179p. [1/1/-].

2012. Halffter, Gonzalo, 1987. Biogeography of the montane entomofauna of Mexico and Central America. *Annual Review of Entomology* 32: 95-114. [1].

2013. Hall, Geoffrey S., ed., 1996. *Methods for the Examination of Organismal Diversity in Soils and Sediments.* Wallingford, Oxon, U.K., and New York: CAB International. 307p. [1/1/1] *A.T.:* soil microbiology.

2014. Hall, J.A., & C.L.J. Frid, 1997. Estuarine sediment remediation: effects on benthic biodiversity. *Estuarine, Coastal and Shelf Science* 44, Suppl. A: 55-61. [1] *A.T.:* England; environmental monitoring.

2015. Hall, Pamela, & K.S. Bawa, 1993. Methods to assess the impact of extraction of non-timber tropical forest products on plant populations. *Economic Botany* 47: 234-247. [1].

2016. Hall, Stephen J.G., & D.G. Bradley, 1995. Conserving livestock breed biodiversity. *TREE* 10: 267-270. [1] *A.T.:* inventories.

2017. Hall, Stephen J.G., & J. Ruane, 1993. Livestock breeds and their conservation: a global overview. *CB* 7: 815-825. [1].

2018. Hall, S.J., & S.P. Greenstreet, 1998. Taxonomic distinctiveness and diversity measures: responses to marine fish communities. *Marine Ecology Progress Series* 166: 227-229.

2019. Halladay, Patricia, & D.A. Gilmour, eds., 1995. *Conserving Biodiversity Outside Protected Areas: The Role of Traditional Agro-ecosystems* [workshop papers]. Gland, Switzerland: IUCN. 228p. [1/1/1].

2020. Hallam, Anthony, 1987. End-Cretaceous mass extinction event: argument for terrestrial causation. *Science* 238: 1237-1242. [1] *A.T.:* stepwise extinction; asteroid impacts.

2021. ———, 1989. The case for sea-level change as a dominant causal factor in mass extinction of marine invertebrates. *PTRSL B* 325: 437-455. [1].

2022. ———, 1991. Why was there a delayed radiation after the end-Paleozoic extinctions? *Historical Biology* 5: 257-262. [1] *A.T.:* mass extinctions; sea-level change; invertebrates.

2023. ———, 1994. *An Outline of Phanerozoic Biogeography.* Oxford, U.K., and New York: OUP. 246p. [1/1/2] *A.T.:* paleobiogeography.

2024. Hallam, Anthony, & P.B. Wignall, 1997. *Mass Extinctions and Their Aftermath.* Oxford, U.K., and New York: OUP. 320p. [1/1/-] *A.T.:*

geologic catastrophes.

2025. Hallegraeff, G.M., 1993. A review of harmful algal blooms and their apparent global increase. *Phycologia* 32: 79-99. [3].

2026. Halley, John M., R.S. Oldham, & J.W. Arntzen, 1996. Predicting the persistence of amphibian populations with the help of a spatial model. *J. of Applied Ecology* 33: 455-470. [1] *A.T.:* population biology; ponds.

2027. Halliday, Tim, 1978. *Vanishing Birds: Their Natural History and Conservation.* New York: Holt, Rinehart & Winston. 296p. [2/2/1] *A.T.:* endangered and threatened birds.

2028. Hallock, Pamela, 1987. Fluctuations in the trophic resource continuum: a factor in global diversity cycles? *Paleoceanography* 2: 457-471. [2] *A.T.:* marine diversity.

2029. Hallock, Pamela, F.E. Müller-Karger, & J.C. Halas, 1993. Anthropogenic nutrients and the degradation of Western Atlantic and Caribbean coral reefs. *Research & Exploration* 9: 358-378. [1] *A.T.:* sedimentation; fishing.

2030. Halloy, S.R.P., 1995. Status of New Zealand biodiversity research and resources: how much do we know? *J. of the Royal Society of New Zealand* 25: 55-80. [1] *A.T.:* genetic resources.

2031. Halpern, Charles B., & T.A. Spies, 1995. Plant species diversity in natural and managed forests of the Pacific Northwest. *Ecological Applications* 5: 913-934. [2] *A.T.:* succession; disturbance.

2032. Hambler, Clive, & M.R. Speight, 1996. Extinction rates in British nonmarine invertebrates since 1900. *CB* 10: 892-896. [1].

2033. Hamer, K.C., J.K. Hill, et al., 1997. Ecological and biogeographical effects of forest disturbance on tropical butterflies of Sumba, Indonesia. *J. of Biogeography* 24: 67-75. *A.T.:* habitat disturbance.

2034. Hamilton, Lawrence S., & H.F. Takeuchi, eds., 1993. *Ethics, Religion and Biodiversity: Relations Between Conservation and Cultural Values* [symposium papers]. Cambridge, U.K.: White Horse. 218p. [1/1/1] *A.T.:* environmental ethics.

2035. Hamilton, L.S., 1995. Mountain cloud forest conservation and research: a synopsis. *Mountain Research and Development* 15: 259-266. [1] *A.T.:* water supply; montane environments.

2036. Hamilton, M.B., 1994. *Ex situ* conservation of wild plant species: time to reassess the genetic assumptions and implications of seed banks. *CB* 8: 39-49. [1].

2037. Hamilton, M.P., & M. Flaxman, 1992. Scientific data visualization and biological diversity: new tools for spatializing multimedia observations of species and ecosystems. *Landscape and Urban Planning* 21: 285-288. [1] *A.T.:* computer programs; bd visualization.

2038. Hamilton, T.H., R.H. Barth, Jr., & I. Rubinoff, 1964. The environmental control of insular variation in bird species abundance. *PNAS* 52:

132-140. [1] *A.T.:* environmental factors.

2039. Hamilton, T.H., I. Rubinoff, et al., 1963. Species abundance: natural regulation of insular variation. *Science* 142: 1575-1577. [1] *A.T.:* Galápagos Islands; land plants.

2040. Hamilton, W.J., III, & K.E.F. Watt, 1970. Refuging. *ARES* 1: 263-286. [1].

2041. Hammer, M., A.M. Jansson, & B.-O. Jansson, 1993. Diversity change and sustainability: implications for fisheries. *Ambio* 22: 97-105. [1] *A.T.:* Baltic Sea; Sweden; fisheries management.

2042. Hammond, Paul C., 1995. Conservation of biodiversity in native prairie communities in the United States. *J. of the Kansas Entomological Society* 68: 1-6. [1].

2043. Hammond, Paul C., & J.C. Miller, 1998. Comparison of the biodiversity of Lepidoptera within three forested ecosystems. *Annals of the Entomological Society of America* 91: 323-328. *A.T.:* indicator taxa.

2044. Hammond, P.M., 1994. Practical approaches to the estimation of the extent of biodiversity in speciose groups. *PTRSL B* 345: 119-136. [1].

2045. Hampson, A.M., & G.F. Peterken, 1998. Enhancing the biodiversity of Scotland's forest resource through the development of a network of forest habitats. *Biodiversity and Conservation* 7: 179-192. *A.T.:* Scottish Natural Heritage; habitat fragmentation.

2046. Hamre, J., 1994. Biodiversity and exploitation of the main fish stocks in the Norwegian-Barents Sea ecosystem. *Biodiversity and Conservation* 3: 473-492. [1] *A.T.:* cod; herring.

2047. Hamrick, J.L., M.J.W. Godt, & S.L. Sherman-Broyles, 1992. Factors influencing levels of genetic diversity in woody plant species. *New Forests* 6: 95-124. [3].

2048. Hancock, C.N., P.G. Ladd, & R.H. Froend, 1996. Biodiversity and management of riparian vegetation in Western Australia. *Forest Ecology and Management* 85: 239-250. [1] *A.T.:* disturbance.

2049. Handel, Steve N., G.R. Robinson, & A.J. Beattie, 1994. Biodiversity resources for restoration ecology. *Restoration Ecology* 2: 230-241. [1] *A.T.:* community structure.

2050. Hanemann, W. Michael, fall 1994. Valuing the environment through contingent valuation. *J. of Economic Perspectives* 8(4): 19-43. [3].

2051. Haney, J. Christopher, J.M. Wunderle, Jr., & W.J. Arendt, 1991. Some initial effects of Hurricane Hugo on endangered and endemic species of West Indian birds. *American Birds* 45: 234-236. [1].

2052. Hanley, Nick, & Clive Spash, 1993. *Preferences, Information and Biodiversity Preservation.* Stirling, U.K.: Discussion Paper in Economics 93-12, Dept. of Economics, University of Stirling. 39p. [1/1/1] *A.T.:* environmental economics; bd valuation.

2053. Hanley, Nick, Clive Spash, & Lorna Walker, 1995. Problems in valuing the benefits of biodiversity protection. *Environmental & Resource Economics* 5: 249-272. [1] *A.T.:* bd valuation; contingent valuation.

2054. Hannah, Lee, Berthe Rakotosamimanana, et al., 1998. Participatory planning, scientific priorities, and landscape conservation in Madagascar. *Environmental Conservation* 25: 30-36. *A.T.:* action plans; landscape approach.

2055. Hannah, Lee, D. Lohse, et al., 1994. A preliminary inventory of human disturbance of world ecosystems. *Ambio* 23: 246-250. [1] *A.T.:* mapping techniques.

2056. Hansen, Andrew J., & Francesco Di Castri, eds., 1992. *Landscape Boundaries: Consequences for Biotic Diversity and Ecological Flows*. New York: Springer. 452p. [1/1/2] *A.T.:* landscape ecology; ecotones.

2057. Hansen, Andrew J., T.A. Spies, et al., 1991. Conserving biodiversity in managed forests: lessons from natural forests. *Bioscience* 41: 382-392. [2] Includes suggested guidelines for multiple-use-oriented management. *A.T.:* Pacific Northwest; disturbance.

2058. Hansen, Stein, 1989. Debt for nature swaps—overview and discussion of key issues. *Ecological Economics* 1: 77-93. [1].

2059. Hanski, Ilkka, 1982. Dynamics of regional distribution: the core and satellite species hypothesis. *Oikos* 38: 210-221. [3] *A.T.:* local extinction.

2060. ———, 1985. Single-species spatial dynamics may contribute to long-term rarity and commonness. *Ecology* 66: 335-343. [2] *A.T.:* immigration; small populations.

2061. ———, 1989. Metapopulation dynamics: does it help to have more of the same? *TREE* 4: 113-114. [2].

2062. ———, 1994. A practical model of metapopulation dynamics. *J. of Animal Ecology* 63: 151-162. [2] *A.T.:* patch dynamics; butterflies.

2063. ———, 1994. Patch-occupancy dynamics in fragmented landscapes. *TREE* 9: 131-135. [2] *A.T.:* habitat fragmentation; metapopulations.

2064. ———, 1994. Spatial scale, patchiness and population dynamics on land. *PTRSL B* 343: 19-25. [2] *A.T.:* metapopulation dynamics.

2065. Hanski, Ilkka, & M.E. Gilpin, 1991. Metapopulation dynamics: brief history and conceptual domain. *Biological J. of the Linnean Society* 42: 3-16. [3].

2066. ———, eds., 1996. *Metapopulation Biology: Ecology, Genetics, and Evolution*. San Diego: Academic Press. 512p. [1/1/2] *A.T.:* population biology.

2067. Hanski, Ilkka, & Mats Gyllenberg, 1993. Two general metapopulation models and the core-satellite species hypothesis. *American Naturalist* 142: 17-41. [2] *A.T.:* patch dynamics.

2068. ———, 1997. Uniting two general patterns in the distribution of species. *Science* 275: 397-400. The two patterns discussed are the species-area curve and the range size-local abundance relation.

2069. Hanski, Ilkka, & P. Hammond, 1995. Biodiversity in boreal forests [editorial]. *TREE* 10: 5-6. [1].

2070. Hanski, Ilkka, & C.D. Thomas, 1994. Metapopulation dynamics and conservation: a spatially explicit model applied to butterflies. *Biological Conservation* 68: 167-180. [2] *A.T.:* Finland; U.K.

2071. Hanski, Ilkka, Atte Moilanen, & Mats Gyllenberg, 1996. Minimum viable metapopulation size. *American Naturalist* 147: 527-541. [2].

2072. Hanski, Ilkka, T. Pakkala, et al., 1995. Metapopulation persistence of an endangered butterfly in a fragmented landscape. *Oikos* 72: 21-28. [2] *A.T.:* Finland.

2073. Hansson, Lennart, 1992. Landscape ecology of boreal forests. *TREE* 7: 299-302. [1] *A.T.:* mosaics; habitat fragmentation.

2074. ———, ed., 1992. *The Ecological Principles of Nature Conservation: Applications in Temperate and Boreal Environments.* London and New York: Elsevier Applied Science. 436p. [1/1/1].

2075. ———, ed., 1997. *Boreal Ecosystems and Landscapes: Structures, Processes and Conservation of Biodiversity.* Copenhagen and Malden, MA: Munksgaard. 203p. [1/1/-] *A.T.:* arctic environments.

2076. ———, 1997. Environmental determinants of plant and bird diversity in ancient oak-hazel woodland in Sweden. *Forest Ecology and Management* 91: 137-143. *A.T.:* species richness.

2077. Hansson, Lennart, & P. Angelstam, 1991. Landscape ecology as a theoretical basis for nature conservation. Landscape *Ecology* 5: 191-201. [1] *A.T.:* habitat fragmentation; Sweden.

2078. Hansson, Lennart, Lenore Fahrig, & Gray Merriam, eds., 1995. *Mosaic Landscapes and Ecological Processes.* London: Chapman & Hall. 356p. [1/1/2] *A.T.:* landscape ecology.

2079. Harch, B.D., R.L. Correll, et al., 1997. Using the Gini coefficient with BIOLOG substrate utilisation data to provide an alternative quantitative measure for comparing bacterial soil communities. *J. of Microbiological Methods* 30: 97-107. *A.T.:* soil microorganisms; bd measures.

2080. Harcourt, A.H., 1996. Is the gorilla a threatened species? How should we judge? *Biological Conservation* 75: 165-176. [1] An essay on the shortcomings of the IUCN criteria for assessing conservation status.

2081. Harcourt, Caroline S., J.A. Sayer, & Clare Billington, eds., 1995. *The Conservation Atlas of Tropical Forests: The Americas.* New York: Simon & Schuster. 335p. [2/1/1].

2082. Harcourt, Wendy, ed., 1994. *Feminist Perspectives on Sustainable Development.* London and Atlantic Highlands, NJ: Zed Books. 255p. [1/2/1].

2083. Hardin, Garrett J., 1968. The tragedy of the commons. *Science* 162: 1243-1248. [3] One of the most celebrated and influential essays on the human population explosion problem, arguing that it "has no technical solution; it requires a fundamental extension in morality."

2084. Hardin, Garrett J., & John Baden, eds., 1977. *Managing the Commons*. San Francisco: W.H. Freeman. 294p. [2/1/2] *A.T.:* environmental policy.

2085. Harding, Lee E., & Emily McCullum, eds., 1994. *Biodiversity in British Columbia: Our Changing Environment*. Ottawa: Canadian Wildlife Service. 426p. [1/1/1].

2086. Harding, S.P., & J.E. Lovelock, 1996. Exploiter-mediated coexistence and frequency-dependent selection in a numerical model of biodiversity. *J. of Theoretical Biology* 182: 109-116. [1] *A.T.:* mathematical models.

2087. Hardon, J.J., 1996. Conservation and use of agro-biodiversity. *Biodiversity Letters* 3: 92-96. [1] *A.T.:* in situ conservation; ex situ conservation.

2088. Hardy, Cheryl D., 1994. Patent protection and raw materials: the Convention on Biological Diversity and its implications for United States policy on the development and commercialization of biotechnology. *University of Pennsylvania J. of International Business Law* 15: 299-326. [1].

2089. Harlan, Jack R., 1975. Our vanishing genetic resources: modern varieties replace ancient populations that have provided genetic variability for plant breeding programs. *Science* 188: 618-621. [1].

2090. ———, 1976. Genetic resources in wild relatives of crops. *Crop Science* 16: 329-333. [1] *A.T.:* germplasm resources; disease resistance.

2091. Harmelin, J.G., 1997. Diversity of bryozoans in a Mediterranean sublittoral cave with bathyal-like conditions: role of dispersal processes and local factors. *Marine Ecology Progress Series* 153: 139-152. *A.T.:* French Mediterranean.

2092. Harmelin, J.G., & J. Vacelet, 1997. Clues to deep-sea biodiversity in a nearshore cave. *Vie et Milieu* 47: 351-354. *A.T.:* sponges; bryozoans; French Mediterranean.

2093. Harner, Richard F., & K.T. Harper, 1976. The role of area, heterogeneity, and favorability in plant species diversity of pinyon-juniper ecosystems. *Ecology* 57: 1254-1263. [1] *A.T.:* New Mexico; Utah; species-area curves.

2094. Harper, J.L., & D.L. Hawksworth, 1994. Biodiversity: measurement and estimation. *PTRSL B* 345: 5-12. [2].

2095. Harrington, Richard, & N.E. Stork, eds., 1995. *Insects in a Changing Environment* [symposium proceedings]. London and San Diego: Academic Press. 535p. [1/1/1].

2096. Harris, H.J., M.S. Milligan, & G.A. Fewless, 1983. Diversity: quantification and ecological evaluation in freshwater marshes. *Biological Conservation* 27: 99-110. [1] *A.T.:* waterbirds.

2097. Harris, Kim S., ed., 1991. *Proceedings of the Native Ecosystems and Rare Species Workshops*. Honolulu: Nature Conservancy of Hawaii. 182p. [1/1/1] *A.T.:* Hawaii; bdc.

2098. Harris, Larry D., 1984. *The Fragmented Forest: Island Biogeogra-*

phy Theory and the Preservation of Biotic Diversity. Chicago: University of Chicago Press. 211p. [2/2/3] *A.T.:* habitat fragmentation.

2099. Harrison, Robert L., 1992. Toward a theory of inter-refuge corridor design. *CB* 6: 293-295. [2].

2100. Harrison, Susan, 1991. Local extinction in a metapopulation context: an empirical evaluation. *Biological J. of the Linnean Society* 42: 73-88. [3] *A.T.:* patch dynamics.

2101. Harrison, Susan, S.J. Ross, & J.H. Lawton, 1992. Beta diversity on geographic gradients in Britain. *J. of Animal Ecology* 61: 151-158. [1] *A.T.:* species turnover; alpha diversity.

2102. Harrop, S., 1995. The GATT 1994, the Biological Diversity Convention and their relationship with macro-biodiversity management. *Biodiversity and Conservation* 4: 1019-1025. [1] *A.T.:* intellectual property; environmental economics.

2103. Hart, C.W., Jr., R.B. Manning, & T.M. Iliffe, 1985. The fauna of Atlantic marine caves: evidence of dispersal by sea floor spreading while maintaining ties to deep waters. *Proc. of the Biological Society of Washington* 98: 288-292. [1].

2104. Hart, M.B., ed., 1996. *Biotic Recovery From Mass Extinction Events* [conference papers]. London: Geological Society. 392p. [1/1/1].

2105. Hart, Terese B., J.A. Hart, & P.G. Murphy, 1989. Monodominant and species-rich forests of the humid tropics: causes for their co-occurrence. *American Naturalist* 133: 613-633. [2].

2106. Harte, John, 1996. Feedbacks, thresholds and synergies in global change: population as a dynamic factor. *Biodiversity and Conservation* 5: 1069-1083. [1] *A.T.:* environmental degradation; global warming.

2107. Hartl, Günter B., & Zdzislaw Pucek, 1994. Genetic depletion in the European bison (*Bison bonasus*) and the significance of electrophoretic heterozygosity for conservation. *CB* 8: 167-174. [1] *A.T.:* Poland.

2108. Hartman, Hyman, & Koichiro Matsuno, eds., 1992. *The Origin and Evolution of the Cell* [conference proceedings]. Singapore and River Edge, NJ: World Scientific Publishing Company. 434p. [1/1/1].

2109. Hartnett, David C., ed., 1995. *Proceedings of the Fourteenth North American Prairie Conference: Prairie Biodiversity*. Manhattan, KS: Kansas State University. 257p. [1/1/1].

2110. Hartvigsen, R., & O. Halvorsen, 1994. Spatial patterns in the abundance and distribution of parasites of freshwater fish. *Parasitology Today* 10: 28-31. [1].

2111. Harvey, Bryan L., & Brad Fraleigh, 1995. Impacts on Canadian agriculture of the Convention on Biological Diversity. *Canadian J. of Plant Science* 75: 17-21. [1] *A.T.:* crop germplasm resources.

2112. Harvey, C.A., & David Pimentel, 1996. Effects of soil and wood depletion on biodiversity. *Biodiversity and Conservation* 5: 1121-1130. [1]

A.T.: soil erosion; bd loss.

2113. Hassell, M.P., H.N. Comins, & R.M. May, 1994. Species coexistence and self-organizing spatial dynamics. *Nature* 370: 290-292. [2] *A.T.:* patch dynamics; metapopulations.

2114. Hastings, Alan, & Susan Harrison, 1994. Metapopulation dynamics and genetics. *ARES* 25: 167-188. [3] *A.T.:* extinction.

2115. Hattori, Tsutomu, Hisayuki Mitsui, et al., 1997. Advances in soil microbial ecology and the biodiversity. *Antonie van Leeuwenhoek* 72: 21-28. *A.T.:* bacteria.

2116. Haufler, J.B., 1994. An ecological framework for planning for forest health. *J. of Sustainable Forestry* 2: 307-316. [1] A.T; forest management; western U.S.

2117. Haufler, J.B., C.A. Mehl, & G.J. Roloff, 1996. Using a coarse-filter approach with species assessment for ecosystem management. *Wildlife Society Bull.* 24: 200-208. [1] *A.T.:* forest ecosystems; Idaho.

2118. Haukos, D.A., & L.M. Smith, 1994. The importance of playa wetlands to biodiversity of the southern High Plains. *Landscape and Urban Planning* 28: 83-98. [1] *A.T.:* Texas; New Mexico.

2119. Hauser, Gertrude, M.A. Little, & D.F. Roberts, eds., 1994. *Man, Culture and Biodiversity: Understanding Interdependencies* [workshop report]. Paris: Biology International Special Issue No. 32, IUBS. 91p. [1/1/1].

2120. Hawkes, John G., 1983. *The Diversity of Crop Plants* [lectures]. Cambridge, MA: Harvard University Press. 184p. [2/1/1] *A.T.:* crop germplasm resources; field crops.

2121. ———, 1990. *The Potato: Evolution, Biodiversity and Genetic Resources.* Washington, DC: Smithsonian Institution Press. 259p. [2/1/3].

2122. ———, 1991. International workshop on dynamic *in situ* conservation of wild relatives of major cultivated plants: summary of final discussion and recommendations. *Israel J. of Botany* 40: 529-536. [1].

2123. Hawkes, M.W., 1992. Seaweed biodiversity and marine conservation in the Pacific Northwest. *Northwest Environmental J.* 8: 146-148. [1] *A.T.:* British Columbia.

2124. Hawkins, Bradford A., 1988. Species diversity in the third and fourth trophic levels: patterns and mechanisms. *J. of Animal Ecology* 57: 137-162. [1] *A.T.:* insects; host-parasite relations.

2125. ———, 1993. Parasitoid species richness, host mortality, and biological control. *American Naturalist* 141: 634-641. [1] *A.T.:* insects; refuge theory.

2126. ———, 1995. Latitudinal body-size gradients for the bees of the eastern United States. *Ecological Entomology* 20: 195-198. [1].

2127. Hawkins, Bradford A., & S.G. Compton, 1992. African fig wasp communities: undersaturation and latitudinal gradients in species richness. *J. of Animal Ecology* 61: 361-372. [2].

2128. Hawkins, Bradford A., & Paul Gross, 1992. Species richness and population limitation in insect parasitoid-host system. *American Naturalist* 139: 417-423. [1] *A.T.:* North America.

2129. Hawkins, Bradford A., & J.H. Lawton, 1995. Latitudinal gradients in butterfly body sizes: is there a general pattern? *Oecologia* 102: 31-36. [1].

2130. Hawkins, Bradford A., M.B. Thomas, & M.E. Hochberg, 1993. Refuge theory and biological control. *Science* 262: 1429-1432. [1] *A.T.:* parasites; insects.

2131. Hawksworth, David L., ed., 1991. *The Biodiversity of Microorganisms and Invertebrates: Its Role in Sustainable Agriculture* [workshop proceedings]. Wallingford, Oxon, U.K.: CAB International. 302p. [1/1/1].

2132. ———, 1991. The fungal dimension of biodiversity: magnitude, significance, and conservation [address]. *Mycological Research* 95: 641-655. [2].

2133. ———, ed., 1995. *Biodiversity: Measurement and Estimation*. London and New York: Chapman & Hall. 140p. [1/1/1].

2134. ———, 1997. Fungi and international biodiversity initiatives. *Biodiversity and Conservation* 6: 661-668. *A.T.:* bd definitions; environmental policy.

2135. Hawksworth, David L., & J.M. Ritchie, 1993. *Biodiversity and Biosystematic Priorities: Microorganisms and Invertebrates*. Wallingford, Oxon, U.K.: CAB International. 120p. [1/1/1].

2136. Hawksworth, David L., P.M. Kirk, & S.D. Clarke, eds., 1997. *Biodiversity Information: Needs and Options* [workshop proceedings]. Wallingford, Oxon, U.K.: CAB International. 194p. [1/1/-] *A.T.:* information resources.

2137. Hay, Mark E., 1986. Associational plant defenses and the maintenance of species diversity: turning competitors into accomplices. *American Naturalist* 128: 617-641. [2] *A.T.:* defense mechanisms; palatable plants; succession.

2138. Hayden, F. Gregory, 1991. Instrumental valuation indicators for natural resources and ecosystems. *J. of Economic Issues* 25: 917-935. [1] *A.T.:* value indicators; valuation.

2139. ———, 1993. Ecosystem valuation: combining economics, philosophy, and ecology. *J. of Economic Issues* 27: 409-420. [1] *A.T.:* holistic modelling.

2140. Hayes, John P., 1991. How mammals become endangered. *Wildlife Society Bull.* 19: 210-215. [1] Results of a survey investigating the causes of endangerment.

2141. Hayes, M.P., & M.R. Jennings, 1986. Decline of ranid frog species in western North America: are bullfrogs (*Rana catesbeiana*) responsible? *J. of Herpetology* 20: 490-509. [2].

2142. Hazevoet, C.J., 1996. Conservation and species list: taxonomic neglect promotes the extinction of endemic birds, as exemplified by taxa from eastern Atlantic islands. *Bird Conservation International* 6: 181-196. [1] *A.T.:* species concept; subspecies.

2143. He, Fangliang, Pierre Legendre, & J.V. LaFrankie, 1996. Spatial pattern of diversity in a tropical rain forest in Malaysia. *J. of Biogeography* 23: 57-74. [1] *A.T.:* trees; spatial structure.

2144. He, Fangliang, Pierre Legendre, et al., 1994. Diversity pattern and spatial scale: a study of a tropical rain forest of Malaysia. *Environmental and Ecological Statistics* 1: 265-286. [1].

2145. Head, Suzanne, & Robert Heinzman, eds., 1990. *Lessons of the Rainforest.* San Francisco: Sierra Club Books. 275p. [2/2/1] *A.T.:* rainforest management.

2146. Heads, Michael J., 1997. Regional patterns of biodiversity in New Zealand: one degree grid analysis of plant and animal distributions. *J. of the Royal Society of New Zealand* 27: 337-354. *A.T.:* mapping.

2147. Heal, Geoffrey M., 1994. *Markets and Biodiversity.* New York: PaineWebber Working Paper Series in Money, Economics and Finance PW-95-17, Columbia University. 10p. [1/1/1] *A.T.:* economic commodities.

2148. Heaney, L.R., 1986. Biogeography of mammals in SE Asia: estimates of rates of colonization, extinction and speciation. *Biological J. of the Linnean Society* 28: 127-165. [2] *A.T.:* Philippines; Sunda Shelf; endemism.

2149. Hebda, Richard J., 1998. Atmospheric change, forests and biodiversity. *Environmental Monitoring and Assessment* 49: 195-212. *A.T.:* British Columbia; climatic change.

2150. Heckman, Charles W., 1998. *The Pantanal of Poconé: Biota and Ecology in the Northern Section of the World's Largest Pristine Wetland.* Boston: Kluwer Academic. 622p. *A.T.:* Brazil.

2151. Hecnar, Stephen J., & R.T. M'Closkey, 1996. Amphibian species richness and distribution in relation to pond water chemistry in south-western Ontario, Canada. *Freshwater Biology* 36: 7-15. [1] *A.T.:* lake ecology.

2152. ———, 1997. The effects of predatory fish on amphibian species richness and distribution. *Biological Conservation* 79: 123-131. *A.T.:* lake ecology; Ontario; predator-prey relations.

2153. Hedges, S. Blair, 1993. Global amphibian declines: a perspective from the Caribbean. *Biodiversity and Conservation* 2: 290-303. [1] *A.T.:* West Indies.

2154. ———, 1996. Historical biogeography of West Indian vertebrates. *ARES* 27: 163-196. [1].

2155. Hedges, S. Blair, C.A. Haas, & L.R. Maxson, 1992. Caribbean biogeography: molecular evidence for dispersal in West Indian terrestrial vertebrates. *PNAS* 89: 1909-1913. [1] *A.T.:* dispersal.

2156. Hedrick, Philip W., & P.S. Miller, 1992. Conservation genetics: techniques and fundamentals. *Ecological Applications* 2: 30-46. [2] *A.T.:*

DNA fingerprinting; inbreeding; outbreeding depression; population bottlenecks; genetic fitness.

2157. Heikkinen, Risto K., 1998. Can richness patterns of rarities be predicted from mesoscale atlas data? A case study of vascular plants in the Kevo Reserve. *Biological Conservation* 83: 133-143. *A.T.:* Finland; hotspots; rare plants.

2158. Heikkinen, Risto K., & Seppo Neuvonen, 1997. Species richness of vascular plants in the subarctic landscape of northern Finland: modelling relationships to the environment. *Biodiversity and Conservation* 6: 1181-1201.

2159. Heinen, Joel T., & Bijaya Kattel, 1992. Parks, people, and conservation: a review of management issues in Nepal's protected areas. *Population and Environment* 14: 49-84. [1].

2160. Heinen, Joel T., & P.B. Yonzon, 1994. A review of conservation issues and programs in Nepal: from a single species focus toward biodiversity protection. *Mountain Research and Development* 14: 61-76. [1] *A.T.:* wildlife conservation; action plans.

2161. Heip, Carlo, & P. Engels, 1974. Comparing species diversity and evenness indices. *J. of the Marine Biological Association* 54: 555-563. [1] *A.T.:* copepods.

2162. Heissenbuttel, John, & W.R. Murray, Aug. 1992. A troubled law in need of revision. *J. of Forestry* 90(8): 13-16. [1] *A.T.:* Endangered Species Act.

2163. Hejl, Sallie Jo, 1992. The importance of landscape patterns to bird diversity: a perspective from the northern Rocky Mountains. *Northwest Environmental J.* 8: 119-138. [1] *A.T.:* habitat fragmentation.

2164. Helfman, Gene S., B.B. Collette, & D.E. Facey, 1997. *The Diversity of Fishes*. Malden, MA: Blackwell Science. 528p. [1/1/-].

2165. Hellriegel, Barbara, Urs Leugger, et al., 1993. Education and science for maintaining biodiversity: critical comments on the symposium of the Swiss UNESCO commission in September 1992 in Basle. *Biodiversity Letters* 1: 131-133. [1] *A.T.:* bd research.

2166. Heltshe, James F., & N.E. Forrester, 1983. Estimating species richness using the jackknife procedure. *Biometrics* 39: 1-11. [2].

2167. Hemley, Ginette, ed., 1994. *International Wildlife Trade: A CITES Sourcebook*. Washington, DC: WWF and Island Press. 166p. [1/1/1].

2168. Hendrickson, John A., 1971. *An Expanded Concept of "Species Diversity."* Philadelphia: Notulae Naturae No. 439, Academy of Natural Sciences. 6p. [1/1/1].

2169. Hendrix, Paul F., ed., 1995. *Earthworm Ecology and Biogeography in North America*. Boca Raton, FL: Lewis Publishers. 244p. [2/1/1].

2170. ———, 1996. Nearctic earthworm fauna in the southern USA: biodiversity and effects on ecosystem processes. *Biodiversity and Con-*

servation 5: 223-234. [1] *A.T.:* soil fauna; biological invasions.

2171. Hendrix, Steven, Jan.-Feb. 1997. Bolivia's outpost of hope. *International Wildlife* 27(1): 12-19. [1] Concerning a joint proposal by Conservation International and an indigenous Indian group to develop an ecotourism camp.

2172. Hengeveld, Rob, 1989. *Dynamics of Biological Invasions.* London and New York: Chapman & Hall. 160p. [1/1/2] *A.T.:* population biology.

2173. ———, 1990. *Dynamic Biogeography.* Cambridge, U.K., and New York: CUP. 249p. [1/2/2].

2174. ———, 1994. Biodiversity—the diversification of life in a nonequilibrium world. *Biodiversity Letters* 2: 1-10. [1] *A.T.:* reserve size; corridors; reserve design.

2175. Hengeveld, Rob, & J. Haeck, 1982. The distribution of abundance. I. Measurements. *J. of Biogeography* 9: 303-316. [2] *A.T.:* species density; northwestern Europe; biogeography.

2176. Henne, Denise, Jan.-Feb. 1995. Taking an ecosystem approach. *Endangered Species Bull.* 20(1): 6-9. [1] *A.T.:* ecosystem management; Fish and Wildlife Service; wildlife conservation.

2177. Hennig, Willi, 1966. *Phylogenetic Systematics.* Urbana, IL: University of Illinois Press. 263p. [2/1/3] This influential work was central to the development of cladistic biogeography.

2178. Henrikson, Lennart, A. Hindar, & E. Thornelof, 1995. Freshwater liming. *Water, Air, and Soil Pollution* 85: 131-142. [1] *A.T.:* Sweden; Norway; acid precipitation; air pollution.

2179. Heppner, John B., 1991. *Faunal Regions and the Diversity of Lepidoptera.* Gainesville, FL: Association for Tropical Lepidoptera. 85p. [1/1/1].

2180. Herbstritt, Robert L., & A.D. Marble, 1996. Current state of biodiversity impact analysis in state transportation agencies. *Transportation Research Record*, no. 1559: 51-66. [1].

2181. Hercock, Marion J., 1997. Appreciating the biodiversity of remnant bushland: an "architectural" approach. *Environmentalist* 17: 249-258. *A.T.:* Western Australia; habitat fragmentation; vegetation structure.

2182. Herkert, J.R., 1994. The effects of habitat fragmentation on midwestern grassland bird communities. *Ecological Applications* 4: 461-471. [2] *A.T.:* Illinois; species-area relationship.

2183. Herman, Kim D., L.A. Masters, et al., 1997. Floristic quality assessment: development and application in the state of Michigan (USA). *Natural Areas J.* 17: 265-279. *A.T.:* plant communities; wetlands.

2184. Hernández Bermejo, J.E., M. Clemente, & V.H. Heywood, eds., 1990. *International Conference on Conservation Techniques in Botanic Gardens* [conference proceedings]. Königstein, Germany: Koeltz Scientific Books. 205p. [1/1/1] *A.T.:* plant conservation.

2185. Herremans, M., 1995. Effects of woodland modification by African elephant *Loxodonta africana* on bird diversity in northern Botswana. *Ecography* 18: 440-454. [1].

2186. Herrera, Lisa S. de, & Vicente Murphy, Sept.-Oct. 1994. Help for Honduras: a country at the cross roads of conservation. *Wildlife Conservation* 97(5): 40-49. [1] *A.T.:* environmental protection.

2187. Hersch Martínez, Paul, 1997. Medicinal plants and regional traders in Mexico: physiographic differences and conservational change. *Economic Botany* 51: 107-120.

2188. Hess, George R., 1994. Conservation corridors and contagious disease: a cautionary note. *CB* 8: 256-262. [2] *A.T.:* simulation models; metapopulations; patch dynamics.

2189. Hessler, Robert R., & H.L. Sanders, 1967. Faunal diversity in the deep-sea. *Deep-Sea Research* 14: 65-78. [2].

2190. Hewitt, N., & K. Miyanishi, 1997. The role of mammals in maintaining plant species richness in a floating *Typha* marsh in southern Ontario. *Biodiversity and Conservation* 6: 1085-1102. *A.T.:* disturbance ecology.

2191. Hey, Ellen, 1995. Increasing accountability for the conservation and sustainable use of biodiversity: an issue of transnational global character. *Colorado J. of International Environmental Law and Policy* 6: 1-29. [1] *A.T.:* international law.

2192. Hey, Jody, 1992. Using phylogenetic trees to study speciation and extinction. *Evolution* 46: 627-640. [2] *A.T.:* phylogenetic analysis.

2193. Heyer, W. Ronald, M.A. Donnelly, et al., eds., 1994. *Measuring and Monitoring Biological Diversity. Standard Methods for Amphibians.* Washington, DC: Smithsonian Institution Press. 364p. [2/1/2].

2194. Heywood, Vernon H., 1989. *The Botanic Gardens Conservation Strategy.* Kew, Richmond, U.K.: IUCN Botanic Gardens Secretariat. 60p. [1/1/1] *A.T.:* plant conservation.

2195. ———, 1993. The new science of synthesis. *Naturopa*, no. 73: 4-5. [1] A short overview of bdc in Europe.

2196. Heywood, Vernon H., & S.D. Davis, 1994-97. *Centres of Plant Diversity: A Guide and Strategy for Their Conservation.* Cambridge, U.K.: WWF and IUCN. 3 vols. [1/1/1] *A.T.:* biogeography.

2197. Heywood, Vernon H., & R.T. Watson, eds., 1995. *Global Biodiversity Assessment.* Cambridge, U.K., and New York: CUP. 1140p. [1/1/2] *A.T.:* bdc.

2198. Hickman, Carole S., 1993. Biological diversity: elements of a paleontological agenda [editorial]. *Palaios* 8: 309-310. [1].

2199. Hietz-Seifert, Ursula, Peter Hietz, & S. Guevara, 1996. Epiphyte vegetation and diversity on remnant trees after forest clearance in southern Veracruz, Mexico. *Biological Conservation* 75: 103-111. [1] *A.T.:* habitat fragmentation.

2200. Higashi, Seigo, Akira Osawa, & Kana Kanagawa, eds., 1993. *Bio-*

diversity and Ecology in the Northernmost Japan. Sapporo, Japan: Hokkaido University Press. 154p. [1/1/1].

2201. Hill, Julie, 1991. *Conserving the World's Biological Diversity: How Can Britain Contribute?* [seminar report] London: Dept. of the Environment. 83p. [1/1/1].

2202. Hill, J., 1993. Biodiversity: a case study of Ghanaian tropical forests. *Swansea Geographer* 30: 123-131. [1].

2203. Hill, J.K., K.C. Hamer, et al., 1995. Effects of selective logging on tropical forest butterflies on Buru, Indonesia. *J. of Applied Ecology* 32: 754-760. [2].

2204. Hill, M.O., 1973. Diversity and evenness: a unifying notation and its consequences. *Ecology* 54: 427-432. [3] Explores several aspects of the concept and measure of diversity. *A.T.:* species-abundance curve.

2205. Hinrichsen, Don, 1987. The forest decline enigma: what underlies extensive dieback on two continents? *Bioscience* 37: 542-546. [1] *A.T.:* Europe; eastern North America; environmental stress.

2206. ———, summer 1996. Coasts in crisis: the earth's most biologically productive habitats are being smothered by development. *Issues in Science and Technology* 12(4): 39-47. [1] *A.T.:* human disturbance; coastal zone management; coral reefs; habitat degradation.

2207. ———, 1997. Coral reefs in crisis. *Bioscience* 47: 554-558. *A.T.:* pollution; habitat degradation; reef management.

2208. Hinsley, S.A., P.E. Bellamy, et al., 1998. Geographical and land use influences on bird species richness in small woods in agricultural landscapes. *Global Ecology and Biogeography Letters* 7: 125-135. *A.T.:* diversity gradients; Europe; species-area relation.

2209. Hinsley, S.A., R. Pakeman, et al., 1996. Influences of habitat fragmentation on bird species distributions and regional population sizes. *PRSL B* 263: 307-313. [1] *A.T.:* England; habitat fragmentation.

2210. Hirsch, P., 1994. Protected areas and people. *Geography Review* 8(2): 37-41. [1] *A.T.:* UNCED; GEF; developing countries.

2211. Hitchings, Susan P., & T.J.C. Beebee, 1997. Genetic substructuring as a result of barriers to gene flow in urban *Rana temporaria* (common frog) populations: implications for biodiversity conservation. *Heredity* 79: 117-127. *A.T.:* small populations; genetic diversity.

2212. Hoage, R.J., ed., 1985. *Animal Extinctions: What Everyone Should Know*. Washington, DC: Smithsonian Institution Press. 192p. [2/1/1].

2213. Hoagland, K. Elaine, & A.Y. Rossman, eds., 1997. *Global Genetic Resources: Access, Ownership, and Intellectual Property Rights* [symposium papers]. Washington, DC: Association of Systematics Collections. 347p. [1/1/-] *A.T.:* germplasm resources.

2214. Hobbs, Richard J., ed., 1992. *Biodiversity of Mediterranean Ecosystems of Australia*. Chipping Norton, NSW: Surrey Beatty. 246p. [1/1/1].

2215. ———, 1992. The role of corridors in conservation: solution or bandwagon? *TREE* 7: 389-392. [2].

2216. ———, 1993. Can revegetation assist in the conservation of biodiversity in agricultural areas? *Pacific Conservation Biology* 1: 29-38. [1].

2217. ———, 1994. Landscape ecology and conservation: moving from description to application. *Pacific Conservation Biology* 1: 170-176. [1].

2218. ———, 1997. Future landscapes and the future of landscape ecology. *Landscape and Urban Planning* 37: 1-9. Reviews and critiques research trends in the field of landscape ecology.

2219. Hobbs, Richard J., & L.F. Huenneke, 1992. Disturbance, diversity, and invasion: implications for conservation. *CB* 6: 324-337. [2] *A.T.:* habitat degradation.

2220. Hobbs, Richard J., & S.E. Humphries, 1995. An integrated approach to the ecology and management of plant invasions. *CB* 9: 761-770. [1] *A.T.:* weeds; ecosystem management.

2221. Hobbs, Richard J., & H.A. Mooney, 1998. Broadening the extinction debate: population deletions and additions in California and Western Australia. *CB* 12: 271-283. *A.T.:* Argues that more attention should be given to population, as distinct from species, extinction.

2222. Hobbs, Richard J., & D.A. Saunders, eds., 1993. *Reintegrating Fragmented Landscapes: Towards Sustainable Production and Nature Conservation*. New York: Springer. 332p. [1/1/1] *A.T.:* habitat fragmentation; Western Australia; landscape ecology.

2223. Hobdy, Robert, 1993. Lana'i—a case study: the loss of biodiversity on a small Hawaiian island. *Pacific Science* 47: 201-210. [1] *A.T.:* human impact.

2224. Hoberg, Eric P., 1992. Congruent and synchronic patterns in biogeography and speciation among seabirds, pinnipeds, and cestodes. *J. of Parasitology* 78: 601-615. [1] *A.T.:* paleobiogeography; tapeworms (cestodes).

2225. Hochberg, Michael E., & B.A. Hawkins, 1993. Predicting parasitoid species richness. *American Naturalist* 142: 671-693. [1] *A.T.:* refuges; host-parasite relations.

2226. Hochberg, Michael E., Jean Clobert, & Robert Barbault, eds., 1996. *Aspects of the Genesis and Maintenance of Biological Diversity* [workshop papers]. Oxford, U.K., and New York: OUP. 316p. [1/1/1] *A.T.:* evolution.

2227. Hockey, P.A.R., & G.M. Branch, 1994. Conserving marine biodiversity on the African coast: implications of a terrestrial perspective. *Aquatic Conservation* 4: 345-362. [1] *A.T.:* marine reserves; reserve selection; coastal zone management

2228. Hocutt, Charles H., & E.O. Wiley, eds., 1986. *The Zoogeography of North American Freshwater Fishes*. New York: Wiley. 866p. [2/1/2].

2229. Hodder, K.H., & J.M. Bullock, 1997. Translocations of native species in the UK: implications for biodiversity. *J. of Applied Ecology* 34:

547-565.

2230. Hodges, John, 1993. Animals too. *Naturopa*, no. 73: 28-29. [1] Concerning threatened livestock breeds in Europe.

2231. Hodgkin, Toby, A.H.D. Brown, et al., eds., 1995. *Core Collections of Plant Genetic Resources*. Chichester, U.K., and New York: Wiley. 269p. [1/1/1] *A.T.:* gene banks; plant collections.

2232. Hodkinson, Ian D., & D. Casson, 1991. A lesser predilection for bugs: Hemiptera (Insecta) diversity in tropical rain forests. *Biological J. of the Linnean Society* 43: 101-109. [1] *A.T.:* Sulawesi.

2233. Hodkinson, Ian D., & Elaine Hodkinson, 1993. Pondering the imponderable: a probability-based approach to estimating insect diversity from repeat faunal samples. *Ecological Entomology* 18: 91-92. [1].

2234. Hoffman, Antoni, 1985. Patterns of family extinction depend on definition and geological time scale. *Nature* 315: 659-662. [1] *A.T.:* mass extinctions.

2235. ———, 1986. Neutral model of Phanerozoic diversification: implications for macroevolution. *Neues Jahrbuch für Geologie und Paläontologie*, Abhandlungen 172: 219-244. [1].

2236. ———, 1989. Mass extinctions: the view of a skeptic. *J. of the Geological Society* 146: 21-35. [1].

2237. ———, 1989. What, if anything, are mass extinctions? *PTRSL B* 325: 253-261. [1].

2238. Hoffman, M. Timm, G.F. Midgley, & R.M. Cowling, 1994. Plant richness is negatively related to energy availability in semi-arid Southern Africa. *Biodiversity Letters* 2: 35-38. [1] *A.T.:* potential evapotranspiration.

2239. Holbrook, Sally J., M.J. Kingsford, et al., 1994. Spatial and temporal patterns in assemblages of temperate reef fish. *American Zoologist* 34: 463-475. [1] *A.T.:* reef ecology.

2240. Holdaway, R.N., 1989. New Zealand's pre-human avifauna and its vulnerability. *New Zealand J. of Ecology* 12, Suppl.: 11-25. [1] *A.T.:* biological invasions.

2241. Holden, John H.W., & J.T. Williams, eds., 1984. *Crop Genetic Resources: Conservation and Evaluation*. London and Boston: Allen & Unwin. 296p. [1/1/1] *A.T.:* plant germplasm resources.

2242. Holden, John H.W., W.J. Peacock, & J.T. Williams, 1993. *Genes, Crops, and the Environment*. Cambridge, U.K., and New York: CUP. 162p. [2/1/1] *A.T.:* germplasm resources.

2243. Holdgate, Martin W., 1994. Ecology, development and global policy. *J. of Applied Ecology* 31: 201-211. [1] An overview of the global relevance of the field of ecology.

2244. ———, 1996. The ecological significance of biological diversity. *Ambio* 25: 409-416. [1] *A.T.:* ecosystem resilience.

2245. Holdridge, Leslie R., 1967. *Life Zone Ecology* (rev. ed.). San José,

Costa Rica: Tropical Science Center. 206p. [1/1/3].

2246. Holl, Karen D., G.C. Daily, & P.R. Ehrlich, 1995. Knowledge and perceptions in Costa Rica regarding environment, population, and biodiversity issues. *CB* 9: 1548-1558. [1] *A.T.:* environmental awareness; bd education.

2247. Holland, D.N., R.J. Lilieholm, et al., 1994. Economic trade-offs of managing forests for timber production and vegetative diversity. *Canadian J. of Forest Research* 24: 1260-1265. [1] *A.T.:* diversity indices.

2248. Holland, Marjorie M., P.G. Risser, & R.J. Naiman, eds., 1991. *Ecotones: The Role of Landscape Boundaries in the Management and Restoration of Changing Environments.* London and New York: Chapman & Hall. 142p. [1/1/1] *A.T.:* landscape ecology.

2249. Hölldobler, Bert, & E.O. Wilson, 1990. *The Ants.* Cambridge, MA: Harvard University Press. 732p. [3/3/3] Winner of a 1991 Pulitzer Prize.

2250. Holling, C.S., 1973. Resilience and stability of ecological systems. *ARES* 4: 1-23. [3] A classic review.

2251. Holloway, J.D., 1991. Biodiversity and tropical agriculture: a biogeographic view. *Outlook on Agriculture* 20: 9-13. [1].

2252. ———, 1993. Lepidoptera in New Caledonia: diversity and endemism in a plant-feeding insect group. *Biodiversity Letters* 1: 92-101. [1].

2253. Holloway, Marguerite, Oct. 1992. Musseling in: exotic species hitch rides in ships' ballast water. *Scientific American* 267(4): 22-23. [1] *A.T.:* biological invasions; zebra mussel; ballast dumping.

2254. ———, July 1993. Sustaining the Amazon. *Scientific American* 269(1): 90-99. [1] Considers various sustainable development strategies for the Amazon rainforest. *A.T.:* deforestation.

2255. Holmes, Bob, 17 June 1995. Noah's new challenge. *New Scientist* 146(1982): 33-37. [1] Concerning the setting of priorities for endangered species preservation.

2256. ———, 18 Oct. 1997. When we were worms. *New Scientist* 156 (2104): 30-35. *A.T.:* Cambrian explosion; hox genes; evolution.

2257. Holmes, John S., J. Marchant, et al., 1998. The British List: new categories and their relevance to conservation. *British Birds* 91: 2-11. *A.T.:* birds; endangerment lists.

2258. Holmes, J.C., 1996. Parasites as threats to biodiversity in shrinking ecosystems. *Biodiversity and Conservation* 5: 975-983. [1] *A.T.:* environmental stress; disease; habitat fragmentation.

2259. Holmquist, Jeff G., J.M. Schmidt-Gengenbach, & B.B. Yoshioka, 1998. High dams and marine-freshwater linkages: effects on native and introduced fauna in the Caribbean. *CB* 12: 621-630. *A.T.:* disturbance; biological invasions.

2260. Holsinger, J.R., 1993. Biodiversity of subterranean amphipod crustaceans: global patterns and zoogeographic implications. *J. of Natural*

History 27: 821-835. [1].

2261. Holsinger, Kent E., 1995. Population biology for policy makers. *Bioscience* 45, Suppl.: S10-S20. [1] *A.T.:* environmental policy; endangered species.

2262. Holst, Jon D., 1992. The unforeseeability factor: federal lands, managing for uncertainty, and the preservation of biological diversity. *Public Land Law Review* 13: 113-136. [1].

2263. Hominick, William M., A.P. Reid, et al., 1996. Entomopathogenic nematodes: biodiversity, geographical distribution and the Convention on Biological Diversity. *Biocontrol Science and Technology* 6: 317-332. *A.T.:* biological control.

2264. Hood, Laura C., 1998. *Frayed Safety Nets: Conservation Planning Under the Endangered Species Act*. Washington, DC: Defenders of Wildlife. 115p.

2265. Hoose, Phillip M., 1981. *Building an Ark: Tools for the Preservation of Natural Diversity Through Land Protection*. Covelo, CA: Island Press. 221p. [2/2/1] *A.T.:* Nature Conservancy.

2266. Hooten, Anthony J., & M.E. Hatziolos, eds., 1995. *Sustainable Financing Mechanisms for Coral Reef Conservation: Proceedings of a Workshop*. Washington, DC: World Bank. 116p. [1/1/1].

2267. Hoover, S.R., & A.J. Parker, 1991. Spatial components of biotic diversity in landscapes of Georgia, USA. *Landscape Ecology* 5: 125-136. [1] *A.T.:* landscape ecology; spatial variation.

2268. Hopper, S.D., & D.J. Coates, 1990. Conservation of genetic resources in Australia's flora and fauna. *Proc. of the Ecological Society of Australia* 16: 567-577. [1].

2269. Hori, M., M.M. Gashagaza, et al., 1993. Littoral fish communities in Lake Tanganyika: irreplaceable diversity supported by intricate interactions among species. *CB* 7: 657-666. [1] *A.T.:* cichlids; mutualism.

2270. Horikoshi, Koki, Masao Fukuda, & Toshiaki Kudo, eds., 1997. *Microbial Diversity and Genetics of Biodegradation*. New York: Karger. 210p. [1/1/-] *A.T.:* bioremediation.

2271. Horn, Henry S., Jan. 1993. Biodiversity in the backyard. *Scientific American* 268(1): 150-152. [1] *A.T.:* inventories; bd education.

2272. Horta, Korinna, 1991. The last big rush for the green gold: the plundering of Cameroon's rainforests. *Ecologist* 21: 142-147. [1] *A.T.:* resource exploitation; indigenous peoples; logging.

2273. Horton, Curtis M., 1995. Protecting biological diversity and cultural diversity under intellectual property law: toward a new international system. *J. of Environmental Law and Litigation* 10: 1-38. [1] *A.T.:* indigenous peoples.

2274. Hotta, Mitsura, ed., 1989. *Diversity and Plant-Animal Interaction in Equatorial Rain Forests: Report of the 1987-1988 Sumatra Research*.

Kagoshima, Japan: Research Center for the South Pacific, Kagoshima University. 203p. [1/1/1] *A.T.:* Indonesia.

2275. Houck, Oliver A., 1997. On the law of biodiversity and ecosystem management. *Minnesota Law Review* 81: 869-979. [2].

2276. Houdijk, A.L.F.M., P.J.M. Verbeek, et al., 1993. Distribution and decline of endangered herbaceous heathland species in relation to the chemical composition of the soil. *Plant and Soil* 148: 137-143. [1] *A.T.:* acidification; soil chemistry; Netherlands.

2277. Houghton, John T., G.J. Jenkins, & J.J. Ephraums, eds., 1990. *Climate Change: The IPCC Scientific Assessment* [panel report]. Cambridge, U.K., and New York: CUP. 364p. [2/2/3].

2278. Houghton, Richard A., 1994. The worldwide extent of land use change. *Bioscience* 44: 305-313. [2] *A.T.:* tropical forests; deforestation.

2279. Houghton, Richard A., & G.M. Woodwell, April 1989. Global climatic change. *Scientific American* 260(4): 36-44. [2] *A.T.:* carbon dioxide; methane.

2280. House, M.R., 1989. Ammonoid extinction events. *PTRSL B* 325: 307-326. [1] *A.T.:* Cretaceous.

2281. Howard, P.C., P. Viskanic, et al., 1998. Complementarity and the use of indicator groups for reserve selection in Uganda. *Nature* 394: 472-475. *A.T.:* species richness.

2282. Howard, W.J., J.M. Ayres, et al., 1995. Mamiraua: a case study of biodiversity conservation involving local people. *Commonwealth Forestry Review* 74: 76-79. [1] *A.T.:* Amazonia; flooded forest; local participation.

2283. Howarth, Francis G., 1980. The zoogeography of specialized cave animals: a bioclimatic model. *Evolution* 34: 394-406. [1] *A.T.:* troglobites.

2284. Howe, H.F., 1994. Managing species diversity in tallgrass prairie: assumptions and implications. *CB* 8: 691-704. [1] *A.T.:* restoration ecology; environmental management.

2285. Howell, E.A., 1986. Woodland restoration: an overview. *Restoration and Management Notes* 4: 13-17. [1].

2286. Howes, Chris, 1997. *The Spice of Life: Biodiversity and the Extinction Crisis.* London: Blandford. 192p. [1/1/-].

2287. Howitt, Richard E., Nov.-Dec. 1995. Economic incentives for growers can benefit biological diversity. *California Agriculture* 49(6): 28-33. [1].

2288. Hoyt, Erich, 1992. *Conserving the Wild Relatives of Crops* (2nd ed.). Rome and Gland, Switzerland: International Board for Plant Genetic Resources, IUCN, and WWF. 52p. [1/1/1] *A.T.:* crop germplasm resources.

2289. Huang, W.M., 1996. Bacterial diversity based on Type II DNA topoisomerase genes. *Annual Review of Genetics* 30: 79-107. [2].

2290. Hubálek, Zdenek, 1982. Coefficients of association and similarity, based on binary (presence-absence) data: an evaluation. *Biological Reviews* 57: 669-689. [1].

2291. Hubbell, S.P., 1995. Towards a theory of biodiversity and biogeography on continuous landscapes. In G.R. Carmichael, G.E. Folk, & J.L. Schnoor, eds., *Preparing for Global Change: A Midwestern Perspective* (Amsterdam: SPB Academic): 171-199. [1] *A.T.:* landscape ecology.

2292. Huber, Peter, 26 Oct. 1992. Biodiversity vs. bioengineering? [opinion] *Forbes* 150(10): 266. [1] Thinks conservationists should restrict their arguments for bdc to non-economic factors.

2293. Hudson, Wendy E., ed., 1991. *Landscape Linkages and Biodiversity*. Washington, DC: Island Press. 196p. [2/2/1] *A.T.:* landscape connectivity; habitat fragmentation.

2294. Huey, Raymond B., 1978. Latitudinal pattern of between-altitude faunal similarity: mountains might be "higher" in the tropics. *American Naturalist* 112: 225-229. [1] *A.T.:* altitudinal distribution.

2295. Hugenholtz, P., & N.R. Pace, 1996. Identifying microbial diversity in the natural environment: a molecular phylogenetic approach. *Trends in Biotechnology* 14: 190-197. [2] *A.T.:* microorganisms; biotechnology.

2296. Hughes, Jennifer B., G.C. Daily, & P.R. Ehrlich, 1997. Population diversity: its extent and extinction. *Science* 278: 689-692. Estimates the average number of distinct populations found per species. *A.T.:* literature reviews.

2297. Hughes, Robert M., & R.F. Noss, May-June 1992. Biological diversity and biological integrity: current concerns for lakes and streams. *Fisheries* 17(3): 11-19. [1] *A.T.:* bd concepts; EPA; bd loss.

2298. Hughes, R.G., 1986. Theories and models of species abundance. *American Naturalist* 128: 879-899. [2].

2299. Hughes, Terence P., 1989. Community structure and diversity of coral reefs: the role of history. *Ecology* 70: 275-279. [2] *A.T.:* disturbance.

2300. Hugueny, Bernard, 1989. West African rivers as biogeographic islands: species richness of fish communities. *Oecologia* 79: 236-243. [1].

2301. Huisman, J.M., R.A. Cowan, & T.J. Entwisle, 1998. Biodiversity of Australian marine microalgae: a progress report. *Botanica Marina* 41: 89-93.

2302. Hull, David L., 1979. The limits of cladism. *Systematic Zoology* 28: 416-440. [1] *A.T.:* phylogeny.

2303. Hulme, Philip E., 1996. Herbivory, plant regeneration, and species coexistence. *J. of Ecology* 84: 609-615. [1] Links herbivory to vegetational diversity. *A.T.:* nutrient cycling.

2304. Human, Kathleen G., & D.M. Gordon, 1997. Effects of Argentine ants on invertebrate biodiversity in Northern California. *CB* 11: 1242-1248. *A.T.:* biological invasions.

2305. Hummel, Monte, ed., 1989. *Endangered Spaces: The Future for Canada's Wilderness*. Toronto: Key Porter Books. 288p. [1/1/1] *A.T.:* protected areas.

2306. Humphries, Christopher J., & L.R. Parenti, 1986. *Cladistic Biogeography*. Oxford, U.K.: Clarendon Press. 98p. [1/1/2].

2307. Humphries, Christopher J., D. Vane-Wright, & P.H. Williams, 1991. Biodiversity reserves: setting new priorities for the conservation of wildlife. *Parks* 2(2): 34-38. [1].

2308. Humphries, Christopher J., P.H. Williams, & R.I. Vane-Wright, 1995. Measuring biodiversity value for conservation. *ARES* 26: 93-111. [2] Focuses on the philosophy and application of character diversity measures.

2309. Hundloe, T.J., 1990. Measuring the value of the Great Barrier Reef. *Australian Parks and Recreation* 26(3): 11-15. [1].

2310. Hunt, Constance Elizabeth, ed., 1997. *Conservation on Private Lands: An Owner's Manual*. Washington, DC: WWF. 353p. [1/1/-] *A.T.:* bdc.

2311. Hunt, Constance Elizabeth, & Verne Huser, 1988. *Down by the River: The Impact of Federal Water Projects and Policies on Biological Diversity*. Washington, DC: Island Press. 266p. [2/1/1] *A.T.:* environmental policy.

2312. Hunt, Lee O., June 1991. Forestry word games: bi-o-di-ver-si-ty [editorial]. *J. of Forestry* 89(6): 39. [1] *A.T.:* bd concepts.

2313. Hunter, Christopher J., 1997. Sustainable bioprospecting: using private contracts and international legal principles and policies to conserve raw medicinal materials. *Boston College Environmental Affairs Law Review* 25: 129-174. *A.T.:* INBio; Merck.

2314. Hunter, Malcolm L., Jr., 1987. Managing forests for spatial heterogeneity to maintain biological diversity. *Trans. of the 52nd North American Wildlife and Natural Resources Conference*: 60-69. [1] *A.T.:* Maine; clearcutting.

2315. ———, 1990. *Wildlife, Forests and Forestry: Principles of Managing Forests for Biological Diversity*. Englewood Cliffs, NJ: Prentice-Hall. 370p. [1/1/2].

2316. Hunter, Malcolm L., Jr., & A. Hutchinson, 1994. The virtues and shortcomings of parochialism: conserving species that are locally rare, but globally common. *CB* 8: 1163-1165. [1] *A.T.:* climatic change; rarity.

2317. Hunter, Malcolm L., Jr., G.L. Jacobson, Jr., & Thompson Webb III, 1988. Paleoecology and the coarse-filter approach to maintaining biological diversity. *CB* 2: 375-385. [2].

2318. Hunter, M.D., & P.W. Price, 1992. Playing chutes and ladders: heterogeneity and the relative roles of bottom-up and top-down forces in natural communities. *Ecology* 73: 724-732. [3] *A.T.:* environmental heterogeneity; predator-prey relations; nutrient availability.

2319. Huntley, Brian J., ed., 1989. *Biotic Diversity in Southern Africa: Concepts and Conservation* [conference papers]. Cape Town: OUP. 380p. [1/1/1].

2320. ———, 1993. Species-richness in north-temperate zone forests. *J. of Biogeography* 20: 163-180. [1] Compares North American and European forests. *A.T.:* Quaternary.

2321. ———, ed., 1994. *Botanical Diversity in Southern Africa* [conference proceedings]. Pretoria: National Botanical Institute. 412p. [1/1/1] *A.T.:* plant diversity.

2322. Huntley, Brian J., & H.J.B. Birks, 1983. *An Atlas of Past and Present Pollen Maps for Europe: 0-13000 Years Ago.* Cambridge, U.K., and New York: CUP. 667p. [1/1/3].

2323. Huntley, Brian J., & Thompson Webb III, 1989. Migration: species' response to climatic variations caused by changes in the earth's orbit. *J. of Biogeography* 16: 5-19. [2] *A.T.:* orbital forcing.

2324. Huntley, Brian J., et al., eds., 1997. *Past and Future Rapid Environmental Changes: The Spatial and Evolutionary Responses of Terrestrial Biota* [workshop proceedings]. Berlin and New York: Springer. 523p. [1/1/-].

2325. Huntsinger, Lynn, & Peter Hopkinson, 1996. Viewpoint: sustaining rangeland landscapes: a social and ecological process. *J. of Range Management* 49: 167-173. [1] *A.T.:* ranching; California.

2326. Huntsman, Gene R., July 1994. Endangered marine finfish: neglected resources or beasts of fiction? *Fisheries* 19(7): 8-15. [1] *A.T.:* U.S.; stock depletion; fisheries; Endangered Species Act.

2327. Hurlbert, Stuart H., 1971. The nonconcept of species diversity: a critique and alternative parameters. *Ecology* 52: 577-586. [3] "Offers a critique of semantic, conceptual, and technical problems in the diversity literature and suggests that ecologists take more direct approaches to the study of species-numbers relations."

2328. Hurlbert, Stuart H., & J.D. Archibald, 1995. No statistical support for sudden (or gradual) extinction of dinosaurs. *Geology* 23: 881-884. [1] *A.T.:* mass extinctions; Cretaceous-Tertiary boundary.

2329. Hurlbut, David, 1994. Fixing the Biodiversity Convention: toward a special protocol for related intellectual property. *Natural Resources J.* 34: 379-409. [1] *A.T.:* North-South relations; patent rights.

2330. Hurst, L.D., 1990. Parasite diversity and the evolution of diploidy, multicellularity and anisogamy. *J. of Theoretical Biology* 144: 429-443. [1].

2331. Hurst, Philip, 1990. *Rainforest Politics: Ecological Destruction in South-East Asia* [case studies]. London and Atlantic Highlands, NJ: Zed Books. 303p. [2/2/2] *A.T.:* deforestation; environmental policy.

2332. Huston, Michael A., 1979. A general hypothesis of species diversity. *American Naturalist* 113: 81-101. [3] Developed on the basis of a model of nonequilibrial interactions among populations.

2333. ———, 1980. Soil nutrients and tree species richness in Costa Rican forests. *J. of Biogeography* 7: 147-157. [1].

2334. ———, 1985. Patterns of species diversity on coral reefs. *ARES* 16: 149-177. [2].

2335. ———, 1992. Biological diversity and human resources. *Impact of Science on Society* 166: 121-130. [1] *A.T.:* resource management; environmental planning.

2336. ———, 1993. Biological diversity, soils, and economics. *Science* 262: 1676-1680. [2] *A.T.:* agricultural productivity.

2337. ———, 1994. *Biological Diversity: The Coexistence of Species on Changing Landscapes*. Cambridge, U.K., and New York: CUP. 681p. [2/2/3].

2338. ———, spring 1995. Biodiversity thrives where crops fail. *Forum for Applied Research and Public Policy* 10(1): 101-105. [1] On the negative relationship between areas of high agricultural productivity and high bd.

2339. ———, 1997. Hidden treatments in ecological experiments: reevaluating the ecosystem function of biodiversity. *Oecologia* 110: 449-460. [2] Identifies complications in the way cause and effect operates at the bd level.

2340. Hut, Piet, Walter Alvarez, et al., 1987. Comet showers as a cause of mass extinctions. *Nature* 329: 118-126. [2] *A.T.:* stepwise extinction.

2341. Hutchings, P.A., 1986. Biological destruction of coral reefs: a review. *Coral Reefs* 4: 239-252. [2].

2342. Hutchins, Michael, & R.J. Wiese, 1991. Beyond genetic and demographic management: the future of the Species Survival Plan and related AAZPA conservation efforts. *Zoo Biology* 10: 285-292. [1].

2343. Hutchins, Michael, Kevin Willis, & R.J. Wiese, 1995. Strategic collection planning: theory and practice. *Zoo Biology* 14: 5-25. [1].

2344. Hutchinson, George Evelyn, 1948. Teleological mechanisms: circular causal systems in ecology. *Annals of the New York Academy of Science* 50: 221-246. [2].

2345. ———, 1957. Concluding remarks. *Cold Spring Harbor Symposia on Quantitative Biology* 22: 415-427. [2] The conceptualization of the niche as a hypervolume.

2346. ———, 1959. Homage to Santa Rosalia, or why are there so many kinds of animals? *American Naturalist* 93: 145-159. [3] One of the best known essays on the subject of biological diversity.

2347. ———, 1961. The paradox of the plankton. *American Naturalist* 95: 137-145. [2] *A.T.:* competitive exclusion principle; phytoplankton.

2348. ———, 1964. The influence of the environment. *PNAS* 51: 930-934. [1] Notes that many elements are available in such low quantities as to limit their functional importance to living things.

2349. Hutchinson, George Evelyn, & R.H. MacArthur, 1959. A theoretical ecological model of size distributions among species of animals. *American Naturalist* 93: 117-125. [2] *A.T.:* size-number relations; food webs.

2350. Hutto, Richard L., 1988. Is tropical deforestation responsible for the reported declines in Neotropical migrant populations? *American Birds* 42: 375-379. [1] *A.T.:* birds.

2351. Huyghe, Patrick, March-April 1993. New-species fever. *Audubon* 95(2): 88-94, 96. [1] Concerning the ongoing discoveries of new species.

2352. Hyatt, Kim D., Oct. 1996. Stewardship for biomass or biodiversity: a perennial issue for salmon management in Canada's Pacific region. *Fisheries* 21(10): 4-5. [1].

2353. Hyde, Kevin D., ed., 1997. *Biodiversity of Tropical Microfungi*. Hong Kong: Hong Kong University Press. 421p. [1/1/-].

2354. Hyman, Libbie H., 1955. How many species? *Systematic Zoology* 4: 142-143. [1] Gives estimates for the numbers of non-arthropod invertebrate species.

2355. Hyndman, David, 1994. Conservation through self-determination: promoting the interdependence of cultural and biological diversity. *Human Organization* 53: 296-302. [1] Identifies the main bdc-promoting institutions and discusses their similarities and differences in approach. *A.T.:* indigenous peoples.

2356. Ibáñez, M., M.R. Alonso, et al., 1997. Distribution of land snails (Mollusca, Gastropoda, Pulmonata) on the island of Gran Canaria (Canary Islands) in relation to protected natural areas. *Biodiversity and Conservation* 6: 627-632.

2357. Ibarzábal, Diana, & J.P. García, 1996. Cuba: an island's approach to marine biodiversity. *Global Biodiversity* 6(1): 35-36. [1].

2358. Idle, E.T., 1995. Conflicting priorities in site management in England. *Biodiversity and Conservation* 4: 929-937. [1] *A.T.:* protected areas; reserve selection.

2359. Ihlenfeldt, H.-D., 1994. Diversification in an arid world: the Mesembryanthemaceae. *ARES* 25: 521-546. [1] *A.T.:* Southern Africa; genetic drift; homeotic genes.

2360. Iida, Shigeo, & Tohru Nakashizuka, 1995. Forest fragmentation and its effect on species diversity in sub-urban coppice forests in Japan. *Forest Ecology and Management* 73: 197-210. [1].

2361. Imbach, A.C., & J.C. Godoy, 1992. Progress in the management of buffer zones in the American tropics: proposals to increase the influence of protected areas. *Parks* 3(1): 19-22. [1].

2362. Imbert, D., A. Rousteau, & P. Labbe, 1998. Hurricanes and biological diversity in tropical forests: the case of Guadeloupe. *Acta Oecologica* 19: 251-262. *A.T.:* disturbance ecology; mangroves.

2363. Imhoff, Marc L., T.D. Sisk, et al., 1997. Remotely sensed indicators of habitat heterogeneity: use of synthetic aperture radar in mapping vegetation and bird habitat. *Remote Sensing of Environment* 60: 217-227. [1] *A.T.:* Northern Territory, Australia.

2364. Ingerpuu, Nele, Kalevi Kull, & Kai Vellak, 1998. Bryophyte vegetation in a wooded meadow: relationships with phanerogam diversity and responses to fertilisation. *Plant Ecology* 134: 163-171. *A.T.:* Estonia.

2365. Ingram, Gordan B., 1992. The remaining islands with primary rain forest: a global resource. *Environmental Management* 16: 585-595. [1] *A.T.:* environmental management; environmental monitoring; environmental policy.

2366. ———, 1995. Conserving habitat and biological diversity: a study of obstacles on Gwaii Haanas, British Columbia. *Forest & Conservation History* 39: 77-89. [1].

2367. Ingram, Gordan B., & J.T. Williams, 1993. Gap analysis for *in situ* conservation of crop genepools: implications of the Convention on Biological Diversity. *Biodiversity Letters* 1: 141-148. [1].

2368. Ingrouille, Martin, 1992. *Diversity and Evolution of Land Plants.* London and New York: Chapman & Hall. 340p. [1/1/1].

2369. International Alliance of Indigenous-Tribal Peoples of the Tropical Forests, & International Work Group for Indigenous Affairs, 1996. *Indigenous Peoples, Forest, and Biodiversity: Indigenous Peoples and the Global Environmental Agenda.* London. 197p. [1/1/1] *A.T.:* environmental policy.

2370. International Work Group for Indigenous Affairs, 1998. *From Principles to Practice: Indigenous Peoples and Biodiversity Conservation in Latin America* [conference proceedings]. Copenhagen. 304p. *A.T.:* protected areas.

2371. IPGRI, 1993. *Diversity for Development: The Strategy of the International Plant Genetic Resources Institute.* Rome. 62p. [1/1/1] *A.T.:* plant germplasm resources.

2372. Irish, Kerry E., & E.A. Norse, 1996. Scant emphasis on marine biodiversity. *CB* 10: 680. [1].

2373. Irwin, Larry L., & T.B. Wigley, Aug. 1992. Conservation of endangered species: the impact on private forestry. *J. of Forestry* 90(8): 27-30, 40. [1].

2374. Ishwaran, Natarajan, 1992. Biodiversity, protected areas and sustainable development. *Nature and Resources* 28(1): 18-25. [1] *A.T.:* environmental policy; UNESCO.

2375. IUCN, 1980. *World Conservation Strategy.* Morges, Switzerland: IUCN/UNEP/WWF. 4 vols. [1/1/2] *A.T.:* wildlife conservation; environmental policy; action plans.

2376. ———, 1991-92. *Protected Areas of the World: A Review of National Systems.* Gland, Switzerland. 4 vols. [1/1/1].

2377. ———, 1994. *Report of the Global Biodiversity Forum.* Gland, Switzerland. 115p. [1/1/1].

2378. ———, 1996. *Report of the Second Global Biodiversity Forum.* Gland, Switzerland. 143p. [1/1/1].

2379. ———, 1996. *Report of the Third Global Biodiversity Forum.* Gland, Switzerland. 166p. [1/1/1].

2380. ———, 1997. *Report of the Fifth Global Biodiversity Forum.* Gland, Switzerland. 211p. [1/1/-].

2381. Iudicello, Suzanne, & Margaret Lytle, 1994. Marine biodiversity and international law: instruments and institutions that can be used to conserve marine biological diversity internationally. *Tulane Environmental Law J.* 8: 123-161. [1] *A.T.:* pollution; environmental disturbance.

2382. IUDZG & IUCN/SSC Captive Breeding Specialist Group, 1993. *The World Zoo Conservation Strategy: The Role of the Zoos and Aquaria of the World in Global Conservation.* Brookfield, IL: Chicago Zoological Society. 76p. [1/1/1].

2383. Ivany, Linda C., 1996. Coordinated stasis or coordinated turnover: exploring intrinsic vs. extrinsic controls on pattern. *Palaeogeography, Palaeoclimatology, Palaeoecology* 127: 239-256. *A.T.:* ecosystem turnover; ecosystem stability; environmental change; macroevolution.

2384. Iwatsuki, Kunio, & P.H. Raven, eds., 1997. *Evolution and Diversification of Land Plants.* Tokyo and New York: Springer. 330p. [1/1/-].

2385. Iwu, Maurice M., 1996. Implementing the Biodiversity Treaty: how to make international co-operative agreements work. *Trends in Biotechnology* 14: 78-83. [1] *A.T.:* CBD.

2386. ———, 1996. Biodiversity prospecting in Nigeria: seeking equity and reciprocity in intellectual property rights through partnership arrangements and capacity building. *J. of Ethnopharmacology* 51: 209-219. [1].

2387. Jablonski, David, 1986. Background and mass extinctions: the alternation of macroevolutionary regimes. *Science* 231: 129-133. [2] *A.T.:* biogeography; marine invertebrates; Cretaceous-Tertiary boundary.

2388. ———, 1991. Extinctions: a paleontological perspective. *Science* 253: 754-757. [2].

2389. ———, 1993. The tropics as a source of evolutionary novelty through geological time. *Nature* 364: 142-144. [2] *A.T.:* latitudinal diversity gradients; origination-extinction dynamics; biogeography.

2390. Jablonski, David, & J.J. Sepkoski, Jr., 1996. Paleobiology, community ecology, and scales of ecological pattern. *Ecology* 77: 1367-1378. [2] *A.T.:* paleobiogeography.

2391. Jablonski, David, D.H. Erwin, & J.H. Lipps, eds., 1996. *Evolutionary Paleobiology: In Honor of James W. Valentine.* Chicago: University of Chicago Press. 484p. [1/1/1].

2392. Jaccard, Paul, 1908. Nouvelles recherches sur la distribution florale. *Bull. de la Société Vaudoise des Sciences Naturelles* 44: 223-270. [3] *A.T.:* similarity indices.

2393. Jackson, D.A., & H.H. Harvey, 1997. Qualitative and quantitative sampling of lake fish communities. *Canadian J. of Fisheries and Aquatic*

Sciences 54: 2807-2813. *A.T.:* Ontario.

2394. Jackson, Fatimah L.C., 1993. Evolutionary and political economic influences on biological diversity in African Americans. *J. of Black Studies* 23: 539-560. [1] *A.T.:* human diversity; human physiological variation.

2395. ———, 1996. The coevolutionary relationship of humans and domesticated plants. *Yearbook of Physical Anthropology* 39: 161-176. [1] *A.T.:* disease.

2396. Jackson, Jeremy B.C., 1991. Adaptation and diversity of reef corals: patterns result from species differences in resource use and life histories and from disturbance. *Bioscience* 41: 475-482. [2].

2397. Jackson, Jeremy B.C., A.F. Budd, & A.G. Coates, eds., 1996. *Evolution & Environment in Tropical America*. Chicago: University of Chicago Press. 425p. [1/1/1] *A.T.:* Quaternary studies.

2398. Jackson, Jeremy B.C., P. Jung, et al., 1993. Diversity and extinction of tropical American mollusks and emergence of the Isthmus of Panama. *Science* 260: 1624-1626. [2] *A.T.:* Pliocene; paleobiogeography.

2399. Jackson, Laura L., J.R. McAuliffe, & B.A. Roundy, 1991. Desert restoration: revegetation trials on abandoned farmland. *Restoration & Management Notes* 9: 71-80. [1] *A.T.:* Arizona.

2400. Jackson, Michael T., 1995. Protecting the heritage of rice biodiversity. *GeoJournal* 35: 267-274. [1] *A.T.:* IRRI; rice germplasm conservation.

2401. Jackson, Michael T., B.V. Ford-Lloyd, & M.L. Parry, eds., 1990. *Climatic Change and Plant Genetic Resources* [workshop papers]. London and New York: Belhaven Press. 190p. [1/1/1].

2402. Jacobsen, B.A., & P. Simonsen, 1993. Disturbance events affecting phytoplankton biomass, composition and species diversity in a shallow, eutrophic, temperate lake. *Hydrobiologia* 249: 9-14. [1] *A.T.:* Denmark; weather events.

2403. Jacobsen, Dean, Rikke Schultz, & Andrea Encalada, 1997. Structure and diversity of stream invertebrate assemblages: the influence of temperature with altitude and latitude. *Freshwater Biology* 38: 247-261. *A.T.:* Ecuador; Denmark.

2404. Jacobson, Susan Kay, 1990. Graduate education in conservation biology. *CB* 4: 431-440. [1].

2405. ———, ed., 1995. *Conserving Wildlife: International Education and Communication* Approaches [case studies]. New York: Columbia University Press. 302p. [1/1/1].

2406. Jacobson, Susan Kay, & Alfredo Figueroa Lopez, 1994. Biological impacts of ecotourism: tourists and nesting turtles in Tortuguero National Park, Costa Rica. *Wildlife Society Bull.* 22: 414-419. [1].

2407. Jaeger, Jean-Jacques, 1994. The evolution of biodiversity among the southwest European Neogene rodent (Mammalia, Rodentia) communities: pattern and process of diversification and extinction. *Palaeogeogra-*

phy, Palaeoclimatology, Palaeoecology 111: 305-336. [1] *A.T.:* paleobiogeography; oceanic change.

2408. Jaenike, J., 1991. Mass extinction of European fungi. *TREE* 6: 174-175. [1].

2409. Jaffe, Walter R., & E.J. Trigo, 1993. Agrobiotechnology in the developing world. Trends, issues, and policy perspectives. *Annals of the New York Academy of Sciences* 700: 111-127. [1] *A.T.:* agricultural production; sustainable agriculture.

2410. Jaffré, Tanguy, Philippe Bouchet, & J.-M. Veillon, 1998. Threatened plants of New Caledonia: is the system of protected areas adequate? *Biodiversity and Conservation* 7: 109-135. *A.T.:* hotspots; endemism; protected areas.

2411. Jahn, Laurence R., & E.W. Schenck, 1991. What sustainable agriculture means for fish and wildlife. *J. of Soil and Water Conservation* 46: 251-254. [1].

2412. Jakobsson, Kristin M., & A.K. Dragun, 1996. *Contingent Valuation and Endangered Species: Methodological Issues and Applications.* Brookfield, VT: Edward Elgar. 269p. [1/1/1].

2413. James, Frances C., 1970. Geographic size variation in birds and its relationship to climate. *Ecology* 51: 365-390. [2] *A.T.:* Bergmann's rule; U.S.

2414. James, Frances C., & Stephen Rathbun, 1981. Rarefaction, relative abundance, and diversity of avian communities. *Auk* 98: 785-800. [2] *A.T.:* diversity indices; species richness.

2415. James, Frances C., C.E. McCulloch, & D.A. Wiedenfeld, 1996. New approaches to the analysis of population trends in land birds. *Ecology* 77: 13-27. [2] *A.T.:* North American Breeding Bird Survey.

2416. James, Matthew J., ed., 1991. *Galápagos Marine Invertebrates: Taxonomy, Biogeography, and Evolution in Darwin's Islands.* New York: Plenum Press. 474p. [1/2/1].

2417. James, Valentine Udoh, 1993. *Africa's Ecology: Sustaining the Biological and Environmental Diversity of a Continent.* Jefferson, NC: McFarland. 293p. [1/1/1] *A.T.:* sustainable development.

2418. Jamieson, G.S., 1993. Marine invertebrate conservation: evaluation of fisheries over-exploitation concerns. *American Zoologist* 33: 551-567. [1] *A.T.:* overfishing; fisheries management.

2419. Jamieson, J.W., 1995. Biological diversity and ethnic identity: changing patterns in the modern world. *Mankind Quarterly* 36: 193-199. [1] *A.T.:* human evolution; human diversity.

2420. Jana, Sakti, & L.N. Pietrzak, 1988. Comparative assessment of genetic diversity in wild and primitive cultivated barley in a center of diversity. *Genetics* 119: 981-990. [1] *A.T.:* eastern Mediterranean.

2421. Janick, Jules, ed., 1989. *The National Plant Germplasm System of the United States.* Portland, OR: Timber Press. 230p. [1/1/1] *A.T.:* National

Plant Germplasm System.

2422. Jansa, L.F., 1993. Cometary impacts into ocean: their recognition and the threshold constraint for biological extinction. *Palaeogeography, Palaeoclimatology, Palaeoecology* 104: 271-286. [1].

2423. Janson, S., & Jan Vegelius, 1981. Measures of ecological association. *Oecologia* 49: 371-376. [2].

2424. Janzen, Daniel H., 1967. Why mountain passes are higher in the tropics. *American Naturalist* 101: 233-249. [2] *A.T.:* altitudinal gradients.

2425. ———, 1970. Herbivores and the number of tree species in tropical forests. *American Naturalist* 104: 501-528. [3] *A.T.:* recruitment.

2426. ———, 1981. The peak in North American ichneumonid species richness lies between 38° and 42°N. *Ecology* 62: 532-537. [1] *A.T.:* latitudinal diversity gradients.

2427. ———, ed., 1983. *Costa Rican Natural History*. Chicago: University of Chicago Press. 816p. [2/1/2].

2428. ———, 1983. No park is an island: increase in interference from outside as park size decreases. *Oikos* 41: 402-410. [2] *A.T.:* Costa Rica; recruitment.

2429. ———, 1987. Insect diversity of a Costa Rican dry forest: why keep it, and how? *Biological Journal of the Linnean Society* 30: 343-356. [1].

2430. ———, 1991. How to save tropical biodiversity: the National Biodiversity Institute of Costa Rica. *American Entomologist* 37: 158-171. [1] *A.T.:* INBio.

2431. ———, 1994. Wildland biodiversity management in the tropics: where are we now and where are we going? *Vida Silvestre Neotropical* 3: 3-15. [1].

2432. Jarvie, James K., & P.F. Stevens, 1998. Interactive keys, inventory, and conservation. *CB* 12: 222-224. Concerning botanical information and constraints on its delivery. *A.T.:* plants; systematists.

2433. Järvinen, Olli, 1982. Conservation of endangered plant populations: single large or several small reserves? *Oikos* 38: 301-307. [1] *A.T.:* SLOSS.

2434. Järvinen, Olli, & Staffan Ulfstrand, 1980. Species turnover of a continental bird fauna: Northern Europe, 1850-1970. *Oecologia* 46: 186-195. [1] *A.T.:* extinction.

2435. Järvinen, Olli, & R.A. Väisänen, 1980. Quantitative biogeography of Finnish land birds as compared with regionality in other taxa. *Annales Zoologici Fennici* 17: 67-85. [1].

2436. Jarvis, Alice, & Tony Robertson, 1997. *Endemic Birds of Namibia: Evaluating Their Status and Mapping Biodiversity Hotspots*. Windhoek, Namibia: Directorate of Environmental Affairs, Ministry of Environment and Tourism. 102p. [1/1/-].

2437. Jarvis, Devra I., & Toby Hodgkin, eds, 1998. *Strengthening the*

Scientific Basis of In Situ Conservation of Agricultural Biodiversity On-Farm: Options for Data Collecting and Analysis [workshop proceedings]. Rome: IPGRI. 104p. *A.T.:* crop germplasm resources.

2438. Jay-Robert, Pierre, J.M. Lobo, & J.-P. Lumaret, 1997. Altitudinal turnover and species richness variation in European montane dung beetle assemblages. *Arctic and Alpine Research* 29: 196-205. *A.T.:* biogeography; Western Europe.

2439. Jazdzewski, Krzysztof, J.M. Weslawski, & Claude de Broyer, 1995. A comparison of the amphipod faunal diversity in two polar fjords: Admiralty Bay, King George Island (Antarctic) and Hornsund, Spitsbergen (Arctic). *Polskie Archiwum Hydrobiologii* 42: 367-384. [1] *A.T.:* habitat heterogeneity.

2440. Jeanrenaud, Sally, & Jean-Paul Jeanrenaud, 1997. *Thinking Politically About Community Forestry and Biodiversity: Insider-driven Initiatives in Scotland.* London: Rural Development Forestry Network, Overseas Development Institute. 40p. [1/1/-].

2441. Jeffries, Mike J., 1997. *Biodiversity and Conservation.* London and New York: Routledge. 208p. [1/1/-].

2442. Jelmert, Anders, & D.O. Oppen-Berntsen, 1996. Whaling and deep-sea biodiversity. *CB* 10: 653-654. [1].

2443. Jenkins, Robin, 1992. *Bringing Rio Home: Biodiversity in Our Food and Farming.* London: S.A.F.E. Alliance. 30p. [1/1/1] *A.T.:* European Economic Community.

2444. Jenks, Daniel T., 1995. The Convention on Biological Diversity —an efficient framework for the preservation of life on Earth? *Northwestern J. of International Law & Business* 15: 636-667. [1] *A.T.:* international cooperation; international law.

2445. Jennings, Michael D., 1995. Gap analysis today: a confluence of biology, ecology, and geography for management of biological resources. *Wildlife Society Bull.* 23: 658-662. [1].

2446. Jennings, Simon, & S.S. Marshall, 1995. Seeking sustainability in the Seychelles. *Biologist* 42: 197, 199-202. [1] *A.T.:* ecotourism; Aldabra.

2447. Jennings, Simon, & N.V.C. Polunin, 1997. Impacts of predator depletion by fishing on the biomass and diversity of non-target reef fish communities. *Coral Reefs* 16: 71-82. *A.T.:* Fiji; predator-prey relations.

2448. Jennings, Simon, S.S. Marshall, & N.V.C. Polunin, 1996. Seychelles' marine protected areas: comparative structure and status of reef fish communities. *Biological Conservation* 75: 201-209. *A.T.:* coral reefs.

2449. Jensen, Deborah B., M.S. Torn, & John Harte, 1993. *In Our Hands: A Strategy for Conserving California's Biological Diversity.* Berkeley: University of California Press. 302p. [1/1/1] *A.T.:* environmental policy.

2450. Jensen, Paul R., & William Fenical, 1994. Strategies for the discovery of secondary metabolites from marine bacteria: ecological perspectives. *Annual Review of Microbiology* 48: 559-584. [2] *A.T.:* symbiotic bacteria; natural products.

2451. ———, 1996. Marine bacterial diversity as a resource for novel microbial products. *J. of Industrial Microbiology & Biotechnology* 17: 346-351. [1] *A.T.:* industrial microbiology.

2452. Jepsen, D.B., 1997. Fish species diversity in sand bank habitats of a Neotropical river. *Environmental Biology of Fishes* 49: 449-460. *A.T.:* Venezuela.

2453. Jha, A.K., 1995. Conserving biodiversity: need for statutory support. *Economic and Political Weekly* 30: 492-495. [1] *A.T.:* environmental law.

2454. Jha, Pramod K., ed., 1996. *Environment and Biodiversity in the Context of South Asia: Proceedings of the Regional Conference on Environment and Biodiversity*. Kathmandu, Nepal: Ecological Society. 410p. [1/1/1].

2455. Jiang, Mingkang, & Dayuan Xue, 1996. Nature reserve construction and its contribution to the biodiversity conservation in China mainland. *J. of Environmental Sciences* (China) 8: 15-20. [1].

2456. Jobin, Benoit, Celine Boutin, & J.-L. Desgranges, 1996. Fauna habitats in the rural Quebec environment: a floristic analysis. *Canadian J. of Botany* 74: 323-336. [1] *A.T.:* rural landscapes.

2457. Johns, Andrew D., 1985. Selective logging and wildlife conservation in tropical rain forest: problems and recommendations. *Biological Conservation* 31: 355-375. [1].

2458. Johns, Andrew D., & J.P. Skorupa, 1987. Responses of rain-forest primates to habitat disturbance: a review. *International J. of Primatology* 8: 157-191. [2].

2459. Johns, Andrew G., 1997. *Timber Production and Biodiversity Conservation in Tropical Rain Forests*. Cambridge, U.K., and New York: CUP. 225p. [1/1/-] *A.T.:* forest management.

2460. Johns, David M., 1993. Landscape-scale restoration: The Wildlands Project. *Restoration & Management Notes* 11: 18-19. [1] *A.T.:* North America; environmental perception; nature reserves.

2461. Johnson, H.B., H.W. Polley, & H.S. Mayeux, 1993. Increasing CO_2 and plant-plant interactions: effects on natural vegetation. *Vegetatio* 104-105: 157-170. [2] *A.T.:* North America; grasslands; atmospheric change.

2462. Johnson, K.H., K.A. Vogt, et al., 1996. Biodiversity and the productivity and stability of ecosystems. *TREE* 11: 372-377. [2] *A.T.:* diversity-stability hypothesis; ecosystem dynamics.

2463. Johnson, Lawrence E., 1991. *A Morally Deep World: An Essay on Moral Significance and Environmental Ethics*. Cambridge, U.K., and New York: CUP. 301p. [2/2/2] *A.T.:* animal rights.

2464. Johnson, Michael P., & P.H. Raven, 1970. Natural regulation of plant species diversity. *Evolutionary Biology* 4: 127-162. [1].

2465. Johnson, Michael P., & P.H. Raven, 1973. Species number and endemism: the Galápagos Archipelago revisited. *Science* 179: 893-895. [1] *A.T.:* plants; island biogeography.

2466. Johnson, Michael P., & D.S. Simberloff, 1974. Environmental determinants of island species numbers in the British Isles. *J. of Biogeography* 1: 149-154. [1].

2467. Johnson, Michael P., L.G. Mason, & P.H. Raven, 1968. Ecological parameters and plant species diversity. *American Naturalist* 102: 297-306. [1] *A.T.:* California; Baja California.

2468. Johnson, Nancy C., & D.A. Wedin, 1997. Soil carbon, nutrients, and mycorrhizae during conversion of dry tropical forest to grassland. *Ecological Applications* 7: 171-182. *A.T.:* fire; biological invasions; nutrient cycling; Costa Rica.

2469. Johnson, Nancy C., J.H. Graham, & F.A. Smith, 1997. Functioning of mycorrhizal associations along the mutualism-parasitism continuum. *New Phytologist* 135: 575-586. [2].

2470. Johnson, Ned K., 1975. Controls of number of bird species on montane islands in the Great Basin. *Evolution* 29: 545-567. [1] *A.T.:* island biogeography.

2471. Johnson, Nels, 1995. *Biodiversity in the Balance: Approaches to Setting Geographic Conservation Priorities*. Washington, DC: Biodiversity Support Program. 115p. [1/1/1] *A.T.:* environmental planning.

2472. Johnson, Ralph G., 1970. Variations in diversity within benthic marine communities. *American Naturalist* 104: 285-300. [1].

2473. Johnson, Rebecca L., & G.V. Johnson, eds., 1990. *Economic Valuation of Natural Resources: Issues, Theory, and Applications*. Boulder, CO: Westview Press. 220p. [1/1/1] *A.T.:* contingent valuation.

2474. Johnson, Stanley P., ed., 1992. *The Earth Summit: The United Nations Conference on Environment and Development (UNCED)*. London: Graham & Trotman. 532p. [1/1/1].

2475. Johnson, Timothy H., 1988. *Biodiversity and Conservation in the Caribbean: Profiles of Selected Islands*. Cambridge, U.K.: International Council for Bird Preservation. 144p. [1/1/1] *A.T.:* endangered and threatened species.

2476. Johnson, Timothy H., & A.J. Stattersfield, 1990. A global review of island endemic birds. *Ibis* 132: 167-180. [1].

2477. Johnston, Brent E., 1993. Forests and UNCED '92: a foundation for the future. *Forestry Chronicle* 69: 539-544. [1] *A.T.:* Canada; forest policy; action plans.

2478. Johnston, Mark, 1998. Tree population studies in low-diversity forests, Guyana. II. Assessments on the distribution and abundance of non-timber forest products. *Biodiversity and Conservation* 7: 73-86. *A.T.:* natural products.

2479. Johnston, Mark, & M. Gillman, 1995. Tree population studies in low-diversity forests, Guyana. I. Floristic composition and stand structure. *Biodiversity and Conservation* 4: 339-362. [1] *A.T.:* forest structure.

2480. Jokiel, P., & F.J. Martinelli, 1992. The vortex model of coral reef biogeography. *J. of Biogeography* 19: 449-458. [1] *A.T.:* vicariance.

2481. Jokimäki, Jukka, & Jukka Suhonen, 1993. Effects of urbanization on the breeding bird species richness in Finland: a biogeographical comparison. *Ornis Fennica* 70: 71-77. [1].

2482. Jolly, Alison, Philippe Oberlé, & Roland Albignac, eds., 1984. *Madagascar (Key Environments* series). Oxford, U.K., and New York: Pergamon Press. 239p. [2/1/1].

2483. Joly, Pierre, & Alain Morand, 1994. Theoretical habitat templets, species traits, and species richness: amphibians in the Upper Rhône River and its floodplain. *Freshwater Biology* 31: 455-468. [1] *A.T.:* patch dynamics; habitat analysis.

2484. Jones, Clive G., & J.H. Lawton, eds., 1995. *Linking Species and Ecosystems* [conference papers]. New York: Chapman & Hall. 387p. [1/1/2].

2485. Jones, Clive G., J.H. Lawton, & M. Shachak, 1994. Organisms as ecosystem engineers. *Oikos* 69: 373-386. [3] A study of species whose activities influence the delivery of resources to other species. *A.T.:* keystone species.

2486. Jones, Gareth E., 1987. *The Conservation of Ecosystems and Species*. London and New York: Croom Helm. 277p. [1/1/1].

2487. Jones, Michael J., 1996. Accounting for biodiversity: a pilot study. *British Accounting Review* 28: 281-303. [1] *A.T.:* natural inventories.

2488. Jones, Peter G., S.E. Beebe, et al., 1997. The use of geographical information systems in biodiversity exploration and conservation. *Biodiversity and Conservation* 6: 947-958. *A.T.:* beans; Colombia; wild relatives; germplasm collections.

2489. Jones, P.J., 1994. Biodiversity in the Gulf of Guinea: an overview. *Biodiversity and Conservation* 3: 772-784. [1].

2490. Jones, Robert F., Sept. 1990. Farewell to Africa. *Audubon* 92(5): 50-104. [1] A study of bd loss in Africa, and efforts at nature conservation there.

2491. Jones, Suzanne, Jan. 1994. Endangered Species Act battles. *Fisheries* 19(1): 22-25. [1] *A.T.:* fisheries management; environmental policy; endangered species.

2492. Jones, T., 1995. Down in the woods they have no names—BioNET International: strengthening systematics in developing countries. *Biodiversity and Conservation* 4: 501-509. [1] *A.T.:* international cooperation.

2493. Jong, W. de, 1997. Developing swidden agriculture and the threat of biodiversity loss. *Agriculture, Ecosystems & Environment* 62: 187-197. *A.T.:* Kalimantan; forest management.

2494. Jordan, Andrew, July-Aug. 1994. Paying the incremental costs of global environmental protection: the evolving role of GEF. *Environment* 36(6): 12-20, 31-36. [1] *A.T.:* international cooperation.

2495. Jordan, Carl F., 1982. Amazon rain forests. *American Scientist* 70: 394-401. [1] *A.T.:* forest ecology; forest management.

2496. ———, ed., 1987. *Amazonian Rain Forests: Ecosystem Disturbance and Recovery* [case studies]. New York: Springer. 133p. [2/1/1] *A.T.:* shifting cultivation; deforestation.

2497. Jordan, R.W., & A.H.L. Chamberlain, 1997. Biodiversity among haptophyte algae. *Biodiversity and Conservation* 6: 131-152. *A.T.:* algal blooms.

2498. Jordan, William R., III, M.E. Gilpin, & J.D. Aber, eds., 1987. *Restoration Ecology: A Synthetic Approach to Ecological Research*. Cambridge, U.K., and New York: CUP. 342p. [2/2/2].

2499. Jordan, William R., III, R.L. Peters, II, & E.B. Allen, 1988. Ecological restoration as a strategy for conserving biological diversity. *Environmental Management* 12: 55-72. [1].

2500. Jörgensen, A.F., & H. Nöhr, 1996. The use of satellite images for mapping of landscape and biological diversity in the Sahel. *International J. of Remote Sensing* 17: 91-109. [1] *A.T.:* remote sensing; Senegal; birds.

2501. Joseph, L., C. Moritz, & A. Hugall, 1995. Molecular support for vicariance as a source of diversity in rainforest. *PRSL B* 260: 177-182. [2] *A.T.:* Queensland; mitochondrial DNA; refuges.

2502. Josephson, Julian, 1982. Why maintain biological diversity? *Environmental Science & Technology* 16: 94A-97A. [1] Presents general arguments.

2503. Joyce, Chris B., & P.M. Wade, eds., 1998. *European Wet Grasslands: Biodiversity, Management, and Restoration*. Chichester, U.K., and New York: Wiley. 340p.

2504. Joyce, Christopher, 19 Oct. 1991. Prospectors for tropical medicines. *New Scientist* 132(1791): 36-40. [1] *A.T.:* INBio; Merck; bioprospecting.

2505. Joyce, Christopher, 1994. *Earthly Goods: Medicine-hunting in the Rainforest*. Boston: Little, Brown. 304p. [2/2/1] *A.T.:* pharmacognosy; medicinal plants.

2506. Joyner, Christopher C., 1995. Biodiversity in the marine environment: resource implications for the law of the sea. *Vanderbilt J. of Transnational Law* 28: 635-687. [1] *A.T.:* Convention on the Law of the Sea.

2507. Judd, S., & I. Hodkinson, 1998. The biogeography and regional biodiversity of the British seed bugs (Hemiptera, Lygaeidae). *J. of Biogeography* 25: 227-249.

2508. Jude, David J., 1997. Round gobies: cyberfish of the third millenium. *Great Lakes Research Review* 3(1): 27-34. *A.T.:* Great Lakes; introduced fish.

2509. Jukofsky, Diane, Sept.-Oct. 1992. Path of the Panther. *Wildlife Conservation* 95(5): 18-24. [1] Describes a Central American conservation project.

2510. ———, July 1993. Can marketing save the rainforest? *E: The Environmental Magazine* 4(4): 32-39. [1] *A.T.:* natural products; consumer goods.

2511. Juma, Calestous, 1989. *The Gene Hunters: Biotechnology and the Scramble for Seeds.* Princeton, NJ: Princeton University Press. 288p. [2/2/1] *A.T.:* plant germplasm resources; Kenya; sustainable agriculture.

2512. Jungius, Hartmut, 1988. The national parks and protected areas concept and its applications to the Arabian Peninsula. *Fauna of Saudi Arabia* 9: 3-11. [1].

2513. Jürgens, Norbert, 1997. Floristic biodiversity and history of African arid regions. *Biodiversity and Conservation* 6: 495-514. *A.T.:* biogeography.

2514. Jusoff, K., & N.M. Majid, 1995. Integrating needs of the local community to conserve forest biodiversity in the state of Kelantan. *Biodiversity and Conservation* 4: 108-114. [1] *A.T.:* Malaysia; local participation.

2515. Just, Jean, 1998. Biodiversity: the scene in Australia and the role of the Australian Biological Resources Study. *International J. for Parasitology* 28: 881-885. *A.T.:* environmental policy.

2516. Just, Raymond A., 1996. Intergenerational standing under the Endangered Species Act: giving back the right to biodiversity after *Lujan v. Defenders of Wildlife. Tulane Law Review* 71: 597-633. [1].

2517. Juste, Javier, & J.E. Fa, 1994. Biodiversity conservation in the Gulf of Guinea islands: taking stock and preparing action. *Biodiversity and Conservation* 3: 759-771. [1] *A.T.:* international cooperation.

2518. Jutro, Peter R., 1991. Biological diversity, ecology, and global climate change. *Environmental Health Perspectives* 96: 167-170. [1] *A.T.:* pollution.

2519. Kadidal, Shayana, 1993. Plants, poverty, and pharmaceutical patents. *Yale Law J.* 103: 223-258. [1] *A.T.:* developing countries.

2520. Kadidal, Shayana, 1997. Subject-matter imperialism? Biodiversity, foreign prior art and the neem patent controversy. *IDEA (The J. of Law and Technology)* 37: 371-403. *A.T.:* developing countries.

2521. ———, 1998. United States patent prior art rules and the neem controversy: a case of subject-matter imperialism? *Biodiversity and Conservation* 7: 27-39. *A.T.:* developing countries; intellectual property.

2522. Kadmon, Ronen, & H.R. Pulliam, 1993. Island biogeography: effect of geographical isolation on species composition. *Ecology* 74: 977-981. [1] *A.T.:* Georgia; South Carolina.

2523. Kaiho, Kunio, 1994. Planktonic and benthic foraminiferal extinction events during the last 100 m. y. *Palaeogeography, Palaeoclimatology,*

Palaeoecology 111: 45-71. [2].

2524. Kaiser, J., 1997. Unique, all-taxa survey in Costa Rica "self-destructs." *Science* 276: 893. [1] Concerning how the All-Taxa Biological Inventory project was scrapped.

2525. Kalin Arroyo, Mary T., P.H. Zedler, & M.D. Fox, eds., 1995. *Ecology and Biogeography of Mediterranean Ecosystems in Chile, California, and Australia*. New York: Springer. 455p. [1/1/1].

2526. Kammer, Thomas W., T.K. Baumiller, & W.I. Ausich, 1997. Species longevity as a function of niche breadth: evidence from fossil crinoids. *Geology* 25: 219-222. *A.T.:* Mississippian; evolution.

2527. Kangas, Jyrki, & Jussi Kuusipalo, 1993. Integrating biodiversity into forest management planning and decision-making. *Forest Ecology and Management* 61: 1-15. [1] *A.T.:* species richness; rarity; vulnerability.

2528. Kangas, Jyrki, & Timo Pukkala, 1996. Operationalization of biological diversity as a decision objective in tactical forest planning. *Canadian J. of Forest Research* 26: 103-111. [1] *A.T.:* forest management.

2529. Kant, Shashi, 1997. Integration of biodiversity conservation in tropical forest and economic development of local communities. *J. of Sustainable Forestry* 4: 33-61. *A.T.:* forest management; local involvement.

2530. Kantvilas, G., & S.J. Jarman, 1993. The cryptogamic flora of an isolated rainforest fragment in Tasmania. *Botanical J. of the Linnean Society* 111: 211-228. [1] *A.T.:* rainforest fragments; lichens; bryophytes.

2531. Kapke, Paul A., H.P. Jorgensen, & M.F. Rothschild, spring 1997. Unique collaborative conservation effort scores a win for America's rarest swine breed. *Diversity* 13(1): 24-25. [1] *A.T.:* livestock breeding; mulefoot pig.

2532. Kapoor-Vijay, Promila, 1992. *Biological Diversity and Genetic Resources: The Programme of the Commonwealth Science Council*. London: Commonwealth Science Council. 145p. [1/1/1] *A.T.:* germplasm resources.

2533. Kapoor-Vijay, Promila, & James White, eds., 1992. *Conservation Biology: A Training Manual for Biological Diversity and Genetic Resources*. London: Commonwealth Science Council. 248p. [1/1/1] *A.T.:* germplasm resources.

2534. Kappelle, Maarten., & M.E. Juárez, 1994. The Los Santos Forest Reserve: a buffer zone vital for the Costa Rican La Amistad Biosphere Reserve. *Environmental Conservation* 21: 166-169. [1] *A.T.:* nature reserves.

2535. Karakassis, I., 1995. S_∞: a new method for calculating macrobenthic species richness. *Marine Ecology Progress Series* 120: 299-303. [1] *A.T.:* sampling; species-area relation.

2536. Kareiva, Peter, 1993. No shortcuts in new maps. *Nature* 365: 292-293. [1] *A.T.:* bd mapping; Great Britain; GAP.

2537. ———, 1994. Diversity begets productivity. *Nature* 368: 686-687. [1] Discusses research by Shahid Naeem. *A.T.:* Ecotron facility; microcosms;

England.

2538. ———, 1996. Diversity and sustainability on the prairie. *Nature* 379: 673-674. [1] *A.T.:* Ecotron facility; microcosms; David Tilman; England.

2539. ———, 1996. Developing a predictive ecology for non-indigenous species and ecological invasions [editorial]. *Ecology* 77: 1651-1652. [2] The introductory essay to a special feature.

2540. Kareiva, Peter, & Uno Wennergren, 1995. Connecting landscape patterns to ecosystem and population processes. *Nature* 373: 299-302. [2] *A.T.:* spatially explicit models; patch dynamics.

2541. Kareiva, Peter M., J.G. Kingsolver, & R.B. Huey, eds., 1993. *Biotic Interactions and Global Change* [conference papers]. Sunderland, MA: Sinauer Associates. 559p. [2/2/2] *A.T.:* climatic change; environmental change.

2542. Karkkainen, Bradley C., 1997. Biodiversity and land. *Cornell Law Review* 83: 1-104. *A.T.:* public lands.

2543. Karlson, Ronald H., & H.V. Cornell, 1998. Scale-dependent variation in local vs. regional effects on coral species richness. *Ecological Monographs* 68: 259-274. *A.T.:* spatial scale; historical influences.

2544. Karp, Angela, & D.S. Ingram, 1995. Biotechnology, biodiversity, and conservation [editorial]. *Bio/technology* 13: 522. [1].

2545. Karp, Angela, P.G. Isaac, & D.S. Ingram, eds., 1998. *Molecular Tools for Screening Biodiversity: Plants and Animals*. London: Chapman & Hall. 498p. *A.T.:* molecular screening.

2546. Karp, Angela, Ole Seberg, & Marcello Buiatti, 1996. Molecular techniques in the assessment of botanical diversity. *Annals of Botany* 78: 143-149. [1] *A.T.:* population genetics; DNA analysis.

2547. Karp, Angela, K.J. Edwards, et al., 1997. Molecular technologies for biodiversity evaluation: opportunities and challenges. *Nature Biotechnology* 15: 625-628. *A.T.:* genetic variation; DNA analysis.

2548. Karpachevskii, L.O., 1996. Soil cover pattern and the diversity of forest phytocenoses. *Eurasian Soil Science* 29: 651-656. [1] *A.T.:* soil formation.

2549. Karr, James R., 1976. On the relative abundance of migrants from the North Temperate zone in tropical habitats. *Wilson Bull.* 88: 433-458. [1] *A.T.:* migrant birds.

2550. ———, 1976. Within- and between-habitat avian diversity in African and Neotropical lowland habitats. *Ecological Monographs* 46: 457-481. [1] *A.T.:* community structure; Panama; Liberia.

2551. ———, 1991. Biological integrity: a long-neglected aspect of water resource management. *Ecological Applications* 1: 66-84. [2] *A.T.:* ecosystem monitoring.

2552. Karr, James R., & K.E. Freemark, 1983. Habitat selection and environmental gradients: dynamics in the "stable tropics." *Ecology* 64: 1481-1494. [2] *A.T.*: birds; nonequilibrium communities.

2553. Karr, James R., & R.R. Roth, 1971. Vegetation structure and avian diversity in several New World areas. *American Naturalist* 105: 423-435. [2].

2554. Kate, Kerry ten, 1995. *Biopiracy or Green Petroleum? Expectations & Best Practice in Bioprospecting.* London: Overseas Development Administration. 61p. [1/1/1].

2555. Kato, M., S. Kawakami, & H. Shimizu, eds., 1991. *Asian Botanical Gardens, Nature Conservation and Genetic Resources* [conference proceedings]. Tokyo: Japan Association of Botanical Gardens. 168p. [1/1/1] *A.T.*: plant conservation.

2556. Kattan, Gustavo H., 1992. Rarity and vulnerability: the birds of the Cordillera Central of Colombia. *CB* 6: 64-70. [1] *A.T.*: cloud forests.

2557. Kattan, Gustavo H., H. Alvarez-López, & M. Giraldo, 1994. Forest fragmentation and bird extinctions: San Antonio eighty years later. *CB* 8: 138-146. [2] *A.T.*: Colombia; Andes; cloud forests; local extinctions.

2558. Katzman, Martin T., & W.G. Cale, Jr., 1990. Tropical forest preservation using economic incentives: a proposal of conservation easements. *Bioscience* 40: 827-832. [1] *A.T.*: forest management.

2559. Kauffman, Erle G., & D.H. Erwin, March 1995. Surviving mass extinctions. *Geotimes* 40(3): 14-17. [1] Surveys recent research on recovery from extinction episodes.

2560. Kauffman, J.B., R.L. Sanford, Jr., et al., 1993. Biomass and nutrient dynamics associated with slash fires in Neotropical dry forests. *Ecology* 74: 140-151. [1] *A.T.*: biomass burning; Brazil; caatinga; nutrient depletion.

2561. Kaufman, Dawn M., 1995. Diversity of New World mammals: universality of the latitudinal gradients of species and bauplans. *J. of Mammalogy* 76: 322-334. [2] *A.T.*: latitudinal diversity gradients; biotic and abiotic factors.

2562. Kaufman, Donald G., & C.M. Franz, 1996. *Biosphere 2000: Protecting Our Global Environment* (2nd ed.). Dubuque, IA: Kendall/Hunt. various pagings. [1/1/1] *A.T.*: human ecology.

2563. Kaufman, J.H., D. Brodbeck, & O.R. Melroy, 1998. Critical biodiversity. *CB* 12: 521-532. *A.T.*: simulations; extinction; niche modelling.

2564. Kaufman, Les, 1992. Catastrophic change in species-rich freshwater ecosystems: the lessons of Lake Victoria. *Bioscience* 42: 846-858. [2] *A.T.*: eutrophication; bd loss; fish.

2565. Kaufman, Les, & Kenneth Mallory, eds., 1993. *The Last Extinction* (2nd ed.) [from a public lecture series]. Cambridge, MA: MIT Press. 242p. [3/3/1] *A.T.*: endangered species.

2566. Kaufmann, R.S., K.L. Smith, Jr., et al., 1994. Epipelagic communities in the northwestern Weddell Sea: results from acoustic, trawl, and trapping surveys. *Antarctic J. of the United States* 28(5): 138-141. [1] *A.T.:* sea ice fauna.

2567. Kavanagh, Rodney P., Stephen Debus, et al., 1995. Distribution of nocturnal forest birds and mammals in northeastern New South Wales: relationships with environmental variables and management history. *Wildlife Research* 22: 359-377. [1] *A.T.:* arboreal marsupials.

2568. Kawanabe, Hiroya, Takayuki Ohgushi, & Masahiko Higashi, eds., 1993. *Symbiosphere: Ecological Complexity for Promoting Biodiversity* [workshop report]. Paris: Biology International Special Issue No. 29, IUBS. 86p. [1/1/1].

2569. Kawasaki, Tsuyoshi, S. Tanaka, et al., eds., 1991. *Long-term Variability of Pelagic Fish Populations and Their Environment* [symposium proceedings]. Oxford, U.K., and New York: Pergamon Press. 402p. [1/1/1] *A.T.:* marine fishes; marine fisheries.

2570. Kay, Charles E., 1994. Aboriginal overkill: the role of Native Americans in structuring Western ecosystems. *Human Nature* 5: 359-398. [1] Argues that activities of Native Americans helped to maximize structural bd.

2571. Kay, E. Alison, ed., 1995. *The Conservation Biology of Molluscs: Proceedings of a Symposium.* Gland, Switzerland: IUCN. 81p. [1/1/1].

2572. Kay, J.J., & E. Schneider, 1994. Embracing complexity: the challenge of the ecosystem approach. *Alternatives* 20(3): 32-39. [2] *A.T.:* systems theory; environmental management.

2573. Kearns, Carol Ann, & D.W. Inouye, 1997. Pollinators, flowering plants, and conservation biology. *Bioscience* 47: 297-307. *A.T.:* bees; endangered plants.

2574. Keast, Allen, ed., 1981. *Ecological Biogeography of Australia.* The Hague and Boston: W. Junk. 3 vols. [1/1/1].

2575. ———, ed., 1990. *Biogeography and Ecology of Forest Bird Communities* [symposium papers]. The Hague: SPB Academic. 410p. [1/1/1].

2576. Keast, Allen, & S.E. Miller, eds., 1996. *The Origin and Evolution of Pacific Island Biotas, New Guinea to Eastern Polynesia: Patterns and Processes.* Amsterdam: SPB Academic Publishers. 531p. [1/1/1].

2577. Keast, Allen, & E.S. Morton, eds., 1980. *Migrant Birds in the Neotropics: Ecology, Behavior, Distribution, and Conservation* [symposium proceedings]. Washington, DC: Smithsonian Institution Press. 576p. [2/2/1].

2578. Keating, K.A., & J.F. Quinn, 1998. Estimating species richness: the Michaelis-Menten model revisited. *Oikos* 81: 411-416. *A.T.:* Montana; vascular plants.

2579. Keating, Michael, 1993. *The Earth Summit's Agenda for Change: A Plain Language Version of Agenda 21 and the Other Rio Agreements.* Geneva: Centre for Our Common Future. 70p. [1/1/1] *A.T.:* UNCED.

2580. Keatinge, J.D.H., L.A. Materon, et al., 1995. The role of rhizobial biodiversity in legume crop productivity in the West Asian Highlands. I.

Rationale, methods and overview. *Experimental Agriculture* 31: 473-483. [1] *A.T.:* bacteria; nitrogen fixation.

2581. Keeley, Jon E., ed., 1993. *Interface Between Ecology and Land Development in California* [symposium proceedings]. Los Angeles: Southern California Academy of Sciences. 297p. [1/1/1] *A.T.:* wildlife conservation; restoration ecology.

2582. Keighery, Greg J., 1995. The ecological consequences of genetic engineering. *Search* 26: 274-276. [1] *A.T.:* Australia; biotechnology; crop plants.

2583. ———, 1996. Phytogeography, biology and conservation of Western Australian Epacridaceae. *Annals of Botany* 77: 347-355. [1] *A.T.:* pollination.

2584. Keiter, Robert B., 1994. Conservation biology and the law: assessing the challenges ahead. *Chicago-Kent Law Review* 69: 911-933. [1] *A.T.:* public lands; environmental law.

2585. ———, 1998. Ecosystems and the law: toward an integrated approach. *Ecological Applications* 8: 332-341. Examines how American law conflicts with the new concepts of ecosystem management.

2586. Kelleher, Graeme, Chris Bleakley, & Susan Wells, eds., 1995. *A Global Representative System of Marine Protected Areas*. Washington, DC: World Bank. 4 vols. [1/1/2] *A.T.:* environmental policy.

2587. Kellenberger, E., 1994. Genetic ecology: a new interdisciplinary science, fundamental for evolution, biodiversity and biosafety evaluations. *Experientia* 50: 429-437. [1] *A.T.:* ecological genetics; genetic exchange.

2588. Keller, A.E., & T.L. Crisman, 1990. Factors influencing fish assemblages and species richness in subtropical Florida lakes and a comparison with temperate lakes. *Canadian J. of Fisheries and Aquatic Sciences* 47: 2137-2146. [1] *A.T.:* acid deposition.

2589. Keller, Gerta, 1989. Extended period of extinctions across the Cretaceous/Tertiary boundary in planktonic foraminifera of continental shelf sections: implications for impact and volcanism theories. *Geological Society of America Bull.* 101: 1408-1419. [2] *A.T.:* mass extinctions; asteroid impacts; Tunisia; Texas; environmental change.

2590. Keller, Gerta, Enriqueta Barrera, et al., 1993. Gradual mass extinction, species survivorship, and long-term environmental changes across the Cretaceous-Tertiary boundary in high latitudes. *Geological Society of America Bull.* 105: 979-997. [2] *A.T.:* mass extinctions; oceanic instability.

2591. Kellert, Stephen R., 1985. Social and perceptual factors in endangered species management. *J. of Wildlife Management* 49: 528-536. [1] *A.T.:* developing countries; environmental perception.

2592. ———, 1993. Attitudes, knowledge, and behavior toward wildlife among the industrial superpowers: United States, Japan, and Germany. *J. of Social Issues* 49(1): 53-69. [1] *A.T.:* wildlife attitudes; environmental perception.

2593. ———, 1993. Values and perceptions of invertebrates. *CB* 7: 845-855. [1] *A.T.:* Connecticut; environmental perception.

2594. ———, 1995. Managing for biological and sociological diversity, or "deja vu, all over again" [edited address]. *Wildlife Society Bull.* 23: 274-278. [1] *A.T.:* environmental perception; wildlife management; societal values.

2595. ———, 1996. *The Value of Life: Biological Diversity and Human Society.* Washington, DC: Island Press/Shearwater Books. 263p. [2/3/1] *A.T.:* nature philosophy; human ecology.

2596. ———, 1997. *Kinship to Mastery: Biophilia in Human Evolution and Development.* Washington, DC: Island Press/Shearwater Books. 256p. [2/2/-] *A.T.:* nature philosophy; human ecology.

2597. Kellert, Stephen R., M. Black, et al., 1996. Human culture and large carnivore conservation in North America. *CB* 10: 977-990. [1] *A.T.:* Rocky Mountains; wildlife management; environmental perception.

2598. Kellert, Stephen R., & E.O. Wilson, eds., 1993. *The Biophilia Hypothesis* [essays]. Washington, DC: Island Press. 484p. [2/3/2] *A.T.:* nature philosophy, human ecology.

2599. Kellman, Martin, 1996. Redefining roles: plant community reorganization and species preservation in fragmented systems. *Global Ecology and Biogeography Letters* 5: 111-116. [1] *A.T.:* habitat fragmentation; forest conservation.

2600. Kellman, Martin, & J. Meave, 1997. Fire in the tropical gallery forests of Belize. *J. of Biogeography* 24: 23-34. *A.T.:* forest fragments.

2601. Kellman, Martin, & Rosanne Tackaberry, 1997. *Tropical Environments: The Functioning and Management of Tropical Ecosystems.* London and New York: Routledge. 380p. [1/1/-].

2602. Kellman, Martin, Rosanne Tackaberry, et al., 1994. Tropical gallery forests. *Research & Exploration* 10: 92-103. [1] *A.T.:* Belize; Venezuela; forest fragments.

2603. Kellogg, Erin, ed., 1992. *Coastal Temperate Rain Forests: Ecological Characteristics, Status, and Distribution Worldwide.* Portland, OR: Ecotrust, and Washington, DC: Conservation International. 64p. [1/1/1].

2604. Kelly, B.J., J.B. Wilson, & A.F. Mark, 1989. Causes of the species-area relation: a study of islands in Lake Manapouri, New Zealand. *J. of Ecology* 77: 1021-1028. [1] *A.T.:* island biogeography.

2605. Kelly, Joyce M., & M.R. Hodge, 1996. The role of corporations in ensuring biodiversity. *Environmental Management* 20: 947-954. [1].

2606. Kelly, J.D., & A.W. English, 1997. Conservation biology and the preservation of biodiversity in Australia: a role for zoos and the veterinary profession. *Australian Veterinary J.* 75: 568-574.

2607. Kelly, J.R., & M.A. Harwell, 1990. Indicators of ecosystem recovery. *Environmental Management* 14: 527-545. [1] *A.T.:* disturbance.

2608. Kelso, John R.M., & C.K. Minns, 1996. Is fish species richness at sites in the Canadian Great Lakes the result of local or regional factors? *Canadian J. of Fisheries and Aquatic Sciences* 53, Suppl.: 175-193. [1] *A.T.:* Ontario.

2609. Kemf, Elizabeth, ed., 1993. *The Law of the Mother: Protecting Indigenous Peoples in Protected Areas.* San Francisco: Sierra Club Books. 296p. [2/1/1] *A.T.:* human ecology.

2610. Kemp, R.H., Gene Namkoong, & F.H. Wadsworth, 1993. *Conservation of Genetic Resources in Tropical Forest Management: Principles and Concepts.* Rome: FAO. 105p. [1/1/1] *A.T.:* rainforest conservation.

2611. Kemp, W.P., & M.M. Cigliano, 1994. Drought and rangeland grasshopper species diversity. *Canadian Entomologist* 126: 1075-1092. [1] *A.T.:* Montana.

2612. Kenchington, Richard A., & M.T. Agardy, 1990. Achieving marine conservation through biosphere reserve planning and management. *Environmental Conservation* 17: 39-44. [1].

2613. Kenchington, Richard A., & C. Bleakley, 1994. Identifying priorities for marine protected areas in the insular Pacific. *Marine Pollution Bull.* 29: 3-9. [1] *A.T.:* environmental management.

2614. Kendall, M.A., & M. Aschan, 1993. Latitudinal gradients in the structure of macrobenthic communities: a comparison of Arctic, temperate and tropical sites. *J. of Experimental Marine Biology and Ecology* 172: 157-169. [2] *A.T.:* Spitsbergen; North Sea; Java.

2615. Kennedy, A.C., & K.L. Smith, 1995. Soil microbial diversity and the sustainability of agricultural soils. *Plant and Soil* 170: 75-86. [2] *A.T.:* agroecosystems.

2616. Kennedy, A.D., 1995. Antarctic terrestrial ecosystem response to global environmental change. *ARES* 26: 683-704. [2] *A.T.:* global warming; ultraviolet radiation; carbon cycle.

2617. Kennedy, Clive R., & A.O. Bush, 1992. Species richness in helminth communities: the importance of multiple congeners. *Parasitology* 104: 189-197. [1] *A.T.:* parasites.

2618. Kennedy, Clive R., & J.-F. Guégan, 1994. Regional versus local helminth parasite richness in British freshwater fish: saturated or unsaturated parasite communities? *Parasitology* 109: 175-185. [2].

2619. Kennedy, Clive R., & T. Pojmanska, 1996. Richness and diversity of helminth parasite communities in the common carp and in three more recently introduced carp species. *J. of Fish Biology* 48: 89-100. [1] *A.T.:* Europe; introduced fish.

2620. Kennett, J.P., & L.D. Stott, 1991. Abrupt deep-sea warming, palaeoceanographic changes and benthic extinctions at the end of the Palaeocene. *Nature* 353: 225-229. [2] *A.T.:* global warming; mass extinctions.

2621. Kenrick, Paul, & P.R. Crane, 1997. *The Origin and Early Diversification of Land Plants: A Cladistic Study*. Washington, DC: Smithsonian Institution Press. 441p. [2/2/2].

2622. ———, 1997. The origin and early evolution of plants on land. *Nature* 389: 33-39. [2] *A.T.:* Silurian; Devonian; diversification.

2623. Kent, M., 1989. Habitat conservation: what we can learn from the island biogeography theory? *Geography Review* 2(4): 2-6. [1] *A.T.:* habitat loss; U.K.

2624. Kent, Shawn, 1992. Biological diversity as a natural resource management policy. *Jesse Marvin Unruh Assembly Fellowship J.* 3: 85-107. [1] *A.T.:* environmental policy; California.

2625. Keogh, J. Scott, 1995. The importance of systematics in understanding the biodiversity crisis: the role of biological educators. *J. of Biological Education* 29: 293-299. [1] *A.T.:* bd education.

2626. Kerner, Hannah M., ed., 1992. *Proceedings of the Symposium on Biodiversity of Northwestern California*. Berkeley: Wildland Resources Center, University of California. 283p. [1/1/1].

2627. Kerr, Jeremy T., 1997. Species richness, endemism, and the choice of areas for conservation. *CB* 11: 1094-1100. *A.T.:* reserve selection; North America.

2628. Kerr, Jeremy T., & Laurence Packer, 1997. Habitat heterogeneity as a determinant of mammal species richness in high-energy regions. *Nature* 385: 252-254. *A.T.:* North America; energy-richness hypothesis.

2629. Kerr, Richard A., 1996. New way to read the record suggests abrupt extinction. *Science* 274: 1303-1304. [1] *A.T.:* ammonites; statistical techniques.

2630. Kerrigan, Richard W., D.B. Carvalho, et al., 1995. Indigenous and introduced populations of *Agaricus bisporus*, the cultivated button mushroom, in eastern and western Canada: implications for population biology, resource management, and conservation of genetic diversity. *Canadian J. of Botany* 73: 1925-1938. [1].

2631. Kerry, Knowles R., & Gotthilf Hempel, eds., 1990. *Antarctic Ecosystems: Ecological Change and Conservation* [symposium proceedings]. Berlin and New York: Springer. 427p. [1/1/1].

2632. Kershaw, Diana R., 1992. *Animal Diversity*. London: Chapman & Hall. 428p. Reviews all animal phyla. [1/1/1].

2633. Kershaw, M., G.M. Mace, & P.H. Williams, 1995. Threatened status, rarity, and diversity as alternative selection measures for protected areas: a test using Afrotropical antelopes. *CB* 9: 324-334. [2] *A.T.:* reserve selection.

2634. Kershaw, M., P.H. Williams, & G.M. Mace, 1994. Conservation of Afrotropical antelopes: consequences and efficiency of using different site selection methods and diversity criteria. *Biodiversity and Conservation* 3:

354-372. [2] *A.T.:* reserve selection.

2635. Kessler, Winifred B., Hal Salwasser, et al., 1992. New perspectives for sustainable natural resources management. *Ecological Applications* 2: 221-225. [2] *A.T.:* ecosystem management; Forest Service.

2636. Ketley, Harriet, 1994. Cultural diversity versus biodiversity. *Adelaide Law Review* 16: 99-160. [1] *A.T.:* Australian aborigines; environmental law.

2637. Kevan, Peter G., C.F. Greco, & Svenja Belaoussoff, 1997. Lognormality of biodiversity and abundance in diagnosis and measuring of ecosystemic health: pesticide stress on pollinators on blueberry heaths. *J. of Applied Ecology* 34: 1122-1136. *A.T.:* New Brunswick; ecosystem measures.

2638. Kevan, Peter G., V.G. Thomas, & Svenja Belaoussoff, 1997. AgrECOLture: defining the ecology in agriculture. *J. of Sustainable Agriculture* 9(2-3): 109-129. *A.T.:* agricultural ecology.

2639. Keystone Center, 1991. *Final Consensus Report on the Keystone Policy Dialogue on Biological Diversity on Federal Lands.* Keystone, CO. 96p. [1/1/1] *A.T.:* public lands; environmental policy.

2640. Khalil, Mohamed H., W.V. Reid, & Calestrous Juma, 1992. *Property Rights, Biotechnology, and Genetic Resources.* Nairobi: African Centre for Technology Studies. 58p. [1/1/1] *A.T.:* germplasm resources.

2641. Khan, M. Latif, S. Menon, & K.S. Bawa, 1997. Effectiveness of the protected area network in biodiversity conservation: a case-study of Meghalaya state. *Biodiversity and Conservation* 6: 853-868. *A.T.:* India; bd databases.

2642. Khan, T.I., 1995. Tropical deforestation and its consequences with reference to biodiversity in India. *Environmental Education and Information* 14: 31-44. [1] *A.T.:* Western Ghats; flowering plants.

2643. ———, 1997. Conservation of biodiversity in western India. *Environmentalist* 17: 283-288.

2644. Khasa, P.D., & B.P. Dancik, 1997. Managing for biodiversity in tropical forests. *J. of Sustainable Forestry* 4: 1-31. *A.T.:* forest management.

2645. ———, 1997. Sustaining tropical forest biodiversity. *J. of Sustainable Forestry* 5: 217-234.

2646. Khasa, P.D., G. Vallee, et al., 1995. Utilization and management of forest resources in Zaire. *Forestry Chronicle* 71: 479-488. [1] *A.T.:* bd loss; sustainable forestry.

2647. Khoshoo, T.N., 1994. India's biodiversity: tasks ahead [editorial]. *Current Science* 67: 577-582. [1].

2648. ———, 1995. Biodiversity, bioproductivity and biotechnology [editorial]. *Ambio* 24: 251-253. [1].

2649. Kidd, A.H., & R.M. Kidd, 1997. Aquarium visitors' perceptions and attitudes toward the importance of marine biodiversity. *Psychological Reports* 81(3, Pt. 2): 1083-1088. *A.T.:* California; environmental perception.

2650. Kienast, Felix, Otto Wildi, & B. Brzeziecki, 1998. Potential impacts of climate change on species richness in mountain forests: an ecological risk assessment. *Biological Conservation* 83: 291-305. *A.T.:* Central Europe; vascular plants.

2651. Kiester, A. Ross, 1971. Species density of North American amphibians and reptiles. *Systematic Zoology* 20: 127-137. [1] *A.T.:* latitudinal diversity gradients.

2652. Kiester, A. Ross, J.M. Scott, et al., 1996. Conservation prioritization using GAP data. *CB* 10: 1332-1342. [2] *A.T.:* gap analysis; Idaho; vertebrates.

2653. Kiflemariam, M., 1993. Dialectics of biological dimensionality and biological diversity. *Coenoses* 8: 149-158. [1] *A.T.:* Ethiopia; developing countries.

2654. Kim, Ke Chung, 1993. Biodiversity, conservation and inventory: why insects matter. *Biodiversity and Conservation* 2: 191-214. [1] *A.T.:* bd monitoring.

2655. ———, 1997. Preserving biodiversity in Korea's demilitarized zone. *Science* 278: 242-243. Describes a movement to turn the DMZ into a permanent resource.

2656. Kim, Ke Chung, & L.V. Knutson, eds., 1986. *Foundations for a National Biological Survey* [conference proceedings]. Lawrence, KS: Association of Systematic Collections. 215p. [1/1/1].

2657. Kim, Ke Chung, & R.D. Weaver, eds., 1994. *Biodiversity and Landscapes: A Paradox of Humanity* [essays]. Cambridge, U.K., and New York: CUP. 431p. [1/1/1] *A.T.:* landscape ecology.

2658. Kimball, Lee A., 1995. The Biodiversity Convention: how to make it work. *Vanderbilt J. of Transnational Law* 28: 763-775. [1].

2659. Kimmerer, William J., 1984. Diversity/stability: a criticism [comment]. *Ecology* 65: 1936-1938. [1].

2660. Kimmins, J.P., 1997. Biodiversity and its relationship to ecosystem health and integrity. *Forestry Chronicle* 73: 229-232. *A.T.:* bd measures; forest management.

2661. Kimura, H., R. Matsumoto, et al., 1997. The Vendian-Cambrian delta C-13 record, North Iran: evidence for overturning of the ocean before the Cambrian explosion. *Earth and Planetary Science Letters* 147: E1-E7. *A.T.:* Precambrian-Cambrian boundary; paleoceanography.

2662. Kimura, Motoo, 1983. *The Neutral Theory of Molecular Evolution.* Cambridge, U.K., and New York: CUP. 367p. [2/2/3] *A.T.:* molecular genetics.

2663. King, Carolyn M., 1984. *Immigrant Killers: Introduced Predators and the Conservation of Birds in New Zealand.* Auckland and New York: OUP. 224p. [1/1/1].

2664. King, S.R., T.J. Carlson, & K. Moran, 1996. Biological diversity, indigenous knowledge, drug discovery and intellectual property rights: creating reciprocity and maintaining relationships. *J. of Ethnopharmacology* 51: 45-57. [1].

2665. Kingdon, Jonathan, 1971-82. *East African Mammals: An Atlas of Evolution.* London and New York: Academic Press. 3 vols. in 7 parts. [2/3/3] Neither an atlas nor an evolutionary study *per se*, this series is nevertheless a classic both for its scope of treatment of the fauna, and Kingdon's inimitable artwork.

2666. ———, 1989. *Island Africa: The Evolution of Africa's Rare Animals and Plants.* Princeton, NJ: Princeton University Press. 287p. [2/2/2] *A.T.:* natural history.

2667. Kinley, Trevor A., & N.J. Newhouse, 1997. Relationship of riparian reserve zone width to bird density and diversity in southeastern British Columbia. *Northwest Science* 71: 75-87. *A.T.:* riparian management; forest management.

2668. Kinnaird, M.F., & T.G. O'Brien, 1996. Ecotourism in the Tangkoko DuaSudara Nature Reserve: opening Pandora's box? *Oryx* 30: 65-73. [1] *A.T.:* Sulawesi; protected areas.

2669. Kinzelbach, Ragnar, 1995. Neozoans in European waters—exemplifying the worldwide process of invasion and species mixing. *Experientia* 51: 526-538. [1] *A.T.:* biological invasions; freshwater.

2670. Kirilenko, A.P., & A.M. Solomon, 1998. Modeling dynamic vegetation response to rapid climate change using bioclimatic classification. *Climatic Change* 38: 15-49. *A.T.:* dispersal; carbon flux.

2671. Kirkpatrick, James B., 1983. An iterative method for establishing priorities for the selection of nature reserves: an example from Tasmania. *Biological Conservation* 25: 127-134. [1].

2672. Kirkpatrick, James B., & M.J. Brown, 1994. A comparison of direct and environmental domain approaches to planning reservation of forest higher plant communities and species in Tasmania. *CB* 8: 217-224. [1] *A.T.:* reserve design.

2673. Kirsop, B.E., 1996. The Convention on Biological Diversity: some implications for microbiology and microbial culture collections. *J. of Industrial Microbiology & Biotechnology* 17: 505-511. [1].

2674. Kirsop, B.E., & D.L. Hawksworth, eds., 1994. *The Biodiversity of Microorganisms and the Role of Microbial Resource Centres.* Braunschweig, Germany: World Federation for Culture Collections. 104p. [1/1/1] *A.T.:* germplasm resources; culture collections.

2675. Kiss, Agnes, ed., 1990. *Living With Wildlife: Wildlife Resource Management With Local Participation in Africa.* Washington, DC: World Bank Technical Paper No. 130. 217p. [1/1/1].

2676. Kitahara, M., & K. Fujii, 1994. Biodiversity and community structure of temperate butterfly species within a gradient of human disturbance: an analysis based on the concept of generalist vs. specialist strategies. *Researches on Population Ecology* 36: 187-199. [1] *A.T.:* Japan.

2677. Kitchell, Jennifer A., D.L. Clark, & A.M. Gombos, Jr., 1986. Biological selectivity of extinction: a link between background and mass extinction. *Palaios* 1: 504-511. [2] *A.T.:* Arctic Ocean; diatoms; molluscs; Cretaceous-Tertiary boundary.

2678. Kitching, I.J., 1996. Identifying complementary areas for conservation in Thailand: an example using owls, hawkmoths, and tiger beetles. *Biodiversity and Conservation* 5: 841-858. [1] *A.T.:* indicator taxa; protected areas.

2679. Kitching, Roger L., 1994. Biodiversity: political responsibilities and agendas for research and conservation. *Pacific Conservation Biology* 1: 279-283. [1].

2680. Kitching, Roger L., J.M. Bergelson, et al., 1993. The biodiversity of arthropods from Australian rainforest canopies: general introduction, methods, sites and ordinal results. *Australian J. of Ecology* 18: 181-191. [1].

2681. Kitzmiller, Jennings H., 1990. Managing genetic diversity in a tree improvement program. *Forest Ecology and Management* 35: 131-149. [1] *A.T.:* Forest Service.

2682. Klatt, Brian J., & L.W. Neal, 1996. Benefits of biodiversity management at landfills. *Waste Age* 27(6): 71-85. [1].

2683. Kleiman, Devra G., 1989. Reintroduction of captive mammals for conservation. Guidelines for reintroducing endangered species into the wild. *Bioscience* 39: 152-161. [2].

2684. Klemm, Cyrille de, 1989. The conservation of biological diversity: state obligations and citizens' duties. *Environmental Policy and Law* 19: 50-57. [1].

2685. [not used]

2686. Klemm, Cyrille de, 1996. *Introductions of Non-native Organisms Into the Natural Environment*. Strasbourg: Council of Europe. 91p. [1/1/1] *A.T.:* bd law; Europe.

2687. ———, 1997. The regulation and management of destructive processes: a new type of instrument for the conservation of biological diversity. *Environmental Policy and Law* 27: 350-354.

2688. Klemm, Cyrille de, & Clare Shine, 1993. *Biological Diversity Conservation and the Law: Legal Mechanisms for Conserving Species and Ecosystems*. Gland, Switzerland: IUCN Environmental Policy and Law Paper No. 29. 292p. [1/1/1] *A.T.:* bd law; international cooperation.

2689. Klicka, John, & R.A. Zink, 1997. The importance of recent ice ages in speciation: a failed paradigm. *Science* 277: 1666-1669. [2] *A.T.:* Pleistocene glaciations; North America; songbirds.

2690. Kline, Gary, spring 1998. Biodiversity and development. *J. of Third World Studies* 15(1): 125-139. Discusses bd decline and some policies that might be implemented to prevent it.

2691. Klinger, Janeen, 1994. Debt-for-nature swaps and the limits to international cooperation on behalf of the environment. *Environmental Politics* 3: 229-246. [1] *A.T.:* environmental policy.

2692. Klinkenborg, Verlyn, Jan.-Feb. 1992. The making of a biopolitician. *Audubon* 94(1): 90-93. [1] Debates the wisdom of commercializing conservation efforts. *A.T.:* Russell Mittermeier.

2693. ———, Jan.-Feb. 1993. Barnyard biodiversity. *Audubon* 95(1): 78-86, 88. [1] Considers the loss of livestock breeds worldwide.

2694. Klopfer, Peter H., & R.H. MacArthur, 1960. Niche size and faunal diversity. *American Naturalist* 94: 293-300. [1] *A.T.:* latitudinal gradients; niche relationships.

2695. Kloppenburg, Jack R., Jr., ed., 1988. *Seeds and Sovereignty: The Use and Control of Plant Genetic Resources*. Durham, NC: Duke University Press. 368p. [1/1/1] *A.T.:* plant germplasm resources.

2696. ———, 1988. *First the Seed: The Political Economy of Plant Biotechnology, 1492-2000*. Cambridge, U.K., and New York: CUP. 349p. [2/3/2].

2697. Kloppenburg, Jack R., Jr., & D.L. Kleinman, 1987. The plant germplasm controversy: analysing empirically the distribution of the world's plant genetic resources. *Bioscience* 37: 190-198. [1].

2698. Knight, D., E. LeDrew, & H. Holden, 1997. Mapping submerged corals in Fiji from remote sensing and *in situ* measurements: applications for integrated coastal management. *Ocean & Coastal Management* 34: 153-170.

2699. Knight, Richard L., 1996. Aldo Leopold, the land ethic, and ecosystem management. *J. of Wildlife Management* 60: 471-474. [1] Relates Leopold's advances to the more modern concept.

2700. Knight, Richard L., & K.J. Gutzwiller, eds., 1995. *Wildlife and Recreationists: Coexistence Through Management and Research*. Washington, DC: Island Press. 372p. [1/1/1] *A.T.:* wildlife conservation.

2701. Knoll, Andrew H., 1989. Evolution and extinction in the marine realm: some constraints imposed by phytoplankton. *PTRSL B* 325: 279-290. [1].

2702. ———, 1992. The early evolution of eukaryotes: a geological perspective. *Science* 256: 622-627. [3].

2703. Knoll, Andrew H., R.K. Bambach, et al., 1996. Comparative earth history and Late Permian mass extinction. *Science* 273: 452-457. [2] *A.T.:* carbon dioxide; paleoceanography.

2704. Knoll, Andrew H., K.J. Niklas, et al., 1984. Character diversification and patterns of evolution in early vascular plants. *Paleobiology* 10: 34-47. [1].

2705. Knopf, Fritz L., 1992. Faunal mixing, faunal integrity, and the biopolitical template for diversity conservation. *Trans. of the 57th North Amer-*

ican Wildlife and Natural Resources Conference: 330-342. [1].

2706. Knopf, Fritz L., & F.B. Samson, 1994. Scale perspectives on avian diversity in Western riparian ecosystems. *CB* 8: 669-676. [1] *A.T.:* corridors.

2707. ———, 1997. *Ecology and Conservation of Great Plains Vertebrates*. New York: Springer. 320p. [1/1/-].

2708. Knopf, Fritz L., & Michael Scott, June-Aug. 1990. Riparian areas —green ribbons of diversity. *Fish and Wildlife News* (June-Aug.): 19. [1] *A.T.:* waterbirds; corridors.

2709. Knowles, P.F., 1969. Centers of plant diversity and conservation of crop germplasm: safflower. *Economic Botany* 23: 324-329. [1].

2710. Knuth, B.A., spring 1995. Biodiversity's future: species to systems. *Forum for Applied Research and Public Policy* 10(1): 88-91. [1].

2711. Knutson, Lloyd V., & A.K. Stoner, eds., 1989. *Biotic Diversity and Germplasm Preservation, Global Imperatives: Invited Papers Presented at a Symposium*. Dordrecht and Boston: Kluwer Academic. 530p. [1/1/1].

2712. Kobayashi, S., & K. Kimura, 1994. The number of species occurring in a sample of a biotic community and its connections with species-abundance relationship and spatial distribution. *Ecological Research* 9: 281-294. [1] *A.T.:* Thailand; species-area relation.

2713. Koch, C.F., & J.P. Morgan, 1988. On the expected distribution of species' ranges. *Paleobiology* 14: 126-138. [1] *A.T.:* stepwise extinction; sampling; distribution patterns.

2714. Koch, Eddie, Dec. 1991. People preserves: apartheid's legacy pits wildlife against settlement. *Scientific American* 265(6): 54. [1] *A.T.:* protected areas; South Africa; park-people relations.

2715. Kochmer, J.P., & R.H. Wagner, 1988. Why are there so many kinds of passerine birds? Because they are small. [letter] *Systematic Zoology* 37: 68-69. [1].

2716. Kofron, Christopher P., & Angela Chapman, 1995. Deforestation and bird species composition in Liberia, West Africa. *Tropical Zoology* 8: 239-256. [1] *A.T.:* habitat fragmentation; tropical rainforests.

2717. Kohli, R.K., Neelima Jerath, & Daizy Batish, eds., 1996. *Some Facets of Biodiversity* [seminar papers and proceedings]. Chandigarh, India: Society of Environmental Scientists & Punjab State Council for Science and Technology. 286p. [1/1/1] *A.T.:* India; bdc.

2718. Kohm, Kathryn A., ed., 1991. *Balancing on the Brink of Extinction: The Endangered Species Act and Lessons for the Future*. Washington, DC: Island Press. 318p. [2/1/1] *A.T.:* environmental law; endangered species.

2719. Kohn, D.D., & D.M. Walsh, 1994. Plant species richness—the effect of island size and habitat diversity. *J. of Ecology* 82: 367-377. [2] *A.T.:* island biogeography; Shetland Islands.

2720. Kohn, Robert E., 1993. Measuring the existence value of wildlife: comment. *Land Economics* 69: 304-308. [1] *A.T.:* environmental economics.

2721. Kolasa, Jurek, & S.T.A. Pickett, eds., 1991. *Ecological Heterogeneity*. New York: Springer. 332p. [1/1/2].

2722. Kolmes, Steven, & K.J. Mitchell, 1990. Information theory and biological diversity. *UMAP J.* 11: 25-62. [1] *A.T.:* mathematical models.

2723. Komen, Joris, April-June 1996. Inventorying biodiversity: an African perspective. *Museum International* 48(2): 31-34. [1] *A.T.:* museum collections.

2724. Koopowitz, Harold, & Hilary Kaye, 1983. *Plant Extinction: A Global Crisis*. Washington, DC: Stonewall Press. 239p. [3/2/1] *A.T.:* plant conservation.

2725. Koopowitz, Harold, A.D. Thornhill, & Mark Andersen, 1994. A general stochastic model for the prediction of biodiversity losses based on habitat conversion. *CB* 8: 425-438. [1] *A.T.:* Ecuador; mathematical models.

2726. Koponen, Timo, 1992. Endangered bryophytes on a global scale. *Biological Conservation* 59: 255-258. [1].

2727. Kopp, Raymond J., 1992. Why existence value *should* be used in cost-benefit analysis. *J. of Policy Analysis and Management* 11: 123-130. [1] *A.T.:* environmental economics.

2728. Koshland, Daniel E., Jr., 1994. The case for diversity [editorial]. *Science* 264: 639. [1] Emphasizes increase in bacterial diversity.

2729. Koslow, J.A., A. Williams, & J.R. Paxton, 1997. How many demersal fish species in the deep sea? A test of a method to extrapolate from local to global diversity. *Biodiversity and Conservation* 6: 1523-1532.

2730. Kotanen, P.M., 1995. Responses of vegetation to a changing regime of disturbance: effects of feral pigs in a Californian coastal prairie. *Ecography* 18: 190-199. [2].

2731. Kothari, Ashish, 1994. *Conserving Life: Implications of the Biodiversity Convention for India*. New Delhi: Kalpavriksh. 86p. [1/1/1] *A.T.:* CBD.

2732. ———, 1997. *Understanding Biodiversity: Life, Sustainability, and Equity*. New Delhi: Orient Longman. 161p. [1/1/-] *A.T.:* India.

2733. ———, 1997. *Conserving India's Agro-biodiversity: Prospects and Policy Implications*. London: Gatekeeper Series No. 65, International Institute for Environment and Development. 18p. [1/1/-] *A.T.:* agricultural ecology.

2734. Kotliakov, V.M., M. Uppenbrink, & V. Metreveli, eds., 1998. *Conservation of the Biological Diversity as a Prerequisite for Sustainable Development in the Black Sea Region* [workshop papers]. Dordrecht and Boston: Kluwer Academic. 518p.

2735. Kotliar, N.B., & J.A. Wiens, 1990. Multiple scales of patchiness and patch structure: a hierarchical framework for the study of heterogeneity. *Oikos* 59: 253-260. [2] *A.T.:* spatial scale; landscape ecology.

2736. Kottelat, Maurice, 1995. Systematic studies and biodiversity: the need for a pragmatic approach [editorial]. *J. of Natural History* 29: 565-569. [1].

2737. Kottelat, Maurice, & Tony Whitten, 1996. *Freshwater Biodiversity in Asia: With Special Reference to Fish.* Washington, DC: World Bank Technical Paper No. 281. 59p. [1/1/1].

2738. Kotwal, P.C., & Sujoy Banerjee, eds., 1998. *Biodiversity Conservation in Managed Forests and Protected Areas* [workshop papers]. Bikaner, India: Agro Botanica. 227p. *A.T.:* India.

2739. Kouki, Jari, Pekka Niemelä, & Matti Viitasaari, 1994. Reversed latitudinal gradient in species richness of sawflies (Hymenoptera, Symphyta). *Annales Zoologici Fennici* 31: 83-88. [1] *A.T.:* latitudinal diversity gradients; aphids; host plants.

2740. Kovács, Margit, ed., 1992. *Biological Indicators in Environmental Protection.* New York: Horwood. 207p. [1/1/1].

2741. Krajewski, Carey, 1991. Phylogeny and diversity [letter]. *Science* 254: 918-919. [1] *A.T.:* phylogenetic systematics; conservation genetics.

2742. ———, 1994. Phylogenetic measures of biodiversity: a comparison and critique. *Biological Conservation* 69: 33-39. [1] *A.T.:* DNA analysis; cranes; phylogenetic analysis.

2743. Krajick, Kevin, & Gary Braasch, Nov. 1995. The secret life of backyard trees. *Discover* 16(11): 92-101. [1] Looks at efforts to examine canopy bd in North American old-growth forests.

2744. Kramer, Randall A., Narendra Sharma, & Mohan Munasinghe, 1995. *Valuing Tropical Forests: Methodology and Case Study of Madagascar.* Washington, DC: World Bank Environment Paper No. 13. 66p. [1/1/1] *A.T.:* forest conservation.

2745. Kramer, Randall A., Carel van Schaik, & Julie Johnson, eds., 1997. *Last Stand: Protected Areas and the Defense of Tropical Biodiversity.* New York: OUP. 242p. [1/1/-].

2746. Krattiger, Anatole F., & W.H. Lesser, 1995. The "facilitator": proposing a new mechanism to strengthen the equitable and sustainable use of biodiversity. *Environmental Conservation* 22: 211-215. [1].

2747. Krattiger, Anatole F., J.A. McNeely, et al., eds., 1994. *Widening Perspectives on Biodiversity* [conference papers]. Gland, Switzerland: IUCN, and Geneva: International Academy of the Environment. 473p. [1/1/1] *A.T.:* bdc.

2748. Kraus, Fred, 1995. The conservation of unisexual vertebrate populations. *CB* 9: 956-959. [1].

2749. Krause, D.C., & M.V. Angel, 1994. Marine biogeography, climate change and societal needs. *Progress in Oceanography* 34: 221-235. [1] *A.T.:* sampling; pelagic environments; environmental monitoring.

2750. Kremen, Claire, 1992. Assessing the indicator properties of species assemblages for natural areas monitoring. *Ecological Applications* 2: 203-

217. [2].

2751. ———, 1994. Biological inventory using target taxa: a case study of the butterflies of Madagascar. *Ecological Applications* 4: 407-422. [2] *A.T.:* environmental gradients; biological indicators.

2752. Kremen, Claire, A.M. Merenlender, & D.D. Murphy, 1994. Ecological monitoring: a vital need for integrated conservation and development programs in the tropics. *CB* 8: 388-397. [2] *A.T.:* Madagascar.

2753. Kremen, Claire, Isaia Raymond, & Kate Lance, 1998. An interdisciplinary tool for monitoring conservation impacts in Madagascar. *CB* 12: 549-563. *A.T.:* ICDPs; biological indicators; ethnobotany; natural products.

2754. Kremen, Claire, R.K. Colwell, et al., 1993. Terrestrial arthropod assemblages: their use in conservation planning. *CB* 7: 796-808. [2] *A.T.:* biological indicators; bd monitoring; ICDPs.

2755. Kretzmann, Maria B., W.G. Gilmartin, et al., 1997. Low genetic variability in the Hawaiian monk seal. *CB* 11: 482-490. *A.T.:* endangered species; population bottlenecks; small populations.

2756. Krever, Vladimir, et al., eds., 1994. *Conserving Russia's Biological Diversity: An Analytical Framework and Initial Investment Portfolio.* Washington, DC: WWF. 209p. [1/1/1].

2757. Kricher, John C., 1997. *A Neotropical Companion: An Introduction to the Animals, Plants, and Ecosystems of the New World Tropics* (2nd ed.). Princeton, NJ: Princeton University Press. 451p. [2/2/1].

2758. Krishnan, Rajaram, J.M. Harris, & N.R. Goodman, eds., 1995. *A Survey of Ecological Economics.* Washington, DC: Island Press. 384p. [2/1/1] *A.T.:* sustainable development.

2759. Krivolutskii, Dmitrii A., 1996. Dynamics of biodiversity and ecosystems under conditions of radioactive contamination. *Doklady, Biological Sciences* 347: 166-168. [1].

2760. Kriwoken, L.K., 1996. Australian biodiversity and marine protected areas. *Ocean & Coastal Management* 33: 113-132. *A.T.:* coastal zone management.

2761. Kruckeberg, Arthur R., & Deborah Rabinowitz, 1985. Biological aspects of endemism in higher plants. *ARES* 16: 447-479. [2] *A.T.:* U.S.; rare species.

2762. Kruess, Andreas, & Teja Tscharntke, 1994. Habitat fragmentation, species loss, and biological control. *Science* 264: 1581-1584. [2] *A.T.:* insects; agricultural landscapes.

2763. Krugman, Stanley L., 1992. Biotechnology and biodiversity—the interrelationships. *Forestry Chronicle* 68: 459-461. [1] *A.T.:* environmental politics; forest biotechnology.

2764. Kruuk, L.E.B., & J.S. Gilchrist, 1997. Mechanisms maintaining species differentiation: predator-mediated selection in a *Bombina* hybrid zone. *PRSL B* 264: 105-110. *A.T.:* gene flow; speciation.

2765. Kubo, Takuya, & Yoh Iwasa, 1995. Inferring the rates of branching and extinction from molecular phylogenies. *Evolution* 49: 694-704. [1] *A.T.:* phylogeny; diversification.

2766. Küchler, August W., & I.S. Zonneveld, eds., 1988. *Vegetation Mapping*. Dordrecht and Boston: Kluwer Academic. 635p. [2/2/1] *A.T.:* plant geography; plant ecology.

2767. Kuliopulos, H., 1990. Amazonian biodiversity. *Science* 248: 1305. [1] A short note featuring a map of Amazonia indicating priority areas for bd preservation attempts.

2768. Kumar, Patnam V.S., 1993. Biotechnology and biodiversity: a dialectical relationship. *J. of Scientific and Industrial Research* 52: 523-532. [1].

2769. Kunin, William E., 1998. Biodiversity at the edge: a test of the importance of spatial "mass effects" in the Rothamsted Park Grass experiments. *PNAS* 95: 207-212. *A.T.:* England; grasslands.

2770. Kunin, William E., & K.J. Gaston, eds., 1997. *The Biology of Rarity: Causes and Consequences of Rare-Common Differences*. London and New York: Chapman & Hall. 280p. [1/1/-] *A.T.:* threatened and endangered species.

2771. Kunz, Thomas H., & P.A. Racey, eds., 1998. *Bat Biology and Conservation* [conference papers]. Washington, DC: Smithsonian Institution Press. 365p.

2772. Kupfer, John A., 1995. Landscape ecology and biogeography. *Progress in Physical Geography* 19: 18-34. [1] *A.T.:* reserve design.

2773. Kushan, Jeffrey P., 1995. Biodiversity: opportunities and obligations. *Vanderbilt J. of Transnational Law* 28: 755-761. [1] *A.T.:* technology transfer; CBD; North-South relations.

2774. Kusler, Jon A., W.J. Mitsch, & J.S. Larson, Jan. 1994. Wetlands. *Scientific American* 270(1): 64B-70. [1] *A.T.:* U.S.

2775. Kuss, Fred R., 1986. A review of major factors influencing plant responses to recreation impacts. *Environmental Management* 10: 637-650. [1].

2776. Kuusinen, Mikko, 1996. Epiphyte flora and diversity on basal trunks of six old-growth forest tree species in southern and middle boreal Finland. *Lichenologist* 28: 443-463. [1] *A.T.:* boreal forests.

2777. Kuusipalo, Jussi, & Jyrki Kangas, 1994. Managing biodiversity in a forestry environment. *CB* 8: 450-460. [1] *A.T.:* Finland; forest management; analytical hierarchy process.

2778. La Pierre, Yvette, Oct. 1994. Poached parklands. *American Horticulturist* 73(10): 19-23. [1] On the effective lack of protection of plants from poaching in national parks.

2779. Laarman, Jan G., & R.A. Sedjo, 1992. *Global Forests: Issues for Six Billion People*. New York: McGraw-Hill. 337p. [1/1/1] *A.T.:* forest management.

2780. Labandeira, Conrad C., & J.J. Sepkoski, Jr., 1993. Insect diversity in the fossil record. *Science* 261: 310-315. [2] *A.T.:* diversification.

2781. Lachavanne, Jean-Bernard, & Raphaelle Juge, eds., 1997. *Biodiversity in Land/Inland Water Ecotones* [conference papers]. Paris: UNESCO, and New York: Parthenon. 308p. [1/1/-] *A.T.:* riparian environments.

2782. Lacher, Thomas E., Jr., & M.I. Goldstein, 1997. Tropical ecotoxicology: status and needs. *Environmental Toxicology and Chemistry* 16: 100-111. *A.T.:* pesticides; Latin America; Caribbean region.

2783. Lack, David, 1969. The numbers of bird species on islands. *Bird Study* 16: 193-209. [1].

2784. Lacroix, G., & L. Abbadie, 1998. Linking biodiversity and ecosystem function: an introduction. *Acta Oecologica* 19: 189-193. *A.T.:* stability.

2785. Lacy, Robert C., 1997. Importance of genetic variation to the viability of mammalian populations. *J. of Mammalogy* 78: 320-335. [2] *A.T.:* genetic drift; inbreeding.

2786. Lacy, William B., 1994. Biodiversity, cultural diversity, and food equity. *Agriculture and Human Values* 11(1): 3-9. [1].

2787. Laderman, Aimlee D., ed., 1998. *Coastally Restricted Forests* [symposia papers]. New York: OUP. 334p.

2788. LaDuke, Winona, winter 1991. The trouble with Wasichu. *Earth Island J.* 6(1): 42. [1] *A.T.:* indigenous peoples; cultural diversity.

2789. Lafferty, Kevin D., & A.M. Kuris, 1996. Biological control of marine pests. *Ecology* 77: 1989-2000. [1] *A.T.:* European green crab; biological invasions.

2790. LaHaye, W.S., R.J. Gutiérrez, & H.R. Akçakaya, 1994. Spotted owl metapopulation dynamics in Southern California. *J. of Animal Ecology* 63: 775-785. [2].

2791. Laikre, Linda, & N. Ryman, 1996. Effects on intraspecific biodiversity from harvesting and enhancing natural populations. *Ambio* 25: 504-509. [1] *A.T.:* fisheries.

2792. Lake, P.S., E.S.G. Schreiber, et al., 1994. Species richness in streams: patterns over time, with stream size and with latitude. *Verhandlungen der Internationalen Vereinigung für Theoretische und Angewandte Limnologie* 25: 1822-1826. [1].

2793. Lalonde, André, & Shahid Akhtar, June 1994. Traditional knowledge research for sustainable development. *Nature and Resources* 30(2): 22-28. [1] Describes how the International Development Research Centre (Canada) has integrated indigenous knowledge into its research programs.

2794. Lamarre, Leslie, Sept.-Oct. 1996. Preserving biodiversity: a delicate balance. *EPRI J.* 21(5): 6-17. [1] Concerning the response of power companies to new ecosystem-level agenda.

2795. Lämås, Tomas, & Clas Fries, 1995. Emergence of a biodiversity concept in Swedish forest policy. *Water, Air, and Soil Pollution* 82: 57-66.

[1] *A.T.:* environmental management.

2796. Lamberson, Roland H., B.R. Noon, et al., 1994. Reserve design for territorial species: the effects of patch size and spacing on the viability of the northern spotted owl. *CB* 8: 185-195. [2] *A.T.:* Pacific Northwest; endangered species.

2797. Lambert, John, Jitendra Srivastava, & Noel Vietmeyer, 1997. *Medicinal Plants: Rescuing a Global Heritage.* Washington, DC: World Bank Technical Paper No. 355. 61p. [1/1/1] *A.T.:* China; India.

2798. Lambshead, P.J.D., 1993. Recent developments in marine benthic biodiversity research. *Océanis* 19(6): 5-24. [2] *A.T.:* marine species diversity.

2799. Lamont, B.B., 1995. Testing the effect of ecosystem composition/ structure on its functioning. *Oikos* 74: 283-295. [1] *A.T.:* ecosystem analysis.

2800. Lancaster, John, March 1992. Searching for survival: saving the parks isn't enough. *J. of Forestry* 90(3): 20-21. [1] Describes the bioreserve concept.

2801. Lande, Russell, 1988. Genetics and demography in biological conservation. *Science* 241: 1455-1460. [3] *A.T.:* genetic drift; inbreeding; small populations; minimum viable population size.

2802. ———, 1995. Mutation and conservation. *CB* 9: 782-791. [2] *A.T.:* inbreeding; genetic variation.

2803. Landen, Laura, 1992. A Thomistic analysis of the Gaia hypothesis: how new is this new look at life on Earth? *Thomist* 56: 1-17. [1].

2804. Landres, P.B., J. Verner, & J.W. Thomas, 1988. Ecological uses of vertebrate indicator species: a critique. *CB* 2: 316-328. [2].

2805. Landsberg, Jill, C.D. James, et al., 1997. *The Effects of Artificial Sources of Water on Rangeland Biodiversity* [report]. Canberra: Environment Australia and CSIRO. 208p. [1/1/-] *A.T.:* Australia.

2806. Lang, C., & O. Reymond, 1993. Empirical relationship between diversity of invertebrate communities and altitude in rivers: application to biomonitoring. *Aquatic Sciences* 55: 188-196. [1] *A.T.:* Switzerland.

2807. Lange, Lene, 1996. Microbial metabolites—an infinite source of novel chemistry. *Pure and Applied Chemistry* 68: 745-748. [1] *A.T.:* biotechnology; natural product chemistry.

2808. Langholz, J., 1996. Economics, objectives, and success of private nature reserves in sub-Saharan Africa and Latin America. *CB* 10: 271-280. [1].

2809. Langner, Linda L., & C.H. Flather, 1994. *Biological Diversity: Status and Trends in the United States.* Fort Collins, CO: General Technical Report GTR RM-244, Rocky Mountain Forest and Range Experiment Station. 24p. [1/1/1] *A.T.:* bdc.

2810. Langreth, Robert, Dec. 1994. The world according to Dan Janzen [interview]. *Popular Science* 245(6): 78-82, 112, 114-115. [1] *A.T.:* Costa Rica.

2811. Lapin, Marc, & B.V. Barnes, 1995. Using the landscape ecosystem approach to assess species and ecosystem diversity. *CB* 9: 1148-1158. [1] *A.T.:* Michigan.

2812. Larkin, P.A., 1996. Concepts and issues in marine ecosystem management. *Reviews in Fish Biology and Fisheries* 6: 139-164. [1] *A.T.:* fisheries management.

2813. LaRoe, Edward T., Jan. 1993. Implementation of an ecosystem approach to endangered species conservation. *Endangered Species Update* 10(3-4): 3-6. [1] *A.T.:* wildlife conservation.

2814. LaRoe, Edward T., G.S. Farris, et al., eds., 1995. *Our Living Resources: A Report to the Nation on the Distribution, Abundance, and Health of U.S. Plants, Animals, and Ecosystems.* Washington, DC: National Biological Service, U.S. Dept. of the Interior. 530p. [2/1/1] *A.T.:* bd; bdc; endangered species.

2815. Larwood, Gilbert P., ed., 1988. *Extinction and Survival in the Fossil Record* [symposium papers]. New York: OUP. 365p. [1/1/1].

2816. LaSalle, John, & I.D. Gauld, eds., 1993. *Hymenoptera and Biodiversity*. Wallingford, Oxon, U.K.: CAB International. 348p. [1/1/1] *A.T.:* tropics.

2817. Laska, M.S., 1997. Structure of understory shrub assemblages in adjacent secondary and old growth tropical wet forests, Costa Rica. *Biotropica* 29: 29-37. *A.T.:* species diversity.

2818. Lasley, B.L., N.M. Loskutoff, & G.B. Anderson, 1994. The limitation of conventional breeding programs and the need and promise of assisted reproduction in nondomestic species. *Theriogenology* 41: 119-132. [1].

2819. Lassen, H.H., 1975. The diversity of freshwater snails in view of the equilibrium theory of island biogeography. *Oecologia* 19: 1-8. *A.T.:* Denmark; lakes and ponds.

2820. Lasserre, Pierre, A.D. McIntyre, et al., eds., 1994. *Marine Laboratory Networks for the Study of the Biodiversity, Function and Management of Marine Ecosystems*. Paris: Biology International Special Issue No. 31, IUBS. 33p. [1/1/1].

2821. Latham, R.E., & R.E. Ricklefs, 1993. Global patterns of tree species richness in moist forests: energy-diversity theory does not account for variation in species richness. *Oikos* 67: 325-333. [2] *A.T.:* evapotranspiration; latitudinal diversity gradients.

2822. Latouche, S., E. Ortona, et al., 1997. Biodiversity of *Pneumocystis carinii hominis*: typing with different DNA regions. *J. of Clinical Microbiology* 35: 383-387. [2] *A.T.:* France; Italy.

2823. Lattin, John D., 1993. Lessons from the spotted owl—the utility of nontraditional data [editorial]. *Bioscience* 43: 666. [1] *A.T.:* Pacific Northwest; forest management; arthropods.

2824. ———, 1993. Arthropod diversity and conservation in old-growth

Northwest forests. *American Zoologist* 33: 578-587. [1] *A.T.:* Pacific Northwest.

2825. Launer, Alan E., & D.D. Murphy, 1994. Umbrella species and the conservation of habitat fragments: a case of a threatened butterfly and a vanishing grassland ecosystem. *Biological Conservation* 69: 145-153. [2] *A.T.:* California.

2826. Laurance, William F., 1991. Ecological correlates of extinction proneness in Australian tropical rain forest mammals. *CB* 5: 79-89. [2] *A.T.:* Queensland.

2827. ———, 1996. Catastrophic declines of Australian rainforest frogs: is unusual weather responsible? *Biological Conservation* 77: 203-212. [1] *A.T.:* montane environments.

2828. Laurance, William F., & R.O. Bierregaard, eds., 1997. *Tropical Forest Remnants: Ecology, Management, and Conservation of Fragmented Communities.* Chicago: University of Chicago Press. 616p. [1/-/2].

2829. Laurance, William F., & C. Gascon, 1997. How to creatively fragment a landscape. *CB* 11: 577-579. *A.T.:* habitat fragmentation; deforestation; ecosystem management; Amazonia.

2830. Laurance, William F., S.G. Laurance, et al., 1997. Biomass collapse in Amazonian forest fragments. *Science* 278: 1117-1118. [2] *A.T.:* habitat fragmentation.

2831. Laurance, William F., K.R. McDonald, & R. Speare, 1996. Epidemic disease and the catastrophic decline of Australian rain forest frogs. *CB* 10: 406-413. [2] *A.T.:* montane environments; endemism.

2832. Laurent, J.P., C. Barnabe, et al., 1997. Impact of clonal evolution on the biological diversity of *Trypanosoma cruzi*. *Parasitology* 114: 213-218. [2] *A.T.:* parasites.

2833. Lauver, Chris L., 1997. Mapping species diversity patterns in the Kansas shortgrass region by integrating remote sensing and vegetation analysis. *J. of Vegetation Science* 8: 387-394.

2834. Lauver, Chris L., & J.L. Whistler, 1993. A hierarchical classification of Landsat TM imagery to identify natural grassland areas and rare species habitat. *Photogrammetric Engineering and Remote Sensing* 59: 627-634. [1] *A.T.:* remote sensing; Kansas.

2835. Lavelle, Patrick, Eric Blanchart, et al., 1993. A hierarchical model for decomposition in terrestrial ecosystems: application to soils of the humid tropics. *Biotropica* 25: 130-150. [2] *A.T.:* savannas; tropical rainforests; microbial activity.

2836. Lavelle, Patrick, C. Lattaud, et al., 1995. Mutualism and biodiversity in soils. *Plant and Soil* 170: 23-33. [1] *A.T.:* microorganisms; soil biota; earthworms.

2837. Law, Richard, & R.D. Morton, 1996. Permanence and the assembly of ecological communities. *Ecology* 77: 762-775. [1] *A.T.:* mathematical models; community structure.

2838. Lawlor, Timothy E., 1983. The peninsular effect on mammalian species diversity in Baja California. *American Naturalist* 121: 432-439. [1].

2839. Lawrence, Patrick L., & J.G. Nelson, eds., 1994. *Lake Erie Biodiversity and Ecosystem Health* [workshop proceedings]. Waterloo, ON: Heritage Resources Centre, University of Waterloo. 130p. [1/1/1] *A.T.:* ecosystem management.

2840. Lawton, John H., 1990. Biological control of plants: a review of generalisations, rules and principles using insects as agents. *FRI Bull.*, no. 155: 3-17. [1] *A.T.:* weeds.

2841. ———, 1990. Species richness and population dynamics of animal assemblages. Patterns in body size: abundance space. *PTRSL B* 330: 283-291. [2] *A.T.:* species abundance.

2842. ———, 1993. Range, population abundance and conservation. *TREE* 8: 409-413. [2] *A.T.:* range size.

2843. ———, 1994. What do species do in ecosystems? *Oikos* 71: 367-374. [2] *A.T.:* Ecotron facility; ecosystem engineers; earthworms.

2844. Lawton, John H., D.E. Bignell, et al., 1998. Biodiversity inventories, indicator taxa and effects of habitat modification in tropical forest. *Nature* 391: 72-76. *A.T.:* habitat disturbance; Cameroon.

2845. Lawton, John H., & R.M. May, eds., 1995. *Extinction Rates*. Oxford, U.K., and New York: OUP. 233p. [2/2/2].

2846. Lawton, John H., & D. Shroder, 1977. Effects of plant type, size of geographical range and taxonomic isolation on number of insect species associated with British plants. *Nature* 265: 137-140. [1].

2847. Laxton, R.R., 1978. The measure of diversity. *J. of Theoretical Biology* 70: 51-67. [1].

2848. Le Floc'h, Edouard, James Aronson, et al., 1998. Biodiversity and ecosystem trajectories: first results from a new LTER in southern France. *Acta Oecologica* 19: 285-293. *A.T.:* ecosystem attributes; land use history.

2849. Le Guenno, Bernard, Oct. 1995. Emerging viruses. *Scientific American* 273(4): 56-64. [1] Discusses how environment modification encourages the spread of disease viruses.

2850. Le Houèrou, Henry N., 1997. Climate, flora and fauna changes in the Sahara over the past 500 million years. *J. of Arid Environments* 37: 619-647. *A.T.:* biogeography.

2851. Leach, Mark K., & T.J. Givnish, 1996. Ecological determinants of species loss in remnant prairies. *Science* 273: 1555-1558. *A.T.:* Wisconsin; habitat fragmentation.

2852. Leader-Williams, N., & S.D. Albon, 1988. Allocation of resources for conservation. *Nature* 336: 533-535. [1] Relates rate of decline of rhinoceros and elephant to lack of funding for conservation efforts. *A.T.:* wildlife conservation.

2853. Leader-Williams, N., J. Harrison, & M.J.B. Green, 1990. Designing protected areas to conserve natural resources. *Science Progress* 74: 189-204. [1] *A.T.:* Emphasizes the benefits of protected areas. *A.T.:* wildlife conservation.

2854. Leakey, Richard E., & Roger Lewin, 1995. *The Sixth Extinction: Patterns of Life and the Future of Humankind.* New York: Doubleday. 271p. [2/3/1] *A.T.:* evolution.

2855. Leal, José, Aug. 1994. Marine species: here today, here tomorrow? *Sea Frontiers* 40(4): 9-11. [1] *A.T.:* marine extinctions.

2856. Leape, Gerald, Sept. 1993. We need laws to preserve biodiversity. *World & I* 8(9): 112, 114-116. [1] *A.T.:* Endangered Species Act; environmental legislation.

2857. Lecard, Marc, May-June 1993. Gather no moss? *Sierra* 78(3): 28-30. [1] Laments accelerating loss of peat bogs.

2858. Ledig, F. Thomas, 1988. The conservation of diversity in forest trees. *Bioscience* 38: 471-479. [1].

2859. ———, 1992. Human impacts on genetic diversity in forest ecosystems. *Oikos* 63: 87-108. [1].

2860. Ledig, F. Thomas, C.I. Millar, & L.A. Riggs, 1990. Conservation of diversity in forest ecosystems. *Forest Ecology and Management* 35: 1-4. [1].

2861. Lee, Keekok, 1998. Biodiversity. In Ruth Chadwick, ed., *Encyclopedia of Applied Ethics* (San Diego: Academic Press), Volume 1: 285-304. *A.T.:* environmental ethics; bd definitions.

2862. Lee, Kenneth E., 1996. Biodiversity of soil organisms: community concepts and ecosystem function [editorial]. *Biodiversity and Conservation* 5: 133-134. [1] *A.T.:* soil structure; trophic relationships.

2863. Lee, Kenneth E., & C.E. Pankhurst, 1992. Soil organisms and sustainable productivity. *Australian J. of Soil Research* 30: 855-892. [2] *A.T.:* soil structure; nutrient cycling.

2864. Lee, Michael S.Y., 1997. Documenting present and past biodiversity: conservation biology meets palaeontology [comment]. *TREE* 12: 132-133. *A.T.:* taxonomy.

2865. Leefers, Larry A., 1990. Managing for biodiversity through public forest land management planning processes. *Michigan Academician* 22: 365-370. [1] *A.T.:* Michigan.

2866. Lefebvre, Gaëtan, & B. Poulin, 1997. Bird communities in Panamanian black mangroves: potential effects of physical and biotic factors. *J. of Tropical Ecology* 13: 97-113.

2867. Leggett, Jeremy K., ed., 1990. *Global Warming: The Greenpeace Report.* Oxford, U.K., and New York: OUP. 554p. [2/2/2] *A.T.:* carbon dioxide.

2868. Leggett, M.E., G.L. Brown, & S.E. Campbell, 1996. Developing assays for studying the effect of introduced microbials on the indigenous soil microflora. *Canadian J. of Soil Science* 76: 441-445. [1] *A.T.:* soil microbiology; microbial ecology.

2869. Leibold, Mathew A., 1996. A graphical model of keystone predators in food webs: trophic regulation of abundance, incidence, and diversity patterns in communities. *American Naturalist* 147: 784-812. [2] *A.T.:* keystone species; bottom-up regulation; top-down regulation.

2870. Leigh, Egbert G., Jr., 1965. On the relation between the productivity, biomass, diversity, and stability of a community. *PNAS* 53: 777-783. [1] *A.T.:* mathematical models.

2871. Leigh, Egbert G., Jr., A.S. Rand, & D.M. Windsor, eds., 1996. *The Ecology of a Tropical Forest: Seasonal Rhythms and Long-Term Changes* (2nd ed.). Washington, DC: Smithsonian Institution Press. 503p. [2/1/2] *A.T.:* Panama.

2872. Leigh, Egbert G., Jr., S.J. Wright, et al., 1993. The decline of tree diversity on newly isolated tropical islands: a test of a null hypothesis and some implications. *Evolutionary Ecology* 7: 76-102. [2] *A.T.:* habitat fragmentation; Panama.

2873. Leipe, Detlef D., 1996. Biodiversity, genomes and DNA sequence databases. *Current Opinion in Genetics & Development* 6: 686-691. [1].

2874. Leitch, E.M., & G. Vasisht, 1998. Mass extinctions and the sun's encounters with spiral arms. *New Astronomy* 3: 51-56.

2875. Leitner, W.A., & M.L. Rosenzweig, 1997. Nested species-area curves and stochastic sampling: a new theory. *Oikos* 79: 503-512.

2876. Leiva, M.J., F.S. Chapin III, & R. Fernández Ales, 1997. Differences in species composition and diversity among Mediterranean grasslands with different history: the case of California and Spain. *Ecography* 20: 97-106.

2877. Lele, S.M., 1991. Sustainable development: a critical review. *World Development* 19: 607-621. [2] An overview emphasizing economic aspects.

2878. Lenski, Richard E., & Michael Travisano, 1994. Dynamics of adaptation and diversification: a 10,000-generation experiment with bacterial populations. *PNAS* 91: 6808-6814. [2] *A.T.:* experimental evolution; macroevolution.

2879. Lenton, Timothy M., 1998. Gaia and natural selection. *Nature* 394: 439-447.

2880. Leon, B., & K.R. Young, 1996. Aquatic plants of Peru: diversity, distribution and conservation. *Biodiversity and Conservation* 5: 1169-1190. [1].

2881. Leopold, Aldo, 1933. *Game Management.* New York and London: C. Scribner's Sons. 481p. [2/-/2] *A.T.:* wildlife conservation.

2882. ———, 1949. *A Sand County Almanac, and Sketches Here and There.* New York: OUP. 226p. [3/-/3] An enduring classic by the originator of the land ethic concept. *A.T.:* natural history; landscapes.

2883. Lescourret, F., & M. Genard, 1994. Habitat, landscape and bird composition in mountain forest fragments. *J. of Environmental Management* 40: 317-328. [1] *A.T.:* French Pyrenees.

2884. Lesica, Peter, 1993. Using plant community diversity in reserve design for pothole prairie on the Blackfeet Indian Reservation, Montana, USA. *Biological Conservation* 65: 69-75. [1].

2885. Lesica, Peter, & F.W. Allendorf, 1992. Are small populations of plants worth preserving? *CB* 6: 135-139. [1].

2886. Lesica, Peter, & T.H. DeLuca, 1996. Long-term harmful effects of crested wheatgrass on Great Plains grassland ecosystems. *J. of Soil and Water Conservation* 51: 408-409. [1] *A.T.:* introduced plants.

2887. Leslie, John, 1992. Design and the anthropic principle. *Biology & Philosophy* 7: 349-354. [1].

2888. ———, 1996. *The End of the World: The Science and Ethics of Human Extinction.* London and New York: Routledge. 310p. [2/2/1] *A.T.:* environmental ethics.

2889. Lessa, Enrique P., Blaire Van Valkenburgh, & R.A. Farina, 1997. Testing hypotheses of differential mammalian extinctions subsequent to the Great American Biotic Interchange. *Palaeogeography, Palaeoclimatology, Palaeoecology* 135: 157-162. *A.T.:* body size.

2890. Lesser, William H., 1994. *Institutional Mechanisms Supporting Trade in Genetic Materials: Issues Under the Biodiversity Convention and GATT/TRIPs.* Geneva: UNEP. 72p. [1/1/1] *A.T.:* CBD; germplasm resources; international trade.

2891. ———, 1997. *The Role of Intellectual Property Rights in Biotechnology Transfer Under the Convention on Biological Diversity.* Ithaca, NY: International Service for the Acquisition of Agri-biotech Applications (ISAAA). 22p. [1/1/-].

2892. ———, 1998. *Sustainable Use of Genetic Resources Under the Convention on Biological Diversity: Exploring Access and Benefit Sharing Issues.* Wallingford, Oxon, U.K., and New York: CAB International. 218p.

2893. Lesser, William H., & A.F. Krattiger, spring 1994. The complexities of negotiating terms for germplasm collections. *Diversity* 10(3): 6-10. [1] *A.T.:* technology transfer; North-South relations.

2894. ———, 1994. What is "genetic technology"? *Biodiversity Letters* 2: 31-34. [1].

2895. Lester, R.T., & J.P. Myers, 1989. Global warming, climate disruption, and biological diversity. *Audubon Wildlife Report 1989/1990*: 177-221. [1] *A.T.:* human impact; deforestation.

2896. Letourneau, Deborah K., 1987. The enemies hypothesis: tritrophic interactions and vegetational diversity in tropical agroecosystems. *Ecology* 68: 1616-1622. [1] *A.T.:* Mexico; parasites; traditional agriculture.

2897. Lévêque, Christian, 1995. Role and consequences of fish diversity in the functioning of African freshwater ecosystems: a review. *Aquatic Living Resources* 8: 59-78. [1] *A.T.:* top-down regulation; African Great Lakes.

2898. ———, 1997. *Biodiversity Dynamics and Conservation: The Freshwater Fish of Tropical Africa.* Cambridge, U.K., and New York: CUP. 438p. [1/1/-].

2899. Lever, Christopher, 1985. *Naturalized Mammals of the World.* London and New York: Longman. 487p. [1/1/1].

2900. ———, 1987. *Naturalized Birds of the World.* New York: Wiley. 615p. [1/2/1].

2901. ———, 1994. *Naturalized Animals: The Ecology of Successfully Introduced Species.* London: T. & A.D. Poyser. 354p. [1/1/1].

2902. ———, 1996. *Naturalized Fishes of the World.* San Diego: Academic Press. 408p. [1/1/1] *A.T.:* introduced fishes.

2903. Levesque, E., 1996. Minimum area and cover-abundance scales as applied to polar desert vegetation. *Arctic and Alpine Research* 28: 156-162. [1] *A.T.:* Ellesmere Island, Northwest Territories; cover assessment; sampling.

2904. Levetin, Estelle, & Karen McMahon, 1996. *Plants and Society.* Dubuque, IA: Wm. C. Brown. 441p. [1/1/1] *A.T.:* economic botany.

2905. Levin, Donald A., Javier Francisco-Ortega, & R.K. Jansen, 1996. Hybridization and the extinction of rare plant species. *CB* 10: 10-16.

2906. Levin, Lisa A., C.L. Huggett, & K.F. Wishner, 1991. Control of deep-sea benthic community structure by oxygen and organic-matter gradients in the eastern Pacific Ocean. *J. of Marine Research* 49: 763-800. [2].

2907. Levin, Simon A., 1992. The problem of pattern and scale in ecology. *Ecology* 73: 1943-1967. [3] A major review.

2908. Levin, Ted, May-June 1996. Immersed in the Everglades. *Sierra* 81(3): 56-63, 86-87. [1] A description of flood control and development threats in the Everglades system.

2909. Levine, Joel S., ed., 1991. *Global Biomass Burning: Atmospheric, Climatic, and Biospheric Implications* [conference proceedings]. Cambridge, MA: MIT Press. 569p. [2/2/2].

2910. ———, ed., 1996. *Biomass Burning and Global Change.* Cambridge, MA: MIT Press. 2 vols. [1/1/-]. *A.T.:* climatic change.

2911. Levinton, Jeffrey S., 1979. A theory of diversity equilibrium and morphological evolution. *Science* 204: 335-336. [1].

2912. Lewin, Roger, 1986. Damage to tropical forests, or why were there so many kinds of animals? *Science* 234: 149-150. [1] *A.T.:* habitat destruction.

2913. Lewins, W.A., & D.N. Joanes, 1984. Bayesian estimation of the number of species. *Biometrics* 40: 323-328. [1].

2914. Lewis, Clifford E., B.F. Swindel, & G.W Tanner, 1988. Species diversity and diversity profiles: concept, measurement, and application to timber and range management. *J. of Range Management* 41: 466-469. [1].

2915. Lewis, Connie D., 1996. *Managing Conflicts in Protected Areas* [case studies]. Gland, Switzerland: IUCN. 100p. [1/1/1].

2916. Lewis, Damien, Dec. 1990. Conflict of interests. *Geographical Magazine* 62(12): 18-22. [1] Concerning bdc strategies that conflict with the interests of local residents, and new strategies that take a more integrated approach.

2917. Lewis, Martin W., 1992. *Green Delusions: An Environmentalist Critique of Radical Environmentalism*. Durham, NC: Duke University Press. 288p. [2/3/2].

2918. Lewis, Thomas A., Nov.-Dec. 1987. Will species die out as the Earth heats up? *International Wildlife* 17(6): 18-21. [1] *A.T.:* global warming.

2919. Lewis Smith, R.I., 1996. Introduced plants in Antarctica: potential impacts and conservation issues. *Biological Conservation* 76: 135-146. [1] *A.T.:* Antarctic Treaty.

2920. Liddle, M.J., 1991. Recreation ecology: effects of trampling on plants and corals. *TREE* 6: 13-17. [1] *A.T.:* recreational activities.

2921. Lidicker, William Z., Jr., ed., 1995. *Landscape Approaches in Mammalian Ecology and Conservation* [conference papers]. Minneapolis: University of Minnesota Press. 215p. [1/1/1] *A.T.:* landscape ecology.

2922. Lieth, Helmut, & Martina Lohmann, eds., 1993. *Restoration of Tropical Forest Ecosystems: Proceedings of a Symposium*. Dordrecht and Boston: Kluwer Academic. 269p. [1/1/1] *A.T.:* restoration ecology; reforestation.

2923. Lieth, Helmut, & M.J.A. Werger, eds., 1989. *Tropical Rain Forest Ecosystems: Biogeographical and Ecological Studies*. Amsterdam and New York: Elsevier Scientific. 713p. [1/1/1].

2924. Lieth, Helmut, & R.H. Whittaker, eds., 1975. *Primary Productivity of the Biosphere*. New York: Springer. 339p. [2/1/2].

2925. Liggett, Christopher, & L.A. Leefers, 1990. Temporal aspects of ecosystem biodiversity: a case study of the aspen types in Michigan. *Michigan Academician* 22: 371-379. [1] *A.T.:* simulation models.

2926. Lindberg, David R., 1991. Marine biotic interchange between northern and southern hemispheres. *Paleobiology* 17: 308-324. [1] *A.T.:* antitropical distributions.

2927. Lindberg, Kreg, & D.E. Hawkins, eds., 1993. *Ecotourism: A Guide for Planners and Managers*. North Bennington, VT: Ecotourism Society. 175p. [1/1/1].

2928. Linddal, Michael, & Arto Naskali, eds., 1993. *Valuing Biodiversity: Proceedings of the Workshop on the Social Costs of and Benefits From Preserving Endangered Species and Biodiversity of the Boreal Forests*. Rovaniemi, Finland: Scandinavian Society of Forest Economics. 158p. [1/1/1].

2929. Linden, Eugene, & Ian McCluskey, 22 June 1992. Rio's legacy. *Time* 139(25): 44-45. [1] News story on a critical phase of the UNCED meetings.

2930. Lindenmayer, David B., T.W. Clark, et al., 1993. Population viability analysis as a tool in wildlife conservation policy: with reference to Australia. *Environmental Management* 17: 745-758. [1] *A.T.:* wildlife management.

2931. Lindenmayer, David B., & H.A. Nix, 1993. Ecological principles for the design of wildlife corridors. *CB* 7: 627-630. [2] *A.T.:* Victoria, Australia; wildlife conservation; marsupials.

2932. Linder, H.P., & J.J. Midgley, 1994. Taxonomy, compositional biodiversity and functional biodiversity of fynbos. *South African J. of Science* 90: 329-333. [1] *A.T.:* South Africa.

2933. Lindsay, Jonathan M., 1993. Overlaps and tradeoffs: coordinating policies for sustainable development in Asia and the Pacific. *J. of Developing Areas* 28: 21-30. [1] *A.T.:* bd loss; deforestation; climatic change; international cooperation.

2934. Lindstrom, Bob, winter 1996. Biodiversity, ecology, and evolution of hot water organisms in Yellowstone National Park: symposium and issues overview. *Park Science* 16(1): 12-13, 19. [1] *A.T.:* thermophiles.

2935. Line, Les, June 1993. Silence of the songbirds. *National Geographic* 183(6): 68-91. [1] Describes declines in songbird populations.

2936. ———, July-Aug. 1995. Gone the way of the dinosaurs. *International Wildlife* 25(4): 16-21. [1] Reports on David Steadman's work on historical extinctions of birds in Tonga.

2937. Lippke, Bruce, & H.L. Fretwell, Jan. 1997. The market incentive for biodiversity. *J. of Forestry* 95(1): 4-7. [1] *A.T.:* forest management; multiple use.

2938. Lips, Karen R., 1998. Decline of a tropical montane amphibian fauna. *CB* 12: 106-117. *A.T.:* Costa Rica.

2939. Lipske, Michael, April-May 1992. Racing to save hot spots of life. *National Wildlife* 30(3): 40-49. [1] Describes operations of the Conservation International Rapid Assessment Program. *A.T.:* Ecuador.

2940. ———, April-May 1996. For backyard biodiversity, try pollinating your yard. *National Wildlife* 34(3): 58, 60. [1] *A.T.:* gardening; Backyard Wildlife Habitat Program.

2941. Littell, Richard, 1992. *Endangered and Other Protected Species: Federal Law and Regulation*. Washington, DC: Bureau of National Affairs. various pagings. [2/1/1] *A.T.:* wildlife conservation.

2942. Little, Charles E., 1990. *Greenways for America*. Baltimore: Johns Hopkins University Press. 237p. [2/2/1].

2943. Liu, J.H., & Peter Hills, 1997. Environmental planning, biodiversity and the development process: the case of Hong Kong's Chinese white dolphins. *J. of Environmental Management* 50: 351-367. *A.T.:* environmental degradation.

2944. Liu, Qinghong, 1995. A model for species diversity monitoring at community level and its applications. *Environmental Monitoring and Assessment* 34: 271-287. [1] *A.T.:* plant communities.

2945. Lloyd, Monte, R.F. Inger, & F.W. King, 1968. On the diversity of reptile and amphibian species in a Bornean rain forest. *American Naturalist* 102: 497-515. [1].

2946. Lloyd, Monte, J.H. Zar, & J.R. Karr, 1968. On the calculation of information-theoretical measures of diversity. *American Midland Naturalist* 79: 257-272. [1] *A.T.:* diversity measures.

2947. [not used]

2948. Lockwood, Jeffrey A., 1987. The moral standing of insects and the ethics of extinction. *Florida Entomologist* 70: 70-89. [1] *A.T.:* environmental ethics; moral values.

2949. Lockwood, Julie L., M.P. Moulton, & S.K. Anderson, 1993. Morphological assortment and the assembly of communities of introduced passeriforms on oceanic islands: Tahiti versus Oahu. *American Naturalist* 141: 398-408. [1].

2950. Loder, N., T.M. Blackburn, & K.J. Gaston, 1997. The slippery slope: towards an understanding of the body size frequency distribution. *Oikos* 78: 195-201. *A.T.:* species richness; frequency distribution slope.

2951. Lodge, D. Jean, 1997. Factors related to diversity of decomposer fungi in tropical forests. *Biodiversity and Conservation* 6: 681-688. *A.T.:* host preference.

2952. Lodge, D. Jean, & Sharon Cantrell, 1995. Fungal communities in wet tropical forests: variation in time and space. *Canadian J. of Botany* 73, Suppl. 1: S1391-S1398. [1] *A.T.:* decomposers; stratification.

2953. Lodge, D.M., 1993. Biological invasions: lessons for ecology. *TREE* 8: 133-137. [2] A general review of the consequences of species introductions. *A.T.:* biotic homogenization.

2954. Lodge, D.M., R.A. Stein, et al., 1998. Predicting impact of freshwater exotic species on native biodiversity: challenges in spatial scaling. *Australian J. of Ecology* 23: 53-67. *A.T.:* Wisconsin; introduced species; ecological effects.

2955. Loehle, Craig, & Gary Wein, 1994. *Landscape Habitat Diversity: An Information Theoretic Measure*. Argonne, IL: Argonne National Laboratory. 27p. [1/1/1].

2956. Loeschcke, Volker, Jürgen Tomiuk, & S.K. Jain, eds., 1994. *Conservation Genetics* [symposium papers]. Basel and Boston: Birkhäuser Verlag. 440p. [1/1/1] *A.T.:* germplasm resources.

2957. Logan, Graham A., J.M. Hayes, et al., 1995. Terminal Proterozoic reorganization of biogeochemical cycles. *Nature* 376: 53-56. [2] *A.T.:* bacteria; atmospheric oxygen.

2958. Lohmann, Larry, 1991. Who defends biological diversity? *Ecologist* 21: 5-13. [1] Describes a study in Thailand suggesting bd is more likely to be conserved as a function of local agenda than corporate- or state-level interference.

2959. Loiselle, Bette A., & J.G. Blake, 1992. Population variation in a tropical bird community: implications for conservation. *Bioscience* 42: 838-845. [2] *A.T.:* stability.

2960. Lombard, A.T., 1995. The problems with multi-species conservation: do hotspots, ideal reserves and existing reserves coincide? *South African J. of Zoology* 30: 145-163. [2] *A.T.:* vertebrates; South Africa; reserve selection.

2961. ———, 1996. Global change, biodiversity and ecosystem functioning [editorial]. *South African J. of Science* 92: 115-116. [1].

2962. Lombard, A.T., P.V. August, & W.R. Siegfried, 1992. A proposed geographic information system for assessing the optimal dispersion of protected areas in South Africa [editorial]. *South African J. of Science* 88: 136-140. [1] *A.T.:* inventories.

2963. Lomolino, Mark V., 1994. An evaluation of alternative strategies for building networks of nature reserves. *Biological Conservation* 69: 243-249. [1] *A.T.:* SLOSS.

2964. ———, 1994. Species richness of mammals inhabiting nearshore archipelagoes: area, isolation, and immigration filters. *J. of Mammalogy* 75: 39-49. [1] *A.T.:* island biogeography.

2965. Lomolino, Mark V., & Rob Channell, 1995. Splendid isolation: patterns of geographic range collapse in endangered mammals. *J. of Mammalogy* 76: 335-347. [1] *A.T.:* GIS.

2966. ———, 1998. Range collapse, re-introductions, and biogeographic guidelines for conservation [comment]. *CB* 12: 481-484.

2967. Lomolino, Mark V., & R. Davis, 1997. Biogeographic scale and biodiversity of mountain forest mammals of western North America. *Global Ecology and Biogeography Letters* 6: 57-76. *A.T.:* immigration; extinction; community structure.

2968. Lomolino, Mark V., J.H. Brown, & Russell Davis, 1989. Island biogeography of montane forest mammals in the American Southwest. *Ecology* 70: 180-194. [2].

2969. Long, Adrian J., M.J. Crosby, et al., 1996. Towards a global map of biodiversity: patterns in the distribution of restricted range birds. *Global Ecology and Biogeography Letters* 5: 281-304. [1] *A.T.:* centers of endemism.

2970. Long, John L., 1981. *Introduced Birds of the World: The Worldwide History, Distribution, and Influence of Birds Introduced to New*

Environments. New York: Universe Books. 528p. [2/2/2].

2971. Longino, J.T., & R.K. Colwell, 1997. Biodiversity assessment using structured inventory: capturing the ant fauna of a tropical rain forest. *Ecological Applications* 7: 1263-1277. *A.T.:* Costa Rica; sampling.

2972. Longuet-Higgins, M.S., 1971. On the Shannon-Weaver index of diversity, in relation to the distribution of species in bird censuses. *Theoretical Population Biology* 2: 271-289. [1].

2973. Loope, Lloyd L., & A.C. Medeiros, June 1995. Strategies for longterm protection of biological diversity in rainforests of Haleakala National Park and East Maui, Hawaii. *Endangered Species Update* 12(6): 1-5. [1] *A.T.:* environmental degradation; biological invasions; environmental management.

2974. Loope, Lloyd L., Ole Hamann, & C.P. Stone, 1988. Comparative conservation biology of oceanic archipelagoes: Hawaii and the Galápagos. *Bioscience* 38: 272-282. [1] *A.T.:* island biology.

2975. Loope, Lloyd L., P.G. Sanchez, et al., 1988. Biological invasions of arid land nature reserves. *Biological Conservation* 44: 95-118. [1].

2976. Loope, Walter L., 1990. Natural forest dynamism and the maintenance of biodiversity: an exploration of agency history and approach. *Michigan Academician* 22: 343-353. [1] *A.T.:* environmental policy; Michigan.

2977. Loranger, G., J.F. Ponge, et al., 1998. Impact of earthworms on the diversity of microarthropods in a vertisol (Martinique). *Biology and Fertility of Soils* 27: 21-26. *A.T.:* Collembola.

2978. Lord, Janice M., & D.A. Norton, 1990. Scale and the spatial concept of fragmentation. *CB* 4: 197-202. [2] *A.T.:* spatial scale.

2979. Loreau, Michel, 1998. Biodiversity and ecosystem functioning: a mechanistic model. *PNAS* 95: 5632-5636. *A.T.:* soil nutrients; plant species richness.

2980. ———, 1998. Separating sampling and other effects in biodiversity experiments. *Oikos* 82: 600-602.

2981. Lorence, D.H., & R.W. Sussman, 1986. Exotic species invasion into Mauritius wet forest remnants. *J. of Tropical Ecology* 2: 147-162. [1].

2982. Loucks, Orie L., 1970. Evolution of diversity, efficiency, and community stability. *American Zoologist* 10: 17-25. [2] *A.T.:* succession.

2983. Louette, Dominique, André Charrier, & Julien Berthaud, 1997. *In situ* conservation of maize in Mexico: genetic diversity and maize; seed management in a traditional community. *Economic Botany* 51: 20-38.

2984. Lövei, Gábor L., 1997. Global change through invasion. *Nature* 388: 627-628. Identifies ratios of native to non-native species in eleven geographical areas.

2985. Lovejoy, Thomas E., 1985. Protecting biological diversity. In Harry Messel, ed., *The Study of Populations* (Sydney: Pergamon): 237-241. [1] Gives general arguments.

2986. ———, 1988. Will unexpectedly the top blow off? [editorial] *Bioscience* 38: 722-726. [1] An address dealing with an array of environmental problems.

2987. ———, 1989. Deforestation and extinction of species. In Daniel B. Botkin, ed., *Changing the Global Environment: Perspectives on Human Involvement* (Boston: Academic Press): 91-98. [1] *A.T.:* Brazil; Madagascar.

2988. ———, Jan.-Feb. 1992. Looking to the next millennium. *National Parks* 66(1-2): 41-44. [1] Concerning increasing challenges for the world's parks.

2989. ———, Jan.-Feb. 1994. People and biodiversity. *Nature Conservancy* 44(1): 28-33. [1] Relates bdc to the ultimate goals of sustainable development.

2990. ———, 1994. The quantification of biodiversity: an esoteric quest or a vital component of sustainable development? *PTRSL B* 345: 81-87. [1] *A.T.:* sustainable development.

2991. ———, 1995. Will expectedly the top blow off? *Bioscience* 45, Suppl.: S3-S6. [1] *A.T.:* environmental policy.

2992. Lovejoy, Thomas E., & R.H. Dwight, winter 1991-1992. Biological diversity and Neptune's realm. *Marine Technology Society J.* 25(4): 7-12. [1] *A.T.:* climatic change; extinction.

2993. Lovelock, James E., 1979 (3rd ed., 1995). *Gaia: A New Look at Life on Earth*. Oxford, U.K., and New York: OUP. 157p. [3/2/3] *A.T.:* biosphere.

2994. ———, 1988. *The Ages of Gaia: A Biography of Our Living Earth*. New York: Norton. 252p. [3/3/2] *A.T.:* biosphere.

2995. ———, 1991. *Healing Gaia: Practical Medicine for the Planet*. New York: Harmony Books. 192p. [2/2/1] *A.T.:* biosphere.

2996. ———, 1992. A numerical model for biodiversity. *PTRSL B* 338: 383-391. [1] Portrays the climate and ecology of an imaginary planet on the basis of an ecosystem model.

2997. Lovett, Jon C., & S.K. Wasser, eds., 1993. *Biogeography and Ecology of the Rain Forests of Eastern Africa*. Cambridge, U.K., and New York: CUP. 341p. [1/2/1].

2998. Lovett, Jon C., J. Hatton, et al., 1997. Assessment of the impact of the lower Kihansi hydropower project. *Biodiversity and Conservation* 6: 915-933. *A.T.:* Tanzania; environmental impact; dams.

2999. Lowe-McConnell, Rosemary H., 1993. Fish faunas of the African Great Lakes: origins, diversity and vulnerability. *CB* 7: 634-643. [1] *A.T.:* cichlids; fish introductions.

3000. Lubchenco, Jane, 1978. Plant species diversity in a marine intertidal community: importance of herbivore food preference and algal competitive abilities. *American Naturalist* 112: 23-39. [2] *A.T.:* environmental policy; science and state.

3001. ———, 1995. The role of science in formulating a biodiversity strategy. *Bioscience* 45, Suppl.: S7-S9. [1].

3002. Lubchenco, Jane, A.M. Olson, et al., 1991. The Sustainable Biosphere Initiative: an ecological research agenda. A report from the Ecological Society of America. *Ecology* 72: 371-412. [3].

3003. Lücking, Robert, 1995. Biodiversity and conservation of foliicolous lichens in Costa Rica. *Mitteilungen der Eidgenössischen Forschungsanstalt für Wald, Schnee und Landschaft* 70: 63-92. [1].

3004. Lucy, T.V., 1995. Empowerment of women for sustainable development. *Social Action* 45: 224-231. [1].

3005. Ludwig, Daniel R., 1995. Assessment and management of wildlife diversity in an urban setting. *Natural Areas J.* 15: 353-361. [1] *A.T.:* faunal inventories; Illinois; resource management.

3006. Ludwig, Donald, Ray Hilborn, & Carl Walters, 1993. Uncertainty, resource exploitation, and conservation: lessons from history. *Science* 260: 17, 36. [3].

3007. Ludyanskiy, Michael L., Derek McDonald, & David MacNeill, 1993. Impact of the zebra mussel, a bivalve invader. *Bioscience* 43: 533-544. [2] *A.T.:* biological invasions; North America.

3008. Lugo, Ariel E., 1988. *Diversity of Tropical Species: Questions That Elude Answers*. Paris: Biology International Special Issue No. 19, IUBS. 37p. [1/1/1].

3009. ———, 1995. Management of tropical biodiversity. *Ecological Applications* 5: 956-961. [1] *A.T.:* ecosystem management; tropical rainforests.

3010. Lugo, Ariel E., J.A. Parrotta, & S. Brown, 1993. Loss in species caused by tropical deforestation and their recovery through management. *Ambio* 22: 106-109. [2] *A.T.:* island biogeography; Caribbean region.

3011. Luke, Timothy W., June 1997. The World Wildlife Fund: ecocolonialism as funding the worldwide "wise use" of nature. *Capitalism, Nature, Socialism* 8(2): 31-62. *A.T.:* environmental ethics.

3012. Lumaret, Roselyne, J.-L. Guillerm, et al., 1997. Plant species diversity and polyploidy in islands of natural vegetation isolated in extensive cultivated lands. *Biodiversity and Conservation* 6: 591-613. *A.T.:* habitat fragmentation; France.

3013. Lund, H. Gyde, ed., 1993. *Integrated Ecological and Resource Inventories: Proceedings, National Workshop*. Washington, DC: USDA Forest Service. 177p. [1/1/1] *A.T.:* ecological surveys; environmental management.

3014. Luning, Klaus, 1990. *Seaweeds: Their Environment, Biogeography, and Ecophysiology*. Edited by Charles Yarish and Hugh Kirkman. New York: Wiley. 527p. [1/1/3].

3015. Lunn, A.G., 1995. Vegetation and species distribution mapping in

Britain, with special reference to north east England. *Bull. of the Society of University Cartographers* 28(2): 13-18. [1].

3016. Lunney, Daniel, ed., 1991. *Conservation of Australia's Forest Fauna*. Mosman, NSW: Royal Zoological Society of New South Wales. 418p. [1/1/1] *A.T.:* wildlife conservation; wildlife management.

3017. Luoma, Jon R., 1987. *A Crowded Ark*. Boston: Houghton Mifflin. 209p. [2/2/1].

3018. ———, Jan.-Feb. 1998. Habitat conservation plans: compromise or capitulation? *Audubon* 100(1): 36-43. An analysis of the pros and cons. *A.T.:* Endangered Species Act; California.

3019. Lupwayi, N.Z., W.A. Rice, & G.W. Clayton, 1998. Soil microbial diversity and community structure under wheat as influenced by tillage and crop rotation. *Soil Biology & Biochemistry* 30: 1733-1741. *A.T.:* bacteria; Alberta.

3020. Lust, N., B. Muys, & L. Nachtergale, 1998. Increase of biodiversity in homeogeneous Scots pine stands by an ecologically diversified management. *Biodiversity and Conservation* 7: 249-260. *A.T.:* Belgium.

3021. Lutz, Ernst, & J.O. Caldecott, eds., 1996. *Decentralization and Biodiversity Conservation* [case studies]. Washington, DC: World Bank. 176p. [1/1/1] *A.T.:* protected areas management.

3022. Lutz, Richard A., & M.J. Kennish, 1993. Ecology of deep-sea hydrothermal vent communities: a review. *Reviews of Geophysics* 31: 211-242. [2].

3023. Lydeard, Charles, 1993. Phylogenetic analysis of species richness: has viviparity increased the diversification of actinopterygian fishes? *Copeia* (2): 514-518. [1].

3024. Lydeard, Charles, & R.L. Mayden, 1995. A diverse and endangered aquatic ecosystem of the southeast United States. *CB* 9: 800-805. [1] Profiles Alabama's unusual aquatic diversity and its endangered state.

3025. Lyman, R. Lee, 1996. Applied zooarchaeology: the relevance of faunal analysis to wildlife management. *World Archaeology* 28: 110-125. [1] *A.T.:* sea otters.

3026. Lynch, Michael, John Conery, & Reinhard Bürger, 1995. Mutation accumulation and the extinction of small populations. *American Naturalist* 146: 489-518. [2] *A.T.:* local extinction; population biology.

3027. Lyons, John, S. Navarro-Pérez, et al., 1995. Index of biotic integrity based on fish assemblages for the conservation of streams and rivers in west-central Mexico. *CB* 9: 569-584. [1].

3028. Ma, Shilai, Lianxian Han, et al., 1995. Faunal resources of the Gaoligongshan region of Yunnan, China: diverse and threatened. *Environmental Conservation* 22: 250-258. [1] *A.T.:* wildlife conservation.

3029. MacArthur, Robert H., 1957. On the relative abundance of bird species. *PNAS* 43: 293-295. [2].

3030. ———, 1960. On the relative abundance of species. *American Naturalist* 94: 25-36. [2] *A.T.:* birds.

3031. ———, 1965. Patterns of species diversity. *Biological Reviews* 40: 510-533. [2].

3032. ———, 1972. *Geographical Ecology: Patterns in the Distribution of Species.* New York: Harper & Row. 269p. [3/2/3] *A.T.:* biogeography.

3033. MacArthur, Robert H., & Richard Levins, 1967. The limiting similarity, convergence, and divergence of coexisting species. *American Naturalist* 101: 377-385. [3] *A.T.:* coefficient of competition; niche relationships; packing.

3034. MacArthur, Robert H., & J.W. MacArthur, 1961. On bird species diversity. *Ecology* 42: 594-598. [2].

3035. MacArthur, Robert H., & E.R. Pianka, 1966. On optimal use of a patchy environment. *American Naturalist* 100: 603-609. [3] *A.T.:* patch utilization.

3036. MacArthur, Robert H., & E.O. Wilson, 1963. An equilibrium theory of insular zoogeography. *Evolution* 17: 373-387. [2] The first published expression of the MacArthur-Wilson theory of island biogeography.

3037. ———, 1967. *The Theory of Island Biogeography.* Princeton, NJ: Princeton University Press. 203p. [2/1/3] One of the century's most important works in theoretical ecology.

3038. MacArthur, Robert H., H.F. Recher, & M.L. Cody, 1966. On the relation between habitat selection and species diversity. *American Naturalist* 100: 319-332. [2].

3039. Macdonald, D.W., F.H. Tattersall, et al., 1995. Reintroducing the European beaver to Britain: nostalgic meddling or restoring biodiversity? *Mammal Review* 25: 161-200. [1].

3040. Macdonald, I.A.W., & G.W. Frame, 1988. The invasion of introduced species into nature reserves in tropical savannas and dry woodlands. *Biological Conservation* 44: 67-93. [1] *A.T.:* biological invasions.

3041. Macdonald, I.A.W., F.J. Kruger, & A.A. Ferrar, eds., 1986. *The Ecology and Management of Biological Invasions in Southern Africa* [symposium proceedings]. Cape Town and New York: OUP. 324p. [1/1/1].

3042. Macdonald, I.A.W., D.M. Graber, et al., 1988. Introduced species in nature reserves in Mediterranean-type climatic regions of the world. *Biological Conservation* 44: 37-66. [1] *A.T.:* biological invasions.

3043. Mace, Georgina M., 1994. Classifying threatened species: means and ends. *PTRSL B* 344: 91-97. [1] *A.T.:* extinction risk.

3044. Mace, Georgina M., & Russell Lande, 1991. Assessing extinction threats: toward a reevaluation of IUCN threatened species categories. *CB* 5: 148-157. [2].

3045. Mace, Georgina M., Andrew Balmford, & J.R. Ginsberg, eds., 1999. *Conservation in a Changing World* [symposium papers]. Cambridge,

U.K., and New York: CUP. 320p.

3046. Machlis, Gary E., 1992. The contribution of sociology to biodiversity research and management. *Biological Conservation* 62: 161-170. [1].

3047. Machlis, Gary E., D.J. Forester, & J.E. McKendry, winter 1994. Gap analysis and national parks: adding the socioeconomic dimension. *Park Science* 14(1): 6-10. [1] *A.T.:* mapping.

3048. ———, 1994. *Biodiversity Gap Analysis: Critical Challenges & Solutions* [workshop report]. Moscow, ID: Idaho Forest, Wildlife and Range Experiment Station, University of Idaho. 61p. [1/1/1] *A.T.:* Idaho.

3049. Machon, N., M. Lefranc, et al., 1997. Allozyme differentiation in *Ulmus* species from France: analysis of differentiation. *Heredity* 78: 12-20. *A.T.:* genetic diversity.

3050. Macia, A., & L. Hernroth, 1995. Maintaining sustainable resources and biodiversity while promoting development—a demanding task for a young nation. *Ambio* 24: 515-517. [1].

3051. Macilwain, C., 1998. Boost to biodiversity research "would strengthen US economy." *Nature* 391: 829. Concerning a recommendation by the President's Council of Advisors to increase funding for bd research.

3052. Macioti, Manfredo, 1992. West to East: America's gifts to the Old World. *Impact of Science on Society* 42: 225-239. [1] *A.T.:* biological invasions; introduced species.

3053. MacIver, D.C., 1998. Atmospheric change and biodiversity. *Environmental Monitoring and Assessment* 49: 177-189. *A.T.:* thermal buffering capacity.

3054. Mack, Richard N., March 1990. Catalog of woes: some of our most troublesome weeds were dispersed through the mail. *Natural History* 99(3): 44-53. [1] *A.T.:* seed catalogs; gardening.

3055. MacKay, W.P., M. Artemio Rebeles, et al., 1991. Impact of the slashing and burning of a tropical rain forest on the native ant fauna (Hymenoptera: Formicidae). *Sociobiology* 18: 257-268. [1] *A.T.:* Mexico.

3056. MacKenzie, Debora, 1992. Seeds of hope. *Tomorrow* 2(1): 46-51. [1] Concerning efforts by the Keystone Dialogue to help farmers in developing countries conserve local crop varieties.

3057. MacKinnon, John R., 1996. *Wild China.* Cambridge, MA: MIT Press. 208p. [1/1/1] *A.T.:* natural history.

3058. ———, 1997. *Protected Areas Systems Review of the Indo-Malayan Realm.* Canterbury, U.K.: Asian Bureau for Conservation. 198p. [1/1/-].

3059. MacKinnon, John R., & Geoff Carey, eds., 1996. *A Biodiversity Review of China.* Hong Kong: WWF International. 529p. [1/1/1].

3060. MacKinnon, John R., & Kathy MacKinnon, 1986. *Review of the Protected Areas System in the Afrotropical Realm.* Gland, Switzerland: IUCN. 259p. [1/1/1].

3061. MacKinnon, John R., Kathy MacKinnon, et al., compilers, 1986.

Managing Protected Areas in the Tropics [conference papers]. Gland, Switzerland: IUCN. 295p. [1/1/1].

3062. Mackintosh, Gay, ed., 1989. *Preserving Communities & Corridors.* Washington, DC: Defenders of Wildlife. 96p. [1/1/1] *A.T.:* wildlife conservation.

3063. Maclean, R.H., & R.W. Jones, 1995. *Aquatic Biodiversity Conservation: A Review of Current Issues and Efforts.* Ottawa: Strategy for International Fisheries Research. 59p. [1/1/1] *A.T.:* fisheries conservation.

3064. MacLeod, Norman, & Gerta Keller, eds., 1996. *Cretaceous-Tertiary Mass Extinctions: Biotic and Environmental Changes.* New York: Norton. 575p. [1/1/1].

3065. MacLeod, Norman, P.F. Rawson, et al., 1997. The Cretaceous-Tertiary biotic transition. *J. of the Geological Society* 154: 265-292. [2].

3066. MacMahon, James A., D.L. Phillips, et al., 1978. Levels of biological organization: an organism-centered approach. *Bioscience* 28: 700-704. [1].

3067. Macpherson, Andrew H., 1965. The origin of diversity in mammals of the Canadian arctic tundra. *Systematic Zoology* 14: 153-173. [1] *A.T.:* biogeography.

3068. Macpherson, E., & C.M. Duarte, 1994. Patterns in species richness, size, and latitudinal range of East Atlantic fishes. *Ecography* 17: 242-248. [1] *A.T.:* latitudinal diversity gradients; latitudinal gradients.

3069. MacRae, R. Andrew, R.A. Fensome, & G.L. Williams, 1996. Fossil dinoflagellate diversity, originations, and extinctions and their significance. *Canadian J. of Botany* 74: 1687-1694. [1].

3070. Mader, H.J., 1984. Animal habitat isolation by roads and agricultural fields. *Biological Conservation* 29: 81-96. [2] *A.T.:* island biogeography; gene flow.

3071. Madin, Laurence P., & Katherine A.C. Madin, fall 1995. Diversity in a vast and stable habitat. *Oceanus* 38(2): 20-24. [1] Describes bd of two midwater regions of the world ocean.

3072. Madson, John, 1994. Rolling forward: toward a new Midwestern land ethic. *Restoration & Management Notes* 12: 5-7. [1] *A.T.:* restoration ecology; prairies.

3073. Maesen, L.J.G. van der, X.M. van der Burgt, & J.M. van Medenbach de Rooy, eds., 1996. *The Biodiversity of African Plants: Proceedings, XIVth AETFAT Congress.* Dordrecht and Boston: Kluwer Academic. 861p. [1/1/1].

3074. Maffei, Maria C., 1993. Evolving trends in the international protection of species. *German Yearbook of International Law* 36: 131-186. [1] *A.T.:* international law; international cooperation.

3075. Maggio, Gregory F., 1997. Recognizing the vital role of local communities in international legal instruments for conserving biodiversity. *UCLA*

J. of Environmental Law & Policy 16: 179-226. *A.T.:* international cooperation; indigenous peoples; environmental policy.

3076. Magnuson, John J., 1990. Long-term ecological research and the invisible present. *Bioscience* 40: 495-501. [2] *A.T.:* research programs.

3077. Magnuson, John J., B.J. Benson, & A.S. McLain, 1994. Insights on species richness and turnover from long-term ecological research: fishes in north temperate lakes. *American Zoologist* 34: 437-451. [1] *A.T.:* survey data; sampling.

3078. Magnussen, S., & T.J.B. Boyle, 1995. Estimating sample size for inference about the Shannon-Weaver and the Simpson indices of species diversity. *Forest Ecology and Management* 78: 71-84. [1] *A.T.:* diversity indices.

3079. Maguire, L.A., & C. Servheen, 1992. Integrating biological and sociological concerns in endangered species management: augmentation of grizzly bear populations. *CB* 6: 426-434. [1] *A.T.:* translocation; Montana.

3080. Magurran, Anne E., 1988. *Ecological Diversity and Its Measurement*. Princeton, NJ: Princeton University Press. 179p. [2/1/3] An indispensable reference work.

3081. Mahar, Dennis J., 1989. *Government Policies and Deforestation in Brazil's Amazon Region*. Washington, DC: World Bank. 56p. [1/1/2] *A.T.:* environmental policy.

3082. Mahathir bin Mohamed, D.S., & José Lutzenberger, summer 1992. Eco-imperialism & bio-monopoly at the Earth Summit. *New Perspectives Quarterly* 9(3): 56-58. [1] *A.T.:* UNCED; North-South relations; Malaysia.

3083. Mahunnah, R.L.A., & K.E. Mshigeni, 1996. Tanzania's policy on biodiversity prospecting and drug discovery programs. *J. of Ethnopharmacology* 51: 221-228. [1].

3084. Maille, P., & R. Mendelsohn, 1993. Valuing ecotourism in Madagascar. *J. of Environmental Management* 38: 213-218. [1].

3085. Majer, J.D., & G. Beeston, 1996. The biodiversity integrity index: an illustration using ants in Western Australia. *CB* 10: 65-73. [1] *A.T.:* bd measures.

3086. Majumdar, Shyamal K., & F.J. Brenner, eds., 1994. *Biological Diversity: Problems and Challenges*. Easton, PA: Pennsylvania Academy of Science. 461p. [1/1/1] *A.T.:* bdc.

3087. Malakoff, David, 1997. Extinction on the high seas. *Science* 277: 486-488. Examines various points of view regarding the vulnerability of marine species.

3088. ———, 1998. Atlantic salmon spawn fight over species protection. *Science* 279: 800. Concerning the decision not to use the Endangered Species Act to protect seven salmon runs in Maine.

3089. Malanson, George P., 1993. *Riparian Landscapes*. Cambridge, U.K., and New York: CUP. 296p. [1/1/2] *A.T.:* landscape ecology.

3090. Malingreau, Jean-Paul, F. Achard, et al., 1996. NOAA-AVHRR based tropical forest mapping for South-East Asia, validated and calibrated with higher spatial resolution imagery. In Giles D'Souza, A.S. Belward, & J.-P. Malingreau, eds., *Advances in the Use of NOAA AVHRR Data for Land Applications* (Dordrecht and Boston: Kluwer): 279-309. [1] *A.T.:* remote sensing.

3091. Mallet, James, 1995. A species definition for the Modern Synthesis [editorial]. *TREE* 10: 294-299. [2] *A.T.:* biological species concept.

3092. Malmqvist, Björn, & Åsa Eriksson, 1995. Benthic insects in Swedish lake-outlet streams: patterns in species richness and assemblage structure. *Freshwater Biology* 34: 285-296. [1].

3093. Maloney, Bernard K., ed., 1998. *Human Activities and the Tropical Rainforest: Past, Present, and Possible Future* [conference papers]. Dordrecht and Boston: Kluwer Academic. 206p. *A.T.:* human ecology.

3094. Malyshev, L.I., & P.L. Nimis, 1997. Climatic dependence of the ecotone between alpine and forest orobiomes in southern Siberia. *Flora* 192: 109-120. *A.T.:* phytogeography.

3095. Mandrak, Nicholas E., 1989. Potential invasion of the Great Lakes by fish species associated with climatic warming. *J. of Great Lakes Research* 15: 306-316. [1] *A.T.:* biological invasions.

3096. Mangel, Marc, & Charles Tier, 1993. A simple direct method for finding persistence times of populations and application to conservation problems. *PNAS* 90: 1083-1086. [1] *A.T.:* conservation biology.

3097. ———, 1994. Four facts every conservation biologist should know about persistence. *Ecology* 75: 607-614. [2] *A.T.:* conservation biology; natural catastrophes.

3098. Mangel, Marc, L.M. Talbot, et al., 1996. Principles for the conservation of wild living resources. *Ecological Applications* 6: 338-362. [2] *A.T.:* conservation biology; wildlife conservation; sustainability.

3099. Mangun, William R., spring 1995. Wildlife habitat: issue of choice. *Forum for Applied Research and Public Policy* 10(1): 84-87. [1] *A.T.:* human disturbance; wildlife conservation.

3100. Mann, Charles C., 1991. Extinction: are ecologists crying wolf? [editorial] *Science* 253: 736-738. [1] Discusses the ongoing battle between "doom and gloom" ecologists and their detractors.

3101. Mann, Charles C., & M.L. Plummer, Jan. 1992. The butterfly problem. *Atlantic* 269(1): 47-70. [1] Discusses contentious effects of the Endangered Species Act on private landowners.

3102. ———, 1993. The high cost of biodiversity [editorial]. *Science* 260: 1868-1871. [1] Considers the goals of the Wildlands Project. *A.T.:* land management; wilderness.

3103. ———, 1995. Are wildlife corridors the right path? [editorial] *Science* 270: 1428-1430. [1] *A.T.:* California.

3104. ———, 1995. *Noah's Choice: The Future of Endangered Species.* New York: Knopf. 302p. [3/3/1] *A.T.:* environmental law.

3105. Mann, D.G., & S.J.M. Droop, 1996. Biodiversity, biogeography and conservation of diatoms. *Hydrobiologia* 336: 19-32. [1].

3106. Mann, L.K., P.D. Parr, et al., 1996. Protection of biota on nonpark public lands: examples from the US Department of Energy Oak Ridge Reservation. *Environmental Management* 20: 207-218. [1] *A.T.:* Tennessee; protected areas.

3107. Manning, Richard, Dec. 1991. The dawn of *Wild Earth*: biodiversity, "big wilderness," and the new environmentalism. *Backpacker* 19(8): 44-47. [1] Describes the "new environmentalism" movement and some related institutions and publications.

3108. Mannion, Antoinette M., 1993. Biotechnology and global change. *Global Environmental Change: Human and Policy Dimensions* 3: 320-329. [1] *A.T.:* international cooperation.

3109. ———, 1994. *Biodiversity and Industry.* Reading, U.K.: Geographical Paper No. 115, Dept. of Geography, University of Reading. 27p. [1/1/1] *A.T.:* biotechnology.

3110. ———, 1995. Biodiversity, biotechnology, and business. *Environmental Conservation* 22: 201-210, 228. [1].

3111. ———, 1997. Agriculture and land transformation. 2. Present trends and future prospects. *Outlook on Agriculture* 26: 151-158. *A.T.:* land use; agricultural development.

3112. Manson, Craig, 1994. Natural communities conservation planning: California's new ecosystem approach to biodiversity. *Environmental Law* 24: 603-615. [1].

3113. Maragos, James E., et al., eds., 1995. *Marine and Coastal Biodiversity in the Tropical Island Pacific Region* [proceedings of two workshops]. Honolulu: Program on Environment, East-West Center, et al. 2 vols. [1/1/1] *A.T.:* Oceania.

3114. Marc, P., & A. Canard, 1997. Maintaining spider biodiversity in agroecosystems as a tool in pest control. *Agriculture, Ecosystems & Environment* 62: 229-235. *A.T.:* biological control; predator-prey relations.

3115. Mares, Michael A., 1992. Neotropical mammals and the myth of Amazonian biodiversity. *Science* 255: 976-979. [2] geographical distribution; macrohabitats.

3116. Mares, Michael A., & D.J. Schmidly, eds., 1991. *Latin American Mammalogy: History, Biodiversity, and Conservation.* Norman: University of Oklahoma Press. 468p. [1/1/1].

3117. Margoluis, Richard, & Nick Salafsky, 1998. *Measures of Success: Designing, Managing, and Monitoring Conservation and Development Projects.* Washington, DC: Island Press. 362p.

3118. Margules, Christopher R., 1992. The Wog Wog habitat fragmentation experiment. *Environmental Conservation* 19: 316-325. [1] *A.T.:* New South Wales; eucalyptus forest.

3119. Margules, Christopher R., & M.P. Austin, eds., 1991. *Nature Conservation: Cost Effective Biological Surveys and Data Analysis* [conference papers]. Canberra: CSIRO Australia. 207p. [1/1/2] *A.T.:* Australia.

3120. ———, 1994. Biological models for monitoring species decline: the construction and use of data bases. *PTRSL B* 344: 69-75. [1] *A.T.:* ecological surveys.

3121. Margules, Christopher R., & K.J. Gaston, 1994. Biological diversity and agriculture [letter]. *Science* 265: 457. [1] *A.T.:* species richness.

3122. Margules, Christopher R., & J.A. Meyers, 1992. Biological diversity and ecosystem fragmentation: an Australian perspective. *Ekistics* 59: 293-300. [1].

3123. Margules, Christopher R., A.J. Higgs, & R.W. Rafe, 1982. Modern biogeographic theory: are there any lessons for nature reserve design? *Biological Conservation* 24: 115-128. [2] *A.T.:* island biogeography.

3124. Margules, Christopher R., A.O. Nicholls, & R.L. Pressey, 1988. Selecting networks of reserves to maximise biological diversity. *Biological Conservation* 43: 63-76. [2] *A.T.:* New South Wales.

3125. Margulies, Rebecca L., 1993. Protecting biodiversity: recognizing international intellectual property rights in plant genetic resources. *Michigan J. of International Law* 14: 322-356. [1].

3126. Margulis, Lynn, 1992. Biodiversity: molecular biological domains, symbiosis and kingdom origins. *Bio Systems* 27: 39-51. [1].

3127. Margulis, Lynn, & J.E. Lovelock, 1974. Biological modulation of the earth's atmosphere. *Icarus* 21: 471-489. [1] "The purpose of this paper is to develop the concept that the earth's atmosphere is actively maintained and regulated by life on the surface." *A.T.:* Gaia hypothesis.

3128. Margulis, Lynn, & Lorraine Olendzenski, eds., 1992. *Environmental Evolution: Effects of the Origin and Evolution of Life on Planet Earth* [essays]. Cambridge, MA: MIT Press. 405p. [2/2/1].

3129. Margulis, Lynn, & Dorion Sagan, 1995. *What Is Life?* New York: Simon & Schuster. 207p. [1/2/1] *A.T.:* biological philosophy; origin of life.

3130. Margulis, Lynn, & K.V. Schwartz, 1998. *Five Kingdoms: An Illustrated Guide to the Phyla of Life on Earth* (3rd ed.). New York: W.H. Freeman. 520p. [3/2/3].

3131. Marinelli, Janet, fall 1992. Backyard biodiversity. *Amicus J.* 14(3): 28-30. [1].

3132. ———, ed., 1994. *Going Native: Biodiversity in Our Own Backyards* [handbook]. Brooklyn, NY: Brooklyn Botanic Garden. 112p. [2/1/1] *A.T.:* gardening.

3133. Markham, Adam, Nigel Dudley, & Sue Stolton, 1993. *Some Like It Hot: Climate Change, Biodiversity, and the Survival of Species.* Gland,

Switzerland: WWF International. 144p. [1/1/1].

3134. Marks, Jonathan M., 1995. *Human Biodiversity: Genes, Race, and History*. New York: Aldine de Gruyter. 321p. [2/2/2] *A.T.:* eugenics.

3135. Marks, Stuart A., 1994. Local hunters and wildlife surveys: a design to enhance participation. *African J. of Ecology* 32: 233-254. [1] Studies a compilation of wildlife sightings made by local hunters and compares to an earlier such compilation. *A.T.:* Zambia.

3136. Marmonier, Pierre, P. Vervier, et al., 1993. Biodiversity in ground waters. *TREE* 8: 392-395. [1] *A.T.:* groundwater colonization; groundwater biota.

3137. Marques, J.C., & D. Bellan-Santini, 1993. Biodiversity in the ecosystem of the Portuguese continental shelf: distributional ecology and the role of benthic amphipods. *Marine Biology* 115: 555-564. [1] *A.T.:* Atlantic.

3138. Marques, J.C., M.A. Pardal, et al., 1997. Analysis of the properties of exergy and biodiversity along an estuarine gradient of eutrophication. *Ecological Modelling* 102: 155-167.

3139. Marquet, P.A., 1994. Diversity of small mammals in the Pacific coastal desert of Peru and Chile and in the adjacent Andean area: biogeography and community structure. *Australian J. of Zoology* 42: 527-542. [1] *A.T.:* Atacama Desert; rodents.

3140. Marroquin-Merino, Victor M., 1995. Wildlife utilization: a new international mechanism for the protection of biological diversity. *Law and Policy in International Business* 26: 303-370. [1] *A.T.:* international law.

3141. Marrs, R.H., 1993. Soil fertility and nature conservation in Europe: theoretical considerations and practical management solutions. *Advances in Ecological Research* 24: 241-300. [2].

3142. Marsden, Stuart J., 1998. Changes in bird abundance following selective logging on Seram, Indonesia. *CB* 12: 605-611. *A.T.:* endemism.

3143. Marsh, Clive W., & R.A. Mittermeier, eds., 1987. *Primate Conservation in the Tropical Rain Forest*. New York: A.R. Liss. 365p. [1/1/1].

3144. Marsh, George Perkins, 1864. *Man and Nature: Or, Physical Geography as Modified by Human Action*. New York: C. Scribner. 560p. The first major work decrying the negative effects of humankind's activities on the environment. [3/-/2].

3145. Marsh, Lindell L., 1994. Conservation planning under the Endangered Species Act: a new paradigm for conserving biological diversity. *Tulane Environmental Law J.* 8: 97-121. [1] *A.T.:* habitat conservation; California.

3146. Marshall, Larry G., 1988. Land mammals and the Great American Interchange. *American Scientist* 76: 380-388. [1] *A.T.:* paleobiogeography.

3147. Marshall, Larry G., S.D. Webb, et al., 1982. Mammalian evolution and the Great American Interchange. *Science* 215: 1351-1357. [2] *A.T.:* paleobiogeography.

3148. Martin, Gary J., & Alison Semple, 1994. Joint ventures in applied ethnobotany. *Nature and Resources* 30(1): 5-17. [1] Describes how ethnobotanists and local residents collaborate on bd-conserving efforts. *A.T.:* local participation; indigenous peoples.

3149. Martin, Paul S., 1973. The discovery of America. *Science* 179: 969-974. [1] Concerning the Late Pleistocene extinctions of American megafauna.

3150. ———, 1990. 40,000 years of extinctions on the "planet of doom." *Global and Planetary Change* 2: 187-201. [1].

3151. Martin, Paul S., & R.G. Klein, eds., 1984. *Quaternary Extinctions: A Prehistoric Revolution.* Tucson: University of Arizona Press. 892p. [2/2/2].

3152. Martin, Paul S., & H.E. Wright, Jr., eds., 1967. *Pleistocene Extinctions: The Search for a Cause* [conference proceedings]. New Haven: Yale University Press. 453p. [2/2/1] The first thorough survey of the subject, including several analyses looking into the possible role of aboriginals in an "overkill" process.

3153. Martin, P., 1996. Oligochaeta and Aphanoneura in ancient lakes: a review. *Hydrobiologia* 334: 63-72. [1] *A.T.:* Lake Baikal; Lake Tanganyika.

3154. Martin, P.H., 1996. Will forest preserves protect temperate and boreal biodiversity from climate change? *Forest Ecology and Management* 85: 335-341. [1] *A.T.:* forest conservation.

3155. Martin, Ronald E., G.J. Vermeij, et al., 1996. Late Permian extinctions [five letters]. *Science* 274: 1549-1552. [1].

3156. Martin, Thomas E., 1981. Species-area slopes and coefficients: a caution on their interpretation. *American Naturalist* 118: 823-837. [1].

3157. Martin, William H., S.G. Boyce, & A.C. Echternacht, eds., 1993. *Biodiversity of the Southeastern United States: Lowland Terrestrial Communities.* New York: Wiley. 502p. [1/1/1].

3158. ———, eds., 1993. *Biodiversity of the Southeastern United States: Upland Terrestrial Communities.* New York: Wiley. 373p. [1/1/1].

3159. Martínez, Neo D., & J.H. Lawton, 1995. Scale and food-web structure—from local to global. *Oikos* 73: 148-154. [1].

3160. Martínez del Rio, Carlos, & Steve Cork, 1997. Exploring nutritional biodiversity: a society is born. *TREE* 12: 9-10. Describes a symposium devoted to animal feeding habits.

3161. Martínez-Murcia, A.J., S.G. Acinas, & F. Rodriguez-Valera, 1995. Evaluation of prokaryotic diversity by restrictase digestion of 16S rDNA directly amplified from hypersaline environments. *FEMS Microbiology Ecology* 17: 247-256. [2] *A.T.:* microbiological analysis; bacteria; archaebacteria.

3162. Martínez Ramos, Miguel, Elena Alvarez-Buylla, et al., 1988. Treefall age determination and gap dynamics in a tropical forest. *J. of Ecology*

76: 700-716. [2] *A.T.:* Mexico; canopy disturbance; understorey growth.

3163. Martius, Chistopher, 1994. Diversity and ecology of termites in Amazonian forests. *Pedobiologia* 38: 407-428. [1].

3164. Marzluff, John M., & K.P. Dial, 1991. Life history correlates of taxonomic diversity. *Ecology* 72: 428-439. [1] *A.T.:* speciose taxa; diversification.

3165. ———, 1991. Does social organization influence diversification? *American Midland Naturalist* 125: 126-134. [1] *A.T.:* speciose taxa.

3166. Marzluff, John M., & Rex Sallabanks, eds., 1998. *Avian Conservation: Research and Managment* [symposium papers]. Washington, DC: Island Press. 563p.

3167. Mascie-Taylor, C.G.N., & Barry Bogin, eds., 1995. *Human Variability and Plasticity.* Cambridge, U.K., and New York: CUP. 241p. [1/1/1].

3168. Mason, Christopher F., 1990. Assessing population trends of scarce birds using information in a county bird report and archive. *Biological Conservation* 52: 303-320. [1] *A.T.:* England; rare birds; bd monitoring.

3169. Masood, Ehsan, 1997. Biodiversity projects face funding challenge. *Nature* 385: 106-107. Describes how GEF has reduced its funding of bd-related projects.

3170. Master, Lawrence L., S.R. Flack, & B.A. Stein, 1998. *Rivers of Life: Critical Watersheds for Protecting Freshwater Biodiversity.* Arlington, VA: Nature Conservancy. 71p.

3171. Mastrorillo, Sylvain, Francis Dauba, et al., 1998. Predicting local fish species richness in the Garonne River Basin. *Comptes Rendus de l'Académie des Sciences Série III* 321: 423-428. *A.T.:* artificial neural networks.

3172. Masundire, Hillary M., & J.Z.Z. Matowanyika, eds., 1995. *Biological Diversity in Southern Africa: The Path Ahead* [workshop report]. Harare, Zimbabwe: IUCN. 68p. [1/1/1].

3173. Matheson, Andrew, S.L. Buchmann, et al., eds., 1996. *The Conservation of Bees.* London: Academic Press. 254p. [1/1/1].

3174. Mathews, Jessica Tuchman, ed., 1991. *Preserving the Global Environment: The Challenge of Shared Leadership* [essays]. New York: Norton. 362p. [2/2/1] *A.T.:* environmental policy; international cooperation.

3175. Matlack, Glenn R., 1994. Plant species migration in a mixed-history forest landscape in eastern North America. *Ecology* 75: 1491-1502. [2] *A.T.:* understorey vegetation; Delaware; Pennsylvania; seed dispersal.

3176. Matson, P.A., W.J. Parton, et al., 1997. Agricultural intensification and ecosystem properties. *Science* 277: 504-509. *A.T.:* environmental degradation; agriculture; sustainable agriculture.

3177. Matsuda, Brent M., 1997. Conservation biology, values, and advocacy [editorial]. *CB* 11: 1449-1450.

3178. Matsuda, H., M. Hori, & P.A. Abrams, 1996. Effects of predator-specific defence on biodiversity and community complexity in two-trophic-level communities. *Evolutionary Ecology* 10: 13-28. [1] *A.T.:* anti-predator behavior; community structure; predator-prey relations.

3179. Matthews, William J., & H.W. Robison, 1998. Influence of drainage connectivity, drainage area and regional species richness on fishes of the Interior Highlands in Arkansas. *American Midland Naturalist* 139: 1-19. *A.T.:* macroecology.

3180. Matthews, William J., & E.G. Zimmerman, Nov.-Dec. 1990. Potential effects of global warming on native fishes of the southern Great Plains and the Southwest. *Fisheries* 15(6): 26-32. [1].

3181. Mattice, Jack, Myra Fraser, et al., 1996. Managing for biodiversity: emerging ideas for the electric utility industry—summary statement. *Environmental Management* 20: 781-788. [1].

3182. Mauchamp, A., 1997. Threats from alien plant species in the Galápagos Islands. *CB* 11: 260-263. *A.T.:* biological invasions; island ecology.

3183. Maunder, M., 1992. Plant reintroduction: an overview. *Biodiversity and Conservation* 1: 51-61. [1].

3184. Maurer, Brian A., 1990. The relationship between distribution and abundance in a patchy environment. *Oikos* 58: 181-189. [1] *A.T.:* birds; boreal forests; simulation studies.

3185. ———, 1993. Biological diversity, ecological integrity, and Neotropical migrants: new perspectives for wildlife management. *General Technical Report RM* GTR-229: 24-31. [1] *A.T.:* birds.

3186. ———, 1994. *Geographical Population Analysis: Tools for the Analysis of Biodiversity.* Oxford, U.K., and Boston: Blackwell Scientific. 130p. [1/1/2].

3187. Maurer, Brian A., & M.-A. Villard, 1994. Population density: geographic variation in abundance of North American birds. *Research & Exploration* 10: 306-317. [1] *A.T.:* population declines; Neotropical migrant species.

3188. Maury-Lechon, Gema, Malcolm Hadley, & Talal Younès, eds., 1984. *The Significance of Species Diversity in Tropical Forest Ecosystems* [meeting report]. Paris: Biology International Special Issue No. 6, IUBS. 76p. [1/1/1].

3189. Mawdsley, N., 1996. Biodiversity loss and ecosystem function in tropical forests [letter]. *TREE* 11: 432. [1].

3190. Maxted, Nigel, B.V. Ford-Lloyd, & J.G. Hawkes, eds., 1997. *Plant Genetic Conservation: The In Situ Approach.* London and New York: Chapman & Hall. 446p. [1/1/-] *A.T.:* plant germplasm resources.

3191. Maxted, Nigel, J.G. Hawkes, et al., 1997. Towards the selection of taxa for genetic conservation. *Genetic Resources and Crop Evolution* 44: 337-348. *A.T.:* genetic erosion; genetic resources.

3192. May, Robert M., 1986. How many species are there? *Nature* 324: 514-515. [1] *A.T.:* species numbers; arthropods.

3193. ———, 1986. The search for patterns in the balance of nature: advances and retreats [lecture]. *Ecology* 67: 1115-1126. [2] *A.T.:* population dynamics; community structure; species numbers; food webs.

3194. ———, 1988. How many species are there on Earth? *Science* 241: 1441-1449. [2] *A.T.:* species numbers; food webs; body size.

3195. ———, 1990. How many species? *PTRSL B* 330: 293-304. [2] *A.T.:* species numbers; diversity assessment; food webs.

3196. ———, 1990. Taxonomy as destiny. *Nature* 347: 129-130. [2] *A.T.:* tuatara; New Zealand; taxonomy.

3197. ———, 1991. A fondness for fungi [editorial]. *Nature* 352: 475-476. [1] Describes David Hawksworth's studies on fungal diversity.

3198. ———, Oct. 1992. How many species inhabit the earth? *Scientific American* 267(4): 42-48. [1] *A.T.:* species numbers.

3199. ———, 1992. Bottoms up for the oceans [editorial]. *Nature* 357: 278-279. [1] *A.T.:* species numbers; North Atlantic.

3200. ———, 1994. Biological diversity: differences between land and sea. *PTRSL B* 343: 105-111. [1] *A.T.:* species numbers; higher taxa.

3201. ———, 1994. Conceptual aspects of the quantification of the extent of biological diversity. *PTRSL B* 345: 13-20. [2] *A.T.:* species numbers; taxonomic distinctiveness.

3202. ———, 1994. Ecological science and the management of protected areas. *Biodiversity and Conservation* 3: 437-448. [1] *A.T.:* edge effects; metapopulations.

3203. May, Robert M., & M.A. Nowak, 1994. Superinfection, metapopulation dynamics, and the evolution of diversity. *J. of Theoretical Biology* 170: 95-114. [1] *A.T.:* pathogens.

3204. Mayden, Richard L., & R.M. Wood, 1995. Systematics, species concepts, and the ESU in biodiversity and conservation biology. In Jennifer L. Nielsen, ed., *Evolution and the Aquatic Ecosystem: Defining Unique Units in Population Conservation* (Bethesda, MD: American Fisheries Society Symposium 17): 58-113. [1].

3205. Mayer, Audrey L., & S.L. Pimm, 1997. Diversity begets diversity. *Current Biology* 7: R430-R432. *A.T.:* seedling survival; tropical rainforests.

3206. Mayr, Ernst, 1963. *Animal Species and Evolution*. Cambridge, MA: Belknap Press of Harvard University Press. 797p. [3/-/3] Mayr's most influential work.

3207. ———, 1976. *Evolution and the Diversity of Life: Selected Essays*. Cambridge, MA, and London: Belknap Press of Harvard University Press. 721p. [2/2/2] A collection of writings by one of the century's leading evolutionists.

3208. ———, 1995. Biodiversity and its molecular foundation [editorial]. *Comptes Rendus de l'Académie des Sciences, Série III* 318: 727-731. [1].

3209. Mays, T.D., & K.D. Mazan, 1996. Legal issues in sharing the benefits of biodiversity prospecting. *J. of Ethnopharmacology* 51: 93-109. [1].

3210. McAdam, J.H., A.C. Bell, et al., 1996. The effects of different hedge restoration strategies on biodiversity. In *Vegetation Management in Forestry, Amenity and Conservation Areas: Managing for Multiple Objectives* (Warwick, U.K.: Association of Applied Biologists): 363-368. [1].

3211. McAllister, D.E., 1991. Estimating the pharmaceutical values of forests, Canadian and tropical. *Canadian Biodiversity* 1(3): 16-25. [1].

3212. McArdle, B.H., 1990. When are rare species not there? *Oikos* 57: 276-277. [2] Concerning the problems associated with monitoring rare species. *A.T.:* sampling.

3213. McCabe, Thomas R., 1994. Assessing values of arctic wildlife and habitat subject to potential petroleum development. *Landscape and Urban Planning* 28: 33-46. [1] *A.T.:* caribou; brown bears; Alaska.

3214. McCall, Robert A., Sean Nee, & P.H. Harvey, 1996. Determining the influence of continental species-richness, island availability and vicariance in the formation of island-endemic bird species. *Biodiversity Letters* 3: 137-150. *A.T.:* island biogeography.

3215. McCallum, H., & A. Dobson, 1995. Detecting disease and parasite threats to endangered species and ecosystems. *TREE* 10: 190-194. [2] *A.T.:* keystone species; local extinction.

3216. McCallum, Rob, & Nikhil Sekhran, 1997. *Race for the Rainforest: Evaluating Lessons From an Integrated Conservation and Development "Experiment" in New Ireland, Papua New Guinea.* Waigani, Papua New Guinea: PNG Biodiversity Conservation and Resource Management Programme. 82p. [1/1/-].

3217. McCarthy, John F., & L.R. Shugart, eds., 1990. *Biomarkers of Environmental Contamination.* Chelsea, MI: Lewis Publishers. 457p. [1/1/2] *A.T.:* biological indicators; environmental monitoring.

3218. McCay, Bonnie J., & J.M. Acheson, eds., 1987. *The Question of the Commons: The Culture and Ecology of Communal Resources* [case studies]. Tucson: University of Arizona Press. 439p. [2/2/2] *A.T.:* human ecology.

3219. McChesney, James D., 1993. Biological and chemical diversity and the search for new pharmaceuticals and other bioactive natural products. In A.D. Kinghorn & M.F. Balandrin, eds., *Human Medicinal Agents From Plants* (Washington, DC: ACS Symposium Series No. 534): 38-47. [1].

3220. McClatchie, S., R.B. Millar, et al., 1997. Demersal fish community diversity off New Zealand: is it related to depth, latitude and regional surface phytoplankton? *Deep-Sea Research, Part I,* 44: 647-667. *A.T.:* diversity gradients; Southwest Pacific.

3221. McCloskey, J.M., & H. Spalding, 1989. A reconnaissance-level inventory of the amount of wilderness remaining in the world. *Ambio* 18: 221-227. [1].

3222. McConnell, Fiona, 1996. *The Biodiversity Convention, A Nego-*

tiating History: A Personal Account of Negotiating the United Nations Convention on Biological Diversity, and After. London and Boston: Kluwer Law International. 223p. [1/1/1] *A.T.:* CBD.

3223. McCoy, Earl D., 1994. "Amphibian decline": a scientific dilemma in more ways than one. *Herpetologica* 50: 98-103. [1]

3224. McCoy, Earl D., & E.F. Connor, 1980. Latitudinal gradients in the species diversity of North American mammals. *Evolution* 34: 193-203. [1].

3225. McCoy, Earl D., & H.R. Mushinsky, 1994. Effects of fragmentation on the richness of vertebrates in the Florida scrub habitat. *Ecology* 75: 446-457. [1].

3226. McCullough, Dale R., 1994. Importance of population data in forest management planning. *Forestry Chronicle* 70: 533-537. [1] *A.T.:* metapopulations.

3227. ———, ed., 1996. *Metapopulations and Wildlife Conservation* [symposium papers]. Washington, DC: Island Press. 429p. [1/2/1].

3228. McDade, Lucinda A., K.S. Bawa, et al., eds., 1994. *La Selva: Ecology and Natural History of a Neotropical Rain Forest.* Chicago: University of Chicago Press. 486p. [2/2/2] *A.T.:* Costa Rica.

3229. McDonald, Kim A., 13 April 1994. Charting biodiversity. *Chronicle of Higher Education* 40(32): A8-A9, A14. [1] Describes the Systematics Agenda 2000 proposal.

3230. McDougall, C.L., & Ronnie Hall, 1995. *Intellectual Property Rights and the Biodiversity Convention: The Impact of GATT.* Luton, Bedfordshire, U.K.: Friends of the Earth. 42p. [1/1/1].

3231. McDowell, R.M., 1996. Volcanism and freshwater fish biogeography in the northeastern North Island of New Zealand. *J. of Biogeography* 23: 139-148. [1].

3232. McEvedy, Imogen, July 1990. Credit where credit's due. *Management Today* (July): 88-89. [1] Describes the debt-for-nature swap concept and gives some examples of its application.

3233. McFadden, Max W., & J.K. Parker, 1994. Human values and biological diversity: are we wasting our time? *Canadian Entomologist* 126: 471-474. [1] *A.T.:* environmental perception.

3234. McGarigal, Kevin, & W.C. McComb, 1995. Relationships between landscape structure and breeding birds in the Oregon Coast Range. *Ecological Monographs* 65: 235-260. [2] *A.T.:* habitat fragmentation.

3235. McGeoch, M.A., & S.L. Chown, 1997. Impact of urbanization on a gall-inhabiting Lepidoptera assemblage: the importance of reserves in urban areas. *Biodiversity and Conservation* 6: 979-993. *A.T.:* South Africa; urban bd; insect conservation.

3236. McGhee, George R., Jr., 1996. *The Late Devonian Mass Extinction: The Frasnian/Famennian Crisis.* New York: Columbia University Press. 303p. [2/1/2].

3237. McGlone, M.S., 1996. When history matters: scale, time, climate and tree diversity. *Global Ecology and Biogeography Letters* 5: 309-314. [1] *A.T.:* energy-diversity theory; glacial cycles.

3238. McGoodwin, James R., 1990. *Crisis in the World's Fisheries: People, Problems, and Politics.* Stanford, CA: Stanford University Press. 235p. [1/2/2] *A.T.:* fisheries management.

3239. McGrady-Steed, Jill, P.M. Harris, & P.J. Morin, 1997. Biodiversity regulates ecosystem predictability. *Nature* 390: 162-165. Describes an experiment with aquatic microorganisms in which increasing bd is linked to greater predictability of ecosystem respiration.

3240. McGuinness, Keith A., 1984. Equations and explanations in the study of species-area curves. *Biological Reviews* 59: 423-440. [2] *A.T.:* random placement hypothesis; habitat diversity hypothesis; equilibrium theory; disturbance hypothesis.

3241. McHarg, Ian L., 1969. *Design With Nature.* Garden City, NY: Natural History Press. 197p. [3/2/2] *A.T.:* human ecology; environmental design.

3242. McIntosh, Heidi J., 1996. National forest management: a new approach based on biodiversity. *J. of Energy, Natural Resources & Environmental Law* 16: 257-317. [1] *A.T.:* Forest Service; ecosystem health.

3243. McKay, David S., E.K. Gibson, Jr., et al., 1996. Search for past life on Mars: possible relic biogenic activity in Martian meteorite ALH84001. *Science* 273: 924-930. [3] Describes the first solid evidence for the existence of past life-forms on the Red Planet.

3244. McKendrick, J.D., R.C. Wilkinson, & R.G.B. Senner, spring 1997. Tundra plant succession and vascular plant species diversity. *Agroborealis* 29(1): 28-30. [1] *A.T.:* Alaska.

3245. McKendry, Jean E., & G.E. Machlis, 1993. The role of geography in extending biodiversity gap analysis. *Applied Geography* 13: 135-152. [1] *A.T.:* socioeconomic factors; GIS.

3246. McKenney, Daniel W., ed., 1994. *Towards a Set of Biodiversity Indicators for Canadian Forests* [workshop proceedings]. Sault Ste. Marie: Canadian Forest Service. 133p. [1/1/1] *A.T.:* forest conservation.

3247. McKenzie, N.L., Lee Belbin, et al., 1989. Selecting representative reserve systems in remote areas: a case study in the Nullarbor Region, Australia. *Biological Conservation* 50: 239-261. [1].

3248. McKibben, Bill, 1989. *The End of Nature.* New York: Random House. 226p. [3/3/2] *A.T.:* greenhouse effect.

3249. McKinney, Michael L., 1998. Is marine biodiversity at less risk? Evidence and implications. *Diversity & Distributions* 4: 3-8. Gives reasons for thinking that this is not the case. *A.T.:* extinction.

3250. McKinney, Michael L., & J.A. Drake, eds., 1998. *Biodiversity Dynamics: Turnover of Populations, Taxa, and Communities.* New York: Columbia University Press. 584p. *A.T.:* evolution; population biology.

3251. McKinney, Michael L., J.L. Lockwood, & D.R. Frederick, 1996. Does ecosystem and evolutionary stability include rare species? *Palaeogeography, Palaeoclimatology, Palaeoecology* 127: 191-207. [1] *A.T.:* coordinated stasis.

3252. McKnight, Bill N., ed., 1993. *Biological Pollution: The Control and Impact of Invasive Exotic Species* [symposium proceedings]. Indianapolis: Indiana Academy of Sciences. 261p. [2/1/1] *A.T.:* North America.

3253. McLaren, D.J., & W.D. Goodfellow, 1990. Geological and biological consequences of giant impacts. *Annual Review of Earth and Planetary Sciences* 18: 123-171. [2] Examines evidence for impacts at several major geological time boundaries. *A.T.:* mass extinctions; gradualistic model.

3254. McLaughlin, Alison, & Pierre Mineau, 1995. The impact of agricultural practices on biodiversity. *Agriculture, Ecosystems & Environment* 55: 201-212. [2] *A.T.:* agricultural ecology; fertilizers.

3255. Mclaughlin, Christopher G., 1998. *Habitat Allometry: Evaluating Changes in Species Richness Using Species-Area Models*. Sault Ste. Marie: Ontario Forest Research Institute. 28p. *A.T.:* mathematical models.

3256. McManis, Charles R., spring 1998. The interface between international intellectual property and environmental protection: biodiversity and biotechnology. *Washington University Law Quarterly* 76(1): 255-279.

3257. McManus, J.W., 1994. The Spratly Islands: a marine park? *Ambio* 23: 181-186. [1] *A.T.:* South China Sea; disputed areas.

3258. ———, 1997. Tropical marine fisheries and the future of coral reefs: a brief review with emphasis on Southeast Asia. *Coral Reefs* 16, Suppl.: S121-S127. *A.T.:* environmental degradation.

3259. McMenamin, Mark A.S., & D.L.S. McMenamin, 1990. *The Emergence of Animals: The Cambrian Breakthrough*. New York: Columbia University Press. 217p. [2/1/2] *A.T.:* diversification; evolution.

3260. ———, 1993. Hypersea and the land ecosystem. *Bio Systems* 31: 145-153. [1] *A.T.:* land eukaryotes.

3261. McMinn, J.W., & D.A. Crossley, eds., 1996. *Biodiversity and Coarse Woody Debris in Southern Forests* [workshop proceedings]. Asheville, NC: General Technical Report SE GTR-94, Southern Research Station. 146p. [1/1/1].

3262. McNab, Brian K., 1971. On the ecological significance of Bergmann's rule. *Ecology* 52: 845-854. [2] *A.T.:* body size; latitudinal gradients.

3263. McNally, Ruth, & Peter Wheale, 1996. Biopatenting and biodiversity: comparative advantages in the new global order. *Ecologist* 26: 222-228. [1] *A.T.:* developing countries.

3264. McNamara, Kenneth J. ed., 1990. *Evolutionary Trends*. Tucson: University of Arizona Press. 368p. [2/2/1] *A.T.:* paleobiology.

3265. McNaughton, S.J., 1977. Diversity and stability of ecological communities: a comment on the role of empiricism in ecology. *American*

Naturalist 111: 515-525. [2].

3266. McNeely, Jeffrey A., 1988. *Economics and Biological Diversity: Developing and Using Economic Incentives to Conserve Biological Resources*. Gland, Switzerland: IUCN. 236p. [1/1/2].

3267. ———, 1989. How to pay for conserving biological diversity. *Ambio* 18: 308-313. [1].

3268. ———, 1990. Climate change and biological diversity: policy implications. In M.M. Boer, R.S. de Groot, & F.A. Eybergen, eds., *Landscape-Ecological Impact of Climatic Change* (Amsterdam: IOS Press): 406-429. [1] *A.T.:* global warming.

3269. ———, 1992. The sinking ark: pollution and the worldwide loss of biodiversity. *Biodiversity and Conservation* 1: 2-18. [1] *A.T.:* environmental degradation.

3270. ———, 1993. Economic incentives for conserving biodiversity: lessons for Africa. *Ambio* 22: 144-150. [1] *A.T.:* environmental policy.

3271. ———, ed., 1993. *Parks for Life: Report of the IVth World Congress on National Parks and Protected Areas*. Gland, Switzerland: IUCN and WWF. 252p. [1/1/1].

3272. ———, 1994. Lessons from the past: forests and biodiversity. *Biodiversity and Conservation* 3: 3-20. [1] *A.T.:* forest management.

3273. ———, 1994. Protected areas for the 21st century: working to provide benefits to society. *Biodiversity and Conservation* 3: 390-405. [1] *A.T.:* protected areas management.

3274. ———, ed., 1995. *Expanding Partnerships in Conservation* [conference papers]. Washington, DC: Island Press. 302p. [1/1/1] *A.T.:* protected areas; environmental planning; citizen participation.

3275. ———, 1995. Keep all the pieces: Systematics 2000 and world conservation. *Biodiversity and Conservation* 4: 510-519. [1] *A.T.:* systematics; bdc; environmental management.

3276. ———, winter 1997. New trends in protecting and managing biodiversity. *Ecodecision*, no. 23: 20-23. *A.T.:* cultural diversity loss; language loss.

3277. McNeely, Jeffrey A., & R.J. Dobias, 1991. Economic incentives for conserving biological diversity in Thailand. *Ambio* 20: 86-90. [1] *A.T.:* environmental policy; bd loss; action plans.

3278. McNeely, Jeffrey A., & K.R. Miller, eds., 1984. *National Parks, Conservation, and Development: The Role of Protected Areas in Sustaining Society* [conference proceedings]. Washington, DC: Smithsonian Institution Press. 825p. [2/1/1].

3279. McNeely, Jeffrey A., & R.B. Norgaard, 1992. Developed country policies and biological diversity in developing countries. *Agriculture, Ecosystems & Environment* 42: 194-204. [1] *A.T.:* North-South relations; agricultural development.

3280. McNeely, Jeffrey A., & D.C. Pitt, eds., 1985. *Culture and Conservation: The Human Dimension in Environmental Planning.* London and Dover, NH: Croom Helm. 308p. [1/1/1].

3281. McNeely, Jeffrey A., & Frank Vorhies, 1997. Policy forum: missing the biodiversity boat. *Environment and Development Economics* 2: 76-81.

3282. McNeely, Jeffrey A., & W.P. Weatherly, 1996. Innovative funding to support biodiversity conservation. *International J. of Social Economics* 23: 98-124. [1] *A.T.:* environmental policy.

3283. McNeely, Jeffrey A., Jeremy Harrison, & P.R. Dingwall, eds., 1994. *Protecting Nature: Regional Reviews of Protected Areas* [conference papers]. Gland, Switzerland: IUCN. 402p. [1/1/1].

3284. McNeely, Jeffrey A., Kal Raustiala, & D.G. Victor, March 1997. Biodiversity since Rio [letters]. *Environment* 39(2): 3-5. Discusses issues raised by implementation of the CBD.

3285. McNeely, Jeffrey A., K.R. Miller, et al., 1990. *Conserving the World's Biological Diversity.* Gland, Switzerland, and Washington, DC: IUCN et al. 193p. [2/1/2].

3286. McNeely, Jeffrey A., K.R. Miller, et al., 1990. Strategies for conserving biodiversity. *Environment* 32(3): 16-20, 36-40. [1] *A.T.:* bd value; environmental management

3287. McNeill, S.E., & P.G. Fairweather, 1993. Single large or several small marine reserves? An experimental approach with seagrass fauna. *J. of Biogeography* 20: 429-440. [1] *A.T.:* SLOSS.

3288. McPherson, Malcolm F., 1985. *Critical Assessment of the Value of and Concern for the Maintenance of Biological Diversity.* Cambridge, MA: Development Discussion Paper No. 212, Harvard Institute for International Development, Harvard University. 89p. [1/1/1].

3289. McRae, M., 1997. Is "good wood" bad for forests? *Science* 275: 1868-1869. [1] Reports on a study of a mahogany-logging operation in Bolivia. *A.T.:* sustainable management.

3290. McRobert, Jencie, 1997. *Biological Diversity in Transport Corridors: Road Drainage Management* [research report]. Vermont South, Victoria, Australia: ARRB Transport Research. 76p. [1/1/-] *A.T.:* Australia.

3291. McRoberts, C.A., & Martin Aberhan, 1997. Marine diversity and sea-level changes: numerical tests for association using Early Jurassic bivalves. *Geologische Rundschau* 86: 160-167.

3292. Meadows, Robin, Sept.-Oct. 1996. Fire management may hinder biodiversity. *California Agriculture* 50(5): 6-8. [1].

3293. Meave, Jorge, & Martin Kellman, 1994. Maintenance of rain forest diversity in riparian forests of tropical savannas: implications for species diversity during Pleistocene drought. *J. of Biogeography* 21: 121-135. [1] *A.T.:* Belize; forest structure; Pleistocene refugia.

3294. Meave, Jorge, Martin Kellman, et al., 1991. Riparian habitats as tropical forest refugia. *Global Ecology and Biogeography Letters* 1: 69-76. [1] *A.T.:* island biogeography; Belize; Venezuela; habitat fragmentation; savannas.

3295. Medail, Frederic, & Pierre Quézel, 1997. Hot-spots analysis for conservation of plant biodiversity in the Mediterranean Basin. *Annals of the Missouri Botanical Garden* 84: 112-127.

3296. Medail, Frederic, & R. Verlaque, 1997. Ecological characteristics and rarity of endemic plants from southeast France and Corsica: implications for biodiversity conservation. *Biological Conservation* 80: 269-281.

3297. Medellín, Rodrigo A., & M. Equihua, 1998. Mammal species richness and habitat use in rainforest and abandoned agricultural fields in Chiapas, Mexico. *J. of Applied Ecology* 35: 13-23.

3298. Medina, E., & J.F. Silva, 1990. Savannas of northern South America: a steady state regulated by water-fire interactions on a background of low nutrient availability. *J. of Biogeography* 17: 403-413. [1] *A.T.:* Venezuela.

3299. Meel, Rosita M. van, 1997. How to augment effectiveness and flexibility by curriculum development in agricultural higher education. A case for a study programme on biological diversity. *European J. of Agricultural Education and Extension* 4: 151-162. *A.T.:* Netherlands; France.

3300. Meffe, Gary K., 1990. Genetic approaches to conservation of rare fishes: examples from North American desert species. *J. of Fish Biology* 37, Suppl. A: 105-112. [1] *A.T.:* hatcheries.

3301. Meffe, Gary K., & C. Ronald Carroll, eds., 1997. *Principles of Conservation Biology* (2nd ed.) [textbook]. Sunderland, MA: Sinauer Associates. 729p. [2/2/2].

3302. Mehrotra, B.N., 1996. Collection of biological materials in biodiversity prospecting in India: problems and solutions. *J. of Ethnopharmacology* 51: 161-165. [1].

3303. Meier, Albert J., S.P. Bratton, & D.C. Duffy, 1995. Possible ecological mechanisms for loss of vernal-herb diversity in logged Eastern deciduous forests. *Ecological Applications* 5: 935-946. [1] *A.T.:* Tennessee; North Carolina.

3304. Mendelsohn, Robert, & M.J. Balick, 1995. The value of undiscovered pharmaceuticals in tropical forests. *Economic Botany* 49: 223-228. [2].

3305. [not used]

3306. Mendes, Maria T.R., 1997. *Biodiversity Conservation in Mozambique and Brazil*. Paris: UNESCO. 32p. [1/1/-].

3307. Menge, Bruce A., & J.P. Sutherland, 1976. Species diversity gradients: synthesis of the roles of predation, competition, and temporal heterogeneity. *American Naturalist* 110: 351-369. [2].

3308. Menge, Bruce A., E.L. Berlow, et al., 1994. The keystone species concept: variation in interaction strength in a rocky intertidal habitat. *Ecological Monographs* 64: 249-286. [2] *A.T.:* mussels; seastars; Oregon.

3309. Menges, E.S., 1990. Population viability analysis for an endangered plant. *CB* 4: 52-62. [2] *A.T.:* louseworts; Maine.

3310. Menges, E.S., & D.R. Gordon, 1996. Three levels of monitoring intensity for rare plant species. *Natural Areas J.* 16: 227-237. [1] *A.T.:* Florida; sampling.

3311. Menke, John, & G.E. Bradford, 1992. Rangelands. *Agriculture, Ecosystems & Environment* 42: 141-163. [1] *A.T.:* rangeland management.

3312. Menon, Shaily, & K.S. Bawa, 1997. Applications of geographic information systems, remote sensing, and a landscape ecology approach to biodiversity conservation in the Western Ghats. *Current Science* 73: 134-145.

3313. Merigoux, S., D. Ponton, & Bernard de Merona, 1998. Fish richness and species-habitat relationships in two coastal streams of French Guiana, South America. *Environmental Biology of Fishes* 51: 25-39.

3314. Merriam, C. Hart, 1892. The geographic distribution of life in North America with special reference to the Mammalia. *Proc. of the Biological Society of Washington* 7: 1-64. [1] *A.T.:* life zones.

3315. ———, 1894. Laws of temperature control of the geographic distribution of terrestrial animals and plants. *National Geographic Magazine* 6: 229-238. [1] *A.T.:* life zones.

3316. Merrifield, J., 1996. A market approach to conserving biodiversity. *Ecological Economics* 16: 217-226. [1] *A.T.:* environmental economics.

3317. Merrill, Samuel B., F.J. Cuthbert, & Gary Oehlert, 1998. Residual patches and their contribution to forest-bird diversity on northern Minnesota aspen clearcuts. *CB* 12: 190-199. *A.T.:* residual timber patches.

3318. Metrick, Andrew, & M.L. Weitzman, 1994. *Patterns of Behavior in Biodiversity Preservation*. Washington, DC: Policy Research Working Paper 1358, World Bank. 28p. [1/1/1] *A.T.:* econometric models; endangered species.

3319. ———, 1996. Patterns of behavior in endangered species preservation. *Land Economics* 72: 1-16. [1] *A.T.:* environmental policy; government spending.

3320. ———, summer 1998. Conflicts and choices in biodiversity preservation. *J. of Economic Perspectives* 12(3): 21-34. *A.T.:* endangered species; optimization analysis.

3321. Metzger, Jean Paul, & Henri Décamps, 1997. The structural connectivity threshold: an hypothesis in conservation biology at the landscape scale. *Acta Oecologica* 18: 1-12.

3322. Metzger, Jean Paul, L.C. Bernacci, & R. Goldenberg, 1997. Pattern of tree species diversity in riparian forest fragments of different widths (SE Brazil). *Plant Ecology* 133: 135-152. *A.T.:* corridors.

3323. Meyer, Carrie A., 1996. NGOs and environmental public goods: institutional alternatives to property rights. *Development and Change* 27:

453-474. [1] *A.T.:* nongovernmental organizations; Costa Rica; environmental policy; INBio.

3324. Meyer, Jean-Yves, & Jacques Florence, 1996. Tahiti's native flora endangered by the invasion of *Miconia calvescens* DC (Melastomataceae). *J. of Biogeography* 23: 775-781. [1] *A.T.:* biological invasions.

3325. Meyer, William B., 1996. *Human Impact on the Earth.* Cambridge, U.K., and New York: CUP. 253p. [2/1/1].

3326. Meyer, William B., & Billie Lee Turner II, eds., 1994. *Changes in Land Use and Land Cover: A Global Perspective* [conference papers]. Cambridge, U.K., and New York: CUP. 537p. [1/2/2].

3327. Meyers, Gary D., 1992. Surveying the lay of the land, air, and water: features of current international environmental and natural resources law, and future prospects for the protection of species habitat to preserve global biological diversity. *Colorado J. of International Environmental Law and Policy* 3: 479-604. [1].

3328. Michener, C.D., 1979. Biogeography of the bees. *Annals of the Missouri Botanical Garden* 66: 277-347. [2].

3329. Midgley, Mary, 1992. Beasts versus the biosphere? *Environmental Values* 1: 113-121. [1] *A.T.:* deep ecology; animal rights; wildlife management.

3330. Mies, Maria, & Vandana Shiva, 1993. *Ecofeminism* [essays]. London and Atlantic Highlands, NJ: Zed Books. 328p. [2/2/2] *A.T.:* human ecology.

3331. Mighetto, Lisa, 1991. *Wild Animals and American Environmental Ethics.* Tucson: University of Arizona Press. 177p. [2/2/1] *A.T.:* animal rights and welfare; wildlife conservation.

3332. Mikitin, Kathleen, 1995. *Issues and Options in the Design of GEF Supported Trust Funds for Biodiversity Conservation.* Washington, DC: Environment Department Paper (Biodiversity Series) No. 11, World Bank. 112p. [1/1/1].

3333. Mikusiński, Grzegorz, & Per Angelstam, 1998. Economic geography, forest distribution, and woodpecker diversity in Central Europe. *CB* 12: 200-208. *A.T.:* bioindicators; socioeconomic factors.

3334. Milberg, P., & Tommy Tyrberg, 1993. Naïve birds and noble savages—a review of man-caused prehistoric extinctions of island birds. *Ecography* 16: 229-250. [1] *A.T.:* overhunting; introduced species; habitat destruction.

3335. Miles, Donald B., 1994. Introduction to the Symposium: contribution of long-term ecological research to current issues in the conservation of biological diversity. *American Zoologist* 34: 367-370. [1].

3336. Milewski, Inka A., 1995. Marine biodiversity: shaping a policy framework. *Natural Areas J.* 15: 61-67. [1] *A.T.:* environmental policy; international cooperation.

3337. Miller, Alan S., 1991. *Gaia Connections: An Introduction to Ecology, Ecoethics, and Economics*. Savage, MD: Rowman & Littlefield. 301p. [2/1/1] *A.T.:* environmental ethics.

3338. Miller, Arnold I., 1997. A new look at age and area: the geographic and environmental expansion of genera during the Ordovician radiation. *Paleobiology* 23: 410-419. *A.T.:* higher taxa; geographical range; paleobiogeography.

3339. ———, 1998. Biotic transitions in global marine diversity. *Science* 281: 1157-1160. *A.T.:* Phanerozoic; mass extinctions; macroevolution.

3340. Miller, Arnold I., & Mike Foote, 1996. Calibrating the Ordovician radiation of marine life: implications for Phanerozoic diversity trends. *Paleobiology* 22: 304-309. [2] *A.T.:* sampling; invertebrates; macroevolution.

3341. Miller, Arnold I., & J.J. Sepkoski, Jr., 1988. Modeling bivalve diversification: the effect of interaction on a macroevolutionary system. *Paleobiology* 14: 364-369. [1] *A.T.:* mass extinctions.

3342. Miller, Brian J., Gerardo Ceballos, & R.P. Reading, 1994. The prairie dog and biotic diversity. *CB* 8: 677-681. [1] *A.T.:* North America; keystone species; grasslands management.

3343. Miller, Brian J., S.H. Anderson, et al., 1988. Biology of the endangered black-footed ferret and the role of captive propagation in its conservation. *Canadian J. of Zoology* 66: 765-773. [1] *A.T.:* Wyoming.

3344. Miller, Douglass R., & A.Y. Rossman, 1995. Systematics, biodiversity, and agriculture. *Bioscience* 45: 680-686. [1] *A.T.:* insect pests; biological control.

3345. Miller, D.J., 1989. Introductions and extinction of fish in the African Great Lakes. *TREE* 4: 56-59. [1] *A.T.:* cichlids; biological invasions.

3346. Miller, Henry I., 1994. *Is the Biodiversity Treaty a Bureaucratic Time Bomb?* Stanford, CA: Essays in Public Policy No. 56, Hoover Institution on War, Revolution and Peace, Stanford University. 11p. [1/1/1] *A.T.:* CBD; environmental law.

3347. Miller, H., D. Kennedy, et al., 1995. Agricultural production, innovation and biological diversity. *Australasian Biotechnology* 5: 238-240. [1].

3348. Miller, Julie Ann, 1989. Diseases for our future: global ecology and emerging viruses. *Bioscience* 39: 509-517. [1] Explores the connection between human interaction and the spread of disease viruses.

3349. ———, 1991. Tropical reforestation, at best, requires patience. *Bioscience* 41: 751-752. [1] Reports on John Terborgh's observations on forest recovery in Peru.

3350. ———, 1994. Charting the biosphere. *Bioscience* 44: 392-393. [1] Describes Systematics Agenda 2000.

3351. Miller, J.C., 1993. Insect natural history, multi-species interactions and biodiversity in ecosystems. *Biodiversity and Conservation* 2: 233-241. [1] *A.T.:* community ecology.

3352. Miller, Kenton R., 1991. Preserving biological diversity. *Evolutionary Trends in Plants* 5: 87-89. [1] *A.T.:* coastal plant communities.

3353. ———, 1994. International cooperation in conserving biological diversity: a world strategy, international convention, and framework for action. *Biodiversity and Conservation* 3: 464-472. [1] *A.T.:* environmental management.

3354. ———, 1996. *Balancing the Scales: Guidelines for Increasing Biodiversity's Chances Through Bioregional Management.* Washington, DC: WRI. 73p. [1/1/1].

3355. Miller, Kenton R., & S.M. Lanou, eds., 1995. *National Biodiversity Planning: Guidelines Based on Early Experiences Around the World* [report]. Washington, DC: WRI. 161p. [1/1/1].

3356. Miller, Kenton R., & J.N. Shores, 1991. *Biodiversity and the Forestry Profession.* Vancouver: Faculty of Forestry, University of British Columbia. 31p. [1/1/1].

3357. Miller, Kenton R., & Laura Tangley, 1991. *Trees of Life: Saving Tropical Forests and Their Biological Wealth.* Boston: Beacon Press. 218p. [2/2/1] *A.T.:* forest conservation; deforestation.

3358. Miller, Robert R., J.D. Williams, & J.E. Williams, Nov.-Dec. 1989. Extinctions of North American fishes during the past century. *Fisheries* 14(6): 22-38. [2] *A.T.:* habitat alteration.

3359. Miller, Ronald I., ed., 1994. *Mapping the Diversity of Nature.* London and New York: Chapman & Hall. 218p. [1/1/1] *A.T.:* bd mapping.

3360. Miller, Ronald I., & R.G. Wiegert, 1989. Documenting completeness, species-area relations, and the species-abundance distribution of a regional flora. *Ecology* 70: 16-22. [1] *A.T.:* Southern Appalachians; vascular plants.

3361. Miller, Ronald I., S.P. Bratton, & P.S. White, 1987. A regional strategy for reserve design and placement based on an analysis of rare and endangered species' distribution patterns. *Biological Conservation* 39: 255-268. [1] *A.T.:* Southern Appalachians; vascular plants.

3362. Miller, Ronald I., S.N. Stuart, & K.M. Howell, 1989. A methodology for analyzing rare species distribution patterns utilizing GIS technology: the rare birds of Tanzania. *Landscape Ecology* 2: 173-189. [1].

3363. Miller, Scott, 1991. Biological diversity and the need to nurture systematic collections [editorial]. *American Entomologist* 37: 76. [1].

3364. Miller, Stanton S., 1992. The road from Rio. *Environmental Science & Technology* 26: 1710-1713. [1] Reviews UNCED. *A.T.:* sustainable development; international cooperation.

3365. Miller, Steven L., 1995. Functional diversity in fungi. *Canadian J. of Botany* 73, Suppl. 1: S50-S57. [1].

3366. Milligan, B.G., J. Leebens-Mack, & A.E. Strand, 1994. Conservation genetics: beyond the maintenance of marker diversity. *Molecular*

Ecology 3: 423-435. [2] *A.T.:* genetic markers.

3367. Milliken, William, & J.A. Ratter, eds., 1998. *Maracá: The Biodiversity and Environment of an Amazonian Rainforest.* Chichester, U.K., and New York: Wiley. 508p. *A.T.:* Brazil.

3368. Mills, Claudia E., & J.T. Carlton, 1998. Rationale for a system of international reserves for the open ocean. *CB* 12: 244-247.

3369. Mills, Edward L., J.H. Leach, et al., 1993. Exotic species in the Great Lakes: a history of biotic crises and anthropogenic introductions. *J. of Great Lakes Research* 19: 1-54. [2] *A.T.:* biological invasions; ballast.

3370. Mills, Edward L., J.H. Leach, et al., 1994. Exotic species and the integrity of the Great Lakes. *Bioscience* 44: 666-676. [1] *A.T.:* biological invasions.

3371. Mills, Judy A., & Christopher Servheen, 1991. *The Asian Trade in Bears and Bear Parts.* Washington, DC: WWF. 113p. [1/1/1] *A.T.:* wild animal trade.

3372. Mills, L. Scott, & P.E. Smouse, 1994. Demographic consequences of inbreeding in remnant populations. *American Naturalist* 144: 412-431. [2] *A.T.:* extinction.

3373. Mills, L. Scott, M.E. Soulé, & D.F. Doak, 1993. The keystone-species concept in ecology and conservation. *Bioscience* 43: 219-224. [2].

3374. Millsap, Brian A., J.A. Gore, et al., 1990. *Setting Priorities for the Conservation of Fish and Wildlife Species in Florida.* Bethesda, MD: Wildlife Monograph No. 111, Wildlife Society. 57p. [1/1/1] *A.T.:* vertebrate conservation.

3375. Milne, Bruce T., 1992. Spatial aggregation and neutral models in fractal landscapes. *American Naturalist* 139: 32-57. [2] *A.T.:* patch dynamics; spatial scale.

3376. Milne, Bruce T., & R.T.T. Forman, 1986. Peninsulas in Maine: woody plant diversity, distance, and environmental patterns. *Ecology* 67: 967-974. [1] *A.T.:* peninsular effect.

3377. Milner-Gulland, E.J., & Ruth Mace, 1998. *Conservation of Biological Resources* [includes case studies]. Malden, MA: Blackwell Science. 404p.

3378. Milton, Suzanne J., W.R.J. Dean, et al., 1994. A conceptual model of arid rangeland degradation: the escalating cost of declining productivity. *Bioscience* 44: 70-76. [2] *A.T.:* range management.

3379. Minckley, Wendell L., & J.E. Deacon, eds., 1991. *Battle Against Extinction: Native Fish Management in the American West* [conference papers]. Tucson: University of Arizona Press. 517p. [1/1/1] *A.T.:* rare fishes.

3380. Mineau, Pierre, & Alison McLaughlin, 1996. Conservation of biodiversity within Canadian agricultural landscapes: integrating habitat for wildlife. *J. of Agricultural & Environmental Ethics* 9: 93-113. [1].

3381. Minelli, Alessandro, 1989. The role of taxonomy in the analysis of natural and agricultural communities. *Agriculture, Ecosystems & Environment* 27: 57-66. [1].

3382. ———, 1994. Systematics and biodiversity [letter]. *TREE* 9: 227. [1].

3383. Mingoti, Sueli A., & G. Meeden, 1992. Estimating the total number of distinct species using presence and absence data. *Biometrics* 48: 863-875. [1] *A.T.:* binary data; bootstrap procedure; jackknife procedure.

3384. Ministry of Environment and Forests (India), 1989. *Wetlands, Mangroves, and Biosphere Reserves: Proceedings of the Indo-US Workshop.* New Delhi. 269p. [1/1/1] *A.T.:* India.

3385. Ministry of State for Population and Environment (Indonesia), 1992. *Indonesian Country Study on Biological Diversity.* Jakarta. 209p. [1/1/1].

3386. Minns, Charles K., 1989. Factors affecting fish species richness in Ontario lakes. *Trans. of the American Fisheries Society* 118: 533-545. [1].

3387. Minns, Charles K., V.W. Cairns, et al., 1994. An index of biotic integrity (IBI) for fish assemblages in the littoral zone of Great Lakes' areas of concern. *Canadian J. of Fisheries and Aquatic Sciences* 51: 1804-1822. [1] *A.T.:* ecosystem health.

3388. Minns, Charles K., J.E. Moore, et al., 1990. Assessing the potential extent of damage to inland lakes in eastern Canada due to acidic deposition. III. Predicted impacts on species richness in seven groups of aquatic biota. *Canadian J. of Fisheries and Aquatic Sciences* 47: 821-830. [2].

3389. Minshall, G. Wayne, R.C. Petersen, Jr., & C.F. Nimz, 1985. Species richness in streams of different size from the same drainage basin. *American Naturalist* 125: 16-38. [1] *A.T.:* Idaho.

3390. Miranda, E.E. de, & C. Mattos, 1992. Brazilian rain forest colonization and biodiversity. *Agriculture, Ecosystems & Environment* 40: 275-296. [1] *A.T.:* deforestation; land use; environmental degradation.

3391. Miranda, Marta, ed., 1998. *All That Glitters Is Not Gold: Balancing Conservation and Development in Venezuela's Frontier Forests.* Washington, DC: WRI. 52p. *A.T.:* sustainable development.

3392. Mishler, Brent D., 1995. Plant systematics and conservation: science and society. *Madroño* 42: 103-113. [1].

3393. Mishra, Charudutt, & G.S. Rawat, 1998. Livestock grazing and biodiversity conservation: comments on Saberwal. *CB* 12: 712-714.

3394. Mistretta, Orlando, L.H. Rieseberg, & T.S. Elias, 1991. Botanic gardens and the preservation of biological diversity. *Evolutionary Trends in Plants* 5: 19-22. [1].

3395. Mitchell, J.C., S.C. Rinehart, et al., 1997. Factors influencing amphibian and small mammal assemblages in central Appalachian forests. *Forest Ecology and Management* 96: 65-76. *A.T.:* Virginia.

3396. Mitchell, Richard S., C.J. Sheviak, & D.J. Leopold, eds., 1990.

Ecosystem Management: Rare Species and Significant Habitats: Proceedings of the 15th Annual Natural Areas Conference. Albany: New York State Museum. 314p. [1/1/1].

3397. Mitchell, Robert C., & R.T. Carson, 1989. *Using Surveys to Value Public Goods: The Contingent Valuation Method.* Washington, DC: Resources for the Future. 463p. [2/2/3] A very frequently referred to source on this subject.

3398. Mitra, Shaibal, Hans Landel, & Stephen Pruett-Jones, 1996. Species richness covaries with mating system in birds. *Auk* 113: 544-551. [2] *A.T.:* sexual selection.

3399. Mittelbach, Gary G., A.M. Turner, et al., 1995. Perturbation and resilience: a long-term, whole-lake study of predator extinction and reintroduction. *Ecology* 76: 2347-2360. [1] *A.T.:* keystone species; predator-prey relations; Michigan.

3400. Mittermeier, Russell A., July-Aug. 1988. Strange and wonderful Madagascar. *International Wildlife* 18(4): 4-13. [1] An overview of bd loss on this island.

3401. ———, 1997. New map of biodiversity "hotspots" aids targeting of conservation efforts. *Diversity* 13(1): 27-29.

3402. Mittermeier, Russell A., & I.A. Bowles, 1993. The Global Environment Facility and biodiversity conservation: lessons to date and suggestions for future action. *Biodiversity and Conservation* 2: 637-655. [1] *A.T.:* UNCED.

3403. ———, 1994. Reforming the approach of the Global Environmental Facility to biodiversity conservation. *Oryx* 28: 101-106. [1].

3404. Mittermeier, Russell A., C.G. Mittermeier, & Patricio Robles Gil, 1997. *Megadiversity: Earth's Biologically Wealthiest Nations.* México, D.F.: CEMEX. 501p. [1/1/-].

3405. Mittermeier, Russell A., T.B. Werner, & A. Lees, 1996. New Caledonia—a conservation imperative for an ancient land. *Oryx* 30: 104-112. [1] *A.T.:* hotspots.

3406. Mittermeier, Russell A., Norman Meyers, et al., 1998. Biodiversity hotspots and major tropical wilderness areas: approaches to setting conservation priorities. *CB* 12: 516-520.

3407. Mittermeier, Russell A., et al., compilers, 1992. *Lemurs of Madagascar: An Action Plan for Their Conservation, 1993-1999.* Gland, Switzerland: IUCN. 58p. [1/1/2].

3408. Miya, Masaki, & Mutsumi Nishida, 1997. Speciation in the open ocean. *Nature* 389: 803-804. *A.T.:* cryptic species; deep-sea fish; genetic diversity.

3409. Mladenoff, David J., R.G. Haight, et al., 1997. Causes and implications of species restoration in altered ecosystems. *Bioscience* 47: 21-31. *A.T.:* wolves; Lakes States.

3410. Mladenoff, David J., M.A. White, et al., 1993. Comparing spatial pattern in unaltered old-growth and disturbed forest landscapes. *Ecological Applications* 3: 294-306. [2] *A.T.:* GIS; Michigan.

3411. Mlot, Christine, 1989. The science of saving endangered species: directions for research in conservation biology [editorial]. *Bioscience* 39: 68-70. [1] *A.T.:* wildlife conservation.

3412. ———, 1989. Blueprint for conserving plant diversity. *Bioscience* 39: 364-368. [1].

3413. ———, 1991. Giving freshwater diversity its due. *Bioscience* 41: 756-757. [1] Discusses relative lack of attention given to freshwater bdc problems.

3414. ———, 1992. Botanists sue Forest Service to preserve biodiversity. *Science* 257: 1618-1619. [1] *A.T.:* environmental litigation.

3415. ———, 1997. Population diversity crowds the ark. *Science News* 152: 260. [1] Reports on studies highlighting population—as distinct from species—loss.

3416. Moe, B., & A. Botnen, 1997. A quantitative study of the epiphytic vegegation on pollarded trunks of *Fraxinus excelsior* at Havra, Osteroy, western Norway. *Plant Ecology* 129: 157-177. *A.T.:* bryophytes; lichens.

3417. Moffat, Anne Simon, 1996. Biodiversity is a boon to ecosystems, not species. *Science* 271: 1497. [1] Concerning David Tilman's work relating stability of individual species populations to ecosystem stability.

3418. ———, 1998. Global nitrogen overload problem grows critical. *Science* 279: 988-989. *A.T.:* fixed nitrogen; pollution; fertilizers.

3419. Moffett, Mark W., 1994. *The High Frontier: Exploring the Tropical Rainforest Canopy.* Cambridge, MA: Harvard University Press. 192p. [2/2/1].

3420. Molau, Ulf, & J.M. Alatalo, 1998. Responses of subarctic-alpine plant communities to simulated environmental change: biodiversity of bryophytes, lichens, and vascular plants. *Ambio* 27: 322-329.

3421. Mönkkönen, M., & D.A. Welsh, 1994. A biogeographical hypothesis on the effects of human caused landscape changes on the forest bird communities of Europe and North America. *Annales Zoologici Fennici* 31: 61-70. [1].

3422. Montgomery, Claire A., & R.A. Pollack, Feb. 1996. Economics and biodiversity: weighing benefits and costs of conservation. *J. of Forestry* 94(2): 34-38. [1].

3423. Montgomery, Claire A., G.M. Brown, Jr., & D.M. Adams, 1994. The marginal cost of species preservation: the northern spotted owl. *J. of Environmental Economics and Management* 26: 111-128. [2] *A.T.:* Pacific Northwest.

3424. Moock, Joyce L., & R.E. Rhoades, eds., 1992. *Diversity, Farmer Knowledge, and Sustainability* [conference papers]. Ithaca, NY: Cornell

University Press. 278p. [1/1/1] *A.T.:* technology transfer.

3425. Mooers, Arne O., & S.B. Heard, 1997. Inferring evolutionary process from phylogenetic tree shape. *Quarterly Review of Biology* 72: 31-54. *A.T.:* macroevolution; phylogeny.

3426. Mooney, Harold A., & Giorgio Bernardi, eds., 1990. *Introduction of Genetically Modified Organisms Into the Environment.* Chichester, U.K., and New York: Wiley. 201p. [1/2/1] *A.T.:* genetic engineering.

3427. Mooney, Harold A., & J.A. Drake, eds., 1986. *Ecology of Biological Invasions of North America and Hawaii* [symposium papers]. New York: Springer. 321p. [1/1/2] *A.T.:* animal introduction; plant introduction.

3428. ———, June 1987. The ecology of biological invasions. *Environment* 29(5): 10-15, 34-37. [1] Relates the study of biological invasions to the release of genetically altered organisms.

3429. Mooney, Harold A., J.H. Cushman, et al., eds., 1996. *Functional Roles of Biodiversity: A Global Perspective.* Chichester, U.K., and New York: Wiley. 493p. [1/1/-].

3430. Mooney, Harold A., Ernesto Medina, et al., eds., 1991. *Ecosystem Experiments.* Chichester, U.K., and New York: Wiley. 268p. [1/1/1]. *A.T.:* environmental change; biosphere.

3431. Moore, H.D.M., W.V. Holt, & G.M. Mace, eds., 1992. *Biotechnology and the Conservation of Genetic Diversity: The Proceedings of a Symposium.* New York: OUP. 240p. [1/1/1].

3432. Moore, Peter D., 1987. Biological diversity: what makes a forest rich? *Nature* 329: 292. [1] *A.T.:* evapotranspiration; species richness.

3433. ———, 1998. Did forests survive the cold in a hotspot? *Nature* 391: 124-127. *A.T.:* Africa; Quaternary glaciations.

3434. Moors, P.J., ed., 1985. *Conservation of Island Birds: Case Studies for the Management of Threatened Island Species* [symposium proceedings]. Cambridge, U.K.: ICBP Technical Publication No. 3, International Council for Bird Preservation. 271p. [1/1/1] *A.T.:* rare birds.

3435. Moraga-Rajel, Jubel, 1992. *Biodiversity Conservation in Chile: Policies and Practices.* Nairobi: African Centre for Technology Studies. 20p. [1/1/1].

3436. Moran, David P., & M.L. Reaka-Kudla, 1991. Effects of disturbance: disruption and enhancement of coral reef cryptofaunal populations by hurricanes. *Coral Reefs* 9: 215-224. [1].

3437. Moran, Dominic, 1994. Debt-swaps for hot-spots: more needed. *Biodiversity Letters* 2: 63-66. [1] *A.T.:* international cooperation.

3438. ———, 1994. Contingent valuation and biodiversity: measuring the user surplus of Kenyan protected areas. *Biodiversity and Conservation* 3: 663-684. [1] *A.T.:* ecotourism.

3439. ———, 1994. *Contingent Valuation and Biodiversity Conservation in Kenyan Protected Areas.* Norwich, U.K.: CSERGE Working Paper GEC

94-16, Centre for Social and Economic Research on the Global Environment. 28p. [1/1/1] *A.T.:* ecotourism.

3440. Moran, Dominic, D.W. Pearce, & Anouk Wendelaar, 1996. Global biodiversity priorities. A cost-effectiveness index for investments. *Global Environmental Change: Human and Policy Dimensions* 6: 103-119. [1] *A.T.:* bd cost; environmental policy.

3441. ———, 1997. Investing in biodiversity: an economic perspective on global priority setting. *Biodiversity and Conservation* 6: 1219-1243. *A.T.:* bd cost.

3442. Morand, S., & R. Poulin, 1998. Density, body mass and parasite species richness of terrestrial mammals. *Evolutionary Ecology* 12: 717-727.

3443. Morat, Philippe, 1993. Our knowledge of the flora of New Caledonia: endemism and diversity in relation to vegetation types and substrates. *Biodiversity Letters* 1: 72-81. [1].

3444. Moravec, J., 1993. Biodiversity changes on an ecosystemic level-phytocoenological approach. *Ecology Bratislava* 12: 317-324. [1] *A.T.:* Czech Republic; biocoenoses; bd monitoring.

3445. Moreira, F.M.S., K. Haukka, & J.P.W. Young, 1998. Biodiversity of rhizobia isolated from a wide range of forest legumes in Brazil. *Molecular Ecology* 7: 889-895. *A.T.:* bacteria.

3446. Morell, Virginia, April 1997. On the origin of (Amazonian) species. *Discover* 18(4): 56-64. Describes attempts to account for the great diversity of Amazonian life. *A.T.:* DNA analysis; rodents.

3447. ———, 1997. Biodiversity in a vial of sugar water. *Science* 278: 390. Describes an elegant bd experiment involving bacteria in a vial of water.

3448. ———, 1997. Counting creatures of the Serengeti, great and small. *Science* 278: 2058-2060. *A.T.:* Tanzania; inventories.

3449. Mori, Scott A., B.M. Boom, & G.T. Prance, 1981. Distribution patterns and conservation of eastern Brazilian coastal forest tree species. *Brittonia* 33: 233-245. [1].

3450. Moritz, Craig, 1994. Applications of mitochondrial DNA analysis in conservation: a critical review. *Molecular Ecology* 3: 401-411. [2] *A.T.:* gene conservation; molecular ecology.

3451. ———, 1994. Defining "evolutionarily significant units" for conservation. *TREE* 9: 373-375. [3].

3452. Moritz, Craig, & D.P. Faith, 1998. Comparative phylogeography and the identification of genetically divergent areas for conservation. *Molecular Ecology* 7: 419-429. *A.T.:* genetic diversity; phylogenetic diversity; Queensland.

3453. Moritz, Craig, & Jiro Kikkawa, eds., 1994. *Conservation Biology in Australia and Oceania.* Chipping Norton, NSW: Surrey Beatty. 404p. [1/1/1] *A.T.:* bdc.

3454. Morowitz, Harold J., 1991. Balancing species preservation and economic considerations. *Science* 253: 752-754. [1] *A.T.:* bd value.

3455. Morris, Douglas W., 1995. Earth's peeling veneer of life [editorial]. *Nature* 373: 25. [1] *A.T.:* human impact; bd loss.

3456. Morris, Douglas W., & Lawrence Heidinga, 1997. Balancing the books on biodiversity [editorial]. *CB* 11: 287-289. *A.T.:* human impact.

3457. Morrone, Juan J., & J.V. Crisci, 1995. Historical biogeography: introduction to methods. *ARES* 26: 373-401. [2].

3458. Morrone, Juan J., L. Katinas, & J.V. Crisci, 1996. On temperate areas, basal clades and biodiversity conservation. *Oryx* 30: 187-194. [1].

3459. Morrone, Juan J., Sergio Roig Juñent, & J.V. Crisci, 1994. South American beetles. *Research & Exploration* 10: 104-115. [1] *A.T.:* biogeography; cladistics.

3460. Morse, Larry E., & Mary Sue Henifin, eds., 1981. *Rare Plant Conservation: Geographical Data Organization* [conference papers]. Bronx, NY: New York Botanical Garden. 377p. [1/1/1] *A.T.:* database management.

3461. Morton, Brian, 1996. Protecting Hong Kong's marine biodiversity: present proposals, future challenges. *Environmental Conservation* 23: 55-65. [1] *A.T.:* coastal zone management; environmental planning; bdc.

3462. Morton, S.R., 1996. Looking after our land: a future for Australia's biodiversity. *Search* 27: 124-126. [1].

3463. Morton, S.R., & C.D. James, 1988. The diversity and abundance of lizards in arid Australia: a new hypothesis. *American Naturalist* 132: 237-256. [2] *A.T.:* termites.

3464. Morton, S.R., D.M.S. Smith, et al., 1995. The stewardship of arid Australia: ecology and landscape management. *J. of Environmental Management* 43: 195-217. [1] *A.T.:* environmental management; sustainable development.

3465. Mosbrugger, Volker, 1992. The evolution of biodiversity—a review of phenomena and problems. *Revue de Paléobiologie* 11: 533-543. [1] *A.T.:* Provides a review of Phanerozoic evolution dwelling on coevolution of biotic and abiotic forces.

3466. Moseley, Bill, April 1993. Daniel Janzen [interview]. *Omni* 15(6): 73-77, 94-95. [1] *A.T.:* Costa Rica.

3467. Mosquin, Theodore, P.G. Whiting, & D.E. McAllister, 1995. *Canada's Biodiversity: The Variety of Life, Its Status, Economic Benefits, Conservation Costs, and Unmet Needs.* Ottawa: Canadian Museum of Nature. 293p. [1/1/1].

3468. Mosseler, Alexander J., 1992. Life history and genetic diversity in red pine: implications for gene conservation in forestry. *Forestry Chronicle* 68: 701-708. [1] *A.T.:* Newfoundland.

3469. Mota, J.F., J. Peñas, et al., 1996. Agricultural development vs. biodiversity conservation: the Mediterranean semiarid vegetation in El Ejido (Almería, southeastern Spain). *Biodiversity and Conservation* 5: 1597-1617.

[1].
3470. Mott, J.J., & P.B. Bridgewater, 1992. Biodiversity conservation and ecologically sustainable development. *Search* 23: 284-287. [1] *A.T.:* Australia; human impact; environmental management.
3471. Mott, Richard N., 1993. The GEF and the Conventions on Climate Change and Biological Diversity. *International Environmental Affairs* 5: 299-312. [1].
3472. Moulton, Michael P., 1993. The all-or-none pattern in introduced Hawaiian passeriforms: the role of competition sustained. *American Naturalist* 141: 105-119. [1].
3473. Mound, Laurence A., & N. Waloff, eds., 1978. *Diversity of Insect Faunas* [symposium papers]. Oxford, U.K.: Blackwell Scientific. 204p. [1/1/1].
3474. Moustafa, Abd El-Raouf, & A. Zayed, 1996. Effect of environmental factors on the flora of alluvial fans in southern Sinai. *J. of Arid Environments* 32: 431-443. [1].
3475. Moyle, Peter B., 1995. Conservation of native freshwater fishes in the Mediterranean-type climate of California, USA: a review. *Biological Conservation* 72: 271-279. [1] *A.T.:* endangered species; environmental management.
3476. Moyle, Peter B., & T. Light, 1996. Biological invasions of fresh water: empirical rules and assembly theory. *Biological Conservation* 78: 149-161. [1].
3477. Moyle, Peter B., & J.E. Williams, 1990. Biodiversity loss in the temperate zone: decline of the native fish fauna of California. *CB* 4: 275-284. [1] *A.T.:* water diversions; introduced species.
3478. Moyle, Peter B., & R.M. Yoshiyama, Feb. 1994. Protection of aquatic biodiversity in California: a five-tiered approach. *Fisheries* 19(2): 6-18. [1] *A.T.:* fish; aquatic conservation.
3479. Mueller-Dombois, Dieter, 1988. Forest decline and dieback—a global ecological problem. *TREE* 3: 310-312. [1].
3480. Mugabe, John, C.V. Barber, et al., 1997. *Access to Genetic Resources: Strategies for Sharing Benefits*. Nairobi: African Centre for Technology Studies. 377p. [1/1/-] *A.T.:* CBD; germplasm resources; international cooperation.
3481. Muir, John, 1912. *The Yosemite*. New York: Century. 284p. [2/-/1] A classic natural history portrait by the celebrated naturalist.
3482. Muller, Frank G., 1996. Transfer payments to developing countries for environmental protection: a viewpoint. *International J. of Environmental Studies* 49: 197-207. [1].
3483. Muller, Richard A., 1988. *Nemesis*. New York: Weidenfeld & Nicolson. 193p. [2/2/1] Concerning asteroid collisions.
3484. Muller, Siemon W., & Alison Campbell, 1954. The relative number

of living and fossil species of animals. *Systematic Zoology* 3: 168-170. [1] *A.T.:* species numbers.

3485. Muller, T., 1994. The role a botanical institute can play in the conservation of the terrestrial biodiversity in a developing country. *Biodiversity and Conservation* 3: 116-125. [1] *A.T.:* Zimbabwe; gene banks.

3486. Mummery, Josephine, & Neal Hardy, 1994. *Australia's Biodiversity: An Overview of Selected Significant Components.* Canberra: Biodiversity Unit, Dept. of the Environment, Sport and Territories. 87p. [1/1/1].

3487. Munasinghe, Mohan, 1992. Biodiversity protection policy: environmental valuation and distribution issues. *Ambio* 21: 227-236. [1] *A.T.:* GEF; sustainable development.

3488. ———, 1993. Environmental economics and biodiversity management in developing countries. *Ambio* 22: 126-135. [1] *A.T.:* Madagascar; Sri Lanka.

3489. Munasinghe, Mohan, & Walter Shearer, eds., 1995. *Defining and Measuring Sustainability: The Biogeophysical Foundations* [case studies]. Washington, DC: World Bank and IUCN. 440p. [1/1/2].

3490. Mungall, Constance, & D.J. McLaren, eds., 1990. *Planet Under Stress: The Challenge of Global Change.* Toronto and New York: OUP. 344p. [2/2/1] *A.T.:* environmental degradation; pollution.

3491. Munro, Neil W.P., & J.H.M. Willison, eds., 1998. *Linking Protected Areas With Working Landscapes Conserving Biodiversity* [conference proceedings]. Wolfville, Nova Scotia: Science and Management of Protected Areas Association. 1018p. *A.T.:* protected areas management.

3492. Munson, Abby, 1995. Should a biosafety protocol be negotiated as part of the Biodiversity Convention? *Global Environmental Change: Human and Policy Dimensions* 5: 7-26. [1] *A.T.:* CBD; genetic engineering.

3493. Munthali, S.M., 1997. Dwindling food-fish species and fishers' preference: problems of conserving Lake Malawi's biodiversity. *Biodiversity and Conservation* 6: 253-261. *A.T.:* fishery management; overfishing.

3494. Murcia, Carolina, 1995. Edge effects in fragmented forests: implications for conservation. *TREE* 10: 58-62. [3].

3495. Murdoch, William W., 1975. Diversity, complexity, stability and pest control. *J. of Applied Ecology* 12: 795-807. [1] *A.T.:* agroecosystems; coevolution.

3496. Murdoch, William W., F.C. Evans, & C.H. Peterson, 1972. Diversity and pattern in plants and insects. *Ecology* 53: 819-829. [2] *A.T.:* Michigan; old fields.

3497. Muriuki, John N., H.M. de Klerk, et al., 1997. Using patterns of distribution and diversity of Kenyan birds to select and prioritize areas for conservation. *Biodiversity and Conservation* 6: 191-210.

3498. Murphy, Dennis D., 1988. Are we studying our endangered butterflies to death? *J. of Research on the Lepidoptera* 26: 236-239. [1].

3499. Murphy, Dennis D., & D.A. Duffus, 1996. Conservation biology and marine biodiversity [editorial]. *CB* 10: 311-312. [1].

3500. Murphy, Dennis D., & B.R. Noon, 1992. Integrating scientific methods with habitat conservation planning: reserve design for the northern spotted owl. *Ecological Applications* 2: 3-17. [2] *A.T.:* Pacific Northwest.

3501. Murphy, Jamie, 13 Oct. 1986. The quiet apocalypse: biologists warn that a mass extinction is happening now. *Time* 128(15): 80. [1] *A.T.:* bd loss.

3502. Murray, David F., 1992. Vascular plant diversity in Alaskan Arctic tundra. *Northwest Environmental J.* 8: 29-52. [1] *A.T.:* environmental change; climatic change.

3503. Murray, E., 1993. The sinister snail. *Endeavour* 17: 78-83. [1] *A.T.:* introduced species; Society Islands; snails.

3504. Murray, M.G., M.J.B. Green, et al., 1997. *Biodiversity Conservation in the Tropics: Gaps in Habitat Protection and Funding Priorities.* Cambridge, U.K.: World Conservation Press. 170p. [1/1/-].

3505. Mwalyosi, R.B.B., 1991. Ecological evaluation for wildlife corridors and buffer zones for Lake Manyara National Park, Tanzania, and its immediate environment. *Biological Conservation* 57: 171-186. [1].

3506. Mwandosya, M.J., & A.K. Semesi, 1997. *Towards a Strategy for the Conservation of Coastal Biological Diversity in Mainland Tanzania.* Dar es Salaam: Centre for Energy, Environment, Science and Technology. 102p. [1/1/-] *A.T.:* environmental management.

3507. Mwinga, D.K.K., 1997. The Biodiversity Convention and *in situ* conservation in Zambia. *Review of European Community & International Environmental Law* 6: 32-41. *A.T.:* CBD; environmental legislation.

3508. Myers, Alan A., 1997. Biogeographic barriers and the development of marine biodiversity. *Estuarine, Coastal, and Shelf Science* 44: 241-248. *A.T.:* Papua New Guinea; New Caledonia; amphipods; tropical lagoons.

3509. Myers, Alan A., & P.S. Giller, eds., 1988. *Analytical Biogeography: An Integrated Approach to the Study of Animal and Plant Distributions.* London and New York: Chapman & Hall. 578p. [2/1/1].

3510. Myers, J.H., C. Higgens, & E. Kovacs, 1989. How many insect species are necessary for the biological control of insects? *Environmental Entomology* 18: 541-547. [1].

3511. Myers, Norman, 1976. An expanded approach to the problem of disappearing species. *Science* 193: 198-202. [1] *A.T.:* bd loss; endangered species.

3512. ———, 1979. *The Sinking Ark: A New Look at the Problem of Disappearing Species.* Oxford, U.K., and New York: Pergamon Press. 307p. [2/2/2] Among the best known works on the subject of species loss. *A.T.:* wildlife conservation; endangered species.

3513. ———, 1980. *Conversion of Tropical Moist Forests.* Washington,

DC: National Academy of Sciences. 205p. [2/1/2] *A.T.:* deforestation; forest surveys.

3514. ———, 1983. *A Wealth of Wild Species: Storehouse for Human Welfare.* Boulder, CO: Westview Press. 274p. [2/1/2] *A.T.:* environmental economics.

3515. ———, 1986. *Tackling Mass Extinction of Species: A Great Creative Challenge.* Berkeley: Dept. of Forestry and Resource Management, College of Natural Resources, University of California. 40p. [1/1/1].

3516. ———, 1987. The impending extinction spasm: synergisms at work. *CB* 1: 14-22. [1].

3517. ———, 1988. Tropical deforestation and remote sensing. *Forest Ecology and Management* 23: 215-225. [1] *A.T.:* bd monitoring.

3518. ———, 1988. Tropical deforestation and climatic change. *Environmental Conservation* 15: 293-298. [1].

3519. ———, 1988. Threatened biotas: "hot-spots" in tropical forests. *Environmentalist* 8: 1-20. [2].

3520. ———, 1988. Threatened biotas: "hot-spots" in tropical forests. *Environmentalist* 8: 187-208. [2] *A.T.:* Costa Rica; Ecuador; Minnesota.

3521. ———, 1989. *Deforestation Rates in Tropical Forests and Their Climatic Implications.* London: Friends of the Earth. 116p. [1/1/2] *A.T.:* greenhouse effect.

3522. ———, 1990. The biodiversity challenge: expanded hot-spots analysis. *Environmentalist* 10: 243-256. [2] *A.T.:* endemism.

3523. ———, 1990. Mass extinctions: what can the past tell us about the present and the future? *Global and Planetary Change* 2: 175-185. [1].

3524. ———, compiler, 1990. *The Wild Supermarket: The Importance of Biological Diversity to Food Security.* Edited by Christopher Hails et al. Gland, Switzerland: WWF. 32p. [1/1/1] *A.T.:* plant germplasm resources; food sources.

3525. ———, 1991. Tropical forests: present status and future outlook. *Climatic Change* 19: 3-32. [2].

3526. ———, 1992. *The Primary Source: Tropical Forests and Our Future* (2nd ed.). New York: Norton. 416p. [3/3/2].

3527. ———, 1993. Questions of mass extinction. *Biodiversity and Conservation* 2: 2-17. [1] *A.T.:* research programs.

3528. ———, 1993. Biodiversity and the precautionary principle. *Ambio* 22: 74-79. [1] *A.T.:* Concerning our debt to future generations. *A.T.:* environmental policy.

3529. ———, 1994. Biodiversity: protected from a greater what? *Biodiversity and Conservation* 3: 411-418. [1].

3530. ———, 1995. The world's forests: need for a policy appraisal [editorial]. *Science* 268: 823-824. [1] *A.T.:* World Commission on Forests and Sustainable Development; bd value.

3531. ———, 1996. Two key challenges for biodiversity: discontinuities and synergisms. *Biodiversity and Conservation* 5: 1025-1034. [1] *A.T.:* extinction.

3532. ———, 1996. Environmental services of biodiversity. *PNAS* 93: 2764-2769. [1] *A.T.:* bd value.

3533. ———, 1996. The biodiversity crisis and the future of evolution. *Environmentalist* 16: 37-47. [1] Expresses concern that the present bd crisis may disrupt evolution.

3534. ———, 1997. Mass extinction and evolution. *Science* 278: 597-598. Treats the possible effect of extinction on the future of evolution.

3535. Myers, Norman, & E.S. Ayensu, 1983. Reduction of biological diversity and species loss. *Ambio* 12: 72-74. [1] Describes the main causes, then suggests some remedies.

3536. Myers, Norman, & Thomas Sisk, A.E. Launer, et al., 1995. Reassessing threats to biodiversity: two replies [letters]. *Bioscience* 45: 379-380. [1] Concerning prioritizing significant centers of bd for conservation.

3537. Myers, Wayne, G.P. Patil, & Kyle Joly, 1997. Echelon approach to areas of concern in synoptic regional monitoring. *Environmental and Ecological Statistics* 4: 131-152. *A.T.:* mammals; Pennsylvania; spatial analysis; geostatistics.

3538. Nabhan, Gary Paul, 1985. Native crop diversity in Aridoamerica: conservation of regional gene pools. *Economic Botany* 39: 387-399. [1].

3539. ———, 1989. *Enduring Seeds: Native American Agriculture and Wild Plant Conservation.* San Francisco: North Point Press. 225p. [2/3/1] *A.T.:* seed banks.

3540. ———, 1995. The dangers of reductionism in biodiversity conservation [editorial]. *CB* 9: 479-481. [1].

3541. ———, 1997. *Cultures of Habitat: On Nature, Culture, and Story.* Washington, DC: Counterpoint. 338p. [2/2/-] *A.T.:* cultural ecology.

3542. Nabhan, Gary Paul, & A.R. Holdsworth, 1998. *State of the Desert Biome: Uniqueness, Biodiversity, Threats and the Adequacy of Protection in the Sonoran Bioregion.* Tucson: Wildlands Project. 81p. *A.T.:* Sonoran Desert.

3543. Nadkarni, Nalini M., 1994. Diversity of species and interactions in the upper tree canopy of forest ecosystems. *American Zoologist* 34: 70-78. [1] *A.T.:* epiphytes; keystone species.

3544. Naeem, Shahid, 1998. Species redundancy and ecosystem reliability. *CB* 12: 39-45.

3545. Naeem, Shahid, & S.B. Li, 1997. Biodiversity enhances ecosystem reliability. *Nature* 390: 507-509. *A.T.:* microcosms; microorganisms.

3546. Naeem, Shahid, K. Hakansson, et al., 1996. Biodiversity and plant productivity in a model assemblage of plant species. *Oikos* 76: 259-264. [2] *A.T.:* England.

3547. Naeem, Shahid, J.H. Lawton, et al., 1995. Biotic diversity and ecosystem processes: using the Ecotron to study a complex relationship. *Endeavour* 19: 58-63. [1] *A.T.:* microcosms.

3548. Naeem, Shahid, L.J. Thompson, et al., 1994. Declining biodiversity can alter the performance of ecosystems. *Nature* 368: 734-737. [3] *A.T.:* ecosystem dynamics; insects.

3549. Naeem, Shahid, L.J. Thompson, et al., 1995. Empirical evidence that declining species diversity may alter the performance of terrestrial ecosystems. *PTRSL B* 347: 249-262. [2] *A.T.:* microcosms; ecosystem processes; community structure.

3550. Næss, Arne, 1973. The shallow and the deep, long-range ecology movement. *Inquiry* 16: 95-100. [2] The first major exposition of the deep ecology concept. *A.T.:* nature philosophy.

3551. Næss, Arne (with David Rothenberg), 1989. *Ecology, Community, and Lifestyle: Outline of an Ecosophy.* Cambridge, U.K., and New York: CUP. 223p. [2/1/2] *A.T.:* nature philosophy; deep ecology.

3552. Næsset, Erik, 1997. Geographical information systems in long-term forest management and planning with special reference to preservation of biological diversity: a review. *Forest Ecology and Management* 93: 121-136.

3553. Nagendra, Harini, & Madhav Gadgil, July 1997. Remote sensing as a tool for estimating biodiversity. *J. of Spacecraft Technology* 7(2): 1-9. *A.T.:* India.

3554. Naiman, Robert J., Henri Décamps, & Michael Pollock, 1993. The role of riparian corridors in maintaining regional biodiversity. *Ecological Applications* 3: 209-212. [2].

3555. Namkoong, Gene, 1992. Biodiversity—issues in genetics, forestry and ethics. *Forestry Chronicle* 68: 438-443. [1] *A.T.:* forest management.

3556. Narayan, L.R.A., 1996. Remote sensing and geographical information systems for conservation and assessment of biological diversity. *International Archives of Photogrammetry and Remote Sensing* 31(B4): 592-597. [1].

3557. Nash, Roderick F., 1967. *Wilderness and the American Mind.* New Haven: Yale University Press. 256p. [3/3/3] *A.T.:* environmental perception.

3558. ———, 1989. *The Rights of Nature: A History of Environmental Ethics.* Madison: University of Wisconsin Press. 290p. [3/3/3] A frequently referred-to source on this subject.

3559. Nathan, R., U.N. Safriel, & H. Shirihai, 1996. Extinction and vulnerability to extinction at distribution peripheries: an analysis of the Israeli breeding avifauna. *Israel J. of Zoology* 42: 361-383. [1].

3560. National Council of the Paper Industry for Air and Stream Improvement, 1996. *The National Gap Analysis Program: Ecological Assumptions and Sensitivity to Uncertainty.* Research Triangle Park, NC: Technical Bull. No. 720, National Council of the Paper Industry for Air and

Stream Improvement. 56p. [1/1/1].

3561. National Research Council, 1972. *Genetic Vulnerability of Major Crops*. Washington, DC: National Academy of Sciences. 307p. [2/1/2] *A.T.:* plant diseases.

3562. ———, 1989. *Evaluation of Biodiversity Projects* [study]. Washington, DC: National Academy Press. 50p. [1/1/1] *A.T.:* research programs.

3563. ———, 1989. *Lost Crops of the Incas: Little-known Plants of the Andes With Promise for Worldwide Cultivation*. Washington, DC: National Academy Press. 415p. [2/1/2].

3564. ———, 1990. *Decline of the Sea Turtles: Causes and Prevention*. Washington, DC: National Academy Press. 259p. [2/1/2].

3565. ———, 1991. *Forest Trees* (*Managing Global Genetic Resources* series). Washington, DC: National Academy Press. 228p. [1/1/1] *A.T.:* tree germplasm resources.

3566. ———, 1991. *The U.S. National Plant Germplasm System* (*Managing Global Genetic Resources* series). Washington, DC: National Academy Press. 171p. [1/2/1].

3567. ———, 1992. *Conserving Biodiversity: A Research Agenda for Development Agencies* [panel report]. Washington, DC: National Academy Press. 127p. [2/1/1].

3568. ———, 1992. *Restoration of Aquatic Ecosystems: Science, Technology, and Public Policy*. Washington, DC: National Academy Press. 552p. [2/1/2].

3569. ———, 1993. *A Biological Survey for the Nation* [study]. Washington, DC: National Academy Press. 205p. [1/1/1] *A.T.:* National Biological Survey.

3570. ———, 1993. *Livestock* (*Managing Global Genetic Resources* series). Washington, DC: National Academy Press. 276p. [1/1/1] *A.T.:* livestock germplasm resources.

3571. ———, 1993. *Sustainable Agriculture and the Environment in the Humid Tropics*. Washington, DC: National Academy Press. 702p. [1/1/1].

3572. ———, 1994. *Assigning Economic Value to Natural Resources* [workshop papers]. Washington, DC: National Academy Press. 185p. [1/1/1] *A.T.:* natural resources valuation.

3573. ———, 1995. *Understanding Marine Biodiversity: A Research Agenda for the Nation* [workshop proceedings]. Washington, DC: National Academy Press. 114p. [1/1/1].

3574. ———, 1995. *Effects of Past Global Change on Life*. Washington, DC: National Academy Press. 250p. [1/1/1].

3575. ———, 1995. *Science and the Endangered Species Act*. Washington, DC: National Academy Press. 271p. [1/1/1].

3576. ———, 1997. *Evaluating Human Genetic Diversity*. Washington, DC: National Academy Press. 91p. [1/1/-].

3577. Nations, James D., & D.I. Komer, April 1983. Rainforests and the hamburger society: can the cycle be broken? *Environment* 25(3): 12-20. [1] Discusses the conversion of rainforest into grasslands for beef cattle production. *A.T.:* Central America.

3578. Nations, James D., et al., eds., 1988. *Biodiversity in Guatemala: Biological Diversity and Tropical Forests Assessment* [study]. Washington, DC: Center for International Development and Environment, WRI. 185p. [1/1/1].

3579. Natural Resources and Environment Program, Thailand Development Research Institute Foundation, 1995. *Proceedings of the Regional Dialogue on Biodiversity and Natural Resources Management in Mainland Southeast Asian Economies.* Bangkok. 130p. [1/1/1].

3580. Nature Conservancy, 1975. *The Preservation of Natural Diversity: A Survey and Recommendations: Final Report.* Arlington, VA. 212, 97p. [1/1/1] *A.T.:* protected areas.

3581. Nature Conservancy Great Lakes Program, 1994. *The Conservation of Biological Diversity in the Great Lakes Ecosystem: Issues and Opportunities.* Chicago. 118p. [1/1/1].

3582. Naveh, Zev, 1994. From biodiversity to ecodiversity: a landscape-ecology approach to conservation and restoration. *Restoration Ecology* 2: 180-189. [1].

3583. ———, 1995. From biodiversity to ecodiversity: new tools for holistic landscape conservation. *International J. of Ecology and Environmental Sciences* 21: 1-16. [1] *A.T.:* Green books; Red lists.

3584. Naveh, Zev, & A.S. Lieberman, 1994. *Landscape Ecology: Theory and Application* (2nd ed.). New York: Springer. 360p. [2/1/2].

3585. Nazarea, Virginia D., 1998. *Cultural Memory and Biodiversity.* Tucson: University of Arizona Press. 189p. *A.T.:* seed banks; plant germplasm resources; traditional farming.

3586. Neave, H.M., R.B. Cunningham, & T.W. Norton, 1996. Biological inventory for conservation evaluation. III. Relationships between birds, vegetation and environmental attributes in southern Australia. *Forest Ecology and Management* 85: 197-218. [1].

3587. Neave, H.M., T.W. Norton, & H.A. Nix, 1996. Biological inventory for conservation evaluation. I. Design of a field survey for diurnal, terrestrial birds in southern Australia. *Forest Ecology and Management* 85: 107-122. *A.T.:* southeastern Australia.

3588. Neave, H.M., R.B. Cunningham, et al., 1992. Evaluation of field sampling strategies for estimating species richness by Monte Carlo methods. *Mathematics and Computers in Simulation* 33: 391-396. [1] *A.T.:* southeastern Australia; GIS.

3589. Neave, H.M., R.B. Cunningham, et al., 1997. Preliminary evaluation of sampling strategies to estimate the species richness of diurnal, terrestrial birds using Monte Carlo simulation. *Ecological Modelling* 95: 17-27. *A.T.:* southeastern Australia.

3590. Nee, Sean, & J.H Lawton, 1996. Body size and biodiversity. *Nature* 380: 672-673. [1] *A.T.:* insects; macroecology.

3591. Nee, Sean, & R.M. May, 1997. Extinction and the loss of evolutionary history. *Science* 278: 692-694. Considers percentage of evolutionary history actually lost when a species and its gene pool goes extinct. *A.T.:* mass extinctions.

3592. Nee, Sean, P.H. Harvey, & R.M. May, 1991. Lifting the veil on abundance patterns. *PRSL B* 243: 161-163. [2] *A.T.:* birds; species abundance; Great Britain; lognormal distribution.

3593. Nee, Sean, A.F. Read, et al., 1991. The relationship between abundance and body size in British birds. *Nature* 351: 312-313. [2].

3594. Negi, Sharad Singh, 1993. *Biodiversity and Its Conservation in India*. New Delhi: Indus. 343p. [1/1/1].

3595. ———, 1996. *Biosphere Reserves in India: Landuse, Biodiversity and Conservation*. New Delhi: Indus. 221p. [1/1/1].

3596. Nei, Masatoshi, 1973. Analysis of gene diversity in subdivided populations. *PNAS* 70: 3321-3323. [3] *A.T.:* heterozygosity; genetic diversity.

3597. Nei, Masatoshi, Takeo Maruyama, & Ranajit Chakraborty, 1975. The bottleneck effect and genetic variability in populations. *Evolution* 29: 1-10. [3] *A.T.:* heterozygosity.

3598. Neitlich, P.N., & B. McCune, 1997. Hotspots of epiphytic lichen diversity in two young managed forests. *CB* 11: 172-182. *A.T.:* Oregon.

3599. Nelson, Gareth J., & P. Ladiges, 1990. Biodiversity and biogeography [editorial]. *J. of Biogeography* 17: 559-560. [1].

3600. Nelson, Gareth J., & N.I. Platnick, 1981. *Systematics and Biogeography: Cladistics and Vicariance*. New York: Columbia University Press. 567p. [2/1/3].

3601. Nelson, J. Gordon, & Rafal Serafin, 1992. Assessing biodiversity: a human ecological approach. *Ambio* 21: 212-218. [1].

3602. Nelson, Mark, T.L. Burgess, et al., 1993. Using a closed ecological system to study earth's biosphere: initial results from Biosphere 2. *Bioscience* 43: 225-236. [1] *A.T.:* microcosms.

3603. Nepal, Sanjay K., & K.E. Weber, 1994. A buffer zone for biodiversity conservation: viability of the concept in Nepal's Royal Chitwan National Park. *Environmental Conservation* 21: 333-341. [1] *A.T.:* protected areas; local participation.

3604. Nettleship, David N., Joanna Burger, & Michael Gochfeld, eds., 1994. *Seabirds on Islands: Threats, Case Studies and Action Plans* [workshop proceedings]. Cambridge, U.K.: ICBP/BirdLife International. 318p. [1/1/1].

3605. Neumayer, E., 1998. Preserving natural capital in a world of uncertainty and scarce financial resources. *International J. of Sustainable Development and World Ecology* 5: 27-42. *A.T.:* environmental economics; sustainable development.

3606. Nevo, Eviatar, 1995. Asian, African and European biota meet at "Evolution Canyon" Israel: local tests of global biodiversity and genetic diversity patterns. *PRSL B* 262: 149-155. [2] *A.T.:* microclimate; arid environments.

3607. Nevo, Eviatar, & Avigdor Beiles, 1991. Genetic diversity and ecological heterogeneity in amphibian evolution. *Copeia* (3): 565-592. [1] *A.T.:* heterozygosity.

3608. New, Timothy R., 1993. Angels on a pin: dimensions of the crisis in invertebrate conservation. *American Zoologist* 33: 623-630. [1] *A.T.:* bd measurement.

3609. ———, 1994. Butterfly ranching: sustainable use of insects and sustainable benefit to habitats. *Oryx* 28: 169-172. [1].

3610. ———, 1995. *An Introduction to Invertebrate Conservation Biology.* Oxford, U.K., and New York: OUP. 194p. [1/1/1].

3611. ———, 1996. Taxonomic focus and quality control in insect surveys for biodiversity conservation. *Australian J. of Entomology* 35: 97-106. [1] *A.T.:* rapid assessment.

3612. ———, 1997. Are Lepidoptera an effective "umbrella group" for biodiversity conservation? *J. of Insect Conservation* 1: 5-12. *A.T.:* flagship species.

3613. ———, 1997. *Butterfly Conservation* (2nd ed.). Oxford, U.K., and New York: OUP. 248p. [2/1/2].

3614. ———, 1998. *Invertebrate Surveys for Conservation.* Oxford, U.K., and New York: OUP. 240p.

3615. New, Timothy R., & I.W.B. Thornton, 1992. The butterflies of Anak Krakatau, Indonesia: faunal development in early succession. *Journal of the Lepidodopterists' Society* 46: 83-96. [1] *A.T.:* island biogeography; colonization.

3616. Newell, Josh, & Emma Wilson, 1996. *The Russian Far East: Forests, Biodiversity Hotspots, and Industrial Development.* Tokyo: Friends of the Earth. 197p. [1/1/1].

3617. Newman, M.E.J., 1997. A model of mass extinction. *J. of Theoretical Biology* 189: 235-252. *A.T.:* mathematical models.

3618. Newmark, William D., 1987. A land-bridge island perspective on mammalian extinctions in western North American parks. *Nature* 325: 430-432. [2] *A.T.:* protected areas.

3619. ———, 1995. Extinction of mammal populations in western North American national parks. *CB* 9: 512-526. [1].

3620. ———, 1996. Insularization of Tanzanian parks and the local extinction of large mammals. *CB* 10: 1549-1556. [1] *A.T.:* island biogeography.

3621. Newsham, K.K., A.H. Fitter, & A.R. Watkinson, 1995. Multifunctionality and biodiversity in arbuscular mycorrhizas. *TREE* 10: 407-411. [2] *A.T.:* fungi; root systems.

3622. Newton, Ian, 1995. The contribution of some recent research on birds to ecological understanding. *J. of Animal Ecology* 64: 675-696. [1] *A.T.:* literature reviews; long-term ecological research.

3623. Newton, Stephen F., & A.V. Newton, 1997. The effect of rainfall and habitat on abundance and diversity of birds in a fenced protected area in the central Saudi Arabian desert. *J. of Arid Environments* 35: 715-735.

3624. Ng, Peter K.L., & H.H. Tan, 1997. Freshwater fishes of Southeast Asia: potential for the aquarium fish trade and conservation issues. *Aquarium Sciences and Conservation* 1: 79-90.

3625. Ng, Peter K.L., L.M. Chou, & T.J. Lam, 1993. The status and impact of introduced freshwater animals in Singapore. *Biological Conservation* 64: 19-24. [1].

3626. Nguyên, Xuân Dung, Quang Huy Do, et al., 1995. Contribution of HRC to the study on the chemistry of natural plants, chemotaxonomy, and biodiversity conservation. *J. of High Resolution Chromatography* 18: 603-606. [1].

3627. Nichols, J.D., & K.H. Pollock, 1983. Estimating taxonomic diversity, extinction rates, and speciation rates from fossil data using capture-recapture models. *Paleobiology* 9: 150-163. [1].

3628. Nichols, William F., K.T. Killingbeck, & P.V. August, 1998. The influence of geomorphological heterogeneity on biodiversity II. A landscape perspective. *CB* 12: 371-379. *A.T.:* vascular plants; Rhode Island; GIS; landscape ecology.

3629. Nicholson, Thomas D., 1991. Preserving the earth's biological diversity: the role of museums. *Curator* 34: 85-108. [1].

3630. Nickens, Eddie, autumn 1995. Is forest management harming songbirds? *American Forests* 101(9): 18-21, 41-42. [1] *A.T.:* avian declines; habitat destruction.

3631. Nielsen, Larry A., spring 1995. Biodiversity: its meaning and its value. *Forum for Applied Research and Public Policy* 10(1): 76-83. [1] *A.T.:* bd loss; human influence.

3632. Niemelä, Jari, Yrjö Haila, & P. Punttila, 1996. The importance of small-scale heterogeneity in boreal forests: variation in diversity in forest-floor invertebrates across the succession gradient. *Ecography* 19: 352-368. [1] *A.T.:* Finland.

3633. Niemi, Gerald J., 1985. Patterns of morphological evolution in bird genera in New World and Old World peatlands. *Ecology* 66: 1215-1228. [1] *A.T.:* Finland; Minnesota; morphological diversity.

3634. Nigh, Ronald B., & J.D. Nations, March 1980. Tropical rainforests.

Bull. of the Atomic Scientists 36(3): 12-19. [1] Discusses the deforestation crisis.

3635. Niklas, Karl J., & B.H. Tiffney, 1994. The quantification of plant biodiversity through time. *PTRSL B* 345: 35-44. [1] *A.T.:* sampling; morphological diversity; taxonomy.

3636. Niklas, Karl J., B.H. Tiffney, & A.H. Knoll, 1983. Patterns in vascular land plant diversification. *Nature* 303: 614-616. [1] *A.T.:* Phanerozoic.

3637. Nilsson, Christer, & F. Götmark, 1992. Protected areas in Sweden: is natural variety adequately represented? *CB* 6: 232-242. [1] *A.T.:* representativeness.

3638. Nilsson, Christer, Alf Ekblad, et al., 1994. A comparison of species richness and traits of riparian plants between a main river channel and its tributaries. *J. of Ecology* 82: 281-295. [2] *A.T.:* Sweden.

3639. Nilsson, Christer, Gunnel Grelsson, et al., 1989. Patterns of plant species richness along riverbanks. *Ecology* 70: 77-84. [2] *A.T.:* Sweden; riparian habiatats.

3640. Nilsson, Sven G., 1978. Fragmented habitats, species richness and conservation practice. *Ambio* 7: 26-27. [1].

3641. ———, 1997. Biodiversity over the last one thousand years in the cultural landscape of southernmost Sweden. *Svensk Botanisk Tidskrift* 91: 85-101. *A.T.:* decomposers; past landscapes.

3642. Nilsson, Sven G., Jan Bengtsson, & Stefan Ås, 1988. Habitat diversity or area *per se*? Species richness of woody plants, carabid beetles and land snails on islands. *J. of Animal Ecology* 57: 685-704. [1] *A.T.:* Sweden; island biogeography; species-area relation.

3643. Nilsson, Sven G., U. Arup, et al., 1995. Tree-dependent lichens and beetles as indicators in conservation forests. *CB* 9: 1208-1216. [1] *A.T.:* Sweden.

3644. Nisbet, E.G., 1991. *Leaving Eden: To Protect and Manage the Earth*. Cambridge, U.K., and New York: CUP. 358p. [2/2/1]. *A.T.:* environmental change; climatic change; environmental management.

3645. Nitecki, Matthew H., ed., 1984. *Extinctions* [symposium papers]. Chicago: University of Chicago Press. 354p. [2/1/1].

3646. Nitta, Eugene T., & J.R. Henderson, 1993. A review of interactions between Hawaii's fisheries and protected species. *Marine Fisheries Review* 55(2): 83-92. [1] *A.T.:* wildlife conservation; longlines.

3647. NOAA, 1991- . *Our Living Oceans: The Annual Report on the Status of U.S. Living Marine Resources*. Washington, DC: National Marine Fisheries Service, NOAA. *A.T.:* fishery resources.

3648. Nøhr, Henning, & A.F. Jorgensen, 1997. Mapping of biological diversity in Sahel by means of satellite image analyses and ornithological surveys. *Biodiversity and Conservation* 6: 545-566. *A.T.:* Senegal; diversity indices; bd monitoring.

3649. Nolte, Kenneth R., & T.E. Fulbright, 1997. Plant, small mammal, and avian diversity following control of honey mesquite. *J. of Range Management* 50: 205-212. *A.T.:* Texas; weed control.

3650. Noon, Barry R., & Kimberley Young, 1991. Evidence of continuing worldwide declines in bird populations: insights from an international conference in New Zealand. *CB* 5: 141-143. [1].

3651. Norgaard, Richard B., 1987. Economics as mechanics and the demise of biological diversity. *Ecological Modelling* 38: 107-121. [1] *A.T.:* socioeconomic evolution; globalization; social organization.

3652. ———, 1988. Sustainable development: a co-evolutionary view. *Futures* 20: 606-620. [2].

3653. Norrena, E.J., 1994. Stewardship of coastal waters and protected spaces. Canada's approach. *Marine Policy* 18: 153-160. [1] *A.T.:* coastal zone management; environmental policy.

3654. Norris, Richard D., 1991. Biased extinction and evolutionary trends. *Paleobiology* 17: 388-399. [2] *A.T.:* morphological diversity; mass extinctions.

3655. Norris, Richard D., winter 1992. So you thought extinction was forever? *Oceanus* 35(4): 96-99. [1] Concerning evolution in marine foraminifera.

3656. Norse, Elliott A., 1987. International lending and the loss of biological diversity. *CB* 1: 259-260. [1].

3657. ———, 1990. *Threats to Biological Diversity in the United States* [report]. Washington, DC: Office of Policy, Planning and Evaluation, U.S. Environmental Protection Agency. 57p. [1/1/1].

3658. ———, 1990. *Ancient Forests of the Pacific Northwest*. Washington, DC: Island Press. 327p. [2/2/2] *A.T.:* old-growth forests; forest conservation.

3659. ———, ed., 1993. *Global Marine Biological Diversity: A Strategy for Building Conservation Into Decision Making*. Washington, DC: Island Press. 383p. [2/2/2].

3660. ———, 1995. Maintaining the world's marine biological diversity. *Bull. of Marine Science* 57: 10-13. [1].

3661. Norse, Elliott A., et al., 1986. *Conserving Biological Diversity in Our National Forests*. Washington, DC: Wilderness Society. 116p. [1/1/1] *A.T.:* forest conservation.

3662. North American Association for Environmental Education & World Wildlife Fund, 1997. *The Biodiversity Debate: Exploring the Issue*. Troy, OH. 42p. [1/1/-] *A.T.:* bdc.

3663. Norton, Bryan G., 1982. Environmental ethics and nonhuman rights. *Environmental Ethics* 4: 17-36. [1].

3664. ———, ed., 1986. *The Preservation of Species: The Value of Biological Diversity* [essays]. Princeton, NJ: Princeton University Press. 305p.

[2/1/2] *A.T.:* wildlife conservation.

3665. ———, spring 1987. Biodiversity & the public lands: the spiral of life. *Wilderness* 50(176): 16-38. [1].

3666. ———, 1987. *Why Preserve Natural Variety?* Princeton, NJ: Princeton University Press. 281p. [2/1/2] *A.T.:* wildlife conservation; bd value.

3667. ———, 1991. *Toward Unity Among Environmentalists.* New York: OUP. 287p. [2/2/2] *A.T.:* environmental policy; nature philosophy.

3668. ———, 1993. Should environmentalists be organicists? *Topoi* 12: 21-30. [1] *A.T.:* environmental philosophy.

3669. Norton, Bryan G., & R.E. Ulanowicz, 1992. Scale and biodiversity policy: a hierarchical approach. *Ambio* 21: 244-249. [2] *A.T.:* spatial scale.

3670. Norton, Bryan G., Michael Hutchins, et al., eds., 1995. *Ethics on the Ark: Zoos, Animal Welfare, and Wildlife Conservation* [workshop papers]. Washington, DC: Smithsonian Institution Press. 330p. [2/2/1] *A.T.:* environmental ethics.

3671. Norton, Tony W., 1996. Conservation of biological diversity in temperate and boreal forest ecosystems. *Forest Ecology and Management* 85: 1-7. [1] *A.T.:* sustainable development.

3672. ———, 1996. Conserving biological diversity in Australia's temperate eucalypt forests. *Forest Ecology and Management* 85: 21-33. [1].

3673. Norton, Tony W., & S.R. Dovers, eds., 1994. *Ecology and Sustainability of Southern Temperate Ecosystems.* East Melbourne, Victoria, Australia: CSIRO. 133p. [1/1/1] *A.T.:* environmental management.

3674. Norton, Trevor A., Michael Melkonian, & R.A. Andersen, 1996. Algal biodiversity. *Phycologia* 35: 308-326. [1].

3675. Norton-Griffiths, Michael, & Clive Southey, 1995. The opportunity costs of biodiversity conservation in Kenya. *Ecological Economics* 12: 125-139. [1] *A.T.:* ecotourism.

3676. Noss, Reed F., 1983. A regional landscape approach to maintain diversity. *Bioscience* 33: 700-706. [2] *A.T.:* landscape ecology.

3677. ———, 1987. Corridors in real landscapes: a reply to Simberloff and Cox. *CB* 1: 159-164. [2].

3678. ———, 1987. From plant communities to landscapes in conservation inventories: a look at The Nature Conservancy (USA). *Biological Conservation* 41: 11-37. [2] *A.T.:* landscape ecology.

3679. ———, 1989. Who will speak for biodiversity? *CB* 3: 202-203. [1].

3680. ———, 1990. Indicators for monitoring biodiversity: a hierarchical approach. *CB* 4: 355-364. [3].

3681. ———, summer 1991. What can wilderness do for biodiversity? *Wild Earth* 1(2): 51-56. [1].

3682. ———, 1991. Wilderness recovery: thinking big in restoration ecology. *Environmental Professional* 13: 225-234. [1].

3683. ———, 1991. Sustainability and wilderness. *CB* 5: 120-122. [1].

3684. Noss, Reed F., & A.Y. Cooperrider, 1994. *Saving Nature's Legacy: Protecting and Restoring Biodiversity.* Washington, DC: Island Press. 416p. [2/2/3].

3685. Noss, Reed F., & L.D. Harris, 1986. Nodes, networks, and MUMs: preserving diversity at all scales. *Environmental Management* 10: 299-309. [2] *A.T.:* network analysis; multiple use; patch dynamics.

3686. ———, 1990. Habitat connectivity and the conservation of biological diversity: Florida as a case study. *Proc. of the Society of American Foresters National Convention 1990*: 131-135. [1] *A.T.:* habitat disturbance.

3687. Noss, Reed F., & R.L. Peters, 1995. *Endangered Ecosystems: A Status Report on America's Vanishing Habitat and Wildlife.* Washington, DC: Defenders of Wildlife. 132p. [1/1/1] *A.T.:* bdc; bd loss.

3688. Noss, Reed F., E.T. LaRoe III, & J.M. Scott, 1995. *Endangered Ecosystems of the United States: A Preliminary Assessment of Loss and Degradation.* Washington, DC: Biological Report No. 28, National Biological Service, U.S. Dept. of the Interior. 58p. [1/1/2].

3689. Noss, Reed F., M.A. O'Connell, & D.D. Murphy, 1997. *The Science of Conservation Planning: Habitat Conservation Under the Endangered Species Act.* Washington, DC: Island Press. 246p. [2/1/-].

3690. Nott, M.P., E. Rogers, & S.L. Pimm, 1995. Extinction rates: modern extinctions in the kilo-death range. *Current Biology* 5: 14-17. [1].

3691. Nout, M.J.R., C.E. Platis, & D.T. Wicklow, 1997. Biodiversity of yeasts from Illinois maize. *Canadian J. of Microbiology* 43: 362-367.

3692. Novacek, Michael J., & Q.D. Wheeler, eds., 1992. *Extinction and Phylogeny.* New York: Columbia University Press. 253p. [1/2/1].

3693. Nowell, Kristin, & Peter Jackson, compilers, 1996. *Wild Cats: Status Survey and Conservation Action Plan.* Gland, Switzerland: IUCN. 382p. [1/1/2].

3694. Noyes, John S., 1989. The diversity of Hymenoptera in the tropics with special reference to *Parasitica* in Sulawesi. *Ecological Entomology* 14: 197-207. [1].

3695. Nunney, L., & K.A. Campbell, 1993. Assessing minimum viable population size: demography meets population genetics. *TREE* 8: 234-239. [2].

3696. Nunney, L., & D.R. Elam, 1994. Estimating the effective population size of conserved populations. *CB* 8: 175-184. [2].

3697. Nürnberg, Gertrud K., 1995. The anoxic factor, a quantitative measure of anoxia and fish species richness in central Ontario lakes. *Trans. of the American Fisheries Society* 124: 677-686. [1] *A.T.:* bd measures; lake ecology.

3698. Nyberg, Per, 1998. Biotic effects in planktonic crustacean communities in acidified Swedish forest lakes after liming. *Water, Air, and Soil*

Pollution 101: 257-288.

3699. N'Yeurt, A.D.R., & G.R. South, 1997. Biodiversity and biogeography of benthic marine algae in the Southwest Pacific, with special reference to Rotuma and Fiji. *Pacific Science* 51: 18-28.

3700. Nyman, Lennart, 1991. *Conservation of Freshwater Fish: Protection of Biodiversity and Genetic Variability in Aquatic Ecosystems.* Goteborg, Sweden: SWEDMAR. 38p. [1/1/1] *A.T.:* fishery conservation.

3701. Obendorf, D.L., & D.M. Spratt, 1995. Wildlife disease and its relevance to conservation biology and biodiversity. *Australasian Biotechnology* 5: 217-219. [1].

3702. Oberdorff, Thierry, J.-F. Guégan, & Bernard Hugueny, 1995. Global scale patterns of fish species richness in rivers. *Ecography* 18: 345-352. [2].

3703. Oberdorff, Thierry, Bernard Hugueny, & J.-F. Guégan, 1997. Is there an influence of historical events on contemporary fish species richness in rivers? Comparisons between Western Europe and North America. *J. of Biogeography* 24: 461-467.

3704. Obrdlik, P., G. Falkner, & E. Castella, 1995. Biodiversity of Gastropoda in European floodplains. *Archiv für Hydrobiologie, Supplement-Band: Large Rivers* 101: 339-356. [1].

3705. O'Brien, Eileen M., 1993. Climatic gradients in woody plant species richness: towards an explanation based on an analysis of Southern Africa's woody flora. *J. of Biogeography* 20: 181-198. [1] *A.T.:* latitudinal gradients.

3706. ———, 1998. Water-energy dynamics, climate, and prediction of woody plant species richness: an interim general model. *J. of Biogeography* 25: 379-398. *A.T.:* Southern Africa.

3707. O'Brien, Stephen J., 1994. Genetic and phylogenetic analyses of endangered species. *Annual Review of Genetics* 28: 467-489. [1] *A.T.:* conservation genetics; population bottlenecks.

3708. ———, 1994. A role for molecular genetics in biological conservation. *PNAS* 91: 5748-5755. [2] *A.T.:* endangered species; population genetics.

3709. O'Brien, Stephen J., D.E. Wildt, & Mitchell Bush, May 1986. The cheetah in genetic peril. *Scientific American* 254(5): 84-92. [1] *A.T.:* population bottlenecks; inbreeding.

3710. O'Brien, Stephen J., D.E. Wildt, et al., 1983. The cheetah is depauperate in genetic variation. *Science* 221: 459-462. [2] *A.T.:* South Africa; population bottlenecks.

3711. O'Connell, Michael A., & R.F. Noss, 1992. Private land management for biodiversity conservation. *Environmental Management* 16: 435-450. [1] *A.T.:* environmental management.

3712. O'Connor, Martin, 1994. Valuing fish in Aotearoa: the treaty, the market, and the intrinsic value of the trout. *Environmental Values* 3: 245-265. [1] *A.T.:* New Zealand; fisheries management.

3713. O'Connor, Raymond J., M.T. Jones, et al., 1996. Spatial partitioning of environmental correlates of avian biodiversity in the conterminous United States. *Biodiversity Letters* 3: 97-110. [1] *A.T.:* species richness.

3714. O'Donnell, A.G., M. Goodfellow, & D.L. Hawksworth, 1994. Theoretical and practical aspects of the quantification of biodiversity among microorganisms. *PTRSL B* 345: 65-73. [1] *A.T.:* bd measures.

3715. Odum, Eugene P., 1985. Biotechnology and the biosphere [letter]. *Science* 229: 1338. [1] Expresses concern over the release of genetically engineered organisms into the environment.

3716. Offerman, Holly L., V.H. Dale, et al., 1995. Effects of forest fragmentation on Neotropical fauna: current research and data availability. *Environmental Reviews* 3: 191-211. [1] *A.T.:* Amazonia.

3717. Office of Technology Assessment, U.S. Congress, 1986. *Assessing Biological Diversity in the United States: Data Considerations.* Washington, DC. 72p. [1/1/1].

3718. ———, 1986. *Grassroots Conservation of Biological Diversity in the United States.* Washington, DC. 67p. [1/1/1].

3719. ———, 1987. *Technologies to Maintain Biological Diversity.* Washington, DC. 334p. [2/1/1].

3720. ———, 1992. *Technologies to Sustain Tropical Forest Resources and Biological Diversity: Combined Summaries.* Washington, DC. 88p. [2/1/1].

3721. Officer, Charles B., & Jake Page, 1996. *The Great Dinosaur Extinction Controversy.* Reading, MA: Addison-Wesley. 209p. [2/2/1].

3722. Officer, Charles B., Anthony Hallam, et al., 1987. Late Cretaceous and paroxysmal Cretaceous-Tertiary extinctions. *Nature* 326: 143-149. [2] *A.T.:* volcanism.

3723. Ogden, John, 1995. The long-term conservation of forest diversity in New Zealand. *Pacific Conservation Biology* 2: 77-90. [1].

3724. Ogutu-Ohwayo, Richard, R.E. Hecky, et al., 1997. Human impacts on the African Great Lakes. *Environmental Biology of Fishes* 50: 117-131. *A.T.:* fish; fisheries; introduced species.

3725. O'Hara, Robert J., 1994. Evolutionary history and the species problem. *American Zoologist* 34: 12-22. [1] *A.T.:* species definitions.

3726. Ohlson, M., L. Söderström, et al., 1997. Habitat qualities versus long-term continuity as determinants of biodiversity in boreal old-growth swamp forests. *Biological Conservation* 81: 221-231. [2] *A.T.:* Sweden.

3727. Ojeda, F., J. Arroyo, & T. Marañón, 1995. Biodiversity components and conservation of Mediterranean heathlands in southern Spain. *Biological Conservation* 72: 61-72. [1].

3728. Okland, Bjorn, 1996. Unlogged forests: important sites for preserving the diversity of mycetophilids (Diptera: Sciaroidea). *Biological Conservation* 76: 297-310. [1] *A.T.:* old-growth forests; Norway.

3729. Okraszewski, J.D., 1990. The biological diversity issue concerns of the private landowner. *Michigan Academician* 22: 337-341. [1] *A.T.:* Michigan.

3730. Oksanen, Jari, 1996. Is the humped relationship between species richness and biomass an artefact due to plot size? *J. of Ecology* 84: 293-295. [2].

3731. Oksanen, M., 1997. The moral value of biodiversity. *Ambio* 26: 541-545. *A.T.:* bd value; environmental ethics.

3732. Okuda, Toshinori, Naoki Kachi, et al., 1997. Tree distribution pattern and fate of juveniles in a lowland tropical rain forest: implications for regeneration and maintenance of species diversity. *Plant Ecology* 131: 155-171. *A.T.:* Malaysia; recruitment.

3733. O'Laughlin, Jay, Aug. 1992. What the law is and what it might become. *J. of Forestry* 90(8): 6-12. [1] *A.T.:* Endangered Species Act.

3734. Oldfield, Margery L., 1989 (reprint of 1984 ed.). *The Value of Conserving Genetic Resources*. Sunderland, MA: Sinauer Associates. 379p. [2/1/1] *A.T.:* germplasm resources.

3735. Oldfield, Margery L., & J.B. Alcorn, 1987. Conservation of traditional agroecosystems. *Bioscience* 37: 199-208. [1] *A.T.:* environmental policy; Mexico; Panama; in situ conservation.

3736. ———, eds., 1991. *Biodiversity: Culture, Conservation, and Ecodevelopment* [symposium papers]. Boulder, CO: Westview Press. 349p. [1/1/1].

3737. Oldfield, S., & C. Sheppard, 1997. Conservation of biodiversity and research needs in the UK Dependent Territories. *J. of Applied Ecology* 34: 1111-1121.

3738. Oleksyn, J., & P.B. Reich, 1994. Pollution, habitat destruction, and biodiversity in Poland. *CB* 8: 943-960. [1] *A.T.:* environmental degradation.

3739. Olembo, R., 1995. Biodiversity and its importance to the biotechnology industry. *Biotechnology and Applied Biochemistry* 21: 1-6. [1].

3740. Oliver, Chadwick D., Sept. 1992. A landscape approach: achieving and maintaining biodiversity and economic productivity. *J. of Forestry* 90(9): 20-25. [1] *A.T.:* landscape ecology.

3741. Oliver, Chadwick D., 1994. Enhancing biodiversity and economic productivity through a systems approach to silviculture. *Proc. of the Fifteenth Annual Forest Vegetation Management Conference*: 104-120. [1] *A.T.:* forest management.

3742. Oliver, Ian, & A.J. Beattie, 1993. A possible method for the rapid assessment of biodiversity. *CB* 7: 562-568. [2].

3743. ———, 1996. Designing a cost-effective invertebrate survey: a test of methods for rapid assessment of biodiversity. *Ecological Applications* 6:

594-607. [2] *A.T.:* Australia.

3744. Oliver, Ian, A.J. Beattie, & Alan York, 1998. Spatial fidelity of plant, vertebrate, and invertebrate assemblages in multiple-use forest in eastern Australia. *CB* 12: 822-835. *A.T.:* assemblage fidelity; species turnover.

3745. Olmsted, Larry L., & J.W. Bolin, 1996. Aquatic biodiversity and the electric utility industry. *Environmental Management* 20: 805-814. [1] *A.T.:* ecosystem management; power plants.

3746. Olney, Peter J.S., G.M. Mace, & Anna Feistner, eds., 1994. *Creative Conservation: Interactive Management of Wild and Captive Animals* [conference papers]. London and New York: Chapman & Hall. 517p. [1/1/1] *A.T.:* wildlife conservation.

3747. Olson, David M., & Eric Dinerstein, 1998. The Global 200: a representation approach to conserving the earth's most biologically valuable ecoregions. *CB* 12: 502-515.

3748. Olson, D., et al., eds., 1997. *Identifying Gaps in Botanical Information for Biodiversity Conservation in Latin America and the Caribbean* [workshop proceedings]. Washington, DC: WWF. various pagings. [1/1/-] *A.T.:* information resources.

3749. Olson, Everett C., 1989. Problems of Permo-Triassic terrestrial vertebrate extinctions. *Historical Biology* 2: 17-35. [1].

3750. Olsson, G.A., L.E. Liljelun, & L. Hedlund, 1991. A strategy for the conservation of biodiversity. *Ambio* 20: 269-270. [1].

3751. Önal, H., 1997. A computationally convenient diversity measure: theory and application. *Environmental & Resource Economics* 9: 409-427. *A.T.:* Indonesia; community structure.

3752. ———, 1997. Trade off between structural diversity and economic objectives in forest management. *American J. of Agricultural Economics* 79: 1001-1012. *A.T.:* diversity measures.

3753. O'Neil, Thomas A., R.J. Steidl, et al., 1995. Using wildlife communities to improve vegetation classification for conserving biodiversity. *CB* 9: 1482-1491. [1] *A.T.:* Oregon.

3754. O'Neill, J., 1997. Managing without prices: the monetary valuation of biodiversity. *Ambio* 26: 546-550. *A.T.:* environmental management; bd valuation.

3755. O'Neill, Robert V., C.T. Hunsaker, et al., 1997. Monitoring environmental quality at the landscape scale. *Bioscience* 47: 513-519. *A.T.:* landscape ecology.

3756. Ono, R. Dana, J.D. Willams, & Anne Wagner, 1983. *Vanishing Fishes of North America*. Washington, DC: Stone Wall Press. 257p. [2/1/1] *A.T.:* endangered fishes.

3757. Opay, Pat, spring 1998. Area de Conservacion de Tortuguero, Costa Rica: challenges to the conservation of biodiversity. *Wild Earth* 8(1): 50-53. *A.T.:* deforestation; poaching.

3758. Opdam, Paul, 1991. Metapopulation theory and habitat fragmentation: a review of Holarctic breeding bird studies. *Landscape Ecology* 5: 93-106. [2].

3759. Opdam, Paul, G. Rijsdijk, & F. Hustings, 1985. Bird communities in small woods in an agricultural landscape: effects of area and isolation. *Biological Conservation* 34: 333-352. [2] *A.T.:* patch dynamics.

3760. Opdam, Paul, R. Foppen, et al., 1995. The landscape ecological approach in bird conservation: integrating the metapopulation concept into spatial planning. *Ibis* 137, Suppl. 1: S139-S146. [1] *A.T.:* Netherlands; habitat fragmentation.

3761. Oppenheimer, Michael, & R.H. Boyle, 1990. *Dead Heat: The Race Against the Greenhouse Effect.* New York: Basic Books. 268p. [3/3/1].

3762. Oregon Biodiversity Project, 1998. *Oregon's Living Landscape: Strategies and Opportunities to Conserve Biodiversity.* Corvallis, OR: Oregon State University Press. 218p. *A.T.:* endangered species.

3763. Organisation for Economic Co-operation and Development (OECD), 1994. *Economic Incentive Measures for the Conservation and Sustainable Use of Biological Diversity: Conceptual Framework and Guidelines for Case Studies.* Paris: OECD Working Papers, Vol. 2, No. 46. 29p. [1/1/1].

3764. ———, 1996. *Investing in Biological Diversity: The Cairns Conference* [proceedings]. Paris. 403p. [1/1/1].

3765. ———, 1996. *Saving Biological Diversity: Economic Incentives.* Paris and Washington, DC. 155p. [1/1/1].

3766. ———, 1997. *Incentive Measures to Promote the Conservation and the Sustainable Use of Biodiversity: Framework for Case Studies.* Paris: OECD Working Papers, Vol. 5, No. 60. 24p. [1/1/-].

3767. Orgel, Leslie E., Oct. 1994. The origin of life on Earth. *Scientific American* 271(4): 76-83. [1] *A.T.:* RNA.

3768. Orians, Gordon H., 1969. The number of bird species in some tropical forests. *Ecology* 50: 783-801. [1] *A.T.:* Costa Rica; species numbers.

3769. ———, 1990. Ecology and conservation biology: mutually supportive sciences. *Physiology and Ecology Japan* 27, Special Number: 151-165. [1] Surveys the whole range of relationships.

3770. ———, 1990. "New Forestry" and the old-growth forests of northwestern North America [interview]. *Northwest Environmental J.* 6: 445-461. [1] Talks with New Forestry advocate Jerry Franklin. *A.T.:* forest management.

3771. ———, 1992. The importance of biodiversity to conservation: a conversation with Walter V. Reid. *Northwest Environmental J.* 8: 235-246. [1] Puts emphasis on Northwest U.S. issues.

3772. ———, 1993. Endangered at what level? *Ecological Applications* 3: 206-208. [1] Considers problems with the Endangered Species Act.

3773. ———, 1997. Biodiversity and terrestrial ecosystem processes.

Science Progress 80: 45-63.

3774. Orians, Gordon H., & P. Lack, 1992. Arable lands. *Agriculture, Ecosystems & Environment* 42: 101-124. [1] Relates the main intensive farming activities to impacts on bd.

3775. Orians, Gordon H., & C.I. Millar, 1992. Forest lands. *Agriculture, Ecosystems & Environment* 42: 125-140. [1] *A.T.:* forest management; agroforestry; Central America.

3776. Orians, Gordon H., Rodolfo Dirzo, & J.H. Cushman, eds., 1996. *Biodiversity and Ecosystem Processes in Tropical Forests.* Berlin and New York: Springer. 229p. [1/1/1].

3777. Orians, Gordon H., G.M. Brown, et al., eds., 1990. *The Preservation and Valuation of Biological Resources* [workshop proceedings]. Seattle: University of Washington Press. 301p. [1/1/1] *A.T.:* germplasm resources.

3778. O'Riordan, Brian, 1992. Fishing out the gene pool. *Appropriate Technology* 18(4): 6-9. [1] *A.T.:* North-South relations; overfishing; fisheries.

3779. Orlove, Benjamin S., & S.B. Brush, 1996. Anthropology and the conservation of biodiversity. *Annual Review of Anthropology* 25: 329-352. [1] *A.T.:* indigenous peoples.

3780. Orme, Mark L., F.B. Samson, & L.H. Suring, 1990. A process for addressing biological diversity within a forest of islands, southeast Alaska. *Proc. of the Society of American Foresters National Convention 1990*: 116-122. [1] *A.T.:* protected areas.

3781. Ormond, Rupert F.G., 1996. Marine biodiversity: causes and consequences [editorial]. *J. of the Marine Biological Association* 76: 151-152. [1].

3782. Ormond, Rupert F.G., J.D. Gage, & M.V. Angel, eds., 1997. *Marine Biodiversity: Patterns and Processes.* New York: CUP. 449p. [1/-/-].

3783. Orr, David W., 1991. Biological diversity, agriculture, and the liberal arts. *CB* 5: 268-270. [1] *A.T.:* agricultural education.

3784. Osborn, J.G., & J.F. Polsenberg, 1996. Meeting of the mangrovellers: the interface of biodiversity and ecosystem function [editorial]. *TREE* 11: 354-356. [1].

3785. Osborne, P.E., & B.J. Tigar, 1992. Interpreting bird atlas data using logistic models: an example from Lesotho, Southern Africa. *J. of Applied Ecology* 29: 55-62. [2].

3786. Osborne, P.L., 1995. Biological and cultural diversity in Papua New Guinea: conservation, conflicts, constraints and compromise. *Ambio* 24: 231-237. [1] A general survey of bd status and conservation efforts.

3787. Osborne, Roy, 1995. The world cycad census and a proposed revision of the threatened species status for cycad taxa. *Biological Conservation* 71: 1-12. [1].

3788. Osemeobo, Gbadebo J., 1993. Impact of land use on biodiversity preservation in Nigerian natural ecosystems: a review. *Natural Resources J.*

33: 1015-1025. [1] *A.T.:* rural land.

3789. Osman, R.W., & R.B. Whitlatch, 1978. Patterns of species diversity: fact or artifact? *Paleobiology* 4: 41-54. [1].

3790. Ostrom, Elinor, 1990. *Governing the Commons: The Evolution of Institutions for Collective Action.* Cambridge, U.K., and New York: CUP. 280p. [2/3/3].

3791. Otte, Daniel, 1976. Species richness patterns of New World desert grasshoppers in relation to plant diversity. *J. of Biogeography* 3: 197-209. [1].

3792. Otte, Daniel, & J.A. Endler, eds., 1989. *Speciation and Its Consequences.* Sunderland, MA: Sinauer Associates. 679p. [2/2/2] *A.T.:* evolution.

3793. Ottenwalder, Jose Alberto, 1996. Conservation of marine biodiversity in the Caribbean: regional challenges. *Global Biodiversity* 6(1): 31-34. [1] *A.T.:* coastal zone management; international cooperation.

3794. Overpeck, J.T., R.S. Webb, & T. Webb III, 1992. Mapping eastern North American vegetation change of the past 18 ka: no-analogs and the future. *Geology* 20: 1071-1074. [2] *A.T.:* modern analogs; paleovegetation.

3795. Owen, Carlton N., & J.M. Sweeney, March 1995. Managing for diversity: Champion International's approach. *J. of Forestry* 93(3): 12-14. [1] *A.T.:* forest management; endangered species.

3796. Owen, James G., 1988. On productivity as a predictor of rodent and carnivore diversity. *Ecology* 69: 1161-1165. [2] *A.T.:* primary productivity; Texas.

3797. ———, 1989. Patterns of herpetofaunal species richness: relation to temperature, precipitation, and variance in elevation. *J. of Biogeography* 16: 141-150. [1] *A.T.:* Texas.

3798. ———, 1990. Patterns of mammalian species richness in relation to temperature, productivity, and variance in elevation. *J. of Mammalogy* 71: 1-13. [1].

3799. Owen-Smith, Norman, 1987. Pleistocene extinctions: the pivotal role of megaherbivores. *Paleobiology* 13: 351-362. [1].

3800. ———, 1989. Megafaunal extinctions: the conservation message from 11,000 years B.P. *CB* 3: 405-412. [1] *A.T.:* Pleistocene extinctions; wildlife conservation; South Africa.

3801. Ozinga, Wim A., Jelte van Andel, & M.P. McDonnell-Alexander, 1997. Nutritional soil heterogeneity and mycorrhiza as determinants of plant species diversity. *Acta Botanica Neerlandica* 46: 237-254. *A.T.:* spatial scale; fungi.

3802. Pacala, S.W., & M.J. Crawley, 1992. Herbivores and plant diversity. *American Naturalist* 140: 243-260. [1].

3803. Pace, N.R., 1997. A molecular view of microbial diversity and the biosphere. *Science* 276: 734-740. [3] *A.T.:* molecular-phylogenetic studies; origin of life.

3804. Pacific Rivers Council, 1993. *Entering the Watershed: An Action Plan to Protect and Restore America's River Ecosystems and Biodiversity* [report to Congress]. Alexandria, VA, and Eugene, OR. 319p. [1/1/1].

3805. Padhi, B.K., & R.K. Mandal, 1994. Improper fish breeding practices and their impact on aquaculture and fish biodiversity [editorial]. *Current Science* 66: 624-626. [1].

3806. Padilla, Michael A., 1996. The mouse that roared: how the National Forest Management Act diversity of species provision is changing public timber harvesting. *UCLA J. of Environmental Law & Policy* 15: 113-150. [1] *A.T.:* old-growth forests; environmental policy.

3807. Padua, Suzana M., 1994. Conservation awareness through an environmental education programme in the Atlantic forest of Brazil. *Environmental Conservation* 21: 145-151. [1] *A.T.:* environmental perception; tamarins.

3808. Page, Roderic D.M., 1990. Temporal congruence and cladistic analysis of biogeography and cospeciation. *Systematic Zoology* 39: 205-226. [2].

3809. Pagel, Mark D., P.H. Harvey, & H.C.J. Godfray, 1991. Species-abundance, biomass, and resource-use distributions. *American Naturalist* 138: 836-850. [2] *A.T.:* body size; population density.

3810. Pagel, Mark D., R.M. May, & A.R. Collie, 1991. Ecological aspects of the geographical distribution and diversity of mammalian species. *American Naturalist* 137: 791-815. [2] *A.T.:* geographical range; latitudinal diversity gradients.

3811. Pagiola, Stefano, John Kellenberg, et al., 1997. *Mainstreaming Biodiversity in Agricultural Development: Toward Good Practice.* Washington, DC: World Bank Environment Paper No. 15. 50p. [1/1/1] *A.T.:* agricultural ecology; agricultural policy.

3812. Pain, Deborah J., & M.W. Pienkowski, eds., 1997. *Farming and Birds in Europe: The Common Agricultural Policy and Its Implications for Bird Conservation.* San Diego: Academic Press. 436p. [1/-/-].

3813. Paine, Robert T., 1966. Food web complexity and species diversity. *American Naturalist* 100: 65-75. [3] A classic treatment of the subject. *A.T.:* predation; rocky intertidal zone.

3814. ———, 1980. Food webs: linkage, interaction strength and community infrastructure. *J. of Animal Ecology* 49: 667-685. [3] *A.T.:* stability.

3815. ———, 1995. A conversation on refining the concept of keystone species. *CB* 9: 962-964. [1].

3816. Palleroni, Norberto J., 1997. Prokaryotic diversity and the importance of culturing. *Antonie van Leeuwenhoek* 72: 3-19. *A.T.:* species concept; culture methods.

3817. Palmer, Margaret A., A.P. Covich, et al., 1997. Biodiversity and ecosystem processes in freshwater sediments. *Ambio* 26: 571-577. *A.T.:*

aquatic habitats; organic carbon.

3818. Palmer, Mary E., 1987. A critical look at rare plant monitoring in the United States. *Biological Conservation* 39: 113-127. [1] *A.T.:* endangered plants; plant conservation.

3819. Palmer, Michael W., 1990. The estimation of species richness by extrapolation. *Ecology* 71: 1195-1198. [2] *A.T.:* mathematical models.

3820. ———, 1992. The coexistence of species in fractal landscapes. *American Naturalist* 139: 375-397. [2] *A.T.:* environmental variability; simulation studies.

3821. ———, 1995. How should one count species? *Natural Areas J.* 15: 124-135. [1] Suggests extrapolation and interpolation as means of estimating species richness.

3822. Palmer, Michael W., & T.A. Maurer, 1997. Does diversity beget diversity? A case study of crops and weeds. *J. of Vegetation Science* 8: 235-240.

3823. Palmer, Michael W., & P.S. White, 1994. Scale dependence and the species-area relationship. *American Naturalist* 144: 717-740. [2] *A.T.:* spatial scale; North Carolina; hotspots.

3824. Palmer, Miguel, & G.X. Pons, 1996. Diversity in Western Mediterranean islets: effects of rat presence on a beetle guild. *Acta Oecologica* 17: 297-305. [1].

3825. Pan American Health Organization, 1996. *Biodiversity, Biotechnology, and Sustainable Development in Health and Agriculture: Emerging Connections.* Washington, DC. 229p. [1/1/1] *A.T.:* Latin America.

3826. Panayotou, Theodore (Todor Panaiotov), 1993. The environment in Southeast Asia: problems and policies. *Environmental Science & Technology* 27: 2270-2274. [1] *A.T.:* sustainable development; resource depletion; environmental degradation.

3827. ———, 1994. Conservation of biodiversity and economic development: the concept of transferable development rights. *Environmental & Resource Economics* 4: 91-110. [1] *A.T.:* developing countries; North-South relations; forest management; forest products.

3828. Panayotou, Theodore (Todor Panaiotov), & P.S. Ashton, 1992. *Not by Timber Alone: Economics and Ecology for Sustaining Tropical Forests.* Washington, DC: Island Press. 282p. [1/1/1].

3829. Pandey, Sanjeeva, 1993. Changes in waterbird diversity due to the construction of Pong Dam Reservoir, Himachal Pradesh, India. *Biological Conservation* 66: 125-130. [1].

3830. Pandey, Sanjeeva, & M.P. Wells, 1997. Ecodevelopment planning at India's Great Himalayan National Park for biodiversity conservation and participatory rural development. *Biodiversity and Conservation* 6: 1277-1292. *A.T.:* protected areas; local participation.

3831. Pangtey, Y.P.S., & R.S. Rawal, eds., 1994. *High Altitudes of the*

Himalaya: Biogeography, Ecology & Conservation. Nainital: Gyanodaya Prakashan. 418p. [1/1/1].

3832. Panjabi, Ranee Khooshie Lal, 1993. International law and the preservation of species: an analysis of the Convention on Biological Diversity signed at the Rio Earth Summit in 1992. *Dickinson J. of International Law* 11: 187-281. [1].

3833. Pankhurst, Clive E., 1997. Biodiversity of soil organisms as an indicator of soil health. In C.E. Pankhurst, B.M. Doube, & V.V.S.R. Gupta, eds., *Biological Indicators of Soil Health* (Wallingford, Oxon, U.K., and New York: CAB International): 297-324. [1].

3834. Pankhurst, Clive E., K. Ophel-Keller, et al., 1996. Biodiversity of soil microbial communities in agricultural systems. *Biodiversity and Conservation* 5: 197-209. [1] *A.T.:* bd estimation; functional diversity; spatial scale.

3835. Panzer, Ron, & M.W. Schwartz, 1998. Effectiveness of a vegetation-based approach to insect conservation. *CB* 12: 693-702. *A.T.:* Illinois; reserve selection; coarse filter approach.

3836. Paoletti, M.G., 1995. Biodiversity, traditional landscapes and agroecosystem management. *Landscape and Urban Planning* 31: 117-128. [1] *A.T.:* agricultural practices.

3837. Paoletti, M.G., David Pimentel, et al., 1992. Agroecosystem biodiversity: matching production and conservation biology. *Agriculture, Ecosystems & Environment* 40: 3-23. [1] *A.T.:* literature reviews; sustainable agriculture.

3838. Paradis, E., 1998. Detecting shifts in diversification rates without fossils. *American Naturalist* 152: 176-187. *A.T.:* phylogeny; primates.

3839. Parish, T., K.H. Lakhani, & T.H. Sparks, 1994. Modelling the relationship between bird population variables and hedgerow and other field margin attributes. I. Species richness of winter, summer and breeding birds. *J. of Applied Ecology* 31: 764-775. [1] *A.T.:* England.

3840. Park, Chris C., 1992. *Tropical Rainforests.* London and New York: Routledge. 188p. [2/1/1] *A.T.:* deforestation; rainforest conservation.

3841. Park, Sang, spring 1995. Biodiversity prospecting: pharmaceutical investment in Costa Rica. *Harvard International Review* 17(2): 52-53, 70-71. [1] Includes discussion of the Merck-INBio agreement.

3842. Parkes, R.J., B.A. Cragg, et al., 1994. Deep bacterial biosphere in Pacific Ocean sediments. *Nature* 371: 410-413. [2] *A.T.:* benthic sediments; microorganisms.

3843. Parkinson, R.W., 1995. Managing biodiversity from a geological perspective. *Bull. of Marine Science* 57: 28-36. [1] *A.T.:* sea-level rise; Florida; barrier islands.

3844. Parks, Catherine G., & D.C. Shaw, 1996. Death and decay: a vital part of living canopies. *Northwest Science* 70, Suppl.: 46-53. [1] *A.T.:* wood decay; dead trees.

3845. Parmesan, Camille, 1996. Climate and species range. *Nature* 382: 765-766. [2] *A.T.:* Describes a study suggesting a global warming effect on the distribution of a butterfly species.

3846. Parr, J.F., R.I. Papendick, et al., 1992. Soil quality: attributes and relationship to alternative and sustainable agriculture. *American J. of Alternative Agriculture* 7: 5-10. [2].

3847. Parrotta, J.A., 1992. The role of plantation forests in rehabilitating degraded tropical ecosystems. *Agriculture, Ecosystems & Environment* 41: 115-133. [2] *A.T.:* environmental degradation; Puerto Rico; agroforestry.

3848. Parson, Edward A., P.M. Haas, & M.A. Levy, Oct. 1992. A summary of the major documents signed at the Earth Summit and the Global Forum. *Environment* 34(8): 12-15, 34-36. [1] *A.T.:* UNCED; CBD.

3849. Parsons, Peter A., 1989. Conservation and global warming: a problem in biological adaptation. *Ambio* 18: 322-325. [1] *A.T.:* environmental stress; greenhouse effect; *Drosophila*.

3850. ———, 1991. Biodiversity conservation under global climatic change: the insect *Drosophila* as a biological indicator? *Global Ecology and Biogeography Letters* 1: 77-83. [1] *A.T.:* environmental stress.

3851. ———, 1993. Stress, extinctions and evolutionary change: from living organisms to fossils. *Biological Reviews* 68: 313-333. [1] *A.T.:* environmental change.

3852. ———, 1994. The energetic cost of stress. Can biodiversity be preserved? *Biodiversity Letters* 2: 11-15. [1] *A.T.:* environmental stress; edge effects.

3853. ———, 1996. Conservation strategies: adaptation to stress and the preservation of genetic diversity. *Biological J. of the Linnean Society* 58: 471-482. [1] *A.T.:* heterozygosity.

3854. Parsons, R.F., & D.G. Cameron, 1974. Maximum plant species diversity in terrestrial communities. *Biotropica* 6: 202-203. [1] Compares several high diversity areas. *A.T.:* South Africa; Australia; fynbos.

3855. Partensky, F., L. Guillou, et al., 1997. Recent advances in the use of molecular techniques to assess the genetic diversity of marine photosynthetic microorganisms. *Vie et Milieu* 47: 367-374. *A.T.:* phytoplankton.

3856. Parthasarathy, N., & R. Karthikeyan, 1997. Plant biodiversity inventory and conservation of two tropical dry evergreen forests on the Coromandel coast, South India. *Biodiversity and Conservation* 6: 1063-1083.

3857. Pastor, John, & Yosef Cohen, 1997. Herbivores, the functional diversity of plants species, and the cycling of nutrients in ecosystems. *Theoretical Population Biology* 51: 165-179.

3858. Patalas, Kazimierz, 1990. Diversity of the zooplankton communities in Canadian lakes as a function of climate. *Verhandlungen der Internationalen Vereinigung für Theoretische und Angewandte Limnologie* 24: 360-368. [1].

3859. Patil, G.P., & C. Taillie, 1982. Diversity as a concept and its measurement. *J. of the American Statistical Association* 77: 548-567. [1] *A.T.:* rarity; diversity indices.

3860. Patlis, Jason M., 1994. Biodiversity, ecosystems and species: where does the Endangered Species Act fit in? *Tulane Environmental Law J.* 8: 33-76. [1] *A.T.:* endangered species.

3861. Patrick, Ruth, 1988. Importance of diversity in the functioning and structure of riverine communities. *Limnology and Oceanography* 33: 1304-1307. [1] *A.T.:* freshwater ecosystems.

3862. Patterson, Alan, Dec. 1990. Debt-for-nature swaps and the need for alternatives. *Environment* 32(10): 5-13, 31-32. [1].

3863. Patterson, B.D., 1994. Accumulating knowledge on the dimensions of biodiversity: systematic perspectives on Neotropical mammals. *Biodiversity Letters* 2: 79-86. [1] *A.T.:* systematic description.

3864. Patterson, Colin, & A.B. Smith, 1989. Periodicity in extinction: the role of systematics. *Ecology* 70: 802-811. [1] *A.T.:* mass extinctions; echinoderms; fish.

3865. Patton, David R., 1997. *Wildlife Habitat Relationships in Forested Ecosystems* (2nd ed.). Portland, OR: Timber Press. 502p. [2/2/-] *A.T.:* wildlife management.

3866. Patton, James L., & M.F. Smith, 1992. mtDNA phylogeny of Andean mice: a test of diversification across ecological gradients. *Evolution* 46: 174-183. [2] *A.T.:* speciation.

3867. Patton, Timothy M., F.J. Rahel, & W.A. Hubert, 1998. Using historical data to assess changes in Wyoming's fish fauna. *CB* 12: 1120-1128. *A.T.:* survey data.

3868. Paul, C.R.C., & S.F. Mitchell, 1994. Is famine a common factor in marine mass extinctions? *Geology* 22: 679-682. [2] *A.T.:* marine fauna.

3869. Paulay, Gustav, 1994. Biodiversity on oceanic islands: its origin and extinction. *American Zoologist* 34: 134-144. [1] *A.T.:* isolation; dispersal; human disturbance.

3870. Paulson, Dennis R., 1992. Northwest bird diversity: from extravagant past and changing present to precarious future. *Northwest Environmental J.* 8: 71-118. [1] *A.T.:* historical changes.

3871. Pauly, Daniel, & V. Christensen, 1995. Primary production required to sustain global fisheries. *Nature* 374: 255-257. [2] *A.T.:* aquatic environments.

3872. Pauly, Daniel, & Purwito Martosubroto, eds., 1996. *Baseline Studies of Biodiversity: The Fish Resources of Western Indonesia*. Jakarta: Directorate General of Fisheries. 312p. [1/1/1] *A.T.:* fishery resources.

3873. Payne, Neil F., 1992. *Techniques for Wildlife Habitat Management of Wetlands*. New York: McGraw-Hill. 549p. [1/2/1].

3874. Payne, Neil F., & F.C. Bryant, 1994. *Techniques for Wildlife*

Habitat Management of Uplands. New York: McGraw-Hill. 840p. [1/1/1].

3875. Pearce, David W., 1991. Deforesting the Amazon: toward an economic solution. *Ecodécision*, no. 1: 40-49. [1].

3876. ———, 1993. *Economic Values and the Natural World.* Cambridge, MA: MIT Press. 129p. [2/2/2] *A.T.:* economic valuation; sustainable development.

3877. ———, 1993. Saving the world's biodiversity [opinion]. *Environment & Planning A* 25: 755-757. [1] Dwells on policies designed to treat bd loss.

3878. ———, 1995. New directions for financing global environmental change. *Global Environmental Change: Human and Policy Dimensions* 5: 27-40. [1] Examines strategies for financing environmental conservation.

3879. Pearce, David W., & Dominic Moran, 1994. *The Economic Value of Biodiversity.* London: Earthscan. 172p. [1/2/2].

3880. Pearce, David W., & Seema Puroshothaman, 1992. *Protecting Biological Diversity: The Economic Value of Pharmaceutical Plants.* Norwich, U.K.: CSERGE Discussion Paper GEC 92-27, Centre for Social and Economic Research on the Global Environment. 17p. [1/1/1] *A.T.:* medicinal plants.

3881. Pearce, David W., & R.K. Turner, 1990. *Economics of Natural Resources and the Environment.* Baltimore: Johns Hopkins University Press. 378p. [2/1/3] *A.T.:* environmental economics.

3882. Pearce, David W., & J.J. Warford, 1993. *World Without End: Economics, Environment, and Sustainable Development.* New York: OUP. 440p. [2/2/2].

3883. Pearce, David W., Anil Markandya, & E.B. Barbier, 1989. *Blueprint for a Green Economy* [report]. London: Earthscan. 192p. [1/1/3] *A.T.:* environmental economics; environmental policy.

3884. Pearce, David W., Samuel Fankhauser, et al., 1992. World economy, world environment. *World Economy* 15: 295-313. [1] Discusses the upcoming UNCED and its probable issues. *A.T.:* North-South relations.

3885. Pearce, David W., et al., eds., 1991. *Blueprint 2: Greening the World Economy.* London: Earthscan. 232p. [1/1/2] *A.T.:* environmental economics; environmental policy.

3886. Pearce, Fred, 28 March 1992. First aid for the Amazon. *New Scientist* 133(1814): 42-46. [1] Concerning World Bank projects.

3887. ———, 11 Sept. 1993. Road to ruin for Britain's wildlife. *New Scientist* 139(1890): 35-38. [1] Discusses threats to the SSSI (Site of Special Scientific Interest) program in Great Britain.

3888. Pearman, P.B., 1995. An agenda for conservation research and its application, with a case-study from Amazonian Ecuador. *Environmental Conservation* 22: 39-43. [1] *A.T.:* deforestation; bdc.

3889. Pearson, David L., 1994. Selecting indicator taxa for the quantitative assessment of biodiversity. *PTRSL B* 345: 75-79. [2].

3890. Pearson, David L., & S.S. Carroll, 1998. Global patterns of species richness: spatial models for conservation planning using bioindicator and precipitation data. *CB* 12: 809-821.

3891. Pearson, David L., & Fabio Cassola, 1992. World-wide species richness patterns of tiger beetles (Coleoptera: Cicindelidae): indicator taxon for biodiversity and conservation studies. *CB* 6: 376-391. [2].

3892. Pechmann, Joseph H.K., & H.M. Wilbur, 1994. Putting declining amphibian populations in perspective: natural fluctuations and human impacts. *Herpetologica* 50: 65-84. [2].

3893. Pechmann, Joseph H.K., D.E. Scott, et al., 1991. Declining amphibian populations: the problem of separating human impacts from natural fluctuations. *Science* 253: 892-895. [3] *A.T.:* South Carolina.

3894. Peck, Sheila, 1998. *Planning for Biodiversity: Issues and Examples.* Washington, DC: Island Press. 221p. *A.T.:* environmental planning.

3895. Pedersen, K., J. Arlinger, et al., 1996. Diversity and distribution of subterranean bacteria in groundwater at Oklo in Gabon, Africa, as determined by 16S rRNA gene sequencing. *Molecular Ecology* 5: 427-436. [2].

3896. Pederson, Judith, ed., 1996. *Exotic Species Workshop: Issues Relating to Aquaculture and Biodiversity* [workshop proceedings]. Cambridge, MA: MITSG No. 96-15, Sea Grant College Program, MIT. 56p. [1/1/1] *A.T.:* marine bd.

3897. Pedrós-Alió, Carlos, 1993. Diversity of bacterioplankton. *TREE* 8: 86-90. [1] *A.T.:* rRNA sequences.

3898. Peet, N.B., & P.W. Atkinson, 1994. The biodiversity and conservation of the birds of São Tomé and Principe. *Biodiversity and Conservation* 3: 851-867. [1] *A.T.:* endemic species; forest management.

3899. Peet, Robert K., 1974. The measurement of species diversity. *ARES* 5: 285-307. [3] *A.T.:* diversity definitions.

3900. Pei, Shengji, ed., 1996. *Banking on Biodiversity: Report on the Regional Consultation on Biodiversity Assessment in the Hindu Kush-Himalayas.* Kathmandu, Nepal: International Centre for Integrated Mountain Development. 485p. [1/1/1].

3901. Pellew, Robin, Jan.-Feb. 1995. Biodiversity conservation—why all the fuss? *RSA Journal* 143(5456): 53-66. [1] An overview of bdc. *A.T.:* U.K.

3902. Peltonen, Anu, & Ilkka Hanski, 1991. Patterns of island occupancy explained by colonization and extinction rates in shrews. *Ecology* 72: 1698-1708. [2] *A.T.:* Finland; body size.

3903. Penev, L.D., 1992. Qualitative and quantitative spatial variation in soil wire-worm assemblages in relation to climatic and habitat factors. *Oikos* 63: 180-192. [1] *A.T.:* Russia; latitudinal gradients.

3904. Peng, Ching-I, & Hang-hung Chou, eds., 1994. *Biodiversity and Terrestrial Ecosystems* [symposium proceedings]. Taipei: Institute of Botany, Academia Sinica. 527p. [1/1/1] *A.T.:* Pacific region; bdc.

3905. Perés, Carlos A., & J.W. Terborgh, 1995. Amazonian nature preserves: an analysis of the defensibility status of existing conservation units and design criteria for the future. *CB* 9: 34-46. [1].

3906. Pérez, Julio E., & M.K. Rylander, 1998. Hybridization and its effect on species richness in natural habitats. *Interciencia* 23: 137-139. Considers aquaculture hybrids and the possible dangers of their release into the environment. *A.T.:* fish.

3907. Perfect, J., 1991. Biodiversity: how important a resource? *Outlook on Agriculture* 20: 5-7. [1] Considers how bd might be used as a model for agricultural production.

3908. Perfecto, Ivette, & Roy Snelling, 1995. Biodiversity and the transformation of a tropical agroecosystem: ants in coffee plantations. *Ecological Applications* 5: 1084-1097. [1] *A.T.:* Costa Rica.

3909. Perfecto, Ivette, & John Vandermeer, 1996. Microclimatic changes and the indirect loss of ant diversity in a tropical agroecosystem. *Oecologia* 108: 577-582. [1] *A.T.:* coffee plantations; Costa Rica.

3910. Perfecto, Ivette, R.A. Rice, et al., 1996. Shade coffee: a disappearing refuge for biodiversity. *Bioscience* 46: 598-608. [1] *A.T.:* northern Latin America; plantations.

3911. Perfecto, Ivette, John Vandermeer, et al., 1997. Arthropod biodiversity loss and the transformation of a tropical agro-ecosystem. *Biodiversity and Conservation* 6: 935-945. *A.T.:* Costa Rica.

3912. Perlman, Dan L., & Glenn Adelson, 1997. *Biodiversity: Exploring Values and Priorities in Conservation.* Malden, MA: Blackwell Science. 182p. [1/1/-] *A.T.:* bdc.

3913. Perrin, William F., 1991. Why are there so many kinds of whales and dolphins? *Bioscience* 41: 460-461. [1].

3914. Perrings, Charles A., 1997. Biodiversity, biospecifics, and ecological services: reply [letter]. *TREE* 12: 67. Concerning bd value.

3915. Perrings, Charles A., & D.W. Pearce, 1994. Threshold effects and incentives for the conservation of biodiversity. *Environmental & Resource Economics* 4: 13-28. [1] *A.T.:* natural capital.

3916. Perrings, Charles A., & Brian Walker, 1997. Biodiversity, resilience and the control of ecological-economic systems: the case of fire-driven rangelands. *Ecological Economics* 22: 73-83. *A.T.:* event-driven systems; disturbance.

3917. Perrings, Charles A., Carl Folke, & Karl-Goran Mäler, 1992. The ecology and economics of biodiversity loss: the research agenda. *Ambio* 21: 201-211. [2] Examines interdisciplinary approaches to researching bd-related problems.

3918. Perrings, Charles A., Karl-Goran Maler, et al., eds., 1995. *Biodiversity Loss: Economic and Ecological Issues.* Cambridge, U.K., and New

York: CUP. 332p. [2/2/2].

3919. Perrings, Charles A., Karl-Goran Maler, et al., eds., 1995. *Biodiversity Conservation: Problems and Policies* [essays]. Dordrecht and Boston: Kluwer. 404p. [1/1/1].

3920. Perry, David A., 1993. Biodiversity and wildlife are not synonymous [editorial]. *CB* 7: 204-205. [1].

3921. ———, 1994. *Forest Ecosystems.* Baltimore: Johns Hopkins University Press. 649p. [2/1/2] *A.T.:* forest ecology.

3922. ———, 1995. Self-organizing systems across scales. *TREE* 10: 241-244. [1] *A.T.:* spatial scale.

3923. Perry, David A., & J.G. Borchers, 1990. Climate change and ecosystems. *Northwest Environmental J.* 6: 293-313. [1] *A.T.:* water stress; greenhouse effect.

3924. Perry, David A., J.G. Borchers, et al., 1990. Species migrations and ecosystem stability during climate change: the belowground connection. *CB* 4: 266-274. [1] *A.T.:* plant-soil links.

3925. Perry, Donald R., Nov. 1984. The canopy of the tropical rain forest. *Scientific American* 251(5): 138-147. [1] *A.T.:* Costa Rica.

3926. Persat, Henri, J.-M. Olivier, & Didier Pont, 1994. Theoretical habitat templets, species traits, and species richness: fish in the Upper Rhône River and its floodplain. *Freshwater Biology* 31: 439-453. [1] *A.T.:* France.

3927. Persiani, A.M., O. Maggi, et al., 1998. Diversity and variability in soil fungi from a disturbed tropical rain forest. *Mycologia* 90: 206-214. *A.T.:* Ivory Coast.

3928. Peterken, G.F., & M. Game, 1984. Historical factors affecting the number and distribution of vascular plant species in the woodlands of central Lincolnshire. *J. of Ecology* 72: 155-182. [2].

3929. Peters, Charles M., A.H. Gentry, & R.O. Mendelsohn, 1989. Valuation of an Amazonian rainforest. *Nature* 339: 655-656. [2] *A.T.:* forest products.

3930. Peters, Robert L., & J.D.S. Darling, 1985. The greenhouse effect and nature reserves—global warming would diminish biological diversity by causing extinctions among reserve species. *Bioscience* 35: 707-717. [2].

3931. Peters, Robert L., & T.E. Lovejoy, eds., 1992. *Global Warming and Biological Diversity* [conference papers]. New Haven: Yale University Press. 386p. [2/2/2] *A.T.:* greenhouse effect.

3932. Peters, Robert L., & J.P. Myers, winter 1991. Preserving biodiversity in a changing climate. *Issues in Science and Technology* 8(2): 66-72. [1].

3933. Peterson, Everett B., N.M. Peterson, & D.F.W. Pollard, 1995. Some principles and criteria to make Canada's protected area systems representative of the nation's forest diversity. *Forestry Chronicle* 71: 497-507. [1] *A.T.:* forest management.

3934. Peterson, George L., et al., eds., 1992. *Valuing Wildlife Resources in Alaska* [workshop papers]. Boulder, CO: Westview Press. 357p. [1/1/1].

3935. Peterson, Melvin N.A., ed., 1992. *Diversity of Oceanic Life: An Evaluative Review* [conference papers]. Honolulu: Ocean Policy Institute, and Washington, DC: Center for Strategic and International Studies. 109p. [1/1/1].

3936. Petraitis, Peter S., R.E. Latham, & R.A. Niesenbaum, 1989. The maintenance of species diversity by disturbance. *Quarterly Review of Biology* 64: 393-418. [2].

3937. Petranka, James W., M.E. Eldridge, & K.E. Haley, 1993. Effects of timber harvesting on Southern Appalachian salamanders. *CB* 7: 363-370. [2] *A.T.:* North Carolina; clearcutting.

3938. Petranu, Adriana, compiler, 1997. *Black Sea Biological Diversity: Romania*. New York: United Nations Publications. 314p. [1/1/-].

3939. Pettersson, M., 1998. Monitoring a freshwater fish population: statistical surveillance of biodiversity. *Environmetrics* 9: 139-150. *A.T.:* Sweden; indicator species.

3940. Pharo, E.J., & A.J. Beattie, 1997. Bryophyte and lichen diversity: a comparative study. *Australian J. of Ecology* 22: 151-162. *A.T.:* southeastern Australia.

3941. Philipp, David P., R.J. Wolotira, Jr., et al., eds., 1995. *Protection of Aquatic Biodiversity: Proceedings of the World Fisheries Congress, Theme 3*. Lebanon, NH: Science Publishers. 282p. [1/1/1] *A.T.:* fishery conservation.

3942. Philippart, J.C., 1995. Is captive breeding an effective solution for the preservation of endemic species? *Biological Conservation* 72: 281-295. [1] *A.T.:* Southwest; fish; genetic diversity.

3943. Philippines Dept. of Environment and Natural Resources, 1997. *Philippine Biodiversity: An Assessment and Plan of Action*. Makati City, Philippines: Bookmark. 298p. [1/1/1].

3944. Phillips, Jeff, May 1992. Treasures and troubles of the Sierra Nevada. *Sunset* 188(5): 88-114. [1] Summarizes Sierra Nevadan bd and current threats to it.

3945. Phillips, J.A., 1998. Marine conservation initiatives in Australia: their relevance to the conservation of macroalgae. *Botanica Marina* 41: 95-103. *A.T.:* protected areas; coastal environments.

3946. Phillips, Kathryn, 1990. Where have all the frogs and toads gone? *Bioscience* 40: 422-424. [2] Describes a workshop sponsored by the NRC. *A.T.:* amphibian declines.

3947. Phillips, O.L., 1997. The changing ecology of tropical forests. *Biodiversity and Conservation* 6: 291-311. *A.T.:* environmental degradation; synergisms.

3948. Phillips, O.L., & A.H. Gentry, 1994. Increasing turnover through time in tropical forests. *Science* 263: 954-958. [2] *A.T.:* turnover rates; recruitment.

3949. Phillips, O.L., P. Hall, et al., 1994. Dynamics and species richness of tropical rain forests. *PNAS* 91: 2805-2809. [2] *A.T.:* productivity; ecosystem dynamics.

3950. Phillips, O.L., P. Hall, et al., 1997. Species richness, tropical forest dynamics, and sampling: response to Sheil. *Oikos* 79: 183-187.

3951. Pianka, Eric R., 1966. Latitudinal gradients in species diversity: a review of concepts. *American Naturalist* 100: 33-46. [2].

3952. ———, 1967. On lizard species diversity: North American flatland deserts. *Ecology* 48: 333-351. [1] Examines eight factors potentially determining species diversity and relates to situation for North American lizards.

3953. ———, 1989. Desert lizard diversity: additional comments and some data. *American Naturalist* 134: 344-364. [1] *A.T.:* Australia; Southern Africa; termites; fire.

3954. Pickett, Steward T.A., & J.N. Thompson, 1978. Patch dynamics and the design of nature reserves. *Biological Conservation* 13: 27-37. [2] *A.T.:* island biogeography; extinction.

3955. Pickett, Steward T.A., & P.S. White, eds., 1985. *The Ecology of Natural Disturbance and Patch Dynamics*. Orlando: Academic Press. 472p. [2/1/3].

3956. Pickett, Steward T.A., Jurek Kolasa, & C.G. Jones, 1994. *Ecological Understanding: The Nature of Theory and the Theory of Nature*. San Diego: Academic Press. 206p. [1/1/2] *A.T.:* environmental philosophy; ecological theory.

3957. Pickett, Steward T.A., R.S. Ostfeld, et al., eds., 1997. *The Ecological Basis of Conservation: Heterogeneity, Ecosystems, and Biodiversity* [conference proceedings]. New York: Chapman & Hall. 466p. [1/-/-] *A.T.:* conservation biology.

3958. Pielou, E.C., 1966. Species-diversity and pattern-diversity in the study of ecological succession. *J. of Theoretical Biology* 10: 370-383. [2] *A.T.:* trees.

3959. ———, 1966. The measurement of diversity in different types of biological collections. *J. of Theoretical Biology* 13: 131-144. [2] *A.T.:* diversity estimation; sampling.

3960. ———, 1966. Shannon's formula as a measure of species diversity: its use and misuse. *American Naturalist* 100: 463-465. [2].

3961. ———, 1986. Assessing the diversity and composition of restored vegetation. *Canadian J. of Botany* 64: 1344-1348. [1] *A.T.:* diversity estimation; diversity comparison.

3962. ———, 1992 (reprint of 1979 ed.). *Biogeography*. Malabar, FL: Krieger. 351p. [2/1/2].

3963. Pienkowski, M.W., E.M. Bignal, et al., 1996. A simplified classification of land-type zones to assist the integration of biodiversity objectives in land-use policies. *Biological Conservation* 75: 11-25. [1] *A.T.:* land use; birds; U.K.; agricultural policy.

3964. Pimbert, Michel P., 1997. Issues emerging in implementing the Convention on Biological Diversity. *J. of International Development* 9: 415-425. *A.T.:* international cooperation.

3965. Pimbert, Michel P., & J.N. Pretty, 1995. *Parks, People and Professionals: Putting "Participation" Into Protected Area Management.* Discussion Paper No. 57, United Nations Research Institute for Social Development. 60p. [1/1/1] *A.T.:* local participation.

3966. Pimentel, David, Ulrich Stachow, et al., 1992. Conserving biological diversity in agricultural/forestry systems. *Bioscience* 42: 354-362. [2] Notes that human-managed ecosystems contain the majority of the world's bd.

3967. Pimentel, David, Christa Wilson, et al., 1997. Economic and environmental benefits of biodiversity. *Bioscience* 47: 747-757. An overview that leads to an estimate of total annual economic benefit of bd. *A.T.:* bd value.

3968. Pimm, Stuart L., 1991. *The Balance of Nature? Ecological Issues in the Conservation of Species and Communities.* Chicago: University of Chicago Press. 434p. [2/2/3] *A.T.:* ecology.

3969. ———, 1996. Lessons from a kill. *Biodiversity and Conservation* 5: 1059-1067. [1] *A.T.:* human impact; Hawaii; bd loss; synergisms.

3970. Pimm, Stuart L., & R.A. Askins, 1995. Forest losses predict bird extinctions in eastern North America. *PNAS* 92: 9343-9347. [1] *A.T.:* deforestation; species-area relation; endemism.

3971. Pimm, Stuart L., & J.L. Gittleman, 1992. Biological diversity: where is it? [editorial] *Science* 255: 940. [1] Includes a discussion of spatial turnover (beta diversity) vs. number of species in a given area (alpha diversity).

3972. Pimm, Stuart L., & J.H. Lawton, 1998. Planning for biodiversity. *Science* 279: 2068-2069. Concerning the problems connected with trying to create reserves over hotspots. *A.T.:* deforestation, endangered and vulnerable species.

3973. Pimm, Stuart L., & A.M. Sugden, 1994. Tropical diversity and global change [editorial]. *Science* 263: 933-934. [1] *A.T.:* forest turnover; global warming.

3974. Pimm, Stuart L., H.L. Jones, & Jared Diamond, 1988. On the risk of extinction. *American Naturalist* 132: 757-785. [2].

3975. Pimm, Stuart L., J.H. Lawton, & J.E. Cohen, 1991. Food web patterns and their consequences. *Nature* 350: 669-674. [2].

3976. Pimm, Stuart L., M.P. Moulton, & L.J. Justice, 1994. Bird extinctions in the central Pacific. *PTRSL B* 344: 27-33. [1]. Examines extinction rates of the past several thousand years.

3977. Pimm, Stuart L., G.J. Russell, et al., 1995. The future of biodiversity. *Science* 269: 347-350. [2] *A.T.:* extinction rates.

3978. Pinchot, Gifford, 1910. *The Fight for Conservation.* New York: Doubleday, Page & Co. 152p. [2/-/1].

3979. Pine, Ronald H., 1994. Systematics and biodiversity [letter]. *TREE* 9: 229. [1].

3980. Pipkin, James, 1996. Biological diversity conservation: a public policy perspective. *Environmental Management* 20: 793-797. [1].

3981. Pitcher, Tony J., & P.J.B. Hart, eds., 1995. *The Impact of Species Changes in African Lakes* [conference papers]. London and New York: Chapman & Hall. 601p. [1/1/1] *A.T.:* introduced fishes; fisheries resources.

3982. Pitkaenen, S., 1997. Correlation between stand structure and ground vegetation: an analytical approach. *Plant Ecology* 131: 109-126. *A.T.:* Finland; understorey.

3983. Place, Susan E., ed., 1993. *Tropical Rainforests: Latin American Nature and Society in Transition.* Wilmington, DE: Scholarly Resources. 229p. [2/1/1] *A.T.:* rainforest conservation.

3984. Planes, S., & R. Galzin, 1997. New perspectives in biogeography of coral reef fish in the Pacific using phylogeography and population genetics approaches. *Vie et Milieu* 47: 375-380.

3985. Planty-Tabacchi, Anne-Marie, Eric Tabacchi, & R.J. Naiman, 1996. Invasibility of species-rich communities in riparian zones. *CB* 10: 598-607. [2] *A.T.:* biological invasions.

3986. Platnick, Norman I., 1981. Widespread taxa and biogeographic congruence. *Advances in Cladistics* 1: 223-227. [1].

3987. ———, 1991. Patterns of biodiversity: tropical vs. temperate [editorial]. *J. of Natural History* 25: 1083-1088. [1].

3988. Platnick, Norman I., & Gareth Nelson, 1978. A method of analysis for historical biogeography. *Systematic Zoology* 27: 1-16. [2] *A.T.:* dispersal; vicariance.

3989. ———, 1984. Composite areas in vicariance biogeography. *Systematic Zoology* 33: 328-335. [1].

3990. Platt, Rutherford H., R.A. Rowntree, & P.C. Muick, eds., 1994. *The Ecological City: Preserving and Restoring Urban Biodiversity* [symposium papers]. Amherst, MA: University of Massachusetts Press. 291p. [2/1/1] *A.T.:* urban ecology.

3991. Pletscher, Daniel H., & R.L. Hutto, 1990. Wildlife management and the maintenance of biological diversity. *Western Wildlands* 17(3): 8-12. [1].

3992. Plotkin, Mark J., 1993. *Tales of a Shaman's Apprentice: An Ethnobotanist Searches for New Medicines in the Amazon Rain Forest.* New York: Viking. 318p. [2/3/2] *A.T.:* medicinal plants.

3993. Plotkin, Mark J., & Lisa Famolare, eds., 1992. *Sustainable Harvest and Marketing of Rain Forest Products* [conference proceedings]. Washington, DC: Island Press. 325p. [2/1/1] *A.T.:* tropical regions; sustainable forestry.

3994. Plowman, K.P., 1994. Stories we tell about fauna. *Memoirs of the Queensland Museum* 36: 185-190. [1] Argues that scientists are not telling the right "stories" about bd to gain the attention of the general public. *A.T.:* environmental perception; Australia; public opinion.

3995. Plucknett, Donald L., & M.E. Horne, 1992. Conservation of genetic resources. *Agriculture, Ecosystems & Environment* 42: 75-92. [1] *A.T.:* in situ conservation; ex situ conservation.

3996. Plucknett, Donald L., N.J.H. Smith, et al., 1983. Crop germplasm conservation and developing countries. *Science* 220: 163-169. [1] *A.T.:* genetic diversity loss; gene banks.

3997. Plucknett, Donald L., N.J.H. Smith, et al., 1987. *Gene Banks and the World's Food.* Princeton, NJ: Princeton University Press. 247p. [2/1/2] *A.T.:* food crops; germplasm resources.

3998. Podolsky, Richard, Oct. 1992. Remote sensing, geographic data and the conservation of biological resources. *Endangered Species Update* 9(12): 1-4. [1] *A.T.:* GIS.

3999. Pojar, J., N. Diaz, et al., 1994. Biodiversity planning and forest management at the landscape scale. *General Technical Report PNW* GTR-336: 55-70. [1] *A.T.:* British Columbia; landscape ecology.

4000. Pokarzhevskii, Andrei D., & D.A. Krivolutskii, 1997. Problems of estimating and maintaining biodiversity of soil biota in natural and agroecosystems: a case study of chernozem soil. *Agriculture, Ecosystems & Environment* 62: 127-133. *A.T.:* Russia.

4001. Polasky, Stephen, & A.R. Solow, 1995. On the value of a collection of species. *J. of Environmental Economics and Management* 29: 298-303. [1] *A.T.:* bd value; valuation methods.

4002. Polasky, Stephen, Holly Doremus, & Bruce Rettig, Oct. 1997. Endangered species conservation on private land. *Contemporary Economic Policy* 15(4): 66-76. Discusses the economic consequences of the conflict between individual and societal conservation priorities.

4003. Polasky, Stephen, A.R. Solow, & James Broadus, 1993. Searching for uncertain benefits and the conservation of biological diversity. *Environmental & Resource Economics* 3: 171-181. [1] *A.T.:* resource allocation.

4004. Polhemus, Dan A., 1993. Conservation of aquatic insects: worldwide crisis or localized threats? *American Zoologist* 33: 588-598. [1].

4005. Polis, Gary A., 1991. Complex trophic interactions in deserts: an empirical critique of food-web theory. *American Naturalist* 138: 123-155. [2] *A.T.:* California.

4006. ———, ed., 1991. *The Ecology of Desert Communities* [conference papers]. Tucson: University of Arizona Press. 456p. [2/2/1].

4007. Polis, Gary A., & S.D. Hurd, 1995. Extraordinarily high spider densities on islands: flow of energy from the marine to terrestrial food webs and the absence of predation. *PNAS* 92: 4382-4386. [2] *A.T.:* Gulf of California.

4008. Pollard, Ernest, & T.J. Yates, 1993. *Monitoring Butterflies for Ecology and Conservation: The British Butterfly Monitoring Scheme*. London and New York: Chapman & Hall. 274p. [1/1/2].

4009. Pollock, M.M., R.J. Naiman, & T.A. Hanley, 1998. Plant species richness in riparian wetlands. A test of biodiversity theory. *Ecology* 79: 94-105. *A.T.:* flood frequency; productivity; Alaska.

4010. Pomeroy, Derek, 1993. Centers of high biodiversity in Africa. *CB* 7: 901-907. [2] *A.T.:* endemism; hot spots.

4011. Pomeroy, Derek, & C. Dranzoa, 1997. Methods of studying the distribution, diversity and abundance of birds in East Africa: some quantitative approaches. *African J. of Ecology* 35: 110-123. *A.T.:* timed species counts; surveys.

4012. Pomeroy, Derek, & Adrian Lewis, 1987. Bird species richness in tropical Africa: some comparisons. *Biological Conservation* 40: 11-28. [1] Compares African patterns with Latin American ones.

4013. Poole, Peter, 1995. *Indigenous Peoples, Mapping & Biodiversity Conservation: An Analysis of Current Activities and Opportunities for Applying Geomatics Technologies*. Landover, MD: Biodiversity Support Program. 83p. [1/1/1] *A.T.:* remote sensing.

4014. Poore, Gary C.B., & G.D.F. Wilson, 1993. Marine species richness [letter]. *Nature* 361: 597-598. [2] *A.T.:* isopods; continental slope benthic communities.

4015. Por, Francis Dov, 1996. Diversity, subservience, and the future of evolution. *Israel J. of Zoology* 42: 455-463. [1] *A.T.:* Focuses on symbionts and parasites. *A.T.:* human influence.

4016. Porembski, Stefan, Jorg Szarzynski, et al., 1996. Biodiversity and vegetation of small-sized inselbergs in a West African rain forest (Tai, Ivory Coast). *J. of Biogeography* 23: 47-55. [1].

4017. Porritt, Jonathon, ed., 1991. *Save the Earth* [essays]. Atlanta: Turner Publications. 208p. [3/2/1] *A.T.:* human ecology; environmental policy.

4018. Porter, Douglas R., & D.A. Salvesen, eds., 1995. *Collaborative Planning for Wetlands and Wildlife: Issues and Examples*. Washington, DC: Island Press. 293p. [2/1/1] *A.T.:* wildlife conservation.

4019. Porter, Sanford D., & D.A. Savignano, 1990. Invasion of polygyne fire ants decimates native ants and disrupts arthropod community. *Ecology* 71: 2095-2106. [2] *A.T.:* biological invasions; Southeast U.S.; competitive replacement.

4020. Portney, Paul R., fall 1994. The contingent valuation debate: why economists should care. *J. of Economic Perspectives* 8(4): 3-17. [2].

4021. Poschlod, Peter, & S. Bonn, 1998. Changing dispersal processes in the Central European landscape since the last Ice Age: an explanation for the actual decrease of plant species richness in different habitats? *Acta Botanica Neerlandica* 47: 27-44. *A.T.:* North America; seed dispersal; mammals.

4022. Posey, Darrell A., Oct. 1996. Protecting indigenous peoples' rights to biodiversity. *Environment* 38(8): 6-9, 37-45. [2].

4023. ———, 1997. Indigenous knowledge, biodiversity, and international rights: learning about forests from the Kayapó Indians of the Brazilian Amazon. *Commonwealth Forestry Review* 76: 53-60.

4024. Poss, Stuart G., & B.B. Collette, 1995. Second survey of fish collections in the United States and Canada. *Copeia* (1): 48-70. [1] *A.T.:* museum collections.

4025. Possingham, Hugh P., 1993. Impact of elevated atmospheric CO_2 on biodiversity: mechanistic population-dynamic perspective. *Australian J. of Botany* 41: 11-21. [1] *A.T.:* climatic change.

4026. Poten, Constance J., Sept. 1991. A shameful harvest: America's illegal wildlife trade. *National Geographic* 180(3): 106-132. [1] *A.T.:* wild animal trade; animal rights; environmental ethics.

4027. Potter, Christopher S., J.I. Cohen, & Dianne Janczewski, eds., 1993. *Perspectives on Biodiversity: Case Studies of Genetic Resource Conservation and Development.* Washington, DC: AAAS Press. 245p. [1/1/1] *A.T.:* germplasm resources.

4028. Potter, C.S., & R.E. Meyer, 1990. The role of soil biodiversity in sustainable dryland farming systems. In R.P. Singh, J.F. Parr, & B.A. Stewart, eds., *Dryland Agriculture: Strategies for Sustainability* (New York: Springer): 241-251. [1].

4029. Potter, D., T.C. Lajeunesse, et al., 1997. Convergent evolution masks extensive biodiversity among marine coccoid picoplankton. *Biodiversity and Conservation* 6: 99-107. [2] *A.T.:* rRNA.

4030. Potter, Van Rensselaer, 1988. *Global Bioethics: Building on the Leopold Legacy.* East Lansing: Michigan State University Press. 203p. [1/1/1].

4031. Potthast, T., 1996. Genetics, evolution and ethics of conservation of nature, or: the invention of biodiversity. *Biologisches Zentralblatt* 115: 177-188. [1].

4032. Poulin, Robert, 1995. Phylogeny, ecology, and the richness of parasite communities in vertebrates. *Ecological Monographs* 65: 283-302. [2] *A.T.:* host-parasite relations; body size.

4033. ———, 1997. Species richness of parasite assemblages: evolution and patterns. *ARES* 28: 341-358. *A.T.:* helminths; community structure.

4034. ———, 1998. Comparison of three estimators of species richness in parasite component communities. *J. of Parasitology* 84: 485-490. *A.T.:* diversity estimators.

4035. Poulin, Robert, & Klaus Rohde, 1997. Comparing the richness of metazoan ectoparasite communities of marine fishes: controlling for host phylogeny. *Oecologia* 110: 278-283. [2].

4036. Poulsen, Axel D., 1997. *Plant Diversity in Forests of Western Uganda and Eastern Zaire: Preliminary Results.* Risskov, Denmark: Dept. of Systematic Botany, University of Aarhus. 76p. [1/1/-] *A.T.:* rainforest diversity.

4037. Poulsen, B.O., & Niels Krabbe, 1998. Avifaunal diversity of five high altitude cloud forests on the Andean western slope of Ecuador: testing a rapid assessment method. *J. of Biogeography* 25: 83-93. *A.T.:* montane environments.

4038. Poulson, Thomas L., & W.J. Platt, 1989. Gap light regimes influence canopy tree diversity. *Ecology* 70: 553-555. [2] *A.T.:* treefall gaps; Michigan.

4039. Pounds, J. Alan, & M.L. Crump, 1994. Amphibian declines and climate disturbance: the case of the golden toad and the harlequin frog. *CB* 8: 72-85. [2] *A.T.:* Costa Rica; El Niño.

4040. Povilitis, Tony, spring 1995. Changes in habitat threaten biodiversity. *Forum for Applied Research and Public Policy* 10(1): 97-100. [1].

4041. Powell, G.V.N., & R.D. Bjork, 1994. Implications of altitudinal migration for conservation strategies to protect tropical biodiversity: a case study of the resplendent quetzal *Pharomacrus mocinno* at Monteverde, Costa Rica. *Bird Conservation International* 4: 161-174. [1] *A.T.:* cloud forests.

4042. ———, 1995. Implications of intratropical migration on reserve design: a case study using *Pharomachrus mocinno*. *CB* 9: 354-362. [1] *A.T.:* quetzals; Costa Rica.

4043. Powell, W., C.A. Orozco-Castillo, et al., 1995. Polymerase chain reaction-based assays for the characterisation of plant genetic resources. *Electrophoresis* 16: 1726-1730. [1] *A.T.:* molecular markers.

4044. Power, Mary E., W.E. Dietrich, & J.C. Finlay, 1996. Dams and downstream aquatic biodiversity: potential food web consequences of hydrologic and geomorphic change. *Environmental Management* 20: 887-895. [1] *A.T.:* river flow variation.

4045. Power, Mary E., David Tilman, et al., 1996. Challenges in the quest for keystones: identifying keystone species is difficult but essential to understanding how loss of species will affect ecosystems. *Bioscience* 46: 609-620. [2].

4046. Powers, Michaelle A., 1993. The United Nations Framework Convention on Biological Diversity: will biodiversity preservation be enhanced through its provisions concerning biotechnology intellectual property rights? *Wisconsin International Law J.* 12: 103-124. [1].

4047. Powledge, Fred, 1998. Biodiversity at the crossroads. *Bioscience* 48: 347-352. Expresses some pessimism about the future of bd.

4048. Prakash, Ishwar, 1997. Ecology of desert mammals. *Current Sci-*

ence 72: 31-34. *A.T.:* India; Pakistan; Thar Desert; human impact.

4049. Pramod, P., R.J.R. Daniels, et al., 1997. Evaluating bird communities of Western Ghats to plan for a biodiversity friendly development. *Current Science* 73: 156-162. *A.T.:* India; bd value.

4050. Prance, Ghillean T., ed., 1982. *Biological Diversification in the Tropics* [symposium proceedings]. New York: Columbia University Press. 714p. [1/1/2] *A.T.:* biogeography.

4051. Prance, Ghillean T., ed., 1986. *Tropical Rain Forests and the World Atmosphere.* Boulder, CO: Westview Press. 105p. [1/1/1] *A.T.:* deforestation; vegetation and climate.

4052. ———, 1994. A comparison of the efficacy of higher taxa and species numbers in the assessment of biodiversity in the Neotropics. *PTRSL B* 345: 89-99. [1] *A.T.:* biogeography; species diversity.

4053. ———, 1995. Systematics, conservation and sustainable development. *Biodiversity and Conservation* 4: 490-500. [1] *A.T.:* Brazil; Zimbabwe.

4054. Prance, Ghillean T., & T.S. Elias, eds., 1977. *Extinction Is Forever: Threatened and Endangered Species of Plants in the Americas and Their Significance in Ecosystems Today and in the Future* [symposium proceedings]. New York: New York Botanical Garden. 437p. [2/1/1].

4055. Prance, Ghillean T., & T.E. Lovejoy, eds., 1985. *Amazonia (Key Environments* series). Oxford, U.K., and New York: Pergamon Press. 442p. [2/1/1] *A.T.:* natural history.

4056. Prance, Ghillean T., W. Balée, et al., 1987. Quantitative ethnobotany and the case for conservation in Amazonia. *CB* 1: 296-310. [1].

4057. Prasad, S.N., B. Prabakaran, & C. Jeganathan, 1996. Visualization in biodiversity research: a case study of Mehao Wildlife Sanctuary, Arunachal Pradesh, northeast India. *Current Science* 71: 1001-1005. [1] *A.T.:* remote sensing; bd visualization.

4058. Preece, R.C., 1998. Impact of early Polynesian occupation on the land snail fauna of Henderson Island, Pitcairn Group (South Pacific). *PTRSL B* 353: 347-368. *A.T.:* human impact; extinction.

4059. Pregill, Gregory K., & S.L. Olson, 1981. Zoogeography of West Indian vertebrates in relation to Pleistocene climatic cycles. *ARES* 12: 75-98. [1] *A.T.:* paleobiogeography.

4060. Prendergast, J.R., & B.C. Eversham, 1995. Butterfly diversity in southern Britain: hotspot losses since 1930. *Biological Conservation* 72: 109-114. [1].

4061. ———, 1997. Species richness covariance in higher taxa: empirical tests of the biodiversity indicator concept. *Ecography* 20: 210-216. [2].

4062. Prendergast, J.R., R.M. Quinn, et al., 1993. Rare species, the coincidence of diversity hotspots and conservation strategies. *Nature* 365: 335-337. [3] *A.T.:* U.K.; in situ conservation.

4063. Prendergast, J.R., S.N. Wood, et al., 1993. Correcting for variation in recording effort in analyses of diversity hotspots. *Biodiversity Letters* 1: 39-53. [1] *A.T.:* Odonata; Hepaticae; U.K.; sampling bias.

4064. Prentice, I.C., 1986. Vegetation responses to past climatic changes. *Vegetatio* 67: 131-141. [2] *A.T.:* response time.

4065. Prescott-Allen, Christine, & Robert Prescott-Allen, 1986. *The First Resource: Wild Species in the North American Economy.* New Haven: Yale University Press. 529p. [2/2/1] *A.T.:* economic biology; animal and plant products.

4066. Prescott-Allen, Robert, & Christine Prescott-Allen, 1982. *What's Wildlife Worth? Economic Contributions of Wild Plants and Animals to Developing Countries.* London: International Institute for Environment and Development. 90p. [1/1/1].

4067. ———, 1988 (1983). *Genes From the Wild: Using Wild Genetic Resources for Food and Raw Materials* (2nd ed.). London: Earthscan. 111p. [1/1/1].

4068. ———, 1990. How many plants feed the world? *CB* 4: 365-374. [1] *A.T.:* genetic diversity; food supply.

4069. Press, Daniel, D.F. Doak, & Paul Steinberg, 1996. The role of local government in the conservation of rare species. *CB* 10: 1538-1548. [1] *A.T.:* California; environmental policy; endangered species.

4070. Pressey, Robert L., 1994. *Ad hoc* reservations: forward or backward steps in developing representative reserve systems? *CB* 8: 662-668. [2] *A.T.:* reserve selection; Australia; environmental planning.

4071. ———, 1995. Conservation reserves in NSW: crown jewels or leftovers? *Search* 26: 47-51. [1] *A.T.:* protected areas; reserve selection.

4072. Pressey, Robert L., & V.S. Logan, 1994. Level of geographical subdivision and its effects on assessments of reserve coverage: a review of regional studies. *CB* 8: 1037-1046. [1].

4073. Pressey, Robert L., C.J. Humphries, et al., 1993. Beyond opportunism: key principles for systematic reserve selection. *TREE* 8: 124-128. [2].

4074. Preston, Brian J., 1995. The role of law in the protection of biological diversity in the Asia-Pacific region. *Environmental and Planning Law J.* 12: 264-277. [1].

4075. Preston, E.M., & C.A. Ribic, 1992. EMAP and other tools for measuring biodiversity, habitat conditions, and environmental trends. *General Technical Report RM* GTR-229: 223-228. [1] *A.T.:* birds.

4076. Preston, F.W., 1948. The commonness, and rarity, of species. *Ecology* 29: 254-283. [2] *A.T.:* species abundance.

4077. ———, 1962. The canonical distribution of commonness and rarity (Parts I & II). *Ecology* 43: 185-215, 410-432. [3] A classic study including a development of the species-area hypothesis. *A.T.:* species abundance.

4078. Preston, G.R., & W.R. Siegfried, 1995. The protection of biological diversity in South Africa: profiles and perceptions of professional practitioners in nature conservation agencies and natural history museums. *South African J. of Wildlife Research* 25: 49-56. [1] *A.T.:* biological invasions.

4079. Price, Andrew, & Sarah Humphrey, eds., 1993. *Application of the Biosphere Reserve Concept to Coastal Marine Areas* [workshop papers]. Gland, Switzerland: IUCN and UNESCO. 114p. [1/1/1] *A.T.:* coastal zone management.

4080. Price, O., J.C.Z. Woinarski, et al., 1995. Patterns of species composition and reserve design for a fragmented estate: monsoon rainforests in the Northern Territory, Australia. *Biological Conservation* 74: 9-19. [1].

4081. Price, P.W., I.R. Diniz, et al., 1995. The abundance of insect herbivore species in the tropics: the high local richness of rare species. *Biotropica* 27: 468-478. [1] *A.T.:* Brazil; cerrado.

4082. Prieto-Samsónov, D.L., R.I. Vázquez-Padrón, et al., 1997. *Bacillus thuringiensis*: from biodiversity to biotechnology. *J. of Industrial Microbiology & Biotechnology* 19: 202-219. *A.T.:* bacteria; transgenic plants.

4083. Primack, Richard B., 1992. Tropical community dynamics and conservation biology. *Bioscience* 42: 818-821. [1] *A.T.:* stability; protected areas.

4084. ———, 1998. *Essentials of Conservation Biology* (2nd ed.) [textbook]. Sunderland, MA: Sinauer Associates. 660p. [2/2/2].

4085. Primack, Richard B., & Pamela Hall, 1991. Species diversity research in Bornean forests with implications for conservation biology and silviculture. *Tropics* 1: 91-111. [1].

4086. ———, 1992. Biodiversity and forest change in Malaysian Borneo. *Bioscience* 42: 829-837. [1] *A.T.:* Sarawak; equilibrium hypothesis; nonequilibrium hypothesis.

4087. Primack, Richard B., & T.E. Lovejoy, eds., 1995. *Ecology, Conservation, and Management of Southeast Asian Rainforests*. New Haven: Yale University Press. 304p. [1/1/1].

4088. Primack, Richard B., & S.L. Miao, 1992. Dispersal can limit local plant distribution. *CB* 6: 513-519. [2] *A.T.:* seed dispersal.

4089. Pring, Allan, 1996. The place of descriptive mineralogy in modern science [editorial]. *Rocks and Minerals* 71: 158-162. [1] Relates mineralogical taxonomy to bd studies.

4090. Pringle, H.J.R., 1995. Pastoralism, nature conservation and ecological sustainability in Western Australia's southern shrubland rangelands. *International J. of Sustainable Development and World Ecology* 2: 26-44. [1].

4091. Prinn, R.G., & B. Fegley, Jr., 1987. Bolide impacts, acid rain, and biospheric traumas at the Cretaceous-Tertiary boundary. *Earth and Planetary Science Letters* 83: 1-15. [2] *A.T.:* asteroid impacts.

4092. Prior-Magee, Julie S., B.C. Thompson, & David Daniel, 1998. Evaluating consistency of categorizing biodiversity management status relative to land stewardship in the gap analysis program. *J. of Environmental Planning and Management* 41: 209-216. *A.T.:* environmental management.

4093. Pritchard, Paul C., March-April 1992. Relevance [opinion]. *National Parks* 66(3-4): 4. [1] States several reasons for supporting nature protection through parks.

4094. Probst, John R., & T.R. Crow, Feb. 1991. Integrating biological diversity and resource management: an essential approach to productive, sustainable ecosystems. *J. of Forestry* 89(2): 12-17. [1].

4095. Prothero, D.R., 1985. North American mammalian diversity and Eocene-Oligocene extinctions. *Paleobiology* 11: 389-405. [1] *A.T.:* faunal turnover.

4096. Pryde, Philip R., 1987. The distribution of endangered fauna in the USSR. *Biological Conservation* 42: 19-37. [1].

4097. ———, 1997. Creating offshore island sanctuaries for endangered species: the New Zealand experience. *Natural Areas J.* 17: 248-254.

4098. Pulliam, H. Ronald, & Bruce Babbitt, 1997. Science and the protection of endangered species [editorial]. *Science* 275: 499-500. *A.T.:* endemic species.

4099. Pulliam, H. Ronald, J.B. Dunning, Jr., & Jianguo Liu, 1992. Population dynamics in complex landscapes: a case study. *Ecological Applications* 2: 165-177. [2] *A.T.:* habitat fragmentation; South Carolina; spatially explicit models; birds.

4100. Pullin, Andrew S., ed., 1995. *Ecology and Conservation of Butterflies*. London and New York: Chapman & Hall. 363p. [1/1/1].

4101. Purdue, Derrick, 1995. Hegemonic trips: world trade, intellectual property and biodiversity. *Environmental Politics* 4: 88-107. [1] *A.T.:* GATT; biotechnology.

4102. Puri, G.S., 1995. Biodiversity and the development of natural resources for the 21st century. *Tropical Ecology* 36: 253-255. [1].

4103. Pushpangadan, P., K. Ravi, & V. Santhosh, eds., 1997. *Conservation and Economic Evaluation of Biodiversity* [workshop papers]. Enfield, NH: Science Publishers. 2 vols. [1/1/-].

4104. Putterman, Daniel M., 1994. Trade and the Biodiversity Convention [editorial]. *Nature* 371: 553-554. [1] *A.T.:* CBD; international cooperation.

4105. ———, 1996. Model material transfer agreements for equitable biodiversity prospecting. *Colorado J. of International Environmental Law and Policy* 7: 149-177. [1] *A.T.:* biotechnology.

4106. Pyle, Richard L., 1996. Exploring the deep coral reef: how much biodiversity are we missing? *Global Biodiversity* 6(1): 3-7. [1].

4107. Pyle, Robert M., M. Bentzien, & Paul Opler, 1981. Insect conservation. *Annual Review of Entomology* 26: 233-258. [1].

4108. Pyrovetsi, Myrto, & E. Papastergiadou, 1992. Biological conservation implications of water-level fluctuations in a wetland of international importance: Lake Kerkini, Macedonia, Greece. *Environmental Conservation* 19: 235-244. [1] *A.T.:* bd loss.

4109. Pysek, Petr, Karel Prach, et al., eds., 1995. *Plant Invasions: General Aspects and Special Problems* [workshop papers]. Amsterdam: SPB Academic Publishing. 263p. [1/1/1].

4110. Quammen, David, 1996. *The Song of the Dodo: Island Biogeography in an Age of Extinctions.* New York: Scribner. 702p. [3/3/1] *A.T.:* Alfred Russel Wallace; endangered species.

4111. Quansah, Nathaniel, 1988. Ethnomedicine in the Maroantsetra region of Madagascar. *Economic Botany* 42: 370-375. [1] *A.T.:* medicinal plants.

4112. Querol Lipcovich, Daniel, 1993. *Genetic Resources: A Practical Guide to Their Conservation.* London and Atlantic Highlands, NJ: Zed Books. 252p. [1/1/1] *A.T.:* plant germplasm resources.

4113. Quinn, James F., & S.P. Harrison, 1988. Effects of habitat fragmentation and isolation on species richness: evidence from biogeographic patterns. *Oecologia* 75: 132-140. [2] *A.T.:* reserve selection; island biogeography; species-area relations.

4114. Quinn, J.S., R.D. Morris, et al., 1996. Design and management of bird nesting habitat: tactics for conserving colonial waterbird biodiversity on artificial islands in Hamilton Harbour, Ontario. *Canadian J. of Fisheries and Aquatic Sciences* 53, Suppl. 1: 45-57. [1].

4115. Quinn, R.M., J.H. Lawton, et al., 1994. The biogeography of scarce vascular plants in Britain with respect to habitat preference, dispersal ability and reproductive biology. *Biological Conservation* 70: 149-157. [1].

4116. Quiros, C.F., S.B. Brush, et al., 1990. Biochemical and folk assessment of variability of Andean cultivated potatoes. *Economic Botany* 44: 254-266. [1] *A.T.:* genetic variation.

4117. Rabb, George B., 1994. The changing roles of zoological parks in conserving biological diversity. *American Zoologist* 34: 159-164. [1].

4118. ———, winter 1997. Global extinction threat. *Defenders* 72(1): 34-36. [1].

4119. Rabb, George B., & T.A. Sullivan, 1995. Coordinating conservation: global networking for species survival. *Biodiversity and Conservation* 4: 536-543. [1] Describes the work of the IUCN's Species Survival Commission.

4120. Rabenold, Kerry N., 1979. A reversed latitudinal diversity gradient in avian communities of Eastern deciduous forests. *American Naturalist* 114: 275-286. [1] *A.T.:* stability.

4121. ———, 1993. Latitudinal gradients in avian species diversity and the role of long-distance migration. *Current Ornithology* 10: 247-273. [1] *A.T.:* biogeography.

4122. Rabenold, Kerry N., P.T. Fauth, et al., 1998. Response of avian communities to disturbance by an exotic insect in spruce-fir forests of the Southern Appalachians. *CB* 12: 177-189. *A.T.:* forest decline; biological invasions.

4123. Rabinowitz, A., Nov.-Dec. 1997. On the road to Myanmar. *Wildlife Conservation* 100(6): 36-43. Describes bdc plans in Myanmar (Burma).

4124. Radcliffe, Samuel J., Aug. 1992. Forestry at the crossroads—integrating economic and social needs with biological concerns. *J. of Forestry* 90(8): 22-26. [1] *A.T.:* endangered species; environmental legislation.

4125. Radmer, Richard J., 1996. Algal diversity and commercial algal products. *Bioscience* 46: 263-270. [1] *A.T.:* biotechnology.

4126. Radomski, Paul J., & T.J. Goeman, July 1995. The homogenizing of Minnesota lake fish assemblages. *Fisheries* 20(7): 20-23. [1] *A.T.:* fish stocking; fisheries management.

4127. Rafe, R.W., M.B. Usher, & R.G. Jefferson, 1985. Birds on reserves: the influence of area and habitat on species richness. *J. of Applied Ecology* 22: 327-335. [1] *A.T.:* seasonal diversity; U.K.

4128. Raghavendra Rao, R., 1994. *Biodiversity in India, Floristic Aspects*. Dehra Dun: Bishen Singh Mahendra Pal Singh. 315p. [1/1/1] *A.T.:* plant ecology.

4129. Rago, Paul J., & J.G. Wiener, 1986. Does pH affect fish species richness when lake area is considered? *Trans. of the American Fisheries Society* 115: 438-447. [1] *A.T.:* Ontario; Wisconsin.

4130. Rahbek, Carsten, 1993. Captive breeding—a useful tool in the preservation of biodiversity? *Biodiversity and Conservation* 2: 426-437. [1] *A.T.:* zoos.

4131. ———, 1995. The elevational gradient of species richness: a uniform pattern? *Ecography* 18: 200-205. [2].

4132. ———, 1997. The relationship among area, elevation, and regional species richness in Neotropical birds. *American Naturalist* 149: 875-902. *A.T.:* Rapoport's rule; elevational gradient.

4133. Raibaut, André, C. Combes, & F. Benoit, 1998. Analysis of the parasitic copepod species richness among Mediterranean fish. *J. of Marine Systems* 15: 185-206.

4134. Rajak, R.C., & M.K. Rai, eds., 1996. *Herbal Medicines, Biodiversity, and Conservation Strategies: Proceedings of the National Seminar*. Dehra Dun: International Book Distributors. 292p. [1/1/1] *A.T.:* medicinal plants; bdc.

4135. Raloff, Janet, 1997. Dying breeds. *Science News* 152: 216-218. [1] Describes the endangerment of many livestock breeds worldwide.

4136. Ram, J., J.S. Singh, & S.P. Singh, 1989. Plant biomass, species diversity and net primary production in a Central Himalayan high altitude grassland. *J. of Ecology* 77: 456-468. [1] *A.T.:* India.

4137. Ramakrishnan, P.S., ed., 1991. *Ecology of Biological Invasion in the Tropics* [workshop proceedings]. New Delhi: International Scientific Publications. 195p. [1/1/1].

4138. Ramakrishnan, P.S., A.N. Purohit, et al., eds., 1997. *Conservation and Management of Biological Resources in Himalaya* [seminar proceedings]. Enfield, NH: Science Publishers. 603p. [1/1/-].

4139. Ramamoorthy, T.P., Robert Bye, et al., eds., 1993. *Biological Diversity of Mexico: Origins and Distribution* [symposium papers]. New York: OUP. 812p. [1/1/1].

4140. Raman, Kannamma S., Jan.-March 1997. Conservation of biodiversity: need of the decade. *Social Action* 47(1): 73-87. Overview of bdc-related problems.

4141. Raman, T.R.S., G.S. Rawat, & A.J.T. Johnsingh, 1998. Recovery of tropical rain forest avifauna in relation to vegetation succession following shifting cultivation in Mizoram, northeast India. *J. of Applied Ecology* 35: 214-231.

4142. Ramesh, B.R., S. Menon, & K.S. Bawa, 1997. A vegetation based approach to biodiversity gap analysis in the Agastyamalai region, Western Ghats, India. *Ambio* 26: 529-536.

4143. Rampino, M.R., & K. Caldeira, 1993. Major episodes of geologic change: correlations, time structure and possible causes. *Earth and Planetary Science Letters* 114: 215-227. [2] Presents evidence of periodic catastrophic events over geological history.

4144. Rana, R.S., et al., eds., 1995. *Plant Germplasm Conservation: Biotechnological Approaches*. New Delhi: National Bureau of Plant Genetic Resources. 259p. [1/1/1] *A.T.:* India.

4145. Rand, Michael, & Martin Kelsey, 1994. Call to action. *American Birds* 48: 36-46. [1] Concerning conservation of the world's bird populations.

4146. Randal, Judith, 1995. Experts warn of health risks from loss of biodiversity [editorial]. *J. of the National Cancer Institute* 87: 714-716. [1] *A.T.:* human health; disease.

4147. Randall, Alan, 1991. The value of biodiversity. *Ambio* 20: 64-68. [1] Employs a "safe minimum standard" approach to rationalize bdc.

4148. Rankin, Colin, & R.M. M'Gonigle, 1991. Legislation for biological diversity: a review and proposal for British Columbia. *University of British Columbia Law Review* 25: 277-333. [1] *A.T.:* bd law.

4149. Rapoport, Eduardo H., 1969. Gloger's rule and pigmentation of Collembola. *Evolution* 23: 622-626. [1].

4150. ———, 1982. *Areography: Geographical Strategies of Species*. Oxford, U.K., and New York: Pergamon Press. 269p. [1/1/2] *A.T.:* biogeography.

4151. Rappole, John H., & M.V. McDonald, 1994. Cause and effect in population declines of migratory birds. *Auk* 111: 652-660. [2] *A.T.:* Latin

America.

4152. Rapport, David J., Robert Costanza, et al., eds., 1998. *Ecosystem Health: Principles and Practice* [includes case studies]. Malden, MA: Blackwell Science. 372p. *A.T.:* environmental management.

4153. Rapson, G.L., K. Thompson, & J.G. Hodgson, 1997. The humped relationship between species richness and biomass—testing its sensitivity to sample quadrat size. *J. of Ecology* 85: 99-100.

4154. Rastogi, Neelkamal, 1996. Crucial role of the landscape "matrix" in determining biodiversity within fragmented habitats [editorial]. *Current Science* 70: 199-200. [1].

4155. Ratcliffe, Philip R., & G.F. Peterken, 1995. The potential for biodiversity in British upland spruce forests. *Forest Ecology and Management* 79: 153-160. [1] *A.T.:* landscape ecology; plantations.

4156. Ratter, J.A., J.F. Ribeiro, & S. Bridgewater, 1997. The Brazilian cerrado vegetation and threats to its biodiversity. *Annals of Botany* 80: 223-230. *A.T.:* human impact.

4157. Raup, David M., 1972. Taxonomic diversity during the Phanerozoic. *Science* 177: 1065-1071. [1].

4158. ———, 1975. Taxonomic diversity estimation using rarefaction. *Paleobiology* 1: 333-342. [1] *A.T.:* echinoderms.

4159. ———, 1976. Species diversity in the Phanerozoic: a tabulation. *Paleobiology* 2: 279-288. [1] *A.T.:* invertebrates.

4160. ———, 1976. Species diversity in the Phanerozoic: an interpretation. *Paleobiology* 2: 289-297. [1] Relates known diversities to volume/area of sedimentary rocks of past ages.

4161. ———, 1979. Size of the Permo-Triassic bottleneck and its evolutionary implications. *Science* 206: 217-218. [2] *A.T.:* mass extinctions.

4162. ———, 1985. Magnetic reversals and mass extinctions. *Nature* 314: 341-343. [1].

4163. ———, 1986. Biological extinction in Earth history. *Science* 231: 1528-1533. [2] Considers both the proposed causes of extinction, and whether it has a Darwinian function.

4164. ———, 1986. *The Nemesis Affair: A Story of the Death of Dinosaurs and the Ways of Science.* New York: Norton. 220p. [3/3/1].

4165. ———, 1989. The case for extraterrestrial causes of extinction. *PTRSL B* 325: 421-435. [1].

4166. ———, 1991. *Extinction: Bad Genes or Bad Luck?* New York: Norton. 210p. [3/3/2].

4167. ———, 1991. A kill curve for Phanerozoic marine species. *Paleobiology* 17: 37-48. [2] *A.T.:* extinction risk; mass extinctions.

4168. ———, 1992. Large-body impact and extinction in the Phanerozoic. *Paleobiology* 18: 80-88. [1] *A.T.:* kill curves; marine species.

4169. ———, 1994. The role of extinction in evolution. *PNAS* 91: 6758-

6763. [1].

4170. Raup, David M., & G.E. Boyajian, 1988. Patterns of generic extinction in the fossil record. *Paleobiology* 14: 109-125. [2] *A.T.:* mass extinctions; Phanerozoic; environmental stress.

4171. Raup, David M., & David Jablonski, eds., 1986. *Patterns and Processes in the History of Life* [workshop papers]. Berlin and New York: Springer. 447p. [1/1/1] *A.T.:* evolution; paleobiology.

4172. Raup, David M., & J.J. Sepkoski, Jr., 1982. Mass extinctions in the marine fossil record. *Science* 215: 1501-1503. [2].

4173. ———, 1984. Periodicity of extinctions in the geologic past. *PNAS* 81: 801-805. [2] Finds a 26-million-year periodicity in major extinction episodes of marine taxa.

4174. ———, 1986. Periodic extinctions of families and genera. *Science* 231: 833-836. [2] More on the 26-million-year event interval.

4175. Raup, David M., S.J. Gould, et al., 1973. Stochastic models of phylogeny and the evolution of diversity. *J. of Geology* 81: 525-542. [2] *A.T.:* reptiles.

4176. Raustiala, Kal, 1997. Domestic institutions and international regulatory cooperation: comparative responses to the Convention on Biological Diversity. *World Politics* 49: 482-509. *A.T.:* environmental policy; U.S.; U.K.

4177. Raustiala, Kal, & D.G. Victor, May 1996. The future of the Convention on Biological Diversity. *Environment* 38(4): 16-20, 37-45. [1].

4178. Raven, John A., 1997. The role of marine biota in the evolution of terrestrial biota: gases and genes. *Biogeochemistry* 39: 139-164. *A.T.:* atmospheric gases.

4179. Raven, Peter H., Feb. 1981. Tropical rainforests: a global responsibility. *Natural History* 90(2): 28-32. [1] *A.T.:* deforestation.

4180. ———, Sept.-Oct. 1985. Disappearing species: a global tragedy. *Futurist* 19(5): 8-14. [1].

4181. ———, 1990. The politics of preserving biodiversity [address]. *Bioscience* 40: 769-774. [1] *A.T.:* environmental policy; human population growth.

4182. ———, Jan.-Feb. 1994. Defining biodiversity. *Nature Conservancy* 44(1): 10-15. [1].

4183. ———, Sept.-Oct. 1995. A time of catastrophic extinction: what we must do. *Futurist* 29(5): 38-41. [1] *A.T.:* bd loss; bd value.

4184. Raven, Peter H., & E.O. Wilson, 1992. A fifty-year plan for biodiversity surveys. *Science* 258: 1099-1100. [2] Concerning the need to increase our knowledge of bd.

4185. Rawat, G.S., 1997. Conservation status of forests and wildlife in the Eastern Ghats, India. *Environmental Conservation* 24: 307-315. *A.T.:* protected areas; bamboo.

4186. Raxworthy, C.J., 1988. Reptiles, rainforest and conservation in Madagascar. *Biological Conservation* 43: 181-211. [1].

4187. Ray, C., 1960. The application of Bergmann's and Allen's rules to the poikilotherms. *J. of Morphology* 106: 85-108. [2].

4188. Ray, G. Carleton, 1991. Coastal-zone biodiversity patterns. *Bioscience* 41: 490-498. [1] *A.T.:* landscape ecology.

4189. ———, 1996. Coastal-marine discontinuities and synergisms: implications for biodiversity conservation. *Biodiversity and Conservation* 5: 1095-1108. [1] *A.T.:* landscape ecology.

4190. Ray, G. Carleton, & J.F. Grassle, 1991. Marine biological diversity. *Bioscience* 41: 453-457. [1] *A.T.:* public awareness.

4191. Ray, G. Carleton, & W.P. Gregg, Jr., 1991. Establishing biosphere reserves for coastal barrier ecosystems. *Bioscience* 41: 301-309. [1] *A.T.:* coastal zone; barrier islands.

4192. Ray, G. Carleton, & M.G. McCormick, 1992. Functional coastal-marine biodiversity. *Trans. of the 57th North American Wildlife and Natural Resource Conference*: 384-397. [1].

4193. Raymond, Anne, & Cheryl Metz, 1995. Laurussian land-plant diversity during the Silurian and Devonian: mass extinction, sampling bias, or both? *Paleobiology* 21: 74-91. [1].

4194. Raymond, Chris, 12 Dec. 1990. Researchers see loss of cultural diversity in destruction of world's rain forests. *Chronicle of Higher Education* 37(15): A5, A8-A9. [1] *A.T.:* indigenous peoples.

4195. Raymond, Ruth D., summer 1996. Genetic resources help ensure world food supply. *Forum for Applied Research and Public Policy* 11(2): 139-142. [1] *A.T.:* CGIAR; crop germplasm resources; CBD.

4196. Razdan, M.K., & E.C. Cocking, eds., 1997. *Conservation of Plant Genetic Resources In Vitro*. Vol. 1. *General Aspects*. Enfield, NH: Science Publishers. 310p. [1/1/-] *A.T.:* plant germplasm resources.

4197. Rea, Val, June 1995. Gender: a vital issue in biodiversity. *Appropriate Technology* 22(1): 8-11. [1] *A.T.:* developing countries.

4198. Ready, Richard C., & R.C. Bishop, 1991. Endangered species and the safe minimum standard. *American J. of Agricultural Economics* 73: 309-312. [1] *A.T.:* game theory.

4199. Reaka-Kudla, Marjorie L., D.E. Wilson, & E.O. Wilson, eds., 1997. *Biodiversity II: Understanding and Protecting Our Biological Resources*. Washington, DC: Joseph Henry Press. 551p. [2/2/2].

4200. Real, Leslie A., 1996. Sustainability and the ecology of infectious disease. *Bioscience* 46: 88-97. [1] *A.T.:* ecosystem management.

4201. Recher, Harry F., 1969. Bird species diversity and habitat diversity in Australia and North America. *American Naturalist* 103: 75-80. [1].

4202. Recher, Harry F., & L. Lim, 1990. A review of current ideas of the extinction, conservation and management of Australia's terrestrial vertebrate fauna. *Proc. of the Ecological Society of Australia* 16: 287-301. [1].

4203. Recher, Harry F., Daniel Lunney, & Irina Dunn, eds., 1986. *A Natural Legacy: Ecology in Australia* (2nd ed.). Sydney and New York: Pergamon Press. 443p. [1/1/1] *A.T.:* nature conservation.

4204. Recher, Harry F., J.D. Majer, & S. Ganesh, 1996. Eucalypts, arthropods and birds: on the relation between foliar nutrients and species richness. *Forest Ecology and Management* 85: 177-195. [1] *A.T.:* New South Wales; canopy species.

4205. Redford, Kent H., 1992. The empty forest. *Bioscience* 42: 412-422. [2] Draws attention to the defaunation problems attending forest disturbance. *A.T.:* hunting.

4206. Redford, Kent H., & J.A. Mansour, eds., 1996. *Traditional Peoples and Biodiversity Conservation in Large Tropical Landscapes* [workshop papers]. Arlington, VA: Nature Conservancy. 267p. [1/1/1] *A.T.:* indigenous peoples; Latin America.

4207. Redford, Kent H., & Christine Padoch, eds., 1992. *Conservation of Neotropical Forests: Working From Traditional Resource Use* [workshop papers]. New York: Columbia University Press. 475p. [1/2/2] *A.T.:* indigenous peoples; Latin America.

4208. Redford, Kent H., & S.E. Sanderson, 1992. The brief, barren marriage of biodiversity and sustainability? [letter] *Bull. of the Ecological Society of America* 73: 36-39. [1].

4209. Redford, Kent H., & A.M. Stearman, 1993. Forest-dwelling native Amazonians and the conservation of biodiversity: interests in common or in collision? *CB* 7: 248-255. [2] *A.T.:* indigenous peoples.

4210. Redford, Kent H., Andrew Taber, & J.A. Simonetti, 1990. There is more to biodiversity than the tropical rain forests. *CB* 4: 328-330. [1] Complains that other ecosystems deserve as much conservation planning attention.

4211. Redmond, Peter G., spring-summer 1997. Protecting the islands in the sky: lessons from conserving the high alpine communities. *Adirondack J. of Environmental Studies* 4(1): 19-23. [1] *A.T.:* New York; montane environments.

4212. Reduron, Jean-Pierre, winter 1997. Biodiversity in an urban setting. *Ecodecision*, no. 23: 64-67.

4213. Reed, Kaye E., & J.G. Fleagle, 1995. Geographic and climatic control of primate diversity. *PNAS* 92: 7874-7876. [1].

4214. Reed, Patrick C., compiler, 1990. *Preparing to Manage Wilderness in the 21st Century: Proceedings of the Conference*. Asheville, NC: General Technical Report SE No. 66, Southeastern Forest Experiment Station, USDA Forest Service. 173p. [1/1/1].

4215. Reed, Timothy M., 1987. Island birds and isolation: Lack revisited. *Biological J. of the Linnean Society* 30: 25-29. [1] *A.T.:* Bahamas; Gulf of

Guinea.

4216. Reeves, Randall R., Stephen Leatherwood, & W.F. Perrin, 1994. *Dolphins, Porpoises, and Whales: 1994-1998 Action Plan for the Conservation of Cetaceans.* Gland, Switzerland: IUCN. 91p. [1/1/2].

4217. Regal, P.J., 1993. The true meaning of "exotic species" as a model for genetically engineered organisms. *Experientia* 49: 225-234. [1].

4218. Regan, Tom, 1983. *The Case for Animal Rights.* Berkeley: University of California Press. 425p. [3/3/3] One of the key works in the literature of this movement. *A.T.:* bioethics; animal welfare.

4219. Regan, Tom, Gary Francione, & Ingrid Newkirk, Jan. 1992. Point/counterpoint: a movement's means create its ends [commentary]. *Animals' Agenda* 12(1): 40-45. Contrasts philosophy of animal welfare and animal rights.

4220. Rege, J.E.O., 1994. International Livestock Center preserves Africa's declining wealth of animal biodiversity. *Diversity* 10(3): 21-25. [1].

4221. Reice, Seth R., 1994. Nonequilibrium determinants of biological community structure. *American Scientist* 82: 424-435. [2] *A.T.:* ecosystem disturbance.

4222. Reichholf, Josef H., 1994. Biodiversity: why are there so many different species? *Universitas* 36: 42-51. [1] *A.T.:* natural selection; mutation; isolation.

4223. Reid, J.W., 1994. Latitudinal diversity patterns of continental benthic copepod species assemblages in the Americas. *Hydrobiologia* 292-293: 341-349. [1].

4224. Reid, Walter V., spring 1992. Toward a national biodiversity policy. *Issues in Science and Technology* 8(3): 59-65. [1].

4225. ———, 1992. Conserving life's diversity: can the extinction crisis be stopped? *Environmental Science & Technology* 26: 1090-1095. [1] *A.T.:* bd loss; environmental policy; international cooperation.

4226. ———, winter 1993. The economic realities of biodiversity. *Issues in Science and Technology* 10(2): 48-55. [1] Focuses on the need to conserve genetic resources.

4227. ———, 1993. Bioprospecting: a force for sustainable development. *Environmental Science & Technology* 27: 1730-1732. [1] *A.T.:* developing countries; environmental policy.

4228. ———, 1994. Formulating a future for diversity. *American Zoologist* 34: 165-171. [1] *A.T.:* bd value; environmental degradation.

4229. ———, July-Aug. 1995. Biodiversity and health: prescription for progress. *Environment* 37(6): 12-15, 35-39. [1] *A.T.:* international cooperation; environmental health.

4230. ———, Sept. 1997. Strategies for conserving biodiversity. *Environment* 39(7): 16-20, 39-43. *A.T.: Global Biodiversity Assessment.*

4231. ———, 1998. Biodiversity hotspots. *TREE* 13: 275-280. *A.T.:*

reserve selection.

4232. Reid, Walter V., & K.R. Miller, 1993 (1989). *Keeping Options Alive: The Scientific Basis for Conserving Biodiversity*. Washington, DC: WRI. 128p. [1/1/2].

4233. Reid, Walter V., & M.C. Trexler, 1991. *Drowning the National Heritage: Climate Change and U.S. Coastal Biodiversity*. Washington, DC: WRI. 48p. [1/1/1].

4234. ———, 1992. Responding to potential impacts of climate change on U.S. coastal biodiversity. *Coastal Management* 20: 117-142. [1] *A.T.*: sea-level change; coastal zone management.

4235. Reid, Walter V., C.V. Barber, & Antonio La Viña, 1995. Translating genetic resource rights into sustainable development: gene cooperatives, the biotrade, and lessons from the Philippines. *Plant Genetic Resources Newsletter*, no. 102: 1-17. [1].

4236. Reid, Walter V., S.A. Laird, et al., eds., 1993. *Biodiversity Prospecting: Using Genetic Resources for Sustainable Development*. Washington, DC: WRI. 341p. [1/1/2] *A.T.*: germplasm resources.

4237. Reid, Walter V., et al., 1993. *Biodiversity Indicators for Policymakers*. Washington, DC: WRI. 42p. [1/1/1] *A.T.*: environmental policy.

4238. Reimink, Ronald L., 1995. Teach biodiversity at the Bell. *American Biology Teacher* 57: 106-107. [1] *A.T.*: bd education.

4239. Reinthal, Peter N., 1993. Evaluating biodiversity and conserving Lake Malawi's cichlid fish fauna. *CB* 7: 712-718. [1] *A.T.*: endemic species.

4240. Reinthal, Peter N., & M.L.J. Stiassny, 1991. The freshwater fishes of Madagascar: a study of an endangered fauna with recommendations for a conservation strategy. *CB* 5: 231-243. [1].

4241. Rejmánek, Marcel, & J.M Randall, 1994. Invasive alien plants in California: 1993 summary and comparison with other areas in North America. *Madroño* 41: 161-177. [1].

4242. Rejmánek, Marcel, P.S. Ward, et al., 1994. Systematics and biodiversity [letter]. *TREE* 9: 228-229. [1].

4243. Remmert, Hermann, ed., 1994. *Minimum Animal Populations*. Berlin and New York: Springer. 156p. [1/1/1] Concerning the viability of small populations. *A.T.*: extinction; population biology.

4244. Remsen, J.V., Jr., 1994. Use and misuse of bird lists in community ecology and conservation. *Auk* 111: 225-227. [1] *A.T.*: sampling; surveys.

4245. ———, 1995. The importance of continued collecting of bird specimens to ornithology and bird conservation. *Bird Conservation International* 5: 145-180. [1] *A.T.*: museum collections.

4246. Remsen, J.V., Jr., & T.A. Parker III, 1983. Contribution of river-created habitats to bird species richness in Amazonia. *Biotropica* 15: 223-231. [2].

4247. ———, 1995. Bolivia has the opportunity to create the planet's richest park for terrestrial biota. *Bird Conservation International* 5: 181-199. [1] *A.T.:* birds.

4248. Renaud, F., D. Clayton, & T. de Meeus, 1996. Biodiversity and evolution in host-parasite associations. *Biodiversity and Conservation* 5: 963-974. [1] *A.T.:* game theory.

4249. Rengifo, Antonio, 1997. Protection of marine biodiversity: a new generation of fisheries agreements. *Review of European Community & International Environmental Law* 6: 313-321.

4250. Rengifo Vásquez, Grimaldo, Dec. 1996. Culture and biodiversity in the Andes. *Development* (Rome) (4): 34-38. [1].

4251. Rennie, John, April 1992. A census of stranglers. *Scientific American* 266(4): 25. [1] Describes life cycle and special conservation needs of strangler figs.

4252. Repetto, Robert C., April 1990. Deforestation in the tropics. *Scientific American* 262(4): 36-42. [1] *A.T.:* environmental policy; developing countries.

4253. Repetto, Robert C., & Malcolm Gillis, eds., 1988. *Public Policies and the Misuse of Forest Resources* [case studies]. Cambridge, U.K., and New York: CUP. 432p. [2/2/2] *A.T.:* developing countries; environmental policy; deforestation.

4254. Rest, Alfred, 1994. The OPOSA decision: implementing the principles of intergenerational equity and responsibility. *Environmental Policy and Law* 24: 314-320. [1] *A.T.:* Philippines; environmental law; logging.

4255. Retallack, G.J., 1995. Permian-Triassic life crisis on land. *Science* 267: 77-80. [2] *A.T.:* Australia; paleobotany.

4256. Reveret, Jean-Pierre, & Catherine Potvin, winter 1997. Loss of biological diversity and international agreements. *Ecodecision,* no. 23: 31-33. [1].

4257. Rex, Michael A., 1997. An oblique slant on deep-sea biodiversity [editorial]. *Nature* 385: 577-578. Reports on paleoecological research casting doubt on theories of ocean-floor environment stability.

4258. Rex, Michael A., & R.J. Etter, 1998. Bathymetric patterns of body size: implications for deep-sea biodiversity. *Deep-Sea Research, Part II,* 45: 103-127. *A.T.:* North Atlantic; gastropods.

4259. Rex, Michael A., C.T. Stuart, et al., 1993. Global-scale latitudinal patterns of species diversity in the deep-sea benthos. *Nature* 365: 636-639. [2] *A.T.:* Atlantic.

4260. Reynolds, J.W., 1994. Earthworms of the world. *Global Biodiversity* 4(1): 11-16. [1].

4261. Reysenbach, Anna-Louise, March 1997. Yellowstone's biological resources [comment]. *Environment* 39(2): 5, 34. Considers the worthiness of conserving microorganisms.

4262. Reznick, David, R.J. Baxter, & John Endler, 1994. Long-term studies of tropical stream fish communities: the use of field notes and museum collections to reconstruct communities of the past. *American Zoologist* 34: 452-462. [1] *A.T.:* Costa Rica; Trinidad.

4263. Rhind, J., 1993. Managing environmental data: the biodiversity map library. *Mapping Awareness and GIS in Europe* 7(2): 3-7. [1] *A.T.:* GIS; World Conservation Monitoring Centre.

4264. Rhoades, Robert E., 1994. Indigenous people and the preservation of biodiversity. *HortScience* 29: 1222-1225. [1].

4265. Rhodes, M.C., & C.W. Thayer, 1991. Mass extinctions: ecological selectivity and primary production. *Geology* 19: 877-880. [1] *A.T.:* Cretaceous-Tertiary boundary; bivalves; brachiopods.

4266. Rhodes, O.E., Jr., & R.K. Chesser, 1994. Genetic concepts for habitat conservation: the transfer and maintenance of genetic variation. *Landscape and Urban Planning* 28: 55-62. [1] *A.T.:* effective population size.

4267. Rhymer, Judith M., & D.S. Simberloff, 1996. Extinction by hybridization and introgression. *ARES* 27: 83-109. [2] *A.T.:* gene flow.

4268. Ricciardi, A., R.J. Neves, & J.B. Rasmussen, 1998. Impending extinctions of North American fresh water mussels (Unionoida) following the zebra mussel (*Dreissena polymorpha*) invasion. *J. of Animal Ecology* 67: 613-619.

4269. Ricciuti, Edward R., Nov.-Dec. 1981. Invaders in paradise. *International Wildlife* 11(6): 30-35. [1] *A.T.:* biological invasions.

4270. Rice, Clifford G., 1997. *Biodiversity Survey Guidelines With Emphasis on Threatened and Endangered Species*. Champaign, IL: USACERL Technical Report 97-39, U.S. Army Corps of Engineers. 34p. [1/1/1] *A.T.:* environmental management.

4271. Rice, Kevin, 1992. Theory and conceptual issues. *Agriculture, Ecosystems & Environment* 42: 9-26. [1] Looks at parallels between bd studies in agricultural systems and natural systems.

4272. Richards, C., L.B. Johnson, & G.E. Host, 1996. Landscape-scale influences on stream habitats and biota. *Canadian J. of Fisheries and Aquatic Sciences* 53, Suppl. 1: 295-311. *A.T.:* Michigan.

4273. Richards, John F., & R.P. Tucker, eds., 1988. *World Deforestation in the Twentieth Century* [symposium papers]. Durham, NC: Duke University Press. 321p. [2/2/1].

4274. Richards, Michael, 1994. Towards valuation of forest conservation benefits in developing countries. *Environmental Conservation* 21: 308-319. [1].

4275. Richards, S.J., K.R. McDonald, & R.A. Alford, 1993. Declines in populations of Australia's endemic tropical rainforest frogs. *Pacific Conservation Biology* 1: 66-77. [2] *A.T.:* Queensland.

4276. Richardson, David M., 1998. Forestry trees as invasive aliens. *CB* 12: 18-26. *A.T.:* biological invasions; pine.

4277. Richardson, David M., & R.M. Cowling, 1993. Biodiversity and ecosystem processes: opportunities in Mediterranean-type ecosystems. *TREE* 8: 79-81. [1].

4278. Richardson, David M., P.A. Williams, & R.J. Hobbs, 1994. Pine invasions in the Southern Hemisphere: determinants of spread and invasibility. *J. of Biogeography* 21: 511-527. [1].

4279. Richardson, David M., B.W. Van Wilgen, et al., 1996. Current and future threats to plant biodiversity on the Cape Peninsula, South Africa. *Biodiversity and Conservation* 5: 607-647. *A.T.:* human impact; threatened plants; endemic plants.

4280. Richardson, J.A., 1998. Wildlife utilization and biodiversity conservation in Namibia: conflicting or complementary objectives? *Biodiversity and Conservation* 7: 549-559. *A.T.:* wildlife conservation.

4281. Richman, Adam D., T.J. Case, & T.D. Schwaner, 1988. Natural and unnatural extinction rates of reptiles on islands. *American Naturalist* 131: 611-630. [1] *A.T.:* South Australia; Baja California.

4282. Richter, Brian D., 1993. Ecosystem-level conservation at the Nature Conservancy: growing needs for applied research in conservation biology. *J. of the North American Benthological Society* 12: 197-200. [1] *A.T.:* research programs.

4283. Richter, D.D., & L.I. Babbar, 1991. Soil diversity in the tropics. *Advances in Ecological Research* 21: 315-389. [2].

4284. Ricklefs, Robert E., 1966. The temporal component of diversity among species of birds. *Evolution* 20: 235-242. [1].

4285. ———, 1987. Community diversity: relative roles of local and regional processes. *Science* 235: 167-171. [3] *A.T.:* community structure; species richness.

4286. Ricklefs, Robert E., & G.W. Cox, 1972. Taxon cycles in the West Indian avifauna. *American Naturalist* 106: 195-219. [1].

4287. Ricklefs, Robert E., & Dolph Schluter, eds., 1993. *Species Diversity in Ecological Communities: Historical and Geographical Perspectives.* Chicago: University of Chicago Press. 414p. [2/2/3] *A.T.:* community ecology; biogeography.

4288. Ridley, Matt, 4 Dec. 1993. Is sex good for anything? *New Scientist* 140(1902): 36-40. [1] *A.T.:* Red Queen hypothesis; sexual selection.

4289. Riede, Klaus, 1993. Monitoring biodiversity: analysis of Amazonian rainforest sounds. *Ambio* 22: 546-548. [1].

4290. Rieley, Jack, & S.E. Page, 1996. The biodiversity, environmental importance and sustainability of tropical peat and peatlands. *Environmental Conservation* 23: 94-95. [1] Reports on conference.

4291. ———, eds., 1997. *Biodiversity and Sustainability of Tropical Peatlands* [symposium proceedings]. Cardigan, Wales: Samara Pub. 370p. [1/1/-].

4292. Riffell, Samuel K., K.J. Gutzwiller, & S.H. Anderson, 1996. Does repeated human intrusion cause cumulative declines in avian richness and abundance? *Ecological Applications* 6: 492-505. [1] *A.T.:* disturbance ecology; Wyoming; wildland management.

4293. Riggs, L.A., 1990. Conserving genetic resources on-site in forest ecosystems. *Forest Ecology and Management* 35: 45-68. [1] *A.T.:* in situ conservation.

4294. Rinker, H. Bruce, M.D. Lowman, & M.W. Moffett, 1995. Africa from the treetops. *American Biology Teacher* 57: 393-401. [1] Describes a study of canopy bd in lowland rainforests between Cameroon and Equatorial Guinea. *A.T.:* bd education.

4295. Rinne, J.N., 1994. Declining Southwestern aquatic habitats and fishes: are they sustainable? *General Technical Report RM* GTR-247: 256-265. [1] *A.T.:* Arizona; human disturbance.

4296. Ripley, J. Douglas, & Michele Leslie, summer 1997. Conserving biodiversity on military lands. *Federal Facilities Environmental J.* 8(2): 93-105. [1] *A.T.:* ecosystem management.

4297. Risch, S.J., D.A. Andow, & M.A. Altieri, 1983. Agroecosystem diversity and pest control: data, tentative conclusions, and new research directions. *Environmental Entomology* 12: 625-629. [2].

4298. Risser, Paul G., 1995. The status of the science examining ecotones. *Bioscience* 45: 318-325. [1].

4299. ———, 1995. Biodiversity and ecosystem function. *CB* 9: 742-746. [1] *A.T.:* ecosystem dynamics.

4300. Rita, H., & E. Ranta, 1993. On analysing species incidence. *Annales Zoologici Fennici* 30: 173-176. [1] *A.T.:* species-area relation.

4301. Ritchie, Mark, Kristin Dawkins, & Mark Vallianatos, 1996. Intellectual property rights and biodiversity: the industrialization of natural resources and traditional knowledge. *St. John's J. of Legal Commentary* 11: 431-453. [1] *A.T.:* developing countries; patents.

4302. Robbins, C.S., J.R. Sauer, et al., 1989. Population declines in North American birds that migrate to the Neotropics. *PNAS* 86: 7658-7662. [3] *A.T.:* Mexico; Neotropical migrant birds.

4303. Robbins, Robert K., 1992. Comparison of butterfly diversity in the Neotropical and Oriental regions. *J. of the Lepidopterists' Society* 46: 298-300. [1].

4304. Roberts, C.M., & N.V.C. Polunin, 1993. Marine reserves: simple solutions to managing complex fisheries? *Ambio* 22: 363-368. [1].

4305. Roberts, D., & H.M. Moore, 1997. Tentacular diversity in deep-sea deposit-feeding holothurians: implications for biodiversity in the deep sea. *Biodiversity and Conservation* 6: 1487-1505. *A.T.:* morphological diversity; biological disturbance hypothesis; stability-time hypothesis; H.L. Sanders.

4306. Roberts, Leslie, 1988. Hard choices ahead on biodiversity [editorial]. *Science* 241: 1759-1761. [1] *A.T.:* rainforests; deforestation.

4307. ———, 1988. Is there life after climate change? *Science* 242: 1010-1012. [1] *A.T.:* greenhouse effect; WWF.

4308. ———, 1989. Does the ozone hole threaten Antarctic life? *Science* 244: 288-289. [1].

4309. ———, 1992. Chemical prospecting: hope for vanishing ecosystems? [editorial] *Science* 256: 1142-1143. [1] *A.T.:* INBio; Merck; Costa Rica; medicinal plants.

4310. Roberts, Mark R., & F.S. Gilliam, 1995. Patterns and mechanisms of plant diversity in forested ecosystems: implications for forest management. *Ecological Applications* 5: 969-977. [1] *A.T.:* diversity theory; research needs.

4311. Robertson, Alistar I., & D.M. Alongi, eds., 1992. *Tropical Mangrove Ecosystems*. Washington, DC: American Geophysical Union. 329p. [1/1/1].

4312. Robertson, John, & Doug Calhoun, 1995. Treaty on Biological Diversity: ownership issues and access to genetic materials in New Zealand. *European Intellectual Property Review* 17: 219-224. [1].

4313. Robichaux, Robert H., G.D. Carr, et al., 1990. Adaptive radiation of the Hawaiian silversword alliance (Compositae-Madiinae): ecological, morphological, and physiological diversity. *Annals of the Missouri Botanical Garden* 77: 64-72. [1].

4314. Robins, C. Richard, 1991. Regional diversity among Caribbean fish species. *Bioscience* 41: 458-459. [1].

4315. Robinson, A.Y., 1991. Sustainable agriculture: the wildlife connection. *American J. of Alternative Agriculture* 6: 161-167. [1] *A.T.:* Fish and Wildlife Service; landscape diversity; agricultural practices.

4316. Robinson, Clare, 1996. Montserrat: biodiversity amidst adversity. *SGM Quarterly* 23: 77-79. [1] *A.T.:* volcanoes.

4317. Robinson, Gaden S., & K.R. Tuck, 1993. Diversity and faunistics of small moths (Microlepidoptera) in Bornean rainforest. *Ecological Entomology* 18: 385-393. [1].

4318. Robinson, George R., & J.F. Quinn, 1992. Habitat fragmentation, species diversity, extinction, and design of nature reserves. In S.K. Jain & L.W. Botsford, eds., *Applied Population Biology* (Dordrecht and Boston: Kluwer Academic): 223-248. [1].

4319. Robinson, George R., R.D. Holt, et al., 1992. Diverse and contrasting effects of habitat fragmentation. *Science* 257: 524-526. [2].

4320. Robinson, John G., 1993. The limits to caring: sustainable living and the loss of biodiversity. *CB* 7: 20-28. [2] *A.T.:* Caring for the Earth; sustainable development.

4321. Robinson, John G., & K.H. Redford, eds., 1991. *Neotropical Wildlife Use and Conservation*. Chicago: University of Chicago Press. 520p.

[1/2/1] *A.T.:* wild animal trade.

4322. Robinson, Michael H., March 1990. Phenomena, comment and notes. *Smithsonian* 20(12): 24-30. [1] Surveys the diversity of tropical ecosystems.

4323. ———, 1992. Global change, the future of biodiversity and the future of zoos. *Biotropica* 24: 345-352. [1] *A.T.:* environmental education.

4324. Robinson, Nicholas A., ed., 1993. *Agenda 21: Earth's Action Plan Annotated.* New York: Oceana Publications. 683p. [1/1/1] *A.T.:* sustainable development; environmental policy.

4325. Robinson, S.K., F.R. Thompson III, et al., 1995. Regional forest fragmentation and the nesting success of migratory birds. *Science* 267: 1987-1990. [3] *A.T.:* Midwest; cowbird parasitism; metapopulations.

4326. Rochefort, L., & F.I. Woodward, 1992. Effects of climate change and a doubling of CO_2 on vegetational diversity. *J. of Experimental Botany* 43: 1169-1180. [1].

4327. Rocheleau, Dianne E., 1995. Gender and biodiversity: a feminist political ecology perspective. *IDS Bull.* 26: 9-16. [1] *A.T.:* Kenya; gender imbalance.

4328. Rodd, Rosemary, 1990. *Biology, Ethics, and Animals.* Oxford, U.K.: Clarendon Press, and New York: OUP. 272p. [3/2/1] *A.T.:* bioethics; animal rights.

4329. Rodda, Gordon H., 1993. How to lie with biodiversity. *CB* 7: 959-960. [1].

4330. Rodda, Gordon H., & T.H. Fritts, 1992. The impact of the introduction of the colubrid snake *Boiga irregularis* on Guam's lizards. *J. of Herpetology* 26: 166-174. [1] *A.T.:* biological invasions.

4331. Roderick, G.K., & R.G. Gillespie, 1998. Speciation and phylogeography of Hawaiian terrestrial arthropods. *Molecular Ecology* 7: 519-531. *A.T.:* natural selection.

4332. Rodriguez, Jon Paul, D.L. Pearson, & R. Barrera, 1998. A test for the adequacy of bioindicator taxa: are tiger beetles (Coleoptera, Cicindelidae) appropriate indicators for monitoring the degradation of tropical forests in Venezuela? *Biological Conservation* 83: 69-76.

4333. Rodriguez, Jon Paul, W.M. Roberts, & Andy Dobson, March-April 1997. Where are endangered species found in the United States? *Endangered Species Update* 14(3-4): 1-4. [1] *A.T.:* geographical distribution; hotspots.

4334. Roelke, Melody E., J.S. Martenson, & S.J. O'Brien, 1993. The consequences of demographic reduction and genetic depletion in the endangered Florida panther. *Current Biology* 3: 340-350. [2].

4335. Roelofs, J.G.M., R. Bobbink, et al., 1996. Restoration ecology of aquatic and terrestrial vegetation on non-calcareous sandy soils in the Netherlands. *Acta Botanica Neerlandica* 45: 517-541. [2].

4336. Roemmich, Dean, & J.A. McGowan, 1995. Climatic warming and the decline of zooplankton in the California Current. *Science* 267: 1324-1326. [2] *A.T.:* nutrient upwelling.

4337. Roff, Derek A., 1992. *The Evolution of Life Histories: Theory and Analysis.* New York: Chapman & Hall. 535p. [2/2/3] *A.T.:* mathematical models; diversification.

4338. Rogers, Adam, 1993. *The Earth Summit: A Planetary Reckoning.* Los Angeles: Global View Press. 351p. [1/1/1] *A.T.:* international cooperation; UNCED.

4339. Rogers, Deborah L., & F.T. Ledig, eds., 1996. *The Status of Temperate North American Forest Genetic Resources* [report]. Davis, CA: Report No. 16, Genetic Resources Conservation Program, Division of Agriculture and Natural Resources, University of California. 85p. [1/1/1].

4340. Rogers, J.W., 1992. Sustainable development patterns: the Chesapeake Bay region. *Water Science and Technology* 26: 2711-2721. [1] *A.T.:* coastal zone management.

4341. Rogers, Kathleen, & J.A. Moore, 1995. Revitalizing the Convention on Nature Protection and Wild Life Preservation in the Western Hemisphere: might awakening a visionary but "sleeping" treaty be the key to preserving biodiversity and threatened natural areas in the Americas? *Harvard International Law J.* 36: 465-508. [1] *A.T.:* environmental policy; international law.

4342. Rogers, R.A., L.A. Rogers, et al., 1991. Native American biological diversity and the biogeographic influence of Ice Age refugia. *J. of Biogeography* 18: 623-630. [1].

4343. Rohde, Klaus, 1978. Latitudinal gradients in species diversity and their causes (in two parts). *Biologisches Zentralblatt* 97: 393-418. [1].

4344. ———, 1992. Latitudinal gradients in species diversity: the search for the primary cause. *Oikos* 65: 514-527. [2] *A.T.:* evolutionary speed.

4345. Rohde, Klaus, Maureen Heap, & David Heap, 1993. Rapoport's rule does not apply to marine teleosts and cannot explain latitudinal gradients in species richness. *American Naturalist* 142: 1-16. [2].

4346. Rohlf, Daniel J., 1989. *The Endangered Species Act: A Guide to Its Protections and Implementation.* Stanford, CA: Stanford Environmental Law Society. 207p. [1/1/1].

4347. ———, 1991. Six biological reasons why the Endangered Species Act doesn't work—and what to do about it. *CB* 5: 273-282. [2] *A.T.:* environmental policy.

4348. Rojas, Martha, 1992. The species problem and conservation: what are we protecting? *CB* 6: 170-178. [2] Considers how different approaches to the species concept affect conservation planning. *A.T.:* species definitions.

4349. Rolston, Holmes, III, 1985. Duties to endangered species. *Bioscience* 35: 718-726. [1] *A.T.:* environmental ethics.

4350. ———, 1988. *Environmental Ethics: Duties to and Values in the Natural World.* Philadelphia: Temple University Press. 391p. [2/2/2].

4351. ———, 1993. Rights and responsibilities on the home planet. *Zygon* 28: 425-439. [1] Distinguishes between individual and political rights and ecological realities. *A.T.:* environmental ethics; enviromental philosophy.

4352. ———, 1994. *Conserving Natural Value.* New York: Columbia University Press. 259p. [1/1/1] Includes a number of instructive case studies. *A.T.:* environmental ethics; environmental philosophy.

4353. ———, 1995. Environmental protection and an equitable international order: ethics after the Earth Summit. *Business Ethics Quarterly* 5: 735-752. [1] *A.T.:* UNCED; environmental economics.

4354. Romeo, John T., J.A. Saunders, & Pedro Barbosa, eds., 1996. *Phytochemical Diversity and Redundancy in Ecological Interactions* [conference proceedings]. New York: Plenum Press. 319p. [1/1/1] *A.T.:* plant ecology.

4355. Rookwood, P., 1995. Landscape planning for biodiversity. *Landscape and Urban Planning* 31: 379-385. [1] *A.T.:* California. *A.T.:* urban planning.

4356. Root, Terry, 1988. Environmental factors associated with avian distributional boundaries. *J. of Biogeography* 15: 489-505. [2] *A.T.:* North America.

4357. ———, 1988. Energy constraints on avian distributions and abundances. *Ecology* 69: 330-339. [2].

4358. Roper, M.M., 1993. Biological diversity of micro-organisms: an Australian perspective. *Pacific Conservation Biology* 1: 21-28. [1] *A.T.:* microbial diversity conservation.

4359. Rose, Deborah B., ed., 1995. *Country in Flames: Proceedings of the 1994 Symposium on Biodiversity and Fire in North Australia.* Canberra: Biodiversity Unit, Dept. of the Environment, Sport and Territories. 127p. [1/1/1].

4360. Rose, Roger, 1992. Economic aspects of conserving biological diversity. *Agriculture & Resources Quarterly* 4: 378-388. [1] *A.T.:* markets.

4361. Rosen, Donn E., 1975. A vicariance model of Caribbean biogeography. *Systematic Zoology* 24: 431-464. [2] *A.T.:* continental drift.

4362. ———, 1978. Vicariant patterns and historical explanation in biogeography. *Systematic Zoology* 27: 159-188. [2] *A.T.:* cladistics.

4363. Rosenberg, Daniel K., B.R. Noon, & C.E. Meslow, 1997. Biological corridors: form, function, and efficacy. *Bioscience* 47: 677-687. *A.T.:* source and sink patches; linear patches.

4364. Rosendal, G.K., 1995. The forest issue in post-UNCED international negotiations: conflicting interests and fora for reconciliation. *Biodiversity and Conservation* 4: 91-107. [1] *A.T.:* international cooperation; environmental policy.

4365. Rosenzweig, Cynthia, & Daniel Hillel, 1993. Agriculture in a greenhouse world. *Research & Exploration* 9: 208-221. [1] Examines ways that global climatic change would both help and hurt agriculture.

4366. Rosenzweig, Michael L., 1987. Habitat selection as a source of biological diversity. *Evolutionary Ecology* 1: 315-330. [1].

4367. ———, 1992. Species diversity gradients: we know more and less than we thought. *J. of Mammalogy* 73: 715-730. [2] *A.T.:* latitudinal diversity gradients; species-area curves.

4368. ———, 1995. *Species Diversity in Space and Time.* Cambridge, U.K., and New York: CUP. 436p. [2/2/3] *A.T.:* biogeography; diversification; paleobiology.

4369. Rossman, Amy Y., & D.F. Farr, 1997. Towards a virtual reality for plant-associated fungi in the United States and Canada. *Biodiversity and Conservation* 6: 739-751. *A.T.:* bd databases.

4370. Rosswall, Thomas, R.G. Woodmansee, & P.G. Risser, eds., 1988. *Scales and Global Change: Spatial and Temporal Variability in Biospheric and Geospheric Processes* [workshop proceedings]. Chichester, U.K., and New York: Wiley. 355p. [1/1/1].

4371. Roth, Dana S., Ivette Perfecto, & B. Rathcke, 1994. The effects of management systems on ground-foraging ant diversity in Costa Rica. *Ecological Applications* 4: 423-436. [1] *A.T.:* plantations; rainforests.

4372. Rothfritz, H., I. Jüttner, et al., 1997. Epiphytic and epilithic diatom communities along environmental gradients in the Nepalese Himalaya: implications for the assessment of biodiversity and water quality. *Archiv für Hydrobiologie* 138: 465-482. *A.T.:* bioindicators.

4373. Rothschild, David, ed., 1997. *Protecting What's Ours: Indigenous Peoples and Biodiversity.* Oakland, CA: South & Meso-American Indian Rights Center. 135p. [1/1/-] *A.T.:* environmental law.

4374. Rothschild, George, 1995. Feed the world! Managing global genetic resources. *Search* 26: 180-182. [1] *A.T.:* plant genetic resources.

4375. Rougier, C., & T. Lam Hoai, 1997. Biodiversity through two groups of microzooplankton in a coastal lagoon (Étang de Thau, France). *Vie et Milieu* 47: 387-394.

4376. Rowan, R., & N. Knowlton, 1995. Intraspecific diversity and ecological zonation in coral-algal symbiosis. *PNAS* 92: 2850-2853. [2].

4377. Rowland, Melanie J., 1992. Bargaining for life: protecting biodiversity through mediated agreements. *Environmental Law* 22: 503-527. [1] *A.T.:* endangered species.

4378. Roy, Jacques, James Aronson, & Francesco Di Castri, eds., 1995. *Time Scales of Biological Responses to Water Constraints: The Case of Mediterranean Biota.* Amsterdam: SPB Academic. 243p. [1/1/1] *A.T.:* moisture availability.

4379. Roy, Kaustuv, 1996. The roles of mass extinction and biotic interaction in large-scale replacements: a reexamination using the fossil record of stromboidean gastropods. *Paleobiology* 22: 436-452.

4380. Roy, Kaustuv, & Mike Foote, 1997. Morphological approaches to measuring biodiversity. *TREE* 12: 277-281. *A.T.:* morphological diversity.

4381. Roy, Kaustuv, David Jablonski, & J.W. Valentine, 1994. Eastern Pacific molluscan provinces and latitudinal diversity gradient: no evidence for "Rapoport's rule." *PNAS* 91: 8871-8874. [2].

4382. ———, 1996. Higher taxa in biodiversity studies: patterns from eastern Pacific marine molluscs. *PTRSL B* 351: 1605-1613. [1].

4383. Royte, Elizabeth, Sept. 1995. On the brink: Hawaii's vanishing species. *National Geographic* 188(3): 2-37. [1].

4384. Rubec, C.D.A., & G.O. Lee, eds., 1997. *Conserving Vitality and Diversity: Proceedings of the World Conservation Congress Workshop on Alien Invasive Species.* Ottawa: IUCN Canada. 96p. [1/1/1] *A.T.:* introduced species.

4385. Rubin, Steven M., & S.C. Fish, 1994. Biodiversity prospecting: using innovative contractual provisions to foster ethnobotanical knowledge, technology, and conservation. *Colorado J. of International Environmental Law and Policy* 5: 23-58. [1] *A.T.:* CBD.

4386. Rubin, Steven M., Jonathan Shatz, & Colleen Deegan, 1994. International conservation finance: using debt swaps and trust funds to foster conservation of biodiversity. *J. of Social, Political, and Economic Studies* 19: 21-43. [1].

4387. Rublúo, A., V. Chávez, & A.P. Martínez, 1993. Strategies for the recovery of endangered orchids and cacti through *in-vitro* culture. *Biological Conservation* 63: 163-169. [1].

4388. Rudel, Thomas K., & Bruce Horowitz, 1993. *Tropical Deforestation: Small Farmers and Land Clearing in the Ecuadorian Amazon.* New York: Columbia University Press. 234p. [2/2/2].

4389. Rudnicky, Tamia C., & M.L. Hunter, 1993. Reversing the fragmentation perspective: effects of clearcut size on bird species richness in Maine. *Ecological Applications* 3: 357-366. [1].

4390. Ruesink, Jennifer L., I.M. Parker, et al., 1995. Reducing the risks of nonindigenous species introductions. *Bioscience* 45: 465-477. [2] *A.T.:* introduced species; biological invasions.

4391. Ruggiero, L.F., G.D. Hayward, & J.R. Squires, 1994. Viability analysis in biological evaluations: concepts of population viability analysis, biological population, and ecological scale. *CB* 8: 364-372. [1].

4392. Ruhl, J.B., 1995. Biodiversity conservation and the ever-expanding web of federal laws regulating nonfederal lands: time for something completely different? *University of Colorado Law Review* 66: 555-673. [2] *A.T.:* environmental policy; environmental law.

4393. Ruitenbeek, H.J., 1992. The rainforest supply price: a tool for evaluating rainforest conservation expenditures. *Ecological Economics* 6: 57-78. [1] *A.T.:* Cameroon; transfers.

4394. Rundel, Philip W., Gloria Montenegro Rizzardini, & F.M. Jaksic,

1998. *Landscape Disturbance and Biodiversity in Mediterranean-Type Ecosystems*. Berlin and New York: Springer. 447p. *A.T.:* landscape ecology.

4395. Runkle, James R., 1989. Synchrony of regeneration, gaps, and latitudinal differences in tree species diversity. *Ecology* 70: 546-547. [1] *A.T.:* treefall gaps.

4396. Rusch, G.M., & M. Oesterheld, 1997. Relationship between productivity, and species and functional group diversity in grazed and nongrazed pampas grassland. *Oikos* 78: 519-526. *A.T.:* primary production; Argentina.

4397. Russ, G.R., 1984. Distribution and abundance of herbivorous grazing fishes in the central Great Barrier Reef (in two parts). *Marine Ecology Progress Series* 20: 23-44. [2].

4398. Russell, Clifford S., 1995. Two propositions about biodiversity. *Vanderbilt J. of Transnational Law* 28: 689-693. [1] Argues against taking a safe minimum standard approach to bdc.

4399. Russell, Dale A., 1995. Biodiversity and time scales for the evolution of extraterrestrial intelligence. In G. Seth Shostak, ed., *Progress in the Search for Extraterrestrial Life* (San Francisco: Astronomical Society of the Pacific Conference Series 74): 143-151. [1].

4400. Russo, Anthony R., 1997. Epifauna living on sublittoral seaweeds around Cyprus. *Hydrobiologia* 344: 169-179.

4401. Rützler, Klaus, & I.C. Feller, March 1996. Caribbean mangrove swamps. *Scientific American* 274(3): 94-99. [1].

4402. Ruzicka, M., 1993. Biotopes mapping, base for research of biodiversity. *Ecology Bratislava* 12: 325-328. [1] *A.T.:* Slovak Republic; bd monitoring.

4403. Ryan, John C., March-April 1992. Conserving biological diversity. *American Forests* 98(3-4): 37-44. [1] *A.T.:* bd loss.

4404. ———, 1992. *Life Support: Conserving Biological Diversity*. Washington, DC: Worldwatch Paper No. 108. 62p. [2/1/1].

4405. Ryder, Graham, David Fastovsky, & Stefan Gartner, eds., 1996. *The Cretaceous-Tertiary Event and Other Catastrophes in Earth History*. Boulder, CO: Special Paper No. 307, Geological Society of America. 569p. [2/1/1] *A.T.:* mass extinction.

4406. Ryder, Oliver A., 1986. Species conservation and systematics: the dilemma of subspecies. *TREE* 1: 9-10. [2].

4407. Ryder, Oliver A., & A.T.C. Feistner, 1995. Research in zoos: a growth area in conservation. *Biodiversity and Conservation* 4: 671-677. [1] *A.T.:* captive breeding.

4408. Ryder, Richard D., 1989. *Animal Revolution: Changing Attitudes Towards Speciesism*. Oxford, U.K., and Cambridge, MA: Basil Blackwell. 385p. [2/2/1] *A.T.:* human-animal relations.

4409. Rydin, Hakan, & Sven-Olov Borgegard, 1988. Plant species richness on islands over a century of primary succession: Lake Hjälmaren. *Ecology* 69: 916-927. [1] *A.T.:* Sweden; succession; island biogeography.

4410. Rydin, Hakan, M. Diekmann, & T. Hallingback, 1997. Biological characteristics, habitat associations, and distribution of macrofungi in Sweden. *CB* 11: 628-640.

4411. Rykken, Jessica J., D.E. Capen, & S.P. Mahabir, 1997. Ground beetles as indicators of land type diversity in the Green Mountains of Vermont. *CB* 11: 522-530. *A.T.:* bioindicators.

4412. Ryman, N., Fred Utter, & Linda Laikre, 1995. Protection of intraspecific biodiversity of exploited fishes. *Reviews in Fish Biology and Fisheries* 5: 417-446. [2] *A.T.:* genetic diversity; fishery resources.

4413. Ryther, John H., 1969. Photosynthesis and fish production in the sea. *Science* 166: 72-76. [2] *A.T.:* primary production.

4414. Ryti, Randall T., 1992. Effect of the focal taxon on the selection of nature reserves. *Ecological Applications* 2: 404-410. [2] *A.T.:* Gulf of California; California; indicator species.

4415. Saberwal, V.K., 1996. Pastoral politics: Gaddi grazing, degradation, and biodiversity conservation in Himachal Pradesh, India. *CB* 10: 741-749. [1] *A.T.:* park-people relations; protected areas.

4416. Sabrosky, Curtis W., 1953. How many insects are there? *Systematic Zoology* 2: 31-36. [1].

4417. Sachs, Carolyn, summer 1992. Reconsidering diversity in agriculture and food systems: an ecofeminist approach. *Agriculture and Human Values* 9(3): 4-10. [1] *A.T.:* agricultural production.

4418. Sachs, Wolfgang, ed., 1993. *Global Ecology: A New Arena of Political Conflict*. London and Atlantic Highlands, NJ: Zed Books. 262p. [2/1/2] *A.T.:* UNCED; environmentalism.

4419. Sætersdal, Magne, H.J.B. Birks, & S.M. Peglar, 1998. Predicting changes in Fennoscandian vascular-plant species richness as a result of future climatic change. *J. of Biogeography* 25: 111-122. *A.T.:* greenhouse effect.

4420. Sætersdal, Magne, J.M. Line, & H.J.B. Birks, 1993. How to maximize biological diversity in nature reserve selection: vascular plants and breeding birds in deciduous woodlands, western Norway. *Biological Conservation* 66: 131-138. [2].

4421. Safford, R.J., & C.G. Jones, 1997. Did organochlorine pesticide use cause declines in Mauritian forest birds? *Biodiversity and Conservation* 6: 1445-1451. *A.T.:* local extinctions.

4422. Safina, Carl, Nov. 1995. The world's imperiled fish. *Scientific American* 273(5): 46-53. [1].

4423. Sagan, Dorion, 1990. *Biospheres: Metamorphosis of Planet Earth*. New York: McGraw-Hill. 198p. [2/2/1] *A.T.:* Gaia hypothesis.

4424. Sage, L., L. Bennasser, et al., 1997. Fungal microflora biodiversity as a function of pollution in Oued Sebou (Morocco). *Chemosphere* 35: 751-

759. *A.T.:* soil pollution.

4425. Sagoff, Mark, 1988. *The Economy of the Earth: Philosophy, Law, and the Environment.* Cambridge, U.K., and New York: CUP. 271p. [2/3/3] *A.T.:* environmental law; environmental policy.

4426. ———, 1993. Biodiversity and the culture of ecology. *Bull. of the Ecological Society of America* 74: 374-381. [1] Relates bd to ideas from the history of ecological studies. *A.T.:* Great Chain of Being.

4427. Sakko, A.L., 1998. The influence of the Benguela upwelling system on Namibia's marine biodiversity. *Biodiversity and Conservation* 7: 419-433.

4428. Sale, Peter F., 1977. Maintenance of high diversity in coral reef fish communities. *American Naturalist* 111: 337-359. [2].

4429. ———, ed., 1991. *The Ecology of Fishes on Coral Reefs.* San Diego: Academic Press. 754p. [2/2/2].

4430. Salo, Jukka, Risto Kalliola, et al., 1986. River dynamics and the diversity of Amazonian lowland forest. *Nature* 322: 254-258. [2] *A.T.:* Peru; geomorphological agents; succession; disturbance.

4431. Salwasser, Hal, 1990. Sustainability as a conservation paradigm [editorial]. *CB* 4: 213-216. [1].

4432. ———, 1990. Conserving biological diversity: a perspective on scope and approaches. *Forest Ecology and Management* 35: 79-90. [1] Concentrates on the matter of how to set goals for bdc. *A.T.:* Forest Service.

4433. ———, 1991. New perspectives for sustaining diversity in U.S. National Forest ecosystems. *CB* 5: 567-569. [1].

4434. ———, Aug. 1994. Ecosystem management: can it sustain diversity and productivity? *J. of Forestry* 92(8): 6-10. [2].

4435. Salwasser, Hal, A.T. Doyle, & W.B. Kessler, 1995. Biodiversity generates new views on old issues. *Forum for Applied Research and Public Policy* 10(2): 121-123. [1] *A.T.:* ecosystem management; environmental planning.

4436. Sample, V. Alaric, ed., 1994. *Remote Sensing and GIS in Ecosystem Management* [case studies]. Washington, DC: Island Press. 369p. [2/1/1].

4437. Samples, Karl C., J.A. Dixon, & M.M. Gowen, 1986. Information disclosure and endangered species valuation. *Land Economics* 62: 306-312. [1].

4438. Samson, Fred B., & F.L. Knopf, 1993. Managing biological diversity. *Wildlife Society Bull.* 21: 509-514. [1] Urges resource managers to attend to basic principles in their efforts at bdc. *A.T.:* biotic integrity.

4439. ———, 1994. Prairie conservation in North America. *Bioscience* 44: 418-421. [2].

4440. ———, eds., 1996. *Prairie Conservation: Preserving America's Most Endangered Ecosystem.* Washington, DC: Island Press. 339p. [2/1/1]

A.T.: ecosystem management; North America.

4441. Samu, F., & G.L. Loevei, 1995. Species richness of a spider community (Araneae): extrapolation from simulated increasing sampling effort. *European J. of Entomology* 92: 633-638. [1].

4442. Samuels, C.L., & J.A. Drake, 1997. Divergent perspectives on community convergence. *TREE* 12: 427-432. *A.T.:* community structure; convergent evolution.

4443. Samuelsson, Johan, Lena Gustafsson, & Torleif Ingelög, 1994. *Dying and Dead Trees: A Review of Their Importance for Biodiversity.* Uppsala, Sweden: Swedish Threatened Species Unit. 109p. [1/1/2] *A.T.:* forest ecology.

4444. Samways, Michael J., 1993. Insects in biodiversity conservation: some perspectives and directives. *Biodiversity and Conservation* 2: 258-282. [1].

4445. ———, 1994. *Insect Conservation Biology.* London and New York: Chapman & Hall. 358p. [1/1/2].

4446. ———, 1996. Insects on the brink of a major discontinuity. *Biodiversity and Conservation* 5: 1047-1058. [1] Speaks of influences on insect species range shifts and population fluctuations.

4447. ———, 1997. Classical biological control and biodiversity conservation: what risks are we prepared to accept? *Biodiversity and Conservation* 6: 1309-1316.

4448. Samways, Michael J., R. Osborn, & F. Carliel, 1997. Effect of a highway on ant (Hymenoptera: Formicidae) species composition and abundance, with a recommendation for roadside verge width. *Biodiversity and Conservation* 6: 903-913. *A.T.:* South Africa.

4449. Sánchez, Vicente, & Calestous Juma, eds., 1994. *Biodiplomacy: Genetic Resources and International Relations.* Nairobi: African Centre for Technology Studies. 370p. [1/1/1] *A.T.:* environmental law; international cooperation.

4450. Sanders, H.L., 1968. Marine benthic diversity: a comparative study. *American Naturalist* 102: 243-282. [3] *A.T.:* rarefaction; stability-time hypothesis; diversity indices.

4451. Sanderson, Michael J., Andy Purvis, & Chris Henze, 1998. Phylogenetic supertrees: assembling the trees of life. *TREE* 13: 105-109. *A.T.:* MRP.

4452. Sandlund, Odd T., & P.J. Schei, eds., 1993. *Proceedings of the Norway/UNEP Expert Conference on Biodiversity.* Trondheim, Norway: Directorate for Nature Management. 190p. [1/1/1] *A.T.:* marine fisheries.

4453. Sandlund, Odd T., & Åslaug Viken, eds., 1997. *Workshop on Freshwater Biodiversity* [summary report]. Trondheim, Norway: Directorate for Nature Management and Norwegian Institute for Nature Research. 46p. [1/1/1] *A.T.:* bdc.

4454. Sandlund, Odd T., Kjetil Hindar, & A.H.D. Brown, eds., 1992. *Conservation of Biodiversity for Sustainable Development*. Oslo: Scandinavian University Press. 324p. [1/1/1].

4455. Sandoy, S., & A.J. Romundstad, 1995. Liming of acidified lakes and rivers in Norway. An attempt to preserve and restore biological diversity in the acidified regions. *Water, Air, and Soil Pollution* 85: 997-1002. [1] *A.T.:* acidification; salmon.

4456. Sands, P., 1996. Microbial diversity and the 1992 Biodiversity Convention. *Biodiversity and Conservation* 5: 473-491. [1] Summarizes the history of regulation of bd, with an emphasis on microbial diversity. *A.T.:* international cooperation.

4457. Sans, Mónica, F.M. Salzano, & Renajit Chakraborty, 1997. Historical genetics in Uruguay: estimates of biological origins and their problems. *Human Biology* 69: 161-170. *A.T.:* human genetics.

4458. Santana C., Eduardo, 1991. Nature conservation and sustainable development in Cuba. *CB* 5: 13-16. [1].

4459. Santiapillai, Charles, & K.R. Ashby, eds., 1987. *Proceedings of the Symposium on the Conservation and Management of Endangered Plants and Animals*. Bogor, Indonesia: Southeast Asian Regional Center for Tropical Biology. 246p. [1/1/1].

4460. Sanyanga, Rudo A., 1995. Management of the Lake Kariba inshore fishery and some thoughts on biodiversity and conservation issues, Zimbabwe. *Environmental Conservation* 22: 111-116. [1].

4461. Sarkar, Sahotra, 1996. Ecological theory and anuran declines. *Bioscience* 46: 199-207. [1] *A.T.:* literature reviews.

4462. Sarre, Philip, 1995. Towards global environmental values: lessons from Western and Eastern experience. *Environmental Values* 4: 115-127. [1].

4463. Sasal, P., S. Morand, & J.-F. Guégan, 1997. Determinants of parasite species richness in Mediterranean marine fishes. *Marine Ecology Progress Series* 149: 61-71.

4464. Sasowsky, Ira D., D.W. Fong, & E.L. White, eds., 1997. *Conservation and Protection of the Biota of Karst* [symposium proceedings]. Charles Town, WV: Karst Waters Institute. 118p. [1/1/-].

4465. Sattaur, Omar, 29 July 1989. The shrinking gene pool. *New Scientist* 123(1675): 37-41. [1] *A.T.:* crop germplasm resources; gene banks.

4466. Saunders, Denis A., & P.J. Curry, 1990. The impact of agricultural and pastoral industries on birds in the southern half of Western Australia: past, present and future. *Proc. of the Ecological Society of Australia* 16: 303-321. [1].

4467. Saunders, Denis A., & R.J. Hobbs, eds., 1991. *The Role of Corridors* (*Nature Conservation 2*) [workshop papers]. Chipping Norton, NSW: Surrey Beatty. 442p. [1/1/2] *A.T.:* Australia.

4468. Saunders, Denis A., J.L. Craig, & E.M. Mattiske, eds., 1996. *The Role of Networks (Nature Conservation 4)* [conference papers]. Chipping Norton, NSW: Surrey Beatty. 684p. [1/1/1] *A.T.:* Australia; citizen participation; conservation.

4469. Saunders, Denis A., R.J. Hobbs, & P.R. Ehrlich, eds., 1993. *The Reconstruction of Fragmented Ecosystems: Global and Regional Perspectives (Nature Conservation 3)* [workshop proceedings]. Chipping Norton, NSW: Surrey Beatty. 326p. [1/1/1] *A.T.:* habitat fragmentation; restoration ecology.

4470. Saunders, Denis A., R.J. Hobbs, & C.R. Margules, 1991. Biological consequences of ecosystem fragmentation: a review. *CB* 5: 18-32. [3] A major, widely cited, review.

4471. Saunders, Denis A., G.W. Arnold, et al., eds., 1987. *The Role of Remnants of Native Vegetation (Nature Conservation 1)* [conference papers]. Chipping Norton, NSW: Surrey Beatty. 410p. [1/1/1] *A.T.:* Australia; New Zealand; vegetation management.

4472. Saunier, Richard E., & R.A. Meganck, eds., 1995. *Conservation of Biodiversity and the New Regional Planning.* Washington, DC: Organization of American States. 150p. [1/1/1].

4473. Savage, Jay M., 1960. Evolution of a peninsular herpetofauna. *Systematic Zoology* 9: 184-212. [1] *A.T.:* peninsular effect; biogeography; Baja California.

4474. ———, 1982. The enigma of the Central American herpetofauna: dispersals or vicariance? *Annals of the Missouri Botanical Garden* 69: 464-547. [2] *A.T.:* biogeography.

4475. ———, 1995. Systematics and the biodiversity crisis. *Bioscience* 45: 673-679. [1] *A.T.:* research programs.

4476. Savidge, Julie A., 1987. Extinction of an island forest avifauna by an introduced snake. *Ecology* 68: 660-668. [2] *A.T.:* Guam; *Boiga irregularis*.

4477. Savitsky, Basil G., & T.E. Lacher, Jr., eds., 1998. *GIS Methodologies for Developing Conservation Strategies: Tropical Forest Recovery and Wildlife Management in Costa Rica.* New York: Columbia University Press. 242p.

4478. Savolainen, V., & J. Goudet, 1998. Rate of gene sequence evolution and species diversification in flowering plants: a re-evaluation. *PRSL B* 265: 603-607.

4479. Sawhill, John C., 1996. Creating biodiversity partnerships: the Nature Conservancy's perspective. *Environmental Management* 20: 789-792. [1].

4480. Sayer, Jeffrey A., 1991. *Rainforest Buffer Zones: Guidelines for Protected Area Managers.* Gland, Switzerland: IUCN. 94p. [1/1/1] *A.T.:* forest conservation.

4481. Sayer, Jeffrey A., & Simon Stuart, 1988. Biological diversity and tropical forests [comment]. *Environmental Conservation* 15: 193-194. [1] Opines that a relatively large proportion of tropical species might be preserved in a relatively small number and area of protected areas.

4482. Sayer, Jeffrey A., & T.C. Whitmore, 1991. Tropical moist forests: destruction and species extinction. *Biological Conservation* 55: 199-213. [2] *A.T.:* deforestation.

4483. Sayer, Jeffrey A., C.S. Harcourt, & N.M. Collins, eds., 1992. *The Conservation Atlas of Tropical Forests: Africa.* New York and London: Simon & Schuster. 288p. [2/2/2].

4484. Schaaf, C.D., 1994. The role of zoological parks in biodiversity conservation in the Gulf of Guinea islands. *Biodiversity and Conservation* 3: 962-968. [1] *A.T.:* Bioko; Equatorial Guinea; in situ conservation.

4485. Schall, Joseph J., & E.R. Pianka, 1978. Geographical trends in numbers of species. *Science* 201: 679-686. [2] *A.T.:* species density; U.S.; Australia; vertebrates.

4486. Scheel, D., T.L.S. Vincent, & G.N. Cameron, 1996. Global warming and the species richness of bats in Texas. *CB* 10: 452-464. [1].

4487. Scheiner, Samuel M., 1992. Measuring pattern diversity. *Ecology* 73: 1860-1867. [2] *A.T.:* bd measures.

4488. Scheiner, Samuel M., & C.A. Istock, 1994. Species enrichment in a transitional landscape, northern Lower Michigan. *Canadian J. of Botany* 72: 217-226. [1] *A.T.:* temperate forests.

4489. Scheiner, Samuel M., & J.M. Rey-Benayas, 1994. Global patterns of plant diversity. *Evolutionary Ecology* 8: 331-347. [1] *A.T.:* bd measures.

4490. Schelhas, John, & Russell Greenburg, eds., 1996. *Forest Patches in Tropical Landscapes.* Washington, DC: Island Press. 426p. [1/2/1] *A.T.:* forest management; forest ecology.

4491. Scheltema, Rudolf S., spring-summer 1996. Describing diversity: too many new species, too few taxonomists. *Oceanus* 39(1): 16-18. [1] *A.T.:* marine fauna.

4492. Schemske, Douglas W., B.C. Husband, et al., 1994. Evaluating approaches to the conservation of rare and endangered plants. *Ecology* 75: 584-606. [2].

4493. Scherf, Beate D., ed., 1995. *World Watch List for Domestic Animal Diversity* (2nd ed.). Rome: Food and Agriculture Organization of the United Nations. 769p. [1/1/1] *A.T.:* animal germplasm resources.

4494. Schieck, Jim, Marie Nietfeld, & J.B. Stelfox, 1995. Differences in bird species richness and abundance among three successional stages of aspen-dominated boreal forests. *Canadian J. of Zoology* 73: 1417-1431. [1] *A.T.:* Alberta.

4495. Schieck, Jim, Kenneth Lertzman, et al., 1995. Effects of patch size on birds in old-growth montane forests. *CB* 9: 1072-1084. [1] *A.T.:* British Columbia.

4496. Schindler, D.W., 1988. Effects of acid rain on freshwater ecosystems. *Science* 239: 149-157. [2] *A.T.:* acidification.

4497. ———, 1992. A view of NAPAP from north of the border. *Ecological Applications* 2: 124-130. [1] *A.T.:* acidification.

4498. Schiøtz, Arne, 1989. Conserving biological diversity: who is responsible? *Ambio* 18: 454-457. [1] Questions the ability of developing countries to conserve bd before becoming more economically solvent.

4499. Schlesinger, R.C., D.T. Funk, et al., 1994. Assessing changes in biological diversity over time. *Natural Areas J.* 14: 235-240. [1] *A.T.:* community structure; temperate forests; Indiana.

4500. Schlickeisen, Rodger, spring 1993. Saving species: a new approach. *Defenders* 68(2): 10-19. [1] *A.T.:* Endangered Species Act.

4501. ———, 1994. Protecting biodiversity for future generations: an argument for a constitutional amendment. *Tulane Environmental Law J.* 8: 181-221. [1].

4502. Schlosser, Isaac J., 1991. Stream fish ecology: a landscape perspective. *Bioscience* 41: 704-712. [2].

4503. Schmidt, K.F., 1997. "No-take" zones spark fisheries debate. *Science* 277: 489-491. *A.T.:* marine fishery stocks; marine reserves.

4504. Schnase, J.L., D.L. Kama, et al., 1997. The Flora of North America digital library: a case study in biodiversity database publishing. *J. of Network and Computer Applications* 20: 87-103.

4505. Schneider, Stephen H., & P.J. Boston, eds., 1991. *Scientists on Gaia* [conference papers]. Cambridge, MA: MIT Press. 433p. [2/2/1].

4506. Schnitzler, A., 1994. Conservation of biodiversity in alluvial hardwood forests of the temperate zone. The example of the Rhine Valley. *Forest Ecology and Management* 68: 385-398. [1] *A.T.:* France.

4507. Schnitzler, A., 1996. Comparison of landscape diversity in forests of the upper Rhine and the middle Loire floodplains (France). *Biodiversity and Conservation* 5: 743-758. [1].

4508. Schnoor, Jerald L., 1993. The Rio Earth Summit. *Environmental Science & Technology* 27: 18-22. [1] *A.T.:* UNCED.

4509. Schoch, William F., & Lawrence Strait, Dec. 1995. Biodiversity and Adirondack fishes. *New York State Conservationist* 50(3): 6-7. [1].

4510. Schoener, Thomas W., 1983. Rate of species turnover decreases from lower to higher organisms: a review of the data. *Oikos* 41: 372-377. [1].

4511. ———, 1991. Extinction and the nature of the metapopulation: a case system. *Acta Oecologica* 12: 53-75. [1] *A.T.:* Bahamas; species; lizards; island biogeography.

4512. Schoener, Thomas W., & D.H. Janzen, 1968. Notes on environmental determinants of tropical versus temperate insect size patterns. *American Naturalist* 102: 207-224. [1].

4513. Schoener, Thomas W., & D.A. Spiller, 1992. Is extinction rate related to temporal variability in population size? An empirical answer for orb spiders. *American Naturalist* 139: 1176-1207. [2].

4514. ———, 1996. Devastation of prey diversity by experimentally introduced predators in the field. *Nature* 381: 691-694. [1] *A.T.:* lizards.

4515. Schofield, E.K., 1989. Effects of introduced plants and animals on island vegetation: examples from the Galápagos Archipelago. *CB* 3: 227-238. [1].

4516. Schonewald-Cox, Christine M., & J.W. Bayless, 1986. The boundary model: a geographical analysis of design and conservation of nature reserves. *Biological Conservation* 38: 305-322. [2].

4517. Schonewald-Cox, Christine M., M. Buechner, et al., 1992. Cross-boundary management between national parks and surrounding lands: a review and discussion. *Environmental Management* 16: 273-282. [1].

4518. Schonewald-Cox, Christine M., S.M. Chambers, et al., eds., 1983. *Genetics and Conservation: A Reference for Managing Wild Animal and Plant Populations.* Menlo Park, CA: Benjamin/Cummings. 722p. [1/1/2] *A.T.:* germplasm resources.

4519. Schoonmaker, Peter K., Bettina von Hagen, & E.C. Wolf, eds., 1997. *The Rain Forests of Home: Profile of a North American Bioregion* [conference papers]. Washington, DC: Island Press. 431p. [2/-/-] *A.T.:* Pacific Northwest.

4520. Schopf, J. William, & Cornelis Klein, eds., 1992. *The Proterozoic Biosphere: A Multidisciplinary Study.* Cambridge, U.K., and New York: CUP. 1348p. [1/2/2].

4521. Schopf, Thomas J.M., 1974. Permo-Triassic extinctions: relation to sea-floor spreading. *J. of Geology* 82: 129-143. [1] *A.T.:* sea-level change.

4522. Schott, Gary W., & S.P. Hamburg, 1997. The seed rain and seed bank of an adjacent native tallgrass prairie and old field. *Canadian J. of Botany* 75: 1-7. *A.T.:* Kansas.

4523. Schreiber, Ramona, spring 1996. Barnacles and blennies: diversity in a dockside habitat. *Mariners Weather Log* 40(1): 24-25. [1].

4524. Schreurs, Miranda A., & Elizabeth Economy, eds., 1997. *The Internationalization of Environmental Protection* [papers from workshops]. Cambridge, U.K., and New York: CUP. 221p. [1/1/-] *A.T.:* environmental law.

4525. Schubert, Andreas, 1993. Conservation of biological diversity in the Dominican Republic. *Oryx* 27: 115-121. [1].

4526. Schulenberg, Thomas S., & Kim Awbrey, eds., 1997. *The Cordillera del Cóndor Region of Ecuador and Peru: A Biological Assessment.* Washington, DC: Conservation International. 231p. [1/-/-] *A.T.:* rapid assessment.

4527. Schulenberg, Thomas S., Kim Awbrey, & Glenda Fabregas, 1997. *A Rapid Assessment of the Humid Forests of South Central Chuquisaca,*

Bolivia. Washington, DC: Conservation International. 84p. [1/1/-].

4528. Schultes, Richard E., 1994. The importance of ethnobotany in environmental conservation. *American J. of Economics and Sociology* 53: 202-206. [1].

4529. Schultes, Richard E., & R.F. Raffauf, 1990. *The Healing Forest: Medicinal and Toxic Plants of the Northwest Amazonia.* Portland, OR: Dioscorides Press. 484p. [2/2/1] *A.T.:* medicinal plants; ethnobotany.

4530. Schulze, Ernst-Detlef, & H.A. Mooney, eds., 1993. *Biodiversity and Ecosystem Function.* Berlin and New York: Springer. 525p. [1/1/2] *A.T.:* bdc.

4531. Schupp, Eugene W., 1992. The Janzen-Connell model for tropical tree diversity: population implications and the importance of spatial scale. *American Naturalist* 140: 526-530. [1].

4532. Schwab, Francis E., & A.R.E. Sinclair, 1994. Biodiversity of diurnal breeding bird communities related to succession in the dry Douglas-fir forests of southeastern British Columbia. *Canadian J. of Forest Research* 24: 2034-2040. [1].

4533. Schwartz, Mark W., 1992. Potential effects of global climate change on the biodiversity of plants. *Forestry Chronicle* 68: 462-471. [1] *A.T.:* carbon dioxide.

4534. ———, 1993. The search for pattern among rare plants: are primitive plants more likely to be rare? *Biological Conservation* 64: 121-127. [1] *A.T.:* North America.

4535. ———, 1994. Conflicting goals for conserving biodiversity: issues of scale and value. *Natural Areas J.* 14: 213-216. [1] *A.T.:* insects; environmental perception.

4536. ———, ed., 1997. *Conservation in Highly Fragmented Landscapes.* New York: Chapman & Hall. 436p. [1/-/-] *A.T.:* Middle West.

4537. Schwarzkopf, Lin, & A.B. Rylands, 1989. Primate species richness in relation to habitat structure in Amazonian rainforest fragments. *Biological Conservation* 48: 1-12. [1].

4538. Schweitzer, Jeff, May-June 1992. Conserving biodiversity in developing countries. *Fisheries* 17(3): 35-38. [1] *A.T.:* international cooperation; USAID.

4539. Schweitzer, Jeff, F.G. Handley, et al., 1991. Summary of the workshop on Drug Development, Biological Diversity, and Economic Growth. *J. of the National Cancer Institute* 83: 1294-1298. [1].

4540. Schwilk, D.W., J.E. Keeley, & W.J. Bond, 1997. The intermediate disturbance hypothesis does not explain fire and diversity patterns in fynbos. *Plant Ecology* 132: 77-84. *A.T.:* South Africa; species-area curves.

4541. Scoble, Malcolm J., Oct.-Dec. 1997. Natural history museums and the biodiversity crisis: the case for Global Taxonomic Facility. *Museum International* 49(4): 55-59.

4542. ———, 1997. The transformation of systematics? *TREE* 12: 465-466. Reports on advances, problems and directions within the field of systematics.

4543. Scott, J. Michael, March 1990. Preserving life on earth: we need a new approach [editorial]. *J. of Forestry* 88(3): 13-14. [1] *A.T.:* endangered species.

4544. ———, 1994. Preserving and restoring avian diversity: a search for solutions. In Joseph R. Jehl, Jr., & N.K. Johnson, eds., *A Century of Avifaunal Change in Western North America* (Camarillo, CA: Cooper Ornithological Society): 340-348. [1].

4545. Scott, J. Michael, & J.W. Carpenter, 1987. Release of captive-reared or translocated endangered birds: what do we need to know? *Auk* 104: 544-545. [1].

4546. Scott, J. Michael, & M.D. Jennings, 1998. Large area mapping of biodiversity. *Annals of the Missouri Botanical Garden* 85: 34-47. *A.T.:* gap analysis.

4547. Scott, J. Michael, M.D. Jennings, & R.G. Wright, 1996. Landscape approaches to mapping biodiversity. *Bioscience* 46: 77-78. [1] *A.T.:* gap analysis.

4548. Scott, J. Michael, Cameron B. Kepler, & Charles Van Riper, 1988. Conservation of Hawaii's vanishing avifauna. *Bioscience* 38: 238-253. [1].

4549. Scott, J. Michael, T.H. Tear, & F.W. Davis, eds., 1996. *Gap Analysis: A Landscape Approach to Biodiversity Planning*. Bethesda, MD: American Society for Photogrammetry and Remote Sensing. 320p. [1/1/1].

4550. Scott, J. Michael, E.D. Ables, et al., 1995. Conservation of biological diversity: perspectives and the future for the wildlife profession [committee report]. *Wildlife Society Bull.* 23: 646-657. [1] *A.T.:* wildlife conservation.

4551. Scott, J. Michael, H. Anderson, et al., 1993. *Gap Analysis: A Geographic Approach to Protection of Biological Diversity*. Bethesda, MD: Wildlife Monographs No. 123, Wildlife Society. 41p. [1/1/3].

4552. Scott, J. Michael, Blair Csuti, et al., 1987. Species richness: a geographic approach to protecting future biological diversity. *Bioscience* 37: 782-788. [2] *A.T.:* GIS; extinction risk.

4553. Scott, J. Michael, Blair Csuti, et al., 1988. Beyond endangered species: an integrated conservation strategy for the preservation of biological diversity. *Endangered Species Update* 5(10): 43-48. [1].

4554. Scott, J. Michael, Blair Csuti, et al., 1989. Status assessment of biodiversity protection. *CB* 3: 85-87. [1].

4555. Scott, J. Michael, J.E. Estes, et al., 1991. An information systems approach to the preservation of biological diversity. In *GIS Applications in Natural Resources* (Fort Collins, CO: GIS World Books), Vol. 1: 283-293. [1].

4556. Scott, Norman J., Jr., 1976. The abundance and diversity of the herpetofaunas of tropical forest litter. *Biotropica* 8: 41-58. [1] *A.T.:* amphibians; lizards; Costa Rica; elevation gradients.

4557. Scott, Ronald R., 1995. Private land use controls and biodiversity preservation in Kentucky. *J. of Natural Resources & Environmental Law* 11: 281-301. [1].

4558. Scott, T., R.B. Standiford, & N. Pratini, Nov.-Dec. 1995. Private landowners critical to saving California biodiversity. *California Agriculture* 49(6): 50-54, 57. [1].

4559. Sebens, Kenneth P., 1994. Biodiversity of coral reefs: what are we losing and why? *American Zoologist* 34: 115-133. [1] *A.T.:* environmental perception.

4560. Seber, George A.F., 1982. *The Estimation of Animal Abundance and Related Parameters* (2nd ed.). New York: Macmillan. 654p. [2/1/3] *A.T.:* mathematical models; population biology.

4561. ———, 1986. A review of estimating animal abundance. *Biometrics* 42: 267-292. [2].

4562. Sedjo, Roger A., July 1989. Forests to offset the greenhouse effect. *J. of Forestry* 87(7): 12-15. [1].

4563. ———, 1992. Property rights, genetic resources, and biotechnological change. *J. of Law & Economics* 35: 199-213. [2].

4564. Seehausen, Ole, J.J.M. van Alphen, & Frans Witte, 1997. Cichlid fish diversity threatened by eutrophication that curbs sexual selection. *Science* 277: 1808-1811. [2] *A.T.:* Lake Victoria; water turbidity.

4565. Seeliger, Ulrich, Clarisse Odebrecht, & Jorge Pablo Castello, eds., 1997. *Subtropical Convergence Environments: The Coast and Sea in the Southwestern Atlantic.* Berlin and New York: Springer. 308p. [1/1/-] *A.T.:* marine environments.

4566. Seger, Jon, 1989. Diversity of little things. *Nature* 337: 305-306. [1] Describes implications of reproductive mode on diversity in thrips. *A.T.:* facultative vivaparity.

4567. Segerström, U., R. Bradshaw, et al., 1994. Disturbance history of a swamp forest refuge in northern Sweden. *Biological Conservation* 68: 189-196. [1] *A.T.:* refugia.

4568. Seidl, Peter R., ed., 1994. *The Use of Biodiversity for Sustainable Development: Investigation of Bioactive Products and Their Commercial Applications* [workshop proceedings]. Rio de Janeiro: Ministério da Ciência e Tecnologia, et al. 123p. [1/1/1] *A.T.:* Amazonia.

4569. Seidl, Peter R., O.R. Gottlieb, & M.A.C. Kaplan, eds., 1995. *Chemistry of the Amazon: Biodiversity, Natural Products, and Environmental Issues* [symposium papers]. Washington, DC: ACS Symposium Series No. 588. 315p. [1/1/1].

4570. Seitz, Alfred, & Volker Loeschcke, eds., 1991. *Species Conserva-*

tion: A Population-Biological Approach [symposium papers]. Basel and Boston: Birkhauser Verlag. 281p. [1/1/1].

4571. Sekgororoane, G.B., & T.G. Dilworth, 1995. Relative abundance, richness, and diversity of small mammals at induced forest edges. *Canadian J. of Zoology* 73: 1432-1437. [1] *A.T.:* edge effect; New Brunswick; ecotones.

4572. Sekhran, Nikhil, & Scott Miller, eds., 1995. *Papua New Guinea Country Study on Biological Diversity.* Waigani, Papua New Guinea: Conservation Resource Centre, Dept. of Environment and Conservation. 438p. [1/1/1].

4573. Selva, S.B., 1994. Lichen diversity and stand continuity in the northern hardwoods and spruce-fir forests of northern New England and western New Brunswick. *Bryologist* 94: 424-429. [2] *A.T.:* endophytes.

4574. Semlitsch, Raymond D., 1998. Biological delineation of terrestrial buffer zones for pond-breeding salamanders. *CB* 12: 1113-1119.

4575. Sepkoski, J. John, Jr., 1976. Species diversity in the Phanerozoic: species-area effects. *Paleobiology* 2: 298-303. [1] *A.T.:* invertebrates.

4576. ———, 1978. A kinetic model of Phanerozoic taxonomic diversity. I. Analysis of marine orders. *Paleobiology* 4: 223-251. [2] *A.T.:* invertebrates.

4577. ———, 1979. A kinetic model of Phanerozoic taxonomic diversity. II. Early Phanerozoic families and multiple equilibria. *Paleobiology* 5: 222-251. [2] *A.T.:* marine faunas; invertebrates.

4578. ———, 1984. A kinetic model of Phanerozoic taxonomic diversity. III. Post-Paleozoic families and mass extinctions. *Paleobiology* 10: 246-267. [2] *A.T.:* marine faunas; invertebrates.

4579. ———, 1988. Alpha, beta, or gamma: where does all the diversity go? *Paleobiology* 14: 221-234. [1].

4580. ———, 1989. Periodicity in extinction and the problem of catastrophism in the history of life. *J. of the Geological Society of London* 146: 7-19. [2] *A.T.:* marine faunas.

4581. ———, 1991. A model of onshore-offshore change in faunal diversity. *Paleobiology* 17: 58-77. [2] *A.T.:* Paleozoic; cladistics; extinction.

4582. ———, 1997. Biodiversity: past, present, and future [address]. *J. of Paleontology* 71: 533-539. [2] *A.T.:* diversification; phylogenetic analysis.

4583. ———, 1998. Rates of speciation in the fossil record. *PTRSL B* 353: 315-326. *A.T.:* recovery from extinction.

4584. Sepkoski, J. John, Jr., R.K. Bambach, et al., 1981. Phanerozoic marine diversity and the fossil record. *Nature* 293: 435-437. [2].

4585. Serafin, Rafal, 1988. Noosphere, Gaia, and the science of the biosphere. *Environmental Ethics* 10: 121-137. [1].

4586. Serena, Melody, ed., 1995. *Reintroduction Biology of Australian and New Zealand Fauna* [conference papers]. Chipping Norton, NSW: Surrey Beatty. 264p. [1/1/1] *A.T.:* endangered species.

4587. Sereno, Paul C., 1997. The origin and evolution of dinosaurs. *Annual Review of Earth and Planetary Sciences* 25: 435-489. *A.T.:* diversification; paleobiogeography.

4588. Serra, M., A. Galiana, & A. Gomez, 1997. Speciation in monogonont rotifers. *Hydrobiologia* 358: 63-70. *A.T.:* intraspecific variation.

4589. Service, Robert F., 1997. Microbiologists explore life's rich, hidden kingdoms. *Science* 275: 1740-1742. *A.T.:* ribosomal sequencing.

4590. Seshadri, Balakrishna, 1997. The dying of biodiversity. *Contemporary Review* 270: 241-246. Describes in general terms the current threats to bd. *A.T.:* human disturbance; bd loss.

4591. Sessions, G., 1987. The deep ecology movement: a review. *Environmental Review* 11: 105-125. [1] An historical look.

4592. Severin, Timothy, 1998. *The Spice Islands Voyage: The Quest for Alfred Wallace, the Man Who Shared Darwin's Discovery of Evolution.* New York: Carroll & Graf. 267p. [2/2/-] *A.T.:* Malay Archipelago.

4593. Seymour, Jane, 12 March 1994. No way to treat a national treasure. *New Scientist* 141(1916): 32-35. [1] Expresses concern over the deteriorating condition of worldwide biological specimen collections. *A.T.:* museums.

4594. Shafer, Craig L., 1990. *Nature Reserves: Island Theory and Conservation Practice.* Washington, DC: Smithsonian Institution Press. 189p. [2/2/2] *A.T.:* protected areas.

4595. ———, 1995. Values and shortcomings of small reserves. *Bioscience* 45: 80-89. [1] *A.T.:* bd education; reserve size.

4596. Shaffer, Mark L., 1981. Minimum population sizes for species conservation. *Bioscience* 31: 131-134. [3].

4597. ———, 1990. Population viability analysis. *CB* 4: 39-40. [1].

4598. Shaffer, Mark L., & F.B. Samson, 1985. Population size and extinction: a note on determining critical population sizes. *American Naturalist* 125: 144-152. [2] *A.T.:* protected areas; reserve design; SLOSS.

4599. Shaltout, K.H., E.F. El-Halawany, & M.M. El-Garawany, 1997. Coastal lowland vegetation of eastern Saudi Arabia. *Biodiversity and Conservation* 6: 1027-1040. *A.T.:* desert plants.

4600. Shand, Hope, 1997. *Human Nature: Agricultural Biodiversity and Farm-based Food Security* [study]. Ottawa: Rural Advancement Foundation International. 94p. [1/1/1] *A.T.:* food supply.

4601. Shands, Henry L., Sept. 1994. Who owns the world's plant genes? [editorial] *Agricultural Research* 42(9): 2. [1] Considers the international origins of agricultural crops.

4602. Shands, William E., & J.S. Hoffman, eds., 1987. *The Greenhouse Effect, Climate Change, and U.S. Forests.* Washington, DC: Conservation Foundation. 304p. [1/1/1].

4603. Shankar, U., S.D. Lama, & K.S. Bawa, 1998. Ecosystem reconstruction through taungya plantations following commercial logging of a dry, mixed deciduous forest in Darjeeling, Himalaya. *Forest Ecology and Management* 102: 131-142. *A.T.:* forest regeneration; India.

4604. Shannon, Daniel E., 1995. A criticism of a false idealism and onward to Hegel: objections to the Gaia hypothesis. *Owl of Minerva* 27: 19-36. [1].

4605. Sharitz, Rebecca R., & J.W. Gibbons, eds., 1989. *Freshwater Wetlands and Wildlife* [symposium proceedings]. Washington, DC: U.S. Department of Energy. 1265p. [1/1/1] *A.T.:* wetlands conservation.

4606. Sharma, Narendra P., ed., 1992. *Managing the World's Forests: Looking for Balance Between Conservation and Development.* Dubuque, IA: Kendall/Hunt. 605p. [1/1/2] *A.T.:* environmental policy.

4607. Sharma, Shalendra D., 1996. Building effective international environmental regimes: the case of the Global Environment Facility. *J. of Environment & Development* 5: 73-86. [1] *A.T.:* developing countries.

4608. Sharp, Robin, 1995. Bird conservation and the U.K. biodiversity action plan. *Ibis* 137, Suppl. 1: S219-S223. [1].

4609. Sharpton, Virgil L., & P.D. Ward, eds., 1990. *Global Catastrophes in Earth History; An Interdisciplinary Conference on Impacts, Volcanism, and Mass Mortality* [conference papers]. Boulder, CO: Special Paper No. 247, Geological Society of America. 631p. [2/1/1] *A.T.:* mass extinction.

4610. Shaw, Scott R., 1995. The biodiversity crisis: a new challenge for entomological teaching. *American Entomologist* 41: 134-135. [1] *A.T.:* bd education.

4611. Sheehan, William, 1986. Response by specialist and generalist natural enemies to agroecosystem diversification: a selective review. *Environmental Entomology* 15: 456-461. [2].

4612. Sheely, D.L., & T.R. Meagher, 1996. Genetic diversity in Micronesian island populations of the tropical tree *Campnosperma brevi-petiolata* (Anacardiaceae). *American J. of Botany* 83: 1571-1579. [1].

4613. Sheil, D., 1996. Species richness, tropical forest dynamics and sampling: questioning cause and effect? *Oikos* 76: 587-590. [1].

4614. ———, 1997. Further notes on species richness, tropical forest dynamics and sampling: a reply to Phillips et al. *Oikos* 79: 188-190.

4615. Shelby, Byron B., & Sandie Arbogast, compilers, 1995. *Management & Biological Conservation* [lectures]. Corvallis, OR: College of Forestry, Oregon State University. 44p. [1/1/1] *A.T.:* ecosystem management.

4616. Sheldon, A.L., 1988. Conservation of stream fishes: patterns of diversity, rarity and risk. *CB* 2: 149-156. [1].

4617. Shen, Susan, 1987. Biological diversity and public policy. *Bioscience* 37: 709-712. [1] Concerning the Office of Technology Assessment's report on bd.

4618. Shepard, W.D., 1993. Desert springs—both rare and endangered.

Aquatic Conservation: Marine and Freshwater Ecosystems 3: 351-359. [3] *A.T.:* biological invasions; western U.S.

4619. Shepherd, Ursula L., 1998. A comparison of species diversity and morphological diversity across the North American latitudinal gradient. *J. of Biogeography* 25: 19-29. *A.T.:* latitudinal diversity gradients; body size; mammals; body shape.

4620. Sherman, Kenneth, 1994. Sustainability, biomass yields, and health of coastal ecosystems: an ecological perspective. *Marine Ecology Progress Series* 112: 277-301. [2] *A.T.:* UNCED; coastal zone management.

4621. Sherman, Kenneth, L.M. Alexander, & B.D. Gold, eds., 1993. *Large Marine Ecosystems: Stress, Mitigation, and Sustainability* [conference papers]. Washington, DC: AAAS Press. 376p. [1/1/1].

4622. Shields, A.O., 1989. World numbers of butterflies. *J. of the Lepidopterists' Society* 43: 178-183. [1].

4623. Shilling, Fraser, 1997. Do habitat conservation plans protect endangered species? *Science* 276: 1662-1663. *A.T.:* California; Endangered Species Act; environmental policy; recovery plans.

4624. Shipp, Diana, ed., 1993. *Loving Them to Death? Sustainable Tourism in Europe's Nature and National Parks*. Grafenau, Germany: Federation of Nature and National Parks of Europe. 96p. [1/1/1] *A.T.:* ecotourism.

4625. Shirayama, Yoshihisa, 1998. Biodiversity and biological impact of ocean disposal of carbon dioxide. *Waste Management* 17: 381-384. *A.T.:* deep-sea species; waste dumping.

4626. Shiva, Vandana, 1990. Biodiversity, biotechnology and profit: the need for a people's plan to protect biological diversity. *Ecologist* 20: 44-47. [1] *A.T.:* action plans; World Bank; genetic resources.

4627. ———, 1991. Biotechnology development and conservation of biodiversity. *Economic and Political Weekly* 26: 2740-2746. [1].

4628. ———, 1993. Farmers' rights, biodiversity and international treaties. *Economic and Political Weekly* 28: 555-560. [1].

4629. ———, 1993. *Monocultures of the Mind: Perspectives on Biodiversity and Biotechnology*. London and Atlantic Highlands, NJ: Zed Books. 184p. [1/1/2] *A.T.:* monoculture; developing countries.

4630. ———, ed., 1994. *Biodiversity Conservation: Whose Resource? Whose Knowledge?* New Delhi: Indian National Trust for Art and Cultural Heritage [seminar papers]. 315p. [1/1/1] *A.T.:* India; intellectual property.

4631. ———, 1995. *Captive Minds, Captive Lives: Essays on Ethical and Ecological Implications of Patents on Life*. Dehra Dun: Research Foundation for Science, Technology and Natural Resource Policy. 212p. [1/1/1] *A.T.:* intellectual property; environmental law.

4632. ———, 1997. *Biopiracy: The Plunder of Nature and Knowledge*. Boston: South End Press. 148p. [2/-/-] *A.T.:* patents.

4633. Shiva, Vandana, & Vanaja Ramprasad, 1993. *Cultivating Diversity:*

Biodiversity Conservation and the Politics of the Seed. Dehra Dun: Research Foundation for Science, Technology and Natural Resource Policy. 130p. [1/1/1] *A.T.:* food crops; seed technology; India.

4634. Shiva, Vandana, P. Anderson, et al., 1991. *Biodiversity: Social and Ecological Perspectives* [essays]. London and Atlantic Highlands, NJ: Zed Books. 123p. [1/1/1] *A.T.:* local participation.

4635. Shiva, Vandana, A.H. Jafri, et al., 1997. *The Enclosure and Recovery of the Commons: Biodiversity, Indigenous Knowledge, and Intellectual Property Rights.* New Delhi: Research Foundation for Science, Technology, and Ecology. 182p. [1/1/1].

4636. Shiva, Vandana, Padmini Krishnan, et al., 1995. *The Seed Keepers.* New Delhi: Navdanya. 156p. [1/1/1] *A.T.:* India; agrobiodiversity conservation.

4637. Shmida, A., & M.J.A. Werger, 1992. Growth form diversity on the Canary Islands. *Vegetatio* 102: 183-199. [1] *A.T.:* endemism; morphological diversity; plants.

4638. Shmida, A., & M.V. Wilson, 1985. Biological determinants of species diversity. *J. of Biogeography* 12: 1-20. [2] *A.T.:* species-area relation; alpha, beta and gamma diversity.

4639. Shorrocks, Bryan, & I.R. Swingland, eds., 1990. *Living in a Patchy Environment.* Oxford, U.K., and New York: OUP. 246p. [1/1/2] *A.T.:* population biology; patch dynamics.

4640. Short, Henry L., & J.B. Hestbeck, 1995. National biotic resource inventories and GAP analysis: problems of scale and unproven assumptions limit a national program. *Bioscience* 45: 535-539. [1] *A.T.:* GIS.

4641. Short, Jeff, & Andrew Smith, 1994. Mammal decline and recovery in Australia. *J. of Mammalogy* 75: 288-297. [2] *A.T.:* wildlife conservation; extinction.

4642. Short, Jeff, S.D. Bradshaw, et al., 1992. Reintroduction of macropods (Marsupialia: Macropodoidea) in Australia: a review. *Biological Conservation* 62: 189-204. [2] *A.T.:* endangered species; wildlife conservation.

4643. Short, J.M., 1997. Recombinant approaches for accessing biodiversity. *Nature Biotechnology* 15: 1322-1323. *A.T.:* biotechnology; recombinant DNA.

4644. Shrader-Frechette, Kristin S., & E.D. McCoy, 1994. Biodiversity, biological uncertainty, and setting conservation priorities. *Biology & Philosophy* 9: 167-195. [1] Asks what criteria conservationists should enlist to make recommendations. *A.T.:* Florida panther; environmental ethics.

4645. Shukla, J., C. Nobre, & P. Sellers, 1990. Amazon deforestation and climate change. *Science* 247: 1322-1325. [2] Looks at the effects of deforestation on climate. *A.T.:* Brazil.

4646. Shulman, Myra J., 1983. Species richness and community predictability in coral reef fish faunas. *Ecology* 64: 1308-1311. [1].

4647. Shvarts, E.A., & M.A. Vaisfeld, 1995. The politics and unpredictable consequences of island transfers for the protection of endangered species: an example from Russia (the European mink, *Mustela lutreola*, on Kunashir Island). *J. of the Royal Society of New Zealand* 25: 314-325. [1] *A.T.:* translocation; Kuril Islands.

4648. Shvarts, E.A., S.V. Pushkaryov, et al., 1995. Geography of mammal diversity and searching for ways to predict global changes in biodiversity. *J. of Biogeography* 22: 907-914. [1] *A.T.:* U.S.S.R.; North America.

4649. Shyamsundar, Priya, & Randall Kramer, 1997. Biodiversity conservation—at what cost? A study of households in the vicinity of Madagascar's Mantadia National Park. *Ambio* 26: 180-184. *A.T.:* local residents; cost-benefit analysis.

4650. Sibanda, B.M.C., & A.K. Omwega, 1996. Some reflections on conservation, sustainable development and equitable sharing of benefits from wildlife in Africa: the case of Kenya and Zimbabwe. *South African J. of Wildlife Research* 26: 175-181. [1] *A.T.:* wildlife conservation; local participation.

4651. Sibley, Charles G., & J.E. Ahlquist, 1990. *Phylogeny and Classification of Birds: A Study in Molecular Evolution.* New Haven: Yale University Press. 976p. [2/2/3].

4652. Sibley, Charles G., & B.L. Monroe, Jr., 1990. *Distribution and Taxonomy of Birds of the World.* New Haven: Yale University Press. 1111p. [2/2/3].

4653. Sibuet, Myriam, & Karine Olu, 1998. Biogeography, biodiversity and fluid dependence of deep-sea cold-seep communities at active and passive margins. *Deep-Sea Research, Part II,* 45: 517-567. *A.T.:* hydrothermal vents; invertebrates.

4654. Siemann, Evan, David Tilman, & John Haarstad, 1996. Insect species diversity, abundance and body size relationships. *Nature* 380: 704-706. [1] *A.T.:* grasslands.

4655. Siepel, Henk, 1996. The importance of unpredictable and short-term environmental extremes for biodiversity in oribatid mites. *Biodiversity Letters* 3: 26-34. [1] *A.T.:* Netherlands; environmental influences.

4656. ———, 1996. Biodiversity of soil microarthropods: the filtering of species. *Biodiversity and Conservation* 5: 251-260. [1] *A.T.:* species loss.

4657. Signor, Philip W., 1990. The geological history of diversity. *ARES* 21: 509-539. [2] *A.T.:* fossil record; Phanerozoic.

4658. ———, 1994. Biodiversity in geological time. *American Zoologist* 34: 23-32. [1] *A.T.:* fossil record; Phanerozoic.

4659. Sihanya, Bernard M., 1994. Technology transfer, intellectual property rights and biosafety: strategies for implementing the Convention on Biodiversity. *AgBiotech News and Information* 6: 53N-60N. [1] *A.T.:* CBD; Kenya.

4660. Silbaugh, John M., & D.R. Betters, 1997. Biodiversity values and measures applied to forest management. *J. of Sustainable Forestry* 5: 235-248. *A.T.:* ecosystem management.

4661. Silva, Nelson Jorge da, Jr., & J.W. Sites, Jr., 1995. Patterns of diversity of Neotropical squamate reptile species with emphasis on the Brazilian Amazon and the conservation potential of indigenous reserves. *CB* 9: 873-901. [1].

4662. Silva, Paul C., 1992. Geographic patterns of diversity in benthic marine algae. *Pacific Science* 46: 429-437. [1] *A.T.:* genetic diversity.

4663. Silver, Leon T., & P.H. Schultz, eds., 1982. *Geological Implications of Impacts of Large Asteroids and Comets on the Earth* [conference papers]. Boulder, CO: Special Paper No. 190, Geological Society of America. 528p. [2/1/1].

4664. Silver, Whendee L., Sandra Brown, & A.E. Lugo, 1996. Effects of changes in biodiversity on ecosystem function in tropical forests. *CB* 10: 17-24. [2] *A.T.:* nutrient cycling.

4665. Simberloff, Daniel S., 1970. Taxonomic diversity of island biotas. *Evolution* 24: 23-47. [1].

4666. ———, 1974. Permo-Triassic extinctions: effects of area on biotic equilibrium. *J. of Geology* 82: 267-274. [1] *A.T.:* invertebrates; marine environment.

4667. ———, 1974. Equilibrium theory of island biogeography and ecology. *ARES* 5: 161-182. [2].

4668. ———, 1988. The contribution of population and community biology to conservation science. *ARES* 19: 473-511. [3] *A.T.:* habitat fragmentation; conservation genetics; island biogeography.

4669. ———, 1995. Habitat fragmentation and population extinction of birds. *Ibis* 137, Suppl. 1: S105-S111. [1] *A.T.:* metapopulations; local extinction.

4670. ———, 1998. Flagships, umbrellas, and keystones: is single-species management passé in the landscape era? *Biological Conservation* 83: 247-257. *A.T.:* bd monitoring.

4671. Simberloff, Daniel S., & L.G. Abele, 1976. Island biogeography theory and conservation practice. *Science* 191: 285-286. [2].

4672. ———, 1982. Refuge design and island biogeographic theory: effects of fragmentation. *American Naturalist* 120: 41-50. [2] *A.T.:* SLOSS.

4673. Simberloff, Daniel S., & William Boecklen, 1991. Patterns of extinction in the introduced Hawaiian avifauna: a reexamination of the role of competition. *American Naturalist* 138: 300-327. [1].

4674. Simberloff, Daniel S., & J. Cox, 1987. Consequences and costs of conservation corridors. *CB* 1: 63-71. [2].

4675. Simberloff, Daniel S., & Nicholas Gotelli, 1984. Effects of insularization on plant species richness in the prairie-forest ecotone. *Biological*

Conservation 29: 27-46. [2] *A.T.:* island biogeography.

4676. Simberloff, Daniel S., & B. Levin, 1985. Predictable sequences of species loss with decreasing island area—land birds in two archipelagoes. *New Zealand J. of Ecology* 8: 11-20. [1] *A.T.:* New Zealand; Cyclades Archipelago; island biogeography.

4677. Simberloff, Daniel S., & J.-L. Martin, 1991. Nestedness of insular avifaunas: simple summary statistics masking complex species patterns. *Ornis Fennica* 68: 178-192. [2] *A.T.:* island biogeography.

4678. Simberloff, Daniel S., J.A. Farr, et al., 1992. Movement corridors: conservation bargains or poor investments? *CB* 6: 493-504. [2] *A.T.:* environmental planning.

4679. Sime, K.R., & A.V.Z. Brower, 1998. Explaining the latitudinal gradient anomaly in ichneumonoid species richness: evidence from butterflies. *J. of Animal Ecology* 67: 387-399. *A.T.:* host-parasite relations.

4680. Simkin, Tom, & R.S. Fiske, eds., 1983. *Krakatau 1883—The Volcanic Eruption and Its Effects*. Washington, DC: Smithsonian Institution Press. 464p. [2/2/2].

4681. Simmons, John B., R.I. Beyer, et al., eds., 1976. *Conservation of Threatened Plants* [conference proceedings]. New York: Plenum Press. 336p. [1/1/1] *A.T.:* botanical gardens.

4682. Simmons, M.T., & R.M. Cowling, 1996. Why is the Cape Peninsula so rich in plant species? An analysis of the independent diversity components. *Biodiversity and Conservation* 5: 551-573. [1] *A.T.:* fynbos; South Africa.

4683. Simmons, Norman M., M.M.R. Freeman, & Julian Inglis, eds., 1987. *Proceedings of the Symposium on Research and Monitoring in Circumpolar Biosphere Reserves*. Edmonton, Alberta: Boreal Institute for Northern Studies. 75p. [1/1/1].

4684. Simon, Detlef L., & Doris Buchenauer, eds., 1993. *Genetic Diversity of European Livestock Breeds: Results of Monitoring by the EAAP Working Group on Animal Genetic Resources*. Wageningen, Netherlands: Wageningen Press. 581p. [1/1/1].

4685. Simon, Hans-Reiner, 1996. From bibliometrics to diversity—a personal view. *J. of Information Science* 22: 457-461. [1] *A.T.:* systematics literature.

4686. Simon, Julian L., 15 May 1986. Disappearing species, deforestation, and data. *New Scientist* 110(1508): 60-63. [1].

4687. Simon, Julian L., & Herman Kahn, eds., 1984. *The Resourceful Earth: A Response to Global 2000*. Oxford, U.K., and New York: Basil Blackwell. 585p. [2/3/1] *A.T.:* environmental economics.

4688. Simon, Noel, 1995. *Nature in Danger: Threatened Habitats and Species*. New York: OUP. 240p. [2/2/1] *A.T.:* nature conservation; endangered species.

4689. Simonetti, Javier A., 1998. Networking and Iberoamerican biodiversity. *TREE* 13: 337. Describes a bd evaluation program operating in Latin American and the Iberian Peninusula.

4690. Simpson, Beryl B., & Joel Cracraft, 1995. Systematics: the science of biodiversity. *Bioscience* 45: 670-672. [1].

4691. Simpson, Beryl B., & Jürgen Haffer, 1978. Speciation patterns in the Amazonian forest biota. *ARES* 9: 497-518. [1].

4692. Simpson, George Gaylord, 1944. *Tempo and Mode in Evolution*. New York: Columbia University Press. 237p. [2/-/3].

4693. ———, 1953. *The Major Features of Evolution*. New York: Columbia University Press. 434p. [2/-/3].

4694. ———, 1960. Notes on the measurement of faunal resemblance. *American J. of Science* 258A: 300-311. [1].

4695. ———, 1964. Species density of North American Recent mammals. *Systematic Zoology* 13: 57-73. [2].

4696. ———, 1980. *Splendid Isolation: The Curious History of South American Mammals*. New Haven: Yale University Press. 266p. [2/2/2] *A.T.:* zoogeography; paleobiogeography.

4697. Simpson, I.A., & Peter Dennis, eds., 1996. *The Spatial Dynamics of Biodiversity: Towards an Understanding of Spatial Patterns & Processes in the Landscape* [conference proceedings]. Aberdeen, Scotland: International Association for Landscape Ecology. 200p. [1/1/1] *A.T.:* landscape ecology.

4698. Simpson, R. David, 1997. Biodiversity prospecting: shopping the wilds is not the key to conservation. *Resources*, no. 126: 12-15. *A.T.:* bd value; environmental economics.

4699. Simpson, R. David, & R.A. Sedjo, Oct. 1994. Commercialization of indigenous genetic resources. *Contemporary Economic Policy* 12(4): 34-44. [1] *A.T.:* technology transfers; bd value.

4700. Simpson, R. David, R.A. Sedjo, & J.W. Reid, 1994. *Valuing Biodiversity: An Application to Genetic Prospecting*. Washington, DC: Discussion Paper 94-20, Resources for the Future. 27p. [1/1/1].

4701. ———, 1996. Valuing biodiversity for use in pharmaceutical research. *J. of Political Economy* 104: 163-185. [2] *A.T.:* bd valuation; bioprospecting.

4702. Sinclair, A.R.E., 1997. Fertility control of mammal pests and the conservation of endangered marsupials. *Reproduction, Fertility, and Development* 9: 1-16. *A.T.:* Australia.

4703. Sinclair, A.R.E., D.S. Hik, et al., 1995. Biodiversity and the need for habitat renewal. *Ecological Applications* 5: 579-587. [1] *A.T.:* habitat loss.

4704. Sinclair, A.R.E., R.P. Pech, et al., 1998. Predicting effects of predation on conservation of endangered prey. *CB* 12: 564-575. *A.T.:* Australia.

4705. Sindermann, Carl J., March-April 1986. Strategies for reducing risks from introductions of aquatic organisms: a marine perspective. *Fisheries* 11(2): 10-15. [1] *A.T.:* environmental management.

4706. Singer, Peter, 1975. *Animal Liberation: A New Ethics for Our Treatment of Animals.* New York: New York Review. 301p. [3/3/3] Perhaps the single most important title in the literature of the animal rights movement.

4707. Singh, J.S., A.S. Raghubanshi, & C.K. Varshney, 1994. Integrated biodiversity research for India [editorial]. *Current Science* 66: 109-112. [1].

4708. Singh, Mahesh P., & Vinita Vishwakarma, 1997. *Forest Environment and Biodiversity.* Delhi: Daya Publishing House. 427p. [1/1/-] *A.T.:* India.

4709. Singh, Shekhar, 1997. *Biodiversity Conservation Through Ecodevelopment Planning and Implementation Lessons From India.* Paris: Working Paper No. 21, South-South Cooperation Programme on Environmentally Sound Socio-Economic Development in the Humid Tropics. 64p. [1/1/-].

4710. Singh, Surendra P., B.S. Adhikari, & D.B. Zobel, 1994. Biomass, productivity, leaf longevity, and forest structure in the central Himalaya. *Ecological Monographs* 64: 401-421. [1] *A.T.:* India; Nepal; elevational gradients.

4711. Sinha, Rajiv K., 1995. Biodiversity conservation through faith and tradition in India: some case studies. *International J. of Sustainable Development and World Ecology* 2: 278-284. [1] *A.T.:* nature worship; sacred groves.

4712. ———, 1995. Sustainable utilisation and conservation of biodiversity by the tribal societies (aborigines) of India: a lesson for modern man. *Environmental Education and Information* 14: 195-204. [1] *A.T.:* indigenous peoples; local knowledge.

4713. ———, 1997. The state of biodiversity in India: some conservation strategies. *Environmental Education and Information* 16: 159-176. *A.T.:* local knowledge; indigenous peoples; bdc.

4714. ———, 1997. *Global Biodiversity: The Library of Life and the Secret of Human Existence on Earth.* Jaipur, India: INA Shree Publishers. 396p. [1/1/-] *A.T.:* bdc; India.

4715. Sinha, Suresh K., 1997. Global change scenario: current and future with reference to land-cover changes and sustainable agriculture: South and South-East Asian context. *Current Science* 72: 846-854.

4716. Sioli, Harold, 1987. The effects of deforestation in Amazonia. *Ecologist* 17: 134-138. [1] *A.T.:* nutrient cycling.

4717. Sisk, Thomas D., & B.R. Noon, fall 1995. Land Use History of North America: an emerging project of the National Biological Service. *Park Science* 15(4): 21. [1].

4718. Sisk, Thomas D., A.E. Launer, et al., 1994. Identifying extinction threats: global analyses of the distribution of biodiversity and the expansion of the human enterprise. *Bioscience* 44: 592-604. [2] *A.T.:* human population growth.

4719. Sitarz, Dan, ed., 1993. *AGENDA 21: The Earth Summit Strategy to Save Our Planet.* Boulder, CO: EarthPress. 321p. [2/2/1] *A.T.:* UNCED; environmental policy; environmental management.

4720. Sites, Jack W., Jr., & K.A. Crandall, 1997. Testing species boundaries in biodiversity studies. *CB* 11: 1289-1297. *A.T.:* species concepts; *Chelydra serpentina*; research methodology.

4721. Sittenfeld, A., & R. Villers, 1993. Exploring and preserving biodiversity in the tropics: the Costa Rican case. *Current Opinion in Biotechnology* 4: 280-285. [1] *A.T.:* INBio.

4722. Skaggs, R.W., & W.J. Boecklen, 1996. Extinctions of montane mammals reconsidered: putting a global-warming scenario on ice. *Biodiversity and Conservation* 5: 759-778. [1] *A.T.:* island biogeography; disperal barriers.

4723. Skelly, David K., & Eli Meir, 1997. Rule-based models for evaluating mechanisms of distributional change. *CB* 11: 531-538. *A.T.:* amphibians; Michigan.

4724. Skelton, Paul H., J.A. Cambray, et al., 1995. Patterns of distribution and conservation status of freshwater fishes in South Africa. *South African J. of Zoology* 30: 71-81. [1] *A.T.:* hotspots; endemism.

4725. Skole, David, & Compton Tucker, 1993. Tropical deforestation and habitat fragmentation in the Amazon: satellite data from 1978 to 1988. *Science* 260: 1905-1910. [3] *A.T.:* remote sensing.

4726. Skov, Flemming, 1997. Stand and neighbourhood parameters as determinants of plant species richness in a managed forest. *J. of Vegetation Science* 8: 573-578. *A.T.:* Denmark.

4727. Skov, Flemming, & F. Borchsenius, 1997. Predicting plant species distribution patterns using simple climatic parameters: a case study of Ecuadorian palms. *Ecography* 20: 347-355. *A.T.:* GIS.

4728. Slack, Nancy G., 1992. Rare and endangered bryophytes in New York State and eastern United States: current status and preservation strategies. *Biological Conservation* 59: 233-241. [1] *A.T.:* Natural Heritage Inventory.

4729. Slesnick, Irwin L., Brad Williamson, et al., 1997. *Biodiversity: Can We Balance Resource Conservation With Economic Growth?* Arlington, VA: National Science Teachers Association. 64p. [1/1/-] *A.T.:* bd education.

4730. Slocombe, D. Scott, 1998. Defining goals and criteria for ecosystem-based management. *Environmental Management* 22: 483-493. *A.T.:* goal-based management.

4731. Slowinski, Joseph B., & Craig Guyer, 1989. Testing the stochasticity of patterns of organismal diversity: an improved null model. *American Naturalist* 134: 907-921. [2] *A.T.:* Markov processes; phylogenetic analysis.

4732. ———, 1993. Testing whether certain traits have caused amplified diversification: an improved method based on a model of random speciation and extinction. *American Naturalist* 142: 1019-1024. [2] *A.T.:* Actinopterygian fishes.

4733. Small, E., 1993. The economic value of plant systematics in Canadian agriculture. *Canadian J. of Botany* 71: 1537-1551. [1] *A.T.:* crop germplasm; systematics research.

4734. Smil, Vaclav, 1997. Some unorthodox perspectives on agricultural biodiversity. The case of legume cultivation. *Agriculture, Ecosystems & Environment* 62: 135-144. *A.T.:* soil microorganisms; agricultural ecology; nitrogen fixation.

4735. Smith, Andrew P., & D.G. Quin, 1996. Patterns and causes of extinction and decline in Australian conilurine rodents. *Biological Conservation* 77: 243-267. [1] *A.T.:* introduced species.

4736. Smith, Andrew P., N. Horning, & D. Moore, 1997. Regional biodiversity planning and lemur conservation with GIS in western Madagascar. *CB* 11: 498-512. *A.T.:* rapid assessment.

4737. Smith, Charles H., 1986. A contribution to the geographical interpretation of biological change. *Acta Biotheoretica* 35: 229-278. [1] *A.T.:* evolution; biogeography; environmental stress.

4738. Smith, Courtland L., Feb. 1994. Connecting cultural and biological diversity in restoring Northwest salmon. *Fisheries* 19(2): 20-26. [1] *A.T.:* fisheries management.

4739. Smith, Daniel S., & Paul C. Hellmund, eds., 1993. *Ecology of Greenways: Design and Function of Linear Conservation Areas.* Minneapolis: University of Minnesota Press. 222p. [2/1/1] *A.T.:* urban ecology.

4740. Smith, Edwin M., 1984. The Endangered Species Act and biological conservation. *Southern California Law Review* 57: 361-413. [1].

4741. Smith, Eric P., & Gerald van Belle, 1984. Nonparametric estimation of species richness. *Biometrics* 40: 119-129. [1] *A.T.:* jackknife procedure; bootstrap procedure.

4742. Smith, Fraser, 1994. *Biodiversity Loss: Ecological and Evolutionary Perspectives.* Norwich, U.K.: CSERGE Discussion Paper GEC 94-17, Centre for Social and Economic Research on the Global Environment. 41p. [1/1/1] *A.T.:* ecosystem resilience.

4743. ———, 1996. Biological diversity, ecosystem stability and economic development. *Ecological Economics* 16: 191-203. [1] *A.T.:* sustainable development; environmental economics.

4744. Smith, Fraser D.M., R.M. May, et al., 1993. How much do we know about the current extinction rate? *TREE* 8: 375-378. [2] *A.T.:* mass extinctions.

4745. Smith, Fraser D.M., R.M. May, et al., 1993. Estimating extinction rates. *Nature* 364: 494-496. [2] Estimates extinction rates for a number of animal groups.

4746. Smith, Joel B., & D.A. Tirpak, eds., 1989. *The Potential Effects of Global Climate Change on the United States*. New York: Hemisphere Publications Corp. 689p. [1/1/2] *A.T.:* greenhouse effect; global warming.

4747. Smith, J.S.C., 1988. Diversity of United States hybrid maize germplasm; isozymic and chromatographic evidence. *Crop Science* 28: 63-69. [2].

4748. Smith, M.A., & A.Y. Blumberg, 1990. Conservation of India's plant genetic resources: USAID's largest biodiversity activity. *Diversity* 6(2): 7-9. [1].

4749. Smith, Nigel J.H., J.T. Williams, et al., 1992. *Tropical Forests and Their Crops*. Ithaca, NY: Comstock Pub. Associates. 568p. [2/1/1] *A.T.:* forest products.

4750. Smith, Paul G.R., & J.B. Theberge, 1986. Evaluating biotic diversity in environmentally significant areas in the Northwest Territories of Canada. *Biological Conservation* 36: 1-18. [1] *A.T.:* bd measures.

4751. Smith, Peter J., 1994. *Genetic Diversity of Marine Fisheries Resources: Possible Impacts of Fishing*. Rome: FAO Fisheries Technical Paper No. 344. 53p. [1/1/1] *A.T.:* fishery conservation.

4752. Smith, P.A., 1994. Autocorrelation in logistic regression modelling of species' distributions. *Global Ecology and Biogeography Letters* 4: 47-61. [1] *A.T.:* U.K.

4753. Smith, Roy, 1993. *International Treaties, Conventions and Agreements Relevant to Biodiversity Conservation*. Bradford, U.K.: Development and Project Planning Centre, University of Bradford. 44p. [1/1/1]. *A.T.:* international cooperation.

4754. Smith, R.C., B.B. Prezelin, et al., 1992. Ozone depletion: ultraviolet radiation and phytoplankton biology in Antarctic waters. *Science* 255: 952-959. [3].

4755. Smith, Stephen V., & R.W. Buddemeier, 1992. Global change and coral reef ecosystems. *ARES* 23: 89-118. [2] *A.T.:* environmental change.

4756. Smith, Thomas B., & R.K. Wayne, eds., 1996. *Molecular Genetic Approaches in Conservation*. New York: OUP. 483p. [1/1/2] *A.T.:* bdc.

4757. Smith, Thomas B., R.K. Wayne, et al., 1997. A role for ecotones in generating rainforest biodiversity. *Science* 276: 1855-1857. *A.T.:* gene flow; birds; Africa; savannas.

4758. Smithson, J.B., & J.M. Lenne, 1996. Varietal mixtures: a viable strategy for sustainable productivity in subsistence agriculture. *Annals of Applied Biology* 128: 127-158. [1] *A.T.:* crop diversity.

4759. Smythe, Katie D., J.C. Bernabo, et al., 1996. Focusing biodiversity research on the needs of decision makers. *Environmental Management* 20: 865-872. [1] *A.T.:* environmental policy.

4760. Snape, William J., III, 1994. Biodiversity and the law: an introduction. *Tulane Environmental Law J.* 8: 5-19. [1] *A.T.:* environmental law.

4761. ———, ed., 1996. *Biodiversity and the Law* [essays]. Washington,

DC: Island Press. 259p. [2/1/1] *A.T.:* environmental law.

4762. Snelgrove, Paul V.R., & J.F. Grassle, fall-winter 1995. The deep sea: desert AND rainforest: debunking the desert analogy. *Oceanus* 38(2): 25-28. [1] *A.T.:* sampling; marine diversity.

4763. ———, fall-winter 1995. What of the deep sea's future diversity? *Oceanus* 38(2): 29. [1] *A.T.:* North Atlantic; waste dumping; benthic ecology.

4764. Snelgrove, Paul V.R., J.F. Grassle, & R.F. Petrecca, 1992. The role of food patches in maintaining high deep-sea diversity: field experiments with hydrodynamically unbiased colonization trays. *Limnology and Oceanography* 37: 1543-1550. [2] *A.T.:* West Indies; benthos.

4765. ———, 1996. Experimental evidence for aging food patches as a factor contributing to high deep-sea macrofaunal diversity. *Limnology and Oceanography* 41: 605-614. [2] *A.T.:* North Atlantic.

4766. Snelgrove, Paul V.R., T.H. Blackburn, et al., 1997. The importance of marine sediment biodiversity in ecosystem processes. *Ambio* 26: 578-583. *A.T.:* benthos.

4767. Snow, Neil, & J.M. Beiswenger, 1997. Biodiversity and human ecology analysis and resolution of fictitious conflict. *American Biology Teacher* 59: 344-348. *A.T.:* land management; bd education.

4768. Snyder, Gary, 1993. Coming in to the watershed: biological and cultural diversity in the California habitat. *Chicago Review* 39(3-4): 75-86. [1] *A.T.:* bioregionalism.

4769. Snyder, Noel F.R., S.R. Derrickson, et al., 1996. Limitations of captive breeding in endangered species recovery. *CB* 10: 338-348.

4770. Sobel, Jack, fall 1993. Conserving biological diversity through marine protected areas: a global challenge. *Oceanus* 36(3): 19-26. [1] *A.T.:* coastal zone management.

4771. ———, 1996. Marine reserves: necessary tools for biodiversity conservation? *Global Biodiversity* 6(1): 8-17. [1].

4772. Soberón M., Jorge, & Jorge Llorente B., 1993. The use of species accumulation functions for the prediction of species richness. *CB* 7: 480-488. [1] *A.T.:* sampling; butterflies; mammals; mathematical models.

4773. Sobolev, N.A., E.A. Shvarts, et al., 1995. Russia's protected areas: a survey and identification of development problems. *Biodiversity and Conservation* 4: 964-983. [1] *A.T.:* NGOs.

4774. Socias i Company, R., & A.J. Felipe, 1992. Almond: a diverse germplasm. *HortScience* 27: 718, 863. [1] *A.T.:* germplasm collections.

4775. Society of American Foresters, 1991. *Task Force Report on Biological Diversity in Forest Ecosystems.* Bethesda, MD. 52p. [1/1/1].

4776. Soejarto, Djaja D., 1996. Biodiversity prospecting and benefit-sharing: perspectives from the field. *J. of Ethnopharmacology* 51: 1-15. [1].

4777. Solbrig, Otto T., June 1991. The origin and function of biodiversity.

Environment 33(5): 16-20, 34-38. [1] A general treatment.

4778. ———, 1991. *Biodiversity: Scientific Issues and Collaborative Research Proposals*. Paris: MAB Digest 9, UNESCO. 77p. [1/1/1].

4779. ———, ed., 1991. *From Genes to Ecosystems: A Research Agenda for Biodiversity* [workshop report]. Cambridge, MA: IUBS. 124p. [1/1/2] *A.T.:* bdc.

4780. ———, 1992. The IUBS-SCOPE-UNESCO program of research in biodiversity. *Ecological Applications* 2: 131-138. [1].

4781. ———, 1992. Biodiversity, global change and scientific integrity [editorial]. *J. of Biogeography* 19: 1-2. [1].

4782. ———, 1993. Biodiversity and economics [editorial]. *Interciencia* 18: 118-119. [1].

4783. ———, 1994. Biodiversity and the world's food crisis. *Developments in Plant and Soil Sciences* 61: 159-168. [1].

4784. Solbrig, Otto T., H.M. van Emden, & P.G.W.J. van Oordt, eds., 1994. *Biodiversity and Global Change* [symposium proceedings]. Wallingford, Oxon, U.K.: CAB International. 227p. [1/1/1].

4785. Solbrig, Otto T., Ernesto Medina, & J.F. Silva, eds., 1996. *Biodiversity and Savanna Ecosystem Processes: A Global Perspective*. Berlin and New York: Springer. 233p. [1/1/1] *A.T.:* ecosystem analysis.

4786. Solé, Ricard V., & S.C. Manrubia, 1997. Criticality and unpredictability in macroevolution. *Physical Review E* 55: 4500-4507. *A.T.:* evolution.

4787. Solé, Ricard V., J. Bascompte, & S.C. Manrubia, 1996. Extinction: bad genes or weak chaos? *PRSL B* 263: 1407-1413. *A.T.:* macroevolution; mathematical models.

4788. Sollins, Phillip, 1998. Factors influencing species composition in tropical lowland rain forest: does soil matter? *Ecology* 79: 23-30. *A.T.:* soil nutrients; soil fertility.

4789. Solomon, Allen M., & H.H. Shugart, eds., 1993. *Vegetation Dynamics & Global Change* [papers from conferences]. New York: Chapman & Hall. 338p. [2/1/1].

4790. Solow, Andrew R., 1993. Measuring biological diversity. *Environmental Science & Technology* 27: 24-26. [1].

4791. ———, 1993. Inferring extinction from sighting data. *Ecology* 74: 962-964. [1].

4792. ———, 1993. A simple test for change in community structure. *J. of Animal Ecology* 62: 191-193. [1] *A.T.:* disturbance measures; mathematical models; diversity indices.

4793. ———, 1994. On the Bayesian estimation of the number of species in a community. *Ecology* 75: 2139-2142. [1].

4794. ———, fall-winter 1995. Estimating biodiversity: calculating unseen richness. *Oceanus* 38(2): 9-10. [1] *A.T.:* marine bd; species-area curves.

4795. ———, Jan. 1997. Biological diversity in the ocean. *Sea Tech-*

nology 38(1): 50-52. [1] A short review of marine bd studies.

4796. Solow, Andrew R., & J.M. Broadus, 1995. Issues in the measurement of biological diversity. *Vanderbilt J. of Transnational Law* 28: 695-702. [1] *A.T.:* bd definitions; bd measurement.

4797. Solow, Andrew R., & Stephen Polasky, 1994. Measuring biological diversity. *Environmental and Ecological Statistics* 1: 95-107. [1].

4798. Solow, Andrew R., Stephen Polasky, & J.M. Broadus, 1993. On the measurement of biological diversity. *J. of Environmental Economics and Management* 24: 60-68. [2] *A.T.:* birds; extinction probabilities.

4799. Somerfield, P.J., F. Olsgard, & M.R. Carr, 1997. A further examination of two new taxonomic distinctness measures. *Marine Ecology Progress Series* 154: 303-306. *A.T.:* bd measures; North Sea; species abundance.

4800. Sommer, Ulrich, 1993. Disturbance-diversity relationships in two lakes of similar nutrient chemistry but contrasting disturbance regimes. *Hydrobiologia* 249: 59-65. [1] *A.T.:* phytoplankton; Germany; intermediate disturbance hypothesis.

4801. ———, 1995. An experimental test of the intermediate disturbance hypothesis using cultures of marine phytoplankton. *Limnology and Oceanography* 40: 1271-1277. [1].

4802. Sosa, Victoria, & Teodoro Platas, 1998. Extinction and persistence of rare orchids in Veracruz, Mexico. *CB* 12: 451-455. *A.T.:* habitat disturbance.

4803. Soulé, Judith D., & J.K. Piper, 1992. *Farming in Nature's Image: An Ecological Approach to Agriculture.* Washington, DC: Island Press. 286p. [2/2/2] *A.T.:* sustainable agriculture.

4804. Soulé, Michael E., 1985. What is conservation biology? *Bioscience* 35: 727-734. [2] Provides a definitional and functional overview.

4805. ———, ed., 1986. *Conservation Biology: The Science of Scarcity and Diversity.* Sunderland, MA: Sinauer Associates. 584p. [2/2/2].

4806. ———, ed., 1987. *Viable Populations for Conservation.* Cambridge, U.K., and New York: CUP. 189p. [2/1/3] *A.T.:* wildlife conservation; conservation biology.

4807. ———, 1990. The real work of systematics. *Annals of the Missouri Botanical Garden* 77: 4-12. [1] Suggests that collective priorities be set for research within the field of systematics.

4808. ———, 1991. Conservation: tactics for a constant crisis. *Science* 253: 744-750. [2] Draws attention to the rate and magnitude of bd loss, and suggests a reappraisal of conservation goals.

4809. ———, 1991. Land use planning and wildlife maintenance: guidelines for conserving wildlife in an urban landscape. *J. of the American Planning Association* 57: 313-323. [1] *A.T.:* California; chaparral; environmental planning.

4810. Soulé, Michael E., & K.A. Kohm, eds., 1989. *Research Priorities for Conservation Biology.* Washington, DC: Island Press. 97p. [2/1/2].

4811. Soulé, Michael E., & M.A. Sanjayan, 1998. Conservation targets: do they help? *Science* 279: 2060-2061. Wonders whether legislated land protection goals might backfire. *A.T.:* bd loss.

4812. Soulé, Michael E., & D.S. Simberloff, 1986. What do genetics and ecology tell us about the design of nature reserves? *Biological Conservation* 35: 19-40. [2] *A.T.:* SLOSS; minimum viable populations; corridors; buffer zones.

4813. Soulé, Michael E., & B.A. Wilcox, eds., 1980. *Conservation Biology: An Evolutionary-Ecological Perspective* [conference papers]. Sunderland, MA: Sinauer Associates. 395p. [2/1/2].

4814. Soulé, Michael E., A.C. Alberts, & D.T. Bolger, 1992. The effects of habitat fragmentation on chaparral plants and vertebrates. *Oikos* 63: 39-47. [2] *A.T.:* California; patch dynamics; birds; rodents.

4815. Sournia, Alain, M.-J. Chrétiennot-Dinet, & Michel Ricard, 1991. Marine phytoplankton: how many species in the world ocean? *J. of Plankton Research* 12: 1093-1099. [1].

4816. Sousa, Wayne P., 1979. Disturbance in marine intertidal boulder fields: the nonequilibrium maintenance of species diversity. *Ecology* 60: 1225-1239. [2] *A.T.:* stability.

4817. ———, 1984. The role of disturbance in natural communities. *ARES* 15: 353-391. [3] A frequently cited review.

4818. Soutar, Andrew, & J.D. Isaacs, 1974. Abundance of pelagic fish during the 19th and 20th centuries as recorded in anaerobic sediment off the Californias. *Fishery Bull.* 72: 257-273. [2] *A.T.:* historical records.

4819. South Pacific Regional Environment Programme, 1993. *Fifth South Pacific Conference on Nature Conservation and Protected Areas.* Apia, Western Samoa. 2 vols. [1/1/1] *A.T.:* Oceania; bdc.

4820. Southerland, M.T., 1995. Conserving biodiversity in highway development projects. *Environmental Professional* 17: 226-242. [1].

4821. ———, winter 1997. Considering biodiversity in integrated natural resource management plans. *Federal Facilities Environmental J.* 7(4): 59-69. [1] *A.T.:* ecosystem management; National Environmental Policy Act.

4822. Southgate, Douglas D., 1998. *Tropical Forest Conservation: An Economic Assessment of the Alternatives in Latin America.* New York: OUP. 175p. *A.T.:* deforestation; forest management.

4823. Southgate, Douglas, & H.L. Clark, 1993. Can conservation projects save biodiversity in South America? *Ambio* 22: 163-166. [1] Criticizes approaches taken by international development and conservation agencies. *A.T.:* developing countries; environmental management.

4824. Southwick, Charles H., 1996. *Global Ecology in Human Perspective.* New York: OUP. 392p. [2/1/1].

4825. Southwood, T.R.E., 1995. Ecological processes and sustainability.

International J. of Sustainable Development and World Ecology 2: 229-239. [1] *A.T.:* environmental policy.

4826. Southwood, T.R.E., & C.E.J. Kennedy, 1983. Trees as islands. *Oikos* 41: 359-371. [1] *A.T.:* colonization; species-area relation.

4827. Southwood, T.R.E., V.C. Moran, & C.E.J. Kennedy, 1982. The richness, abundance and biomass of the arthropod communities on trees. *J. of Animal Ecology* 51: 635-649. [1] *A.T.:* Great Britain; South Africa.

4828. Spackman, Susan C., & J.W. Hughes, 1995. Assessment of minimum stream corridor width for biological conservation: species richness and distribution along mid-order streams in Vermont, USA. *Biological Conservation* 71: 325-332. [1] *A.T.:* birds; mammals; plants.

4829. Sparrow, Helen R., T.D. Sisk, et al., 1994. Techniques and guidelines for monitoring Neotropical butterflies. *CB* 8: 800-809. [1] *A.T.:* sampling; bd monitoring.

4830. Spash, Clive L., & Nick Hanley, 1995. Preferences, information and biodiversity preservation. *Ecological Economics* 12: 191-208. [2] *A.T.:* environmental perception; contingent valuation; preference evaluation.

4831. Specht, Alison, & R.L. Specht, 1994. Biodiversity of overstorey trees in relation to canopy productivity and stand density in the climatic gradient from warm temperate to tropical Australia. *Biodiversity Letters* 2: 39-45. [1] *A.T.:* GIS; Madagascar; Conservation International.

4832. Specht, Jeff, March 1996. Mapping Earth's endangered biodiversity. *GIS World* 9(3): 42-46. [1].

4833. Spellerberg, Ian F., 1991. *Monitoring Ecological Change.* Cambridge, U.K., and New York: CUP. 334p. [1/1/2] *A.T.:* ecological monitoring.

4834. ———, 1992. *Evaluation and Assessment for Conservation: Ecological Guidelines for Determining Priorities for Nature Conservation.* London and New York: Chapman & Hall. 260p. [1/1/1] *A.T.:* conservation evaluation.

4835. ———, ed., 1996. *Conservation Biology.* Harlow, U.K.: Longman. 242p. [1/1/1].

4836. Spellerberg, Ian F., & J.W.D. Sawyer, 1996. Standards for biodiversity: a proposal based on biodiversity standards for forest plantations. *Biodiversity and Conservation* 5: 447-459. [1].

4837. Spellerberg, Ian F., F.B. Goldsmith, & M.G. Morris, eds., 1991. *The Scientific Management of Temperate Communities for Conservation* [conference papers]. Oxford, U.K., and Boston: Blackwell Scientific. 566p. [1/1/1] *A.T.:* bdc; environmental management.

4838. Sperling, Louise, 1997. Assessing crop diversity after war: the Seeds of Hope initiative in Rwanda. *Diversity* 13(1): 36-39. [1] *A.T.:* crop germplasm resources.

4839. Speth, James G., June 1990. Coming to terms: toward a North-

South compact for the environment. *Environment* 32(5): 16-20, 40-43. [1] *A.T.:* international cooperation.

4840. Spiller, D.A., & T.W. Schoener, 1998. Lizards reduce spider species richness by excluding rare species. *Ecology* 79: 503-516. *A.T.:* Bahamas; predator-prey relations.

4841. Spitzer, Karel, Josef Jaroš, et al., 1997. Effect of small-scale disturbance on butterfly communities of an Indochinese montane rainforest. *Biological Conservation* 80: 9-15. *A.T.:* Vietnam.

4842. Spoel, S. van der, & A.C. Pierrot-Bults, eds., 1979. *Zoogeography and Diversity of Plankton*. New York: Halsted Press. 410p. [1/1/1] *A.T.:* zooplankton.

4843. Sponsel, Leslie E., ed., 1995. *Indigenous Peoples and the Future of Amazonia: An Ecological Anthropology of an Endangered World*. Tucson: University of Arizona Press. 312p. [2/1/1] *A.T.:* human ecology.

4844. Spratt, D.M., 1997. Endoparasite control strategies: implications for biodiversity of native fauna. *International J. for Parasitology* 27: 173-180. *A.T.:* parasitic diseases; domestic animals.

4845. Springuel, Irina, M. Sheded, & K.J. Murphy, 1997. The plant biodiversity of the Wadi Allaqi Biosphere Reserve (Egypt): impact of Lake Nasser on a desert wadi ecosystem. *Biodiversity and Conservation* 6: 1259-1275. *A.T.:* periodic flooding.

4846. Srinivasan, M.C., 1994. Microbial biodiversity and its relevance to screening for novel industrially useful enzymes. *Current Science* 66: 137-140. [1] *A.T.:* industrial microbiology; biotechnology.

4847. Srivastava, Jitendra P., John Lambert, & Noel Vietmeyer, 1996. *Medicinal Plants: An Expanding Role in Development*. Washington, DC: World Bank Technical Paper No. 320. 21p. [1/1/1] *A.T.:* environmental policy.

4848. Srivastava, Jitendra P., N.J.H. Smith, & D.A. Forno, eds., 1996. *Biodiversity and Agricultural Intensification: Partners for Development and Conservation*. Washington, DC: World Bank. 128p. [1/1/1] *A.T.:* agricultural ecology; sustainable agriculture.

4849. ———, 1996. *Biodiversity and Agriculture: Implications for Conservation and Development*. Washington, DC: World Bank Technical Paper No. 321. 26p. [1/1/1] *A.T.:* agricultural ecology; sustainable agriculture.

4850. Stacey, Pete B., & Mark Taper, 1992. Environmental variation and the persistence of small populations. *Ecological Applications* 2: 18-29. [2] *A.T.:* woodpeckers; Southwest; habitat fragmentation.

4851. Staddon, W.J., L.C. Duchesne, & J.T. Trevors, 1996. Conservation of forest soil microbial diversity: the impact of fire and research needs. *Environmental Reviews* 4: 267-275. [1].

4852. Stahl, Peter W., 1996. Holocene biodiversity: an archaeological perspective from the Americas. *Annual Review of Anthropology* 25: 105-126.

[1] *A.T.:* environmental change.

4853. Stähler, Frank, 1994. Biological diversity: the international management of genetic resources and its impact on biotechnology. *Ecological Economics* 11: 227-236. [1] *A.T.:* developing countries; international cooperation.

4854. Staley, James T., 1997. Biodiversity: are microbial species threatened? *Current Opinion in Biotechnology* 8: 340-345. *A.T.:* endangered species.

4855. Stanford, J.A., & F.R. Hauer, 1992. Mitigating the impacts of stream and lake regulation in the Flathead River catchment, Montana, USA: an ecosystem perspective. *Aquatic Conservation* 2: 35-63. [1] *A.T.:* flow regulation; fish.

4856. Stangel, Peter W., M.R. Lennartz, & M.H. Smith, 1992. Genetic variation and population structure of red-cockaded woodpeckers. *CB* 6: 283-292. [2] *A.T.:* heterozygosity; endangered species; small populations.

4857. Stanley, Steven M., 1975. A theory of evolution above the species level. *PNAS* 72: 646-650. [2] *A.T.:* macroevolution; higher taxa.

4858. ———, 1979. *Macroevolution, Pattern and Process.* San Francisco: W.H. Freeman. 332p. [2/1/3].

4859. ———, June 1984. Mass extinctions in the ocean. *Scientific American* 250(6): 64-72. [1].

4860. ———, 1984. Temperature and biotic crises in the marine realm. *Geology* 12: 205-208. [1] *A.T.:* climatic change; Cenozoic; benthos.

4861. ———, 1987. *Extinction.* New York: Scientific American Library. 242p. [3/2/1].

4862. ———, 1988. Paleozoic mass extinctions: shared patterns suggest global cooling as a common cause. *American J. of Science* 288: 334-352. [1] *A.T.:* glaciation; marine biotas.

4863. ———, 1989. *Earth and Life Through Time* (2nd ed.). New York: W.H. Freeman. 689p. [2/1/1].

4864. Starr, June, & K.C. Hardy, 1993. Not by seeds alone: the Biodiversity Treaty and the role for native agriculture. *Stanford Environmental Law J.* 12: 85-123. [1] *A.T.:* agricultural practices; CBD; traditional agriculture.

4865. Stattersfield, Alison J., M.J. Crosby, et al., 1998. *Endemic Bird Areas of the World: Priorities for Biodiversity Conservation.* Cambridge, U.K.: BirdLife International. 846p. *A.T.:* bird conservation.

4866. Steadman, David W., 1995. Prehistoric extinctions of Pacific island birds: biodiversity meets zooarchaeology. *Science* 267: 1123-1131. [2] *A.T.:* human disturbance; equatorial Pacific.

4867. Stebbins, G. Ledyard, 1981. Why are there so many species of flowering plants? *Bioscience* 31: 573-577. [1].

4868. Stebbins, G. Ledyard, & G.F. Hrusa, 1995. The North Coast biodiversity arena in central California: a new scenario for research and teaching processes of evolution. *Madroño* 42: 269-294. [1].

4869. Steele, John H., 1991. Marine functional diversity: ocean and land ecosystems may have different time scales for their responses to change. *Bioscience* 41: 470-474. [1].

4870. Steffek, Jozef, 1995. Ecosociological evaluation of the mollusc biodiversity in Slovakian biosphere reserves. *Ecology Bratislava* 14, Suppl. 2: 3-10. [1] *A.T.:* MAB.

4871. Steffek, Jozef, et al., 1995. Significance of the national ecological network of Slovakia for the European Ecological Network—EECONET. *Ecology Bratislava* 14, Suppl. 1: 205-212. [1] *A.T.:* corridors; Slovakia.

4872. Stehli, Francis G., & S.D. Webb, eds., 1985. *The Great American Biotic Interchange*. New York: Plenum Press. 532p. [1/2/2] *A.T.:* paleobiogeography.

4873. Stehli, Francis G., & John W. Wells, 1971. Diversity and age patterns in hermatypic corals. *Systematic Zoology* 20: 115-126. [2] *A.T.:* generic age.

4874. Stehli, Francis G., R.G. Douglas, & N.D. Newell, 1969. Generation and maintenance of gradients in taxonomic diversity. *Science* 164: 947-949. [1] *A.T.:* latitudinal diversity gradients; invertebrates.

4875. Stehli, Francis G., A.L. McAlester, & C.E. Helsley, 1967. Taxonomic diversity of Recent bivalves and some implications for geology. *Geological Society of America Bull.* 78: 455-465. [1] *A.T.:* geographic differentiation; polar wandering.

4876. Steinberg, Paul F., 1998. Consensus by design, policy by default: implementing the Convention on Biological Diversity. *Society & Natural Resources* 11: 375-385. Studies the worldwide allocation of conservation aid by donor agencies. *A.T.:* AID; GEF.

4877. Stelljes, Kathryn B., Nov. 1995. Curator minds his peas: there are nearly 3,000 different kinds in the national collection. *Agricultural Research* 43(11): 18-19. [1] *A.T.:* germplasm resources.

4878. Stelljes, Kathryn B., Marcia Wood, et al., Oct. 1995. Genebanks: treasure houses of uncommon foods. *Agricultural Research* 43(10): 4-8. [1].

4879. Stevens, George C., 1986. Dissection of the species-area relationship among wood-boring insects and their host plants. *American Naturalist* 128: 35-46. [1].

4880. ———, 1989. The latitudinal gradient in geographical range: how so many species coexist in the tropics. *American Naturalist* 133: 240-256. [2] *A.T.:* Rapoport's rule.

4881. ———, 1992. The elevational gradient in altitudinal range: an extension of Rapoport's latitudinal rule to altitude. American Naturalist 140: 893-911. [2].

4882. ———, 1996. Extending Rapoport's rule to Pacific marine fishes. *J. of Biogeography* 23: 149-154. [1] Extends the rule to ocean depth relationships.

4883. Stevens, Thomas H., T.A. More, & R.J. Glass, 1993. Measuring the existence value of wildlife: reply. *Land Economics* 69: 309-312. [1].

4884. Stevens, Thomas H., Jaime Echeverria, et al., 1991. Measuring the existence value of wildlife: what do CVM estimates really show? *Land Economics* 67: 390-400. [2] *A.T.:* contingent valuation; New England; bd value.

4885. Stevens, Victoria, 1995. *Wildlife Diversity in British Columbia: Distribution and Habitat Use of Amphibians, Reptiles, Birds, and Mammals in Biogeoclimatic Zones.* Victoria, BC: Ministry of Forests Research, Province of British Columbia. 288p. [1/1/1].

4886. Stevens, William K., 18 Dec. 1990. Legions of plants thrive where they don't belong. *New York Times* 140(48453): C1, C12. [1] Describes how seeds and plants have been transported and relocated all around the world.

4887. ———, 1995. *Miracle Under the Oaks: The Revival of Nature in America.* New York: Pocket Books. 332p. [2/2/1] *A.T.:* restoration ecology.

4888. Stiassny, Melanie L.J., Sept. 1996. An overview of freshwater biodiversity: with some lessons from African fishes. *Fisheries* 21(9): 7-13. [1] Focuses on bd losses in freshwater environments.

4889. Stiles, F. Gary, & D.A. Clark, 1989. ICBP: Conservation of tropical rain forest birds: a case study from Costa Rica. *American Birds* 43: 420-428. [1].

4890. Stinner, D.H., B.R. Stinner, & E. Martsolf, 1997. Biodiversity as an organizing principle in agroecosystem management: case studies of holistic resource management practitioners in the USA. *Agriculture, Ecosystems & Environment* 62: 199-213. *A.T.:* landscape planning.

4891. Stocker, R., P.W. Leadley, & C. Korner, 1997. Carbon and water fluxes in a calcareous grassland under elevated CO_2. *Functional Ecology* 11: 222-230. [2] *A.T.:* evapotranspiration.

4892. Stocking, Michael A., & Scott Perkin, 1992. Conservation-with-development: an application of the concept in the Usambara Mountains, Tanzania. *Trans. of the Institute of British Geographers* 17: 337-349. [1] *A.T.:* local participation; bdc.

4893. Stoddart, David R., 1992. Biogeography of the tropical Pacific. *Pacific Science* 46: 276-293. [1] *A.T.:* atolls; coral reefs.

4894. Stohlgren, Thomas J., G.W. Chong, et al., 1997. Rapid assessment of plant diversity patterns: a methodology for landscapes. *Environmental Monitoring and Assessment* 48: 25-43. *A.T.:* environmental indicators; Rocky Mountain National Park.

4895. Stohlgren, Thomas J., M.B. Coughenour, et al., 1997. Landscape analysis of plant diversity. *Landscape Ecology* 12: 155-170. *A.T.:* Rocky Mountain National Park; ecosystem management.

4896. Stohlgren, Thomas J., J.F. Quinn, et al., 1995. Status of biotic inventories in US national parks. *Biological Conservation* 71: 97-106. [1].

4897. Stokland, Jogeir N., 1997. Representativeness and efficiency of bird and insect conservation in Norwegian boreal forest reserves. *CB* 11: 101-111. *A.T.:* beetles.

4898. Stolzenburg, William, Jan.-Feb. 1992. Detectives of diversity. *Nature Conservancy* 42(1): 22-27. [1] Describes work of the Natural Heritage Network.

4899. ———, July-Aug. 1992. Biodiversity's burning issue. *Nature Conservancy* 42(9): 30-31. [1] Concerning the implications of the greenhouse effect.

4900. ———, Sept.-Oct. 1992. Sacred peaks: common grounds. *Nature Conservancy* 42(5): 16-23. [1] Concerning traditional knowledge and its application to bdc.

4901. ———, March-April 1993. Indonesia: Wallace's wonderland. *Nature Conservancy* 43(2): 16-22. [1].

4902. ———, July-Aug. 1995. The guardian of Eden. *Nature Conservancy* 45(4): 24-29. [1] A report on the world of park guards. *A.T.:* poaching.

4903. Stoms, David M., 1992. Effects of habitat map generalization in biodiversity assessment. *Photogrammetric Engineering and Romote Sensing* 58: 1587-1591. [1] *A.T.:* GIS; gap analysis; California.

4904. ———, 1994. Scale dependence of species richness maps. *Professional Geographer* 46: 346-358. [1] *A.T.:* GIS; gap analysis; Idaho; spatial scale.

4905. Stoms, David M., & J.E. Estes, 1993. A remote sensing research agenda for mapping and monitoring biodiversity. *International J. of Remote Sensing* 14: 1839-1860. [1].

4906. Stone, Charles P., & L.L. Loope, 1987. Reducing negative effects of introduced animals on native biotas in Hawaii. *Environmental Conservation* 14: 245-258. [1] *A.T.:* biological invasions.

4907. Stone, Charles P., & D.B. Stone, eds., 1989. *Conservation Biology in Hawai'i*. Honolulu: University of Hawaii Cooperative National Park Resources Study Unit. 252p. [1/1/1].

4908. Stone, Charles P., C.W. Smith, & J.T. Tunison, eds., 1992. *Alien Plant Invasions in Native Ecosystems of Hawai'i: Management and Research* [symposium papers]. Honolulu: Cooperative National Park Resources Studies Unit, University of Hawaii. 887p. [1/1/1].

4909. Stone, Christopher D., 1987. *Earth and Other Ethics: The Case for Moral Pluralism*. New York: Harper & Row. 280p. [2/2/2] *A.T.:* environmental ethics.

4910. ———, 1993. *The Gnat Is Older than Man: Global Environment and Human Agenda*. Princeton, NJ: Princeton University Press. 341p. [2/2/2] *A.T.:* environmental policy; environmental law.

4911. [not used]

4912. Stone, Christopher D., 1995. What to do about biodiversity:

property rights, public goods, and the Earth's biological riches. *Southern California Law Review* 68: 577-620. [1] *A.T.:* genetic resources.

4913. ———, 1997. Stemming the loss of biological diversity: the institutional and ethical contours. *Review of European Community & International Environmental Law* 6: 231-238. *A.T.:* CBD.

4914. Stone, David, 1994. *Biodiversity of Indonesia: Tanah Air.* Singapore: Archipelago Press. 208p. [1/1/1].

4915. Stone, David, Kristina Ringwood, & Frank Vorhies, 1997. *Business and Biodiversity: A Guide for the Private Sector* [report]. Geneva: World Business Council for Sustainable Development, and IUCN. 64p. [1/1/-] *A.T.:* sustainable development.

4916. Stone, Jeffrey K., M.A. Sherwood, & G.C. Carroll, 1996. Canopy microfungi: function and diversity. *Northwest Science* 70, Suppl.: 37-45. [1] *A.T.:* Pacific Northwest; endophytes.

4917. Stone, L., 1995. Biodiversity and habitat destruction: a comparative study of model forest and coral reef ecosystems. *PRSL B* 261: 381-388. [1] *A.T.:* Red Sea; habitat fragmentation; Israel.

4918. Stone, L., E. Eilam, et al., 1996. Modelling coral reef biodiversity and habitat destruction. *Marine Ecology Progress Series* 134: 299-302. [1] *A.T.:* mathematical models.

4919. Stone, Paul A., H.L. Snell, & H.M. Snell, 1994. Behavioral diversity as biological diversity: introduced cats and lava lizard wariness. *CB* 8: 569-573. [1].

4920. Stone, Richard, 1995. Taking a new look at life through a functional lens. *Science* 269: 316-317. [1] Concerning the evolution of the keystone species concept.

4921. Stork, Nigel E., 1988. Insect diversity: facts, fiction and speculation. *Biological J. of the Linnean Society* 35: 321-337. [2] *A.T.:* rainforest canopy; Central America.

4922. ———, 1993. How many species are there? *Biodiversity and Conservation* 2: 215-232. [1].

4923. Stork, Nigel E., & K.J. Gaston, 11 Aug. 1990. Counting species one by one. *New Scientist* 127(1729): 43-47. [1] Concerning the reasons for our ignorance as to the total number of species.

4924. Stork, Nigel E., M.J. Samways, & H.A.C. Eeley, 1996. Inventorying and monitoring biodiversity [editorial]. *TREE* 11: 39-40. [1].

4925. Stout, Jean, & John Vandermeer, 1975. Comparison of species richness for stream-inhabiting insects in tropical and mid-latitude streams. *American Naturalist* 109: 263-280. [2].

4926. Straalen, N.M. van, 1994. Biodiversity of ecotoxicological responses in animals. *Netherlands J. of Zoology* 44: 112-129. [1] *A.T.:* isopods; life history studies.

4927. Strahl, Stuart D., 1992. Furthering avian conservation in Latin

America. *Auk* 109: 680-682. [1] *A.T.:* international cooperation.

4928. Strathdee, A.T., & J.S. Bale, 1998. Life on the edge: insect ecology in Arctic environments. *Annual Review of Entomology* 43: 85-106.

4929. Strittholt, James R., & R.E.J. Boerner, 1995. Applying biodiversity gap analysis in a regional nature reserve design for the Edge of Appalachia, Ohio (USA). *CB* 9: 1492-1505. [1] *A.T.:* rule-based models; remote sensing; plant communities.

4930. Strong, Donald R., Jr., E.D. McCoy, & J.R. Rey, 1977. Time and the number of herbivore species: the pests of sugarcane. *Ecology* 58: 167-175. [1] *A.T.:* insect pests; parasite communities.

4931. Stuart, Simon N., R.J. Adams, & M.D. Jenkins, 1990. *Biodiversity in Sub-Saharan Africa and Its Islands: Conservation, Management, and Sustainable Use*. Gland, Switzerland: IUCN. 242p. [1/1/1].

4932. Stuessy, Tod F., 1993. The role of creative monography in the biodiversity crisis. *Taxon* 42: 313-321. [1] Suggests that more attention should be put on producing large-scale systematic treatises. *A.T.:* plants.

4933. Stuessy, Tod F., & S.H. Sohmer, eds., 1996. *Sampling the Green World: Innovative Concepts of Collection, Preservation, and Storage of Plant Diversity* [symposium papers]. New York: Columbia University Press. 289p. [1/1/1] *A.T.:* plant collections.

4934. Stumpel, Anton H.P., 1992. Successful reproduction of introduced bullfrogs *Rana catesbeiana* in northwestern Europe: a potential threat to indigenous amphibians. *Biological Conservation* 60: 61-62. [1] *A.T.:* biological invasions.

4935. Subramanian, Arvind, 1992. Genetic resources, biodiversity and environmental protection—an analysis, and proposals towards a solution. *J. of World Trade* 26: 105-109. [1] *A.T.:* property rights.

4936. Suchanek, Thomas H., 1994. Temperate coastal marine communities: biodiversity and threats. *American Zoologist* 34: 100-114. [1] *A.T.:* coastal environment; human impact.

4937. Sugihara, George, 1980. Minimal community structure: an explanation of species abundance patterns. *American Naturalist* 116: 770-787. [2].

4938. Sukopp, Herbert, & Ulrich Sukopp, 1993. Ecological long-term effects of cultigens becoming feral and of naturalization of non-native species. *Experientia* 49: 210-218. [1].

4939. Sun, Marjorie, 1990. How do you measure the Lovejoy Effect? *Science* 247: 1174-1176. Profiles the work of Thomas Lovejoy. [1] *A.T.:* Amazonia.

4940. Sun, Wen Quan, 1992. Quantifying species diversity of streetside trees in our cities. *J. of Arboriculture* 18: 91-93. [1] *A.T.:* urban bd; urban forestry.

4941. Sunquist, Fiona, Sept.-Oct. 1988. Zeroing in on keystone species. *International Wildlife* 18(5): 18-23. [1].

4942. Supardiyono, E.K., & D. Smith, 1997. Microbial diversity: *ex situ* conservation of Indonesian microorganisms. *World J. of Microbiology & Biotechnology* 13: 359-366. *A.T.:* fungi.

4943. Supkoff, D.M., D.B. Joley, & J.J. Marois, 1988. Effect of introduced biological control organisms on the density of *Chondrilla juncea* in California. *J. of Applied Ecology* 25: 1089-1095. [1] *A.T.:* weeds.

4944. Suter, Werner, 1998. Involving conservation biology in biodiversity strategy and action planning. *Biological Conservation* 83: 235-237. *A.T.:* action plans.

4945. Sutherland, William J., & S.R. Baillie, 1993. Patterns in the distribution, abundance and variation of bird populations [editorial]. *Ibis* 135: 209-210. [1].

4946. Sutherland, William J., & D.A. Hill, eds., 1995. *Managing Habitats for Conservation*. Cambridge, U.K., and New York: CUP. 399p. [2/1/1] *A.T.:* environmental management.

4947. Svavarsson, Jörundur, 1997. Diversity of isopods (Crustacea): new data from the Arctic and Atlantic Oceans. *Biodiversity and Conservation* 6: 1571-1579. *A.T.:* vertical distribution.

4948. Swain, Hilary M., 1995. Reconciling rarity and representation: a review of listed species in the Indian River Lagoon. *Bull. of Marine Science* 57: 252-266. [1] *A.T.:* Florida.

4949. Swain, Roger B., April 1992. Gene soup: diversity is easier to celebrate than to maintain. *Horticulture* 70(4): 23-30. [1] *A.T.:* plant germplasm resources.

4950. Swaine, M.D., & T.C. Whitmore, 1988. On the definition of ecological species groups in tropical rain forests. *Vegetatio* 75: 81-86. [2].

4951. Swaminathan, Monkombu S., June 1994. Biotechnology for beginners. *UNESCO Courier* 47(6): 8-10. [1] *A.T.:* transgenic plants; genetic engineering; bioremediation.

4952. ———, ed., 1995. *Farmers' Rights and Plant Genetic Resources: Recognition & Reward: A Dialogue* [conference proceedings]. Madras: Macmillan India. 440p. [1/1/1].

4953. ———, ed., 1996. *Agrobiodiversity and Farmers' Rights* [conference proceedings]. Delhi: Konark Publishers. 303p. [1/1/1] *A.T.:* food crops; germplasm resources.

4954. Swaminathan, Monkombu S., & S. Jana, eds., 1992. *Biodiversity, Implications for Global Food Security* [contributed papers]. Madras: Macmillan India. 326p. [1/1/1] *A.T.:* germplasm resources.

4955. Swaney, James A., & P.I. Olson, 1992. The economics of biodiversity: lives and lifestyles. *J. of Economic Issues* 26: 1-25. [1].

4956. Swanson, Bradley J., 1998. Autocorrelated rates of change in animal populations and their relationship to precipitation. *CB* 12: 801-808. *A.T.:* mammals; birds.

4957. Swanson, F.J., & J.F. Franklin, 1992. New forestry principles from ecosystem analysis of Pacific Northwest forests. *Ecological Applications* 2: 262-274. [2] *A.T.:* forest management; riparian environments.

4958. Swanson, Timothy M., 1992. Economics of a biodiversity convention. *Ambio* 21: 250-257. [1] *A.T.:* international cooperation; bdc incentives.

4959. ———, 1994. The economics of extinction revisited and revised: a generalised framework for the analysis of the problems of endangered species and biodiversity losses. *Oxford Economic Papers* 46, Suppl.: 800-821. [1] *A.T.:* environmental economics.

4960. ———, 1994. *The International Regulation of Extinction*. Washington Square, NY: New York University Press. 289p. [2/2/1] *A.T.:* environmental policy; international cooperation.

4961. ———, ed., 1995. *Intellectual Property Rights and Biodiversity Conservation: An Interdisciplinary Analysis of the Values of Medicinal Plants*. Cambridge, U.K., and New York: CUP. 271p. [1/1/1].

4962. ———, ed., 1995. *The Economics and Ecology of Biodiversity Decline: The Forces Driving Global Change* [symposium papers]. Cambridge, U.K., and New York: CUP. 162p. [1/2/1].

4963. ———, 1996. The reliance of Northern economies on Southern biodiversity: biodiversity as information. *Ecological Economics* 17: 1-8. [1] *A.T.:* North-South relations; industrial research and development.

4964. ———, 1997. *Global Action for Biodiversity: An International Framework for Implementing the Convention on Biological Diversity*. London: Earthscan. 191p. [1/1/-] *A.T.:* international cooperation.

4965. ———, 1997. What is the public interest in biodiversity conservation for agriculture? *Outlook on Agriculture* 26: 7-12.

4966. Swanson, Timothy M., & E.B. Barbier, eds., 1992. *Economics for the Wilds: Wildlife, Diversity, and Development*. Washington, DC: Island Press. 226p. [1/1/1].

4967. Swanson, Timothy M., Patrick Bolton, & Alan Manning, April 1993. Regulating endangered species; discussion. *Economic Policy*, no. 16: 183-205. [1] *A.T.:* environmental policy; CBD; commons; extinction.

4968. Swindel, Benee F., L.F. Conde, & J.E. Smith, 1987. Index-free diversity orderings: concept, measurement, and observed response to clearcutting and site-preparation. *Forest Ecology and Management* 20: 195-208. [1] *A.T.:* species diversity; plantations; Southeast; forest management.

4969. Swindel, Benee F., J.E. Smith, & R.C. Abt, 1991. Methodology for predicting species diversity in managed forests. *Forest Ecology and Management* 40: 75-85. [1] *A.T.:* pine; plantations.

4970. Swindells, Philip, 1 July 1988. A call to conserve. *American Nurseryman* 168(1): 227-236. [1] A plea to the nursery industry to help save endangered plants.

4971. Swingland, Ian R., 1993. The ecology of stability in Southeast

Asia's forests: biodiversity and common resource property. *Global Ecology and Biogeography Letters* 3: 290-296. [1] *A.T.:* commons; resource management.

4972. Sylvan, Richard, 1984, 1985. A critique of deep ecology (in two parts). *Radical Philosophy*, no. 40: 2-12, and no. 41: 10-22. [1].

4973. Sylvan, Richard, & David Bennett, 1994. *The Greening of Ethics.* Tucson: University of Arizona Press. 269p. [1/2/1] *A.T.:* environmental ethics; Australia.

4974. Sylvester-Bradley, P.C., 1968. The science of diversity. *Systematic Zoology* 17: 176-181. [1] *A.T.:* systematics.

4975. Symstad, A.J., David Tilman, et al., 1998. Species loss and ecosystem functioning: effects of species identity and community composition. *Oikos* 81: 389-397. *A.T.:* community structure; productivity.

4976. Synge, Hugh, ed., 1981. *The Biological Aspects of Rare Plant Conservation* [conference proceedings]. Chichester, U.K., and New York: Wiley. 558p. [1/2/1].

4977. Systematics Agenda 2000, 1994. *Systematics Agenda 2000: Charting the Biosphere: Technical Report.* 34p. [1/1/1] *A.T.:* classification.

4978. Szaro, Robert C., Dec. 1992. Biodiversity and biological realities. *General Technical Report INT* GTR-291: 5-8. [1] *A.T.:* forest management.

4979. Szaro, Robert C., & D.W. Johnston, eds., 1996. *Biodiversity in Managed Landscapes: Theory and Practice* [symposium papers]. New York: OUP. 778p. [1/1/1] *A.T.:* bdc.

4980. Szaro, Robert C., & R.M. King, 1990. Sampling intensity and species richness: effects on delineating Southwestern riparian plant communities. *Forest Ecology and Management* 33-34: 335-349. [1].

4981. Szaro, Robert C., K.E. Severson, & D.R. Patton, coordinators, 1988. *Management of Amphibians, Reptiles, and Small Mammals in North America* [symposium proceedings]. Fort Collins, CO: General Technical Report RM GTR-166, USDA Forest Service. 458p. [1/1/1] *A.T.:* wildlife conservation.

4982. Szentandrasi, Susanne, Stephen Polasky, et al., 1995. Conserving biological diversity and the Conservation Reserve Program. *Growth and Change* 26: 383-404. [1] *A.T.:* wildlife conservation; environmental policy.

4983. Taba, S., & S.A. Eberhart, 1997. Cooperation between U.S. and CIMMYT leads to rescue of thousands of Latin American maize landraces. *Diversity* 13(1): 9-11. *A.T.:* Latin American Maize Project; international cooperation.

4984. Taberlet, P., & J. Bouvet, 1994. Mitochondrial DNA polymorphism, phylogeography, and conservation genetics of the brown bear *Ursus arctos* in Europe. *PRSL B* 255: 195-200. [2].

4985. Tacconi, Luca, 1997. An ecological economic approach to forest and biodiversity conservation: the case of Vanuatu. *World Development* 25:

1995-2008. *A.T.:* environmental policy; local participation.

4986. ———, 1997. Property rights and participatory biodiversity conservation: lessons from Malekula Island, Vanuatu. *Land Use Policy* 14: 151-161. *A.T.:* protected areas; local participation.

4987. Tacconi, Luca, & Jeff Bennett, 1995. Biodiversity conservation: the process of economic assessment and establishment of a protected area in Vanuatu. *Development and Change* 26: 89-110. [1] *A.T.:* logging.

4988. ———, 1995. Economic implications of intergenerational equity for biodiversity conservation. *Ecological Economics* 12: 209-223. [1] *A.T.:* environmental economics.

4989. Tackaberry, Rosanne, & Martin Kellman, 1996. Patterns of tree species richness along peninsular extensions of tropical forests. *Global Ecology and Biogeography Letters* 5: 85-90. [1] *A.T.:* Belize; peninsular effect; Venezuela.

4990. Tackaberry, Rosanne, & N. Brokaw, et al., 1997. Estimating species richness in tropical forest: the missing species extrapolation technique. *J. of Tropical Ecology* 13: 449-458. *A.T.:* Belize.

4991. Takacs, David, 1996. *The Idea of Biodiversity: Philosophies of Paradise*. Baltimore: Johns Hopkins University Press. 393p. [2/2/1] *A.T.:* environmental philosophy.

4992. Takeuchi, K., M. Ide, et al., 1995. Relationships of landform and biological diversity in landscape ecology. *Trans. of the Japanese Geomorphological Union* 16: 215-225. [1] *A.T.:* geomorphology.

4993. Tamisier, Alain, & Patrick Grillas, 1994. A review of habitat changes in the Camargue: an assessment of the effects of the loss of biological diversity on the wintering waterfowl community. *Biological Conservation* 70: 39-47. [2] *A.T.:* France; wetlands.

4994. Tangley, Laura, 1985. A national biological survey. *Bioscience* 35: 686-690. [1] Comments on the absence of a full catalog of U.S. species.

4995. ———, 1988. Preparing for climate change. *Bioscience* 38: 14-18. [1] *A.T.:* greenhouse effect.

4996. ———, 1988. Studying (and saving) the tropics. *Bioscience* 38: 375-385. [1] Describes work of the Organization for Tropical Studies.

4997. ———, 1988. Research priorities for conservation. *Bioscience* 38: 444-448. [1] *A.T.:* conservation biology; research programs.

4998. ———, 1990. Cataloging Costa Rica's diversity. *Bioscience* 40: 633-636. [1].

4999. ———, 1992. *Computers and Conservation Priorities: Mapping Biodiversity*. Washington, DC: Lessons from the Field 1, Conservation International. 28p. [1/1/1] *A.T.:* GIS; Papua New Guinea.

5000. ———, 1996. The case of the missing migrants. *Science* 274: 1299-1300. [1] Relates decline in migratory bird populations to a trend away from traditional coffee plantation systems.

5001. ———, 1996. Ground rules emerge for marine bioprospectors. *Bioscience* 46: 245-249. [1] *A.T.:* international cooperation; resource development.

5002. Tanner, Jason E., T.P. Hughes, & J.H. Connell, 1994. Species coexistence, keystone species and succession: a sensitivity analysis. *Ecology* 75: 2204-2219. [2] *A.T.:* matrix models; Great Barrier Reef.

5003. Tansley, A.G., 1935. The use and abuse of vegetational concepts and terms. *Ecology* 16: 284-307. [2] *A.T.:* succession; ecosystem concept.

5004. Tappan, Helen, & A.R. Loeblich, Jr., 1988. Foraminiferal evolution, diversification, and extinction. *J. of Paleontology* 62: 695-714. [1].

5005. Tarasofsky, Richard G., 1997. The relationship between the TRIPs Agreement and the Convention on Biological Diversity: towards a pragmatic approach. *Review of European Community & International Environmental Law* 6: 148-156. *A.T.:* international trade; intellectual property.

5006. Tarlock, A. Dan, 1993. Local government protection of biodiversity: what is its niche? *University of Chicago Law Review* 60: 555-613. [2] *A.T.:* California; Texas.

5007. ———, 1995. Biodiversity federalism. *Maryland Law Review* 54: 1315-1353. [1] *A.T.:* environmental policy.

5008. Taylor, Barbara L., 1995. The reliability of using population viability analysis for risk classification of species. *CB* 9: 551-558. [2] *A.T.:* sea lions; extinction risk.

5009. Taylor, David, 1997. Seeing the forests for more than the trees. *Environmental Health Perspectives* 105: 1186-1191. Looks at the environmental health effects of global deforestation.

5010. Taylor, Paul D., & G.P. Larwood, eds., 1990. *Major Evolutionary Radiations* [symposium proceedings]. New York: OUP. 437p. [1/1/1] *A.T.:* paleobiology; diversification.

5011. Taylor, Paul W., 1986. *Respect for Nature: A Theory of Environmental Ethics*. Princeton, NJ: Princeton University Press. 329p. [2/2/2].

5012. Taylor, P.D., L. Fahrig, et al., 1993. Connectivity is a vital element of landscape structure. *Oikos* 68: 571-573. [2] *A.T.:* microorganisms; neighborhood effect.

5013. Taylor, Robert J., & P.J. Regal, 1978. The peninsular effect on species diversity and the biogeography of Baja California. *American Naturalist* 112: 583-593. [1] *A.T.:* heteromyid rodents.

5014. Taylor, Victoria J., & Nigel Dunstone, eds., 1996. *The Exploitation of Mammal Populations* [symposium papers]. London and New York: Chapman & Hall. 415p. [1/1/1] *A.T.:* wildlife conservation; animal welfare.

5015. Tear, Timothy H., J.M. Scott, et al., 1993. Status and prospects for success of the Endangered Species Act: a look at recovery plans [editorial]. *Science* 262: 976-977. [2].

5016. Teitel, Martin, 1992. *Rain Forest in Your Kitchen: The Hidden*

Connection Between Extinction and Your Supermarket. Washington, DC: Island Press. 112p. [2/2/1] *A.T.:* consumerism; bdc.

5017. Teketay, Demel, 1995. Floristic composition of Dakata Valley, southeast Ethiopia: an implication for the conservation of biodiversity. *Mountain Research and Development* 15: 183-186. [1].

5018. Temple, Stanley A., ed., 1978. *Endangered Birds: Management Techniques for Preserving Threatened Species* [symposium proceedings]. Madison: University of Wisconsin Press. 466p. [2/1/1].

5019. ———, 1986. The problem of avian extinctions. *Current Ornithology* 3: 453-485. [1].

5020. ———, 1996. Ecological principles, biodiversity, and the electric utility industry. *Environmental Management* 20: 873-878. [1] *A.T.:* carbon release.

5021. Templeton, Alan R., 1994. Biodiversity at the molecular genetic level: experiences from disparate macroorganisms. *PTRSL B* 345: 59-64. [1] *A.T.:* genetic diversity; population viability.

5022. Tenenbaum, David, Feb. 1995. To know, to use, to save. *World & I* 10(2): 208-215. [1] Concerning INBio's attempt to organize a national biological survey of Costa Rica.

5023. ———, Aug.-Sept. 1996. Weeds from hell. *Technology Review* 99(6): 32-40. [1] *A.T.:* plant invasions; U.S.; weed control.

5024. ———, Jan. 1997. In praise of superfly. *Technology Review* 100(1): 19-21. [1] Describes an experiment with houseflies designed to examine the pedigree breeding strategy. *A.T.:* genetic diversity.

5025. Tepedino, V.J., S.D. Sipes, et al., 1997. The need for "extended care" in conservation: examples from studies of rare plants in the western United States. *Acta Horticulturae*, no. 437: 245-248. *A.T.:* pollinators.

5026. Terborgh, John, 1974. Preservation of natural diversity: the problem of extinction-prone species. *Bioscience* 24: 715-722. [2].

5027. ———, 1977. Bird species diversity on an Andean elevational gradient. *Ecology* 58: 1007-1019. [2] *A.T.:* Peru.

5028. ———, 1985. The vertical component of plant species diversity in temperate and tropical forests. *American Naturalist* 126: 760-776. [1] *A.T.:* eastern U.S.

5029. ———, 1988. The big things that run the world—a sequel to E.O. Wilson. *CB* 2: 402-403. [2] *A.T.:* predator-prey relations; stability.

5030. ———, 1989. *Where Have All the Birds Gone? Essays on the Biology and Conservation of Birds That Migrate to the American Tropics.* Princeton, NJ: Princeton University Press. 207p. [2/2/2].

5031. ———, May 1992. Why American songbirds are vanishing. *Scientific American* 266(5): 98-104. [1] *A.T.:* deforestation; avian declines.

5032. ———, 1992. Maintenance of diversity in tropical forests. *Biotropica* 24: 283-292. [2] *A.T.:* Brazil; habitat fragmentation; forest conservation.

5033. ———, 1992. *Diversity and the Tropical Rain Forest*. New York: Scientific American Library. 242p. [3/2/2].

5034. Terborgh, John, R.B. Foster, & Percy Núñez-V., 1996. Tropical tree communities: a test of the nonequilibrium hypothesis. *Ecology* 77: 561-567. *A.T.:* Peru; rainforest ecology.

5035. Tew, T.E., T.J. Crawford, et al., eds., 1997. *The Role of Genetics in Conserving Small Populations*. Peterborough, U.K.: JNCC. 203p. [1/1/-].

5036. Texas Center for Policy Studies, 1993. *Biodiversity Protection in the Texas-Mexico Border Region*. Austin. 43p. [1/1/1].

5037. Thackway, Richard, 1997. Significant trends in nature conservation in Australia. *Natural Areas J.* 17: 233-240. *A.T.:* environmental policy.

5038. Thapar, Valmik, 1997. *Land of the Tiger: A Natural History of the Indian Subcontinent*. Berkeley: University of California Press. 288p. [2/2/-] A book prepared to accompany the television series "Land of the Tiger."

5039. Thauer, Rolf, 1997. Biodiversity and unity in biochemistry. *Antonie van Leeuwenhoek* 71: 21-32. [2].

5040. Thelander, Carl G., ed., 1994. *Life on the Edge: A Guide to California's Endangered Natural Resources: Wildlife*. Santa Cruz and Berkeley: BioSystems Books. 550p. [1/1/1] *A.T.:* endangered species.

5041. Thiery, R.G., 1982. Environmental instability and community diversity. *Biological Reviews* 57: 691-710. [1].

5042. Thiollay, Jean-Marc, 1989. Area requirements for the conservation of rain forest raptors and game birds in French Guiana. *CB* 3: 128-137. [1] *A.T.:* reserve size.

5043. Thiollay, Jean-Marc, 1990. Comparative diversity of temperate and tropical forest bird communities: the influence of habitat heterogeneity. *Acta Oecologica* 11: 887-911. [1] *A.T.:* France; French Guiana.

5044. ———, 1992. Influence of selective logging on bird species diversity in a Guianan rain forest. *CB* 6: 47-63. [2] *A.T.:* French Guiana.

5045. ———, 1995. The role of traditional agroforests in the conservation of rain forest bird diversity in Sumatra. *CB* 9: 335-353. [1].

5046. ———, 1997. Distribution and abundance patterns of bird community and raptor populations in the Andaman Archipelago. *Ecography* 20: 67-82. *A.T.:* habitat preference.

5047. ———, 1997. Disturbance, selective logging and bird diversity: a Neotropical forest study. *Biodiversity and Conservation* 6: 1155-1173. *A.T.:* French Guiana.

5048. Thirgood, J.V., 1989. Man's impact on the forests of Europe. *J. of World Forest Resource Management* 4: 127-167. [1] Provides a historical perspective.

5049. Thomas, C.D., 1990. Fewer species [letter]. *Nature* 347: 237. [1] Questions and revises downward Erwin's estimate that as many as thirty million species of arthropods may exist. *A.T.:* insects; species numbers.

5050. ———, 1994. Extinction, colonization, and metapopulations: environmental tracking by rare species. *CB* 8: 373-378. [2] *A.T.:* local extinction; persistence.

5051. Thomas, C.D., & J.C.G. Abery, 1995. Estimating rates of butterfly decline from distribution maps: the effect of scale. *Biological Conservation* 73: 59-65. [1] *A.T.:* mapping; spatial scale; U.K.

5052. Thomas, C.D., & H.C. Mallorie, 1985. Rarity, species richness and conservation: butterflies of the Atlas Mountains in Morocco. *Biological Conservation* 33: 95-117. [2].

5053. Thomas, C.D., J.A. Thomas, & M.S. Warren, 1992. Distributions of occupied and vacant butterfly habitats in fragmented landscapes. *Oecologia* 92: 563-567. [2] *A.T.:* U.K.; island biogeography.

5054. Thomas, F., F. Cézilly, et al., 1997. Parasitism and ecology of wetlands: a review. *Estuaries* 20: 646-654. *A.T.:* coastal waters.

5055. Thomas, F., F. Renaud, et al., 1995. Differential mortality of two closely related host species induced by one parasite. *PRSL B* 260: 349-352. [1] *A.T.:* invertebrates.

5056. Thomas, Gregory A., 1996. Conserving aquatic biodiversity: a critical comparison of legal tools for augmenting streamflows in California. *Stanford Environmental Law J.* 15: 3-58. [1] *A.T.:* water transfers; public trust doctrine.

5057. Thomas, Heather Smith, 1996. Why some westerners fear "protection" of biodiversity. *Rangelands* 18: 179-181. [1].

5058. Thomas, J.A., & M.G. Morris, 1994. Patterns, mechanisms and rates of extinction among invertebrates in the United Kingdom. *PTRSL B* 344: 47-54. [1].

5059. Thomas, J.C., & W.J. Bond, 1997. Genetic variation in an endangered cedar (*Widdringtonia cedarbergensis*) versus two congeneric species. *South African J. of Botany* 63: 133-140. *A.T.:* Southern Africa.

5060. Thomas, J.D., 1993. Biological monitoring and tropical biodiversity in marine environments: a critique with recommendations, and comments on the use of amphipods as bioindicators. *J. of Natural History* 27: 795-806. [1] *A.T.:* pollution monitoring.

5061. Thomas, J.W., & Hal Salwasser, 1989. Bringing conservation biology into a position of influence in natural resource management. *CB* 3: 123-127. [1] Argues that conservation biologists should focus on bdc as a primary management goal.

5062. Thomas, J.W., L.F. Ruggiero, et al., 1988. Management and conservation of old-growth forests in the United States. *Wildlife Society Bull.* 16: 252-262. [1] *A.T.:* Pacific Northwest; environmental legislation.

5063. Thomas, V.G., & P.G. Kevan, 1993. Basic principles of agroecology and sustainable agriculture. *J. of Agricultural & Environmental*

Ethics 6: 1-19. [1].

5064. Thompson, D.B.A., & A. Brown, 1992. Biodiversity in montane Britain: habitat variation, vegetation diversity and some objectives for conservation. *Biodiversity and Conservation* 1: 179-208. [1].

5065. Thompson, H., & D. Kennedy, 1996. Ecological economics of biodiversity and tropical rainforest deforestation. *J. of Interdisciplinary Economics* 7: 169-190. [1].

5066. Thompson, Ian D., & D.A. Welsh, 1993. Integrated resource management in boreal forest ecosystems—impediments and solutions. *Forestry Chronicle* 69: 32-39. [1] *A.T.:* Canada; ecosystem management.

5067. Thompson, J.N., 1996. Evolutionary ecology and the conservation of biodiversity. *TREE* 11: 300-303. [1].

5068. Thompson, Ken, & J.G. Hodgson, 1996. More on the biogeography of scarce vascular plants. *Biological Conservation* 75: 299-302. [1] *A.T.:* British Isles.

5069. Thompson, Peter A., 1975. The collection, maintenance, and environmental importance, of the genetic resources of wild plants. *Environmental Conservation* 2: 223-228. [1].

5070. Thorne-Miller, Boyce, 1999. *The Living Ocean: Understanding and Protecting Marine Biodiversity* (2nd ed.). Washington, DC: Island Press. 214p. [2/2/1].

5071. Thornton, I.W.B., R.A. Zann, & S. van Balen, 1993. Colonization of Rakata (Krakatau Is.) by non-migrant land birds from 1883 to 1992 and implications for the value of island equilibrium theory. *J. of Biogeography* 20: 441-452. [1] *A.T.:* island biogeography; immigration curves; extinction curves.

5072. Thornton, Robert D., July-Aug. 1997. The No Surprises policy is essential to attract private dollars for the protection of biodiversity. *Endangered Species Update* 14(7-8): 65-66. [1].

5073. Thorpe, John E., G.A.E. Gall, et al., 1995. *Conservation of Fish and Shellfish Resources: Managing Diversity*. London and San Diego: Academic Press. 206p. [1/1/1] *A.T.:* germplasm resources; fishery management.

5074. Thorpe, John E., J.F. Koonce, et al., 1981. Assessing and managing man's impact on fish genetic resources. *Canadian J. of Fisheries and Aquatic Sciences* 38: 1899-1907. [1] *A.T.:* genetic diversity; stock structure.

5075. Thrupp, Lori Ann, 1997. *Linking Biodiversity and Agriculture: Challenges and Opportunities for Sustainable Food Security*. Washington, DC: WRI. 19p. [1/1/1] *A.T.:* food supply; international cooperation.

5076. ———, 1998. *Cultivating Diversity: Agrobiodiversity and Food Security*. Washington, DC: WRI. 80p. *A.T.:* international cooperation; food supply.

5077. Thrush, S.F., J.E. Hewitt, et al., 1998. Disturbance of the marine benthic habitat by commercial fishing: impacts at the scale of the fishery.

Ecological Applications 8: 866-879. *A.T.:* New Zealand.

5078. Ticco, Paul C., 1995. The use of marine protected areas to preserve and enhance marine biological diversity: a case study approach. *Coastal Management* 23: 309-314. [1] *A.T.:* coastal zone management.

5079. Tickell, Oliver, 4 Dec. 1993. Poor management could cost the earth. *New Scientist* 140(1902): 4-5. [1] Reports on criticism of the GEF. *A.T.:* World Bank; environmental policy.

5080. ———, 17 Jan. 1998. Paradise postponed. *New Scientist* 157 (2117): 18-19. Describes effort in Europe to reinstate more bd-friendly farming practices.

5081. Tiedje, James M., R.K. Colwell, et al., 1989. The planned introduction of genetically engineered organisms: ecological considerations and recommendations. *Ecology* 70: 298-315. [3] *A.T.:* transgenic organisms.

5082. Tietenberg, Thomas H., 1996. *Environmental and Natural Resource Economics* (4th ed.). New York: HarperCollins College. 614p. [2/1/2] *A.T.:* environmental economics; environmental policy.

5083. Tiffney, Bruce H., 1985. The Eocene North Atlantic land-bridge: its importance in Tertiary and modern phytogeography of the Northern Hemisphere. *J. of the Arnold Arboretum* 66: 243-273. [2].

5084. Tiggelen, John, summer 1995-1996. Satellites reveal a bleak landscape. *Ecos*, no. 86: 4. [1] Discusses changes in Australia's land cover.

5085. Tilford, David S., 1998. Saving the blueprints: the international legal regime for plant resources. *Case Western Reserve J. of International Law* 30: 373-446. *A.T.:* CBD; North-South relations; plant germplasm resources.

5086. Tilman, David, 1993. Species richness of experimental productivity gradients: how important is colonization limitation? *Ecology* 74: 2179-2191. [2] *A.T.:* grasslands; litter mass.

5087. ———, 1994. Competition and biodiversity in spatially structured habitats. *Ecology* 75: 2-16. [3] *A.T.:* spatial competition; grasslands.

5088. ———, 1996. Biodiversity: population versus ecosystem stability. *Ecology* 77: 350-363. [3] Finds that bd acts to stabilize communities and ecosystems, but not population levels. *A.T.:* diversity-stability hypothesis.

5089. ———, 1997. Community invasibility, recruitment limitation, and grassland biodiversity. *Ecology* 78: 81-92. [2] *A.T.:* community structure.

5090. Tilman, David, & J.A. Downing, 1994. Biodiversity and stability in grasslands. *Nature* 367: 363-365. [3] *A.T.:* primary productivity; drought; diversity-stability hypothesis.

5091. Tilman, David, & A. El Haddi, 1992. Drought and biodiversity in grasslands. *Oecologia* 89: 257-264. [2] *A.T.:* extinction.

5092. Tilman, David, C.L. Lehman, & C.E. Bristow, 1998. Diversity-stability relationships: statistical inevitability or ecological consequence? *American Naturalist* 151: 277-282. *A.T.:* competition; stochastic processes.

5093. Tilman, David, C.L. Lehman, & K.T. Thomson, 1997. Plant diversity and ecosystem productivity: theoretical considerations. *PNAS* 94: 1857-1861. [2] *A.T.:* soil nutrients; mathematical models.

5094. Tilman, David, David Wedin, & Johannes Knops, 1996. Productivity and sustainability influenced by biodiversity in grassland ecosystems. *Nature* 379: 718-720. [3] *A.T.:* soil nutrients; diversity-productivity hypothesis.

5095. Tilman, David, Johannes Knops, et al., 1997. The influence of functional diversity and composition on ecosystem processes. *Science* 277: 1300-1302. [2] *A.T.:* ecosystem management.

5096. Tilman, David, R.M. May, et al., 1994. Habitat destruction and the extinction debt. *Nature* 371: 65-66. [2] *A.T.:* habitat fragmentation; mathematical models.

5097. Timberlake, Lloyd, 1988. *Africa in Crisis: The Causes, the Cures of Environmental Bankruptcy* (new ed.). London: Earthscan. 203p. [3/3/2] *A.T.:* environmental degradation.

5098. Tinker, Catherine, 1995. A "new breed" of treaty: the United Nations Convention on Biological Diversity. *Pace Environmental Law Review* 13: 191-218. [1] *A.T.:* international law.

5099. ———, 1995. Responsibility for biological diversity conservation under international law. *Vanderbilt J. of Transnational Law* 28: 777-821. [1] *A.T.:* CBD; precautionary principle.

5100. Tisdell, Clem, 1991. *Economics of Environmental Conservation: Economics for Environmental & Ecological Management.* Amsterdam and New York: Elsevier. 233p. [1/1/1].

5101. ———, 1994. Conservation, protected areas and the global economic system: how debt, trade, exchange rates, inflation and macroeconomic policy affect biological diversity. *Biodiversity and Conservation* 3: 419-436. [1] *A.T.:* political lobbying; debt-for-nature swaps.

5102. ———, 1995. Issues in biodiversity conservation including the role of local communities. *Environmental Conservation* 22: 216-222. [1] *A.T.:* safe minimum standard; local participation; China; incentive measures.

5103. ———, spring 1996. Ecotourism, economics, and the environment: observations from China. *J. of Travel Research* 34(4): 11-19. [1].

5104. ———, 1998. *Biodiversity, Conservation, and Sustainable Development: Principles and Practices With Asian Examples* [case studies]. Aldershot, U.K., and Brookfield, VT: Edward Elgar. 320p.

5105. Tisdell, Clem, & Kartik Roy, 1997. Sustainability of land use in north-east India: issues involving economics, the environment and biodiversity. *International J. of Social Economics* 24: 160-177. *A.T.:* sustainable agriculture.

5106. Tisdell, Clem, & Xiang Zhu, 1996. Reconciling economic development, nature conservation and local communities: strategies for biodiversity conservation in Xishuangbanna, China. *Environmentalist* 16: 203-

211. [1] *A.T.:* Agenda 21; local participation; sustainable development.

5107. Titus, Timothy R., May-June 1992. Biodiversity: the need for a national policy. *Fisheries* 17(3): 31-34. [1] *A.T.:* environmental policy; environmental management.

5108. Tiwari, B.K., S.K. Barik, & R.S. Tripathi, 1998. Biodiversity value, status, and strategies for conservation of sacred groves of Meghalaya, India. *Ecosystem Health* 4: 20-32. *A.T.:* disturbance ecology.

5109. Tobey, James A., 1993. Toward a global effort to protect the Earth's biological diversity. *World Development* 21: 1931-1945. [1] *A.T.:* international cooperation; environmental economics; North-South relations.

5110. ———, Feb.-March 1996. Economic incentives for biodiversity. *OECD Observer*, no. 198: 25-28. [1] *A.T.:* environmental economics.

5111. Tobias, Michael, 1998. *Nature's Keepers: On the Front Lines of the Fight to Save Wildlife in America*. New York: Wiley. 238p. *A.T.:* wild animal trade; poaching; wildlife conservation.

5112. Tobin, Richard J., 1990. *The Expendable Future: U.S. Politics and the Protection of Biological Diversity*. Durham, NC: Duke University Press. 325p. [2/2/2] *A.T.:* environmental policy.

5113. Tokeshi, Mutsunori, 1990. Niche apportionment or random assortment: species abundance patterns revisited. *J. of Animal Ecology* 59: 1129-1146. [2] *A.T.:* chironomids; epiphytes.

5114. ———, 1993. Species abundance patterns and community structure. *Advances in Ecological Research* 24: 111-186. [2].

5115. Tol, G. van, H.F. van Dobben, et al., 1998. Biodiversity of Dutch forest ecosystems as affected by receding groundwater levels and atmospheric deposition. *Biodiversity and Conservation* 7: 221-228. *A.T.:* desiccation; nitrogen deposition.

5116. Tolba, Mostafa K., 1989. Our biological heritage under siege. *Bioscience* 39: 725-728. [1].

5117. Tomlinson, Timothy R., & Olayiwola Akerele, eds., 1998. *Medicinal Plants: Their Role in Health and Biodiversity* [symposium proceedings]. Philadelphia: University of Pennsylvania Press. 221p.

5118. Tonhasca, A., Jr., 1994. Diversity indices in the analysis of biological communities. *Ciência e Cultura* (São Paulo) 46: 138-140. [1] *A.T.:* carabid beetles; Ohio.

5119. Tonn, William M., J.J. Magnuson, et al., 1990. Intercontinental comparison of small-lake fish assemblages: the balance between local and regional processes. *American Naturalist* 136: 345-375. [2] *A.T.:* Finland; Wisconsin.

5120. Tonteri, T., 1994. Species richness of boreal understorey forest vegetation in relation to site type and successional factors. *Annales Zoologici Fennici* 31: 53-60. [1] *A.T.:* Finland; environmental gradients.

5121. Toro, Taryn, July-Aug. 1993. A Cold War legacy. *Wildlife Conser-*

vation 96(4): 66-71. [1] Describes how abandoned military posts in East Germany have been made nature reserves.

5122. Torsvik, Vigdis, Jostein Goksøyr, & Frida Lise Daae, 1990. High diversity in DNA of soil bacteria. *Applied and Environmental Microbiology* 56: 782-787. [3].

5123. Torsvik, Vigdis, R. Sorheim, & J. Goksoyr, 1996. Total bacterial diversity in soil and sediment communities—a review. *J. of Industrial Microbiology & Biotechnology* 17: 170-178. [2] *A.T.:* soil microorganisms.

5124. Torsvik, Vigdis, Kåre Salte, et al., 1990. Comparison of phenotypic diversity and DNA heterogeneity in a population of soil bacteria. *Applied and Environmental Microbiology* 56: 776-781. [2] *A.T.:* genetic diversity; soil microorganisms.

5125. Townsend, C.R., M.R. Scarsbrook, & Sylvain Doledec, 1997. The intermediate disturbance hypothesis, refugia, and biodiversity in streams. *Limnology and Oceanography* 42: 938-949. *A.T.:* invertebrates.

5126. Tracy, C. Richard, & P.F. Brussard, 1994. Preserving biodiversity: species in landscapes [letter]. *Ecological Applications* 4: 205-207. [1].

5127. Tracy, C. Richard, & T.L. George, 1992. On the determinants of extinction. *American Naturalist* 139: 102-122. [2] *A.T.:* extinction risk; U.K.; birds; body size.

5128. Traina, Frank, & Susan Darley-Hill, eds., 1995. *Perspectives in Bioregional Education*. Troy, OH: North American Association for Environmental Education. 162p. [1/1/1] *A.T.:* environmental education.

5129. Tramer, Elliot J., 1969. Bird species diversity: components of Shannon's formula. *Ecology* 50: 927-929. [1] *A.T.:* phytoplankton.

5130. ———, 1974. On latitudinal gradients in avian diversity. *Condor* 76: 123-130. [1] *A.T.:* North America; alpha diversity; gamma diversity.

5131. Tranvik, Lars J., & L.A. Hansson, 1997. Predator regulation of aquatic microbial abundance in simple food webs of sub-Antarctic lakes. *Oikos* 79: 347-356. *A.T.:* South Georgia Island; copepods.

5132. Travaini, A., M. Delibes, et al., 1997. Diversity, abundance or rare species as a target for the conservation of mammalian carnivores: a case study in southern Spain. *Biodiversity and Conservation* 6: 529-535. *A.T.:* wildlife management.

5133. Traverse, Alfred, 1990. Plant evolution in relation to world crises and the apparent resilience of Kingdom Plantae. *Global and Planetary Change* 2: 203-211. [1] *A.T.:* greenhouse effect.

5134. Travis, S.E., J. Maschinski, & P. Keim, 1996. An analysis of genetic variation in *Astragalus cremnophylax* var. *cremnophylax*, a critically endangered plant, using AFLP markers. *Molecular Ecology* 5: 735-745. [2] *A.T.:* heterozygosity; conservation genetics.

5135. Trebino, Hernan J., E.J. Chaneton, & R.J.C. Leon, 1996. Flooding, topography, and successional age as determinants of species diversity in oldfield vegetation. *Canadian J. of Botany* 74: 582-588. [1] *A.T.:* Argentina; grasslands.

5136. Trémolières, Michèle, J.M. Sánchez-Pérez, et al., 1998. Impact of river management history on the community structure, species composition and nutrient status in the Rhine alluvial hardwood forest. *Plant Ecology* 135: 59-78.

5137. Trevors, J.T., 1998. Bacterial biodiversity in soil with an emphasis on chemically-contaminated soils. *Water, Air, and Soil Pollution* 101: 45-67. *A.T.:* soil pollution; microorganisms.

5138. Trinder-Smith, T.H., R.M. Cowling, & H.P. Linder, 1996. Profiling a besieged flora: endemic and threatened plants of the Cape Peninsula, South Africa. *Biodiversity and Conservation* 5: 575-589. [1] *A.T.:* fynbos.

5139. Trinder-Smith, T.H., A.T. Lombard, & M.D. Picker, 1996. Reserve scenarios for the Cape Peninsula: high-, middle- and low-road options for conserving the remaining biodiversity. *Biodiversity and Conservation* 5: 649-669. [1] *A.T.:* fynbos; South Africa.

5140. Tripp, R., fall 1996. Biodiversity and modern crop varieties: sharpening the debate. *Agriculture and Human Values* 13(4): 48-63. [1] *A.T.:* genetic resources.

5141. Triquet, A.M., G.A. McPeek, & W.C. McComb, 1990. Songbird diversity in clearcuts with and without a riparian buffer strip. *J. of Soil and Water Conservation* 45: 500-503. [1] *A.T.:* Kentucky.

5142. Troumbis, A.Y., & P.G. Dimitrakopoulos, 1998. Geographic coincidence of diversity threatspots for three taxa and conservation planning in Greece. *Biological Conservation* 84: 1-6. *A.T.:* endangered species.

5143. Trüper, H.G., 1992. Prokaryotes: an overview with respect to biodiversity and environmental importance. *Biodiversity and Conservation* 1: 227-236. [1].

5144. Tryon, Rolla M., 1986. The biogeography of species, with special reference to ferns. *Botanical Review* 52: 117-156. [1] *A.T.:* geographical range; speciation.

5145. Tsai, Tzong-Bing, 1995. One step further towards biodiversity conservation. *Tulsa Law J.* 30: 657-670. [1].

5146. Tudge, Colin, 1992. *Last Animals at the Zoo: How Mass Extinction Can Be Stopped.* Washington, DC: Island Press. 266p. [3/3/1] *A.T.:* zoos; conservation biology.

5147. Tuluhan Yilmaz, K., 1998. Ecological diversity of the Eastern Mediterranean region of Turkey and its conservation. *Biodiversity and Conservation* 7: 87-96.

5148. Tuomisto, Hanna, 1998. What satellite imagery and large scale field studies can tell about biodiversity patterns in Amazonian forests. *Annals of the Missouri Botanical Garden* 85: 48-62. *A.T.:* remote sensing; Peru; ecological heterogeneity.

5149. Tuomisto, Hanna, & Kalle Ruokolainen, 1997. The role of ecological knowledge in explaining biogeography and biodiversity in Amazonia. *Biodiversity and Conservation* 6: 347-357.

5150. Tuomisto, Hanna, Kalle Ruokolainen, et al., 1995. Dissecting Amazonian biodiversity. *Science* 269: 63-66. [2] *A.T.:* ecological heterogeneity; biogeography.

5151. Turner, Billie Lee, W.C. Clark, et al., eds., 1990. *The Earth as Transformed by Human Action: Global and Regional Changes in the Biosphere Over the Past 300 Years* [symposium papers]. Cambridge, U.K., and New York: CUP. 713p. [2/2/2] *A.T.:* human ecology; environmental change.

5152. Turner, Ian M., 1996. Species loss in fragments of tropical rain forest: a review of the evidence. *J. of Applied Ecology* 33: 200-209. [2] *A.T.:* habitat fragmentation; local extinction.

5153. ———, 1997. A tropical flora summarized: a statistical analysis of the vascular plant diversity of Malaya. *Flora* 192: 157-163.

5154. Turner, Ian M., K.S. Chua, et al., 1996. A century of plant species loss from an isolated fragment of lowland tropical rain forest. *CB* 10: 1229-1244. [2] *A.T.:* Singapore; historical records.

5155. Turner, Ian M., C.H. Diong, et al., eds., 1996. *Biodiversity and the Dynamics of Ecosystems* [workshop proceedings]. Kyoto: International Network for DIVERSITAS in Western Pacific and Asia. 383p. [1/1/1].

5156. Turner, Ian M., H.T.W. Tan, et al., 1994. A study of plant species extinction in Singapore: lessons for the conservation of biological diversity. *CB* 8: 705-712. [1] *A.T.:* species loss.

5157. Turner, John R.G., C.M. Gatehouse, & C.A. Corey, 1987. Does solar energy control organic diversity? Butterflies, moths and the British climate. *Oikos* 48: 195-205. [2] *A.T.:* species-energy hypothesis; seasonal diversity.

5158. Turner, John R.G., J.J. Lennon, & J.A. Lawrenson, 1988. British bird species distributions and the energy theory. *Nature* 335: 539-541. [2] *A.T.:* species-energy hypothesis; latitudinal diversity gradients.

5159. Turner, Monica G., ed., 1987. *Landscape Heterogeneity and Disturbance.* New York: Springer. 239p. [1/1/2] *A.T.:* landscape ecology.

5160. ———, 1989. Landscape ecology: the effect of pattern on process. *ARES* 20: 171-197. [3] *A.T.:* spatial pattern.

5161. Turner, Monica G., & R.H. Gardner, eds., 1991. *Quantitative Methods in Landscape Ecology: The Analysis and Interpretation of Landscape Heterogeneity.* New York: Springer. 536p. [2/1/2].

5162. Turner, Monica G., G.J. Arthaud, et al., 1995. Usefulness of spatially explicit population models in land management. *Ecological Applications* 5: 12-16. [2].

5163. Turner, Monica G., R.H. Gardner, et al., 1995. Ecological dynamics at broad scales: ecosystems and landscapes. *Bioscience* 45, Suppl.: S29-S35.

[1] *A.T.:* landscape ecology.

5164. Turner, Terence, 1993. The role of indigenous peoples in the environmental crisis: the example of the Kayapó of the Brazilian Amazon. *Perspectives in Biology and Medicine* 36: 526-545. [1].

5165. Tuxill, John D., & Chris Bright, June-July 1998. Protecting nature's biodiversity: mending strands in the web of life. *Futurist* 32(5): 46-51. An overview of the increasing ratio of species extinctions, and international attempts to deal with the problem.

5166. Tuxill, John D., & J.A. Peterson, 1998. *Losing Strands in the Web of Life: Vertebrate Declines and the Conservation of Biological Diversity.* Washington, DC: Worldwatch Paper No. 141. 88p. *A.T.:* wildlife conservation; bd loss.

5167. Tyler, Peter A., 1996. Endemism in fresh water algae. *Hydrobiologia* 336: 127-135. [1] *A.T.:* Australia.

5168. Udall, James R., & K.Y. Craft, Sept.-Oct. 1991. Launching the natural ark. *Sierra* 76(5): 80-89. [1] Describes the expansion of conservation efforts to the level of habitats and ecosystems.

5169. Udvardy, Miklos D.F., 1975. *A Classification of the Biogeographical Provinces of the World.* Morges, Switzerland: IUCN Occasional Paper No. 18. 48p. [1/1/1].

5170. Uetz, George W., 1974. Species diversity: a review. *Biologist* 56: 111-129. [1].

5171. Uhl, Christopher, April 1983. You *can* keep a good forest down: how much can Amazonian rain forests take and still recover? *Natural History* 92(4): 70-79. [1] *A.T.:* habitat disturbance; deforestation.

5172. Ulfstrand, Staffan, 1992. Biodiversity—how to reduce its decline [editorial]. *Oikos* 63: 3-5. [1].

5173. Ulgiati, S., M.T. Brown, et al., 1995. Emergy-based indices and ratios to evaluate the sustainable use of resources. *Ecological Engineering* 5: 519-531. [1].

5174. UNCED, 1995. *Convention on Biological Diversity, and the Bern Convention: The Next Steps.* Strasbourg: Council of Europe Publishing. 149p. [1/1/1] *A.T.:* Europe.

5175. Underwood, A.J., & M.G. Chapman, eds., 1995. *Coastal Marine Ecology of Temperate Australia.* Sydney: UNSW Press. 341p. [1/1/1].

5176. UNDP, UNEP, & World Bank, 1994. *Global Environment Facility: Independent Evaluation of the Pilot Phase.* Washington, DC: World Bank. 192p. [1/1/1] *A.T.:* UNDP; UNEP; environmental policy.

5177. ———, 1995. *Implementing the Convention on Biological Diversity: Toward a Strategy for World Bank Assistance.* Washington, DC: World Bank. 29p. [1/1/1].

5178. UNEP, 1990. *Elements for Possible Inclusion in a Global Framework Legal Instrument on Biological Diversity.* Nairobi. 23p. [1/1/1] *A.T.:*

environmental law.

5179. ———, 1992. *Global Support for the Preparation of Biodiversity Country Studies*. Nairobi. 40p. [1/1/1].

5180. ———, 1995. *The UNEP Biodiversity Programme and Implementation Strategy: A Framework for Supporting Global Conservation and Sustainable Use of Biodiversity*. Nairobi. 74p. [1/1/1].

5181. ———, 1997. *The Biodiversity Agenda: Decisions From the Third Meeting of the Conference of the Parties to the Convention on Biological Diversity* (2nd ed.) [conference proceedings]. New York: United Nations. 116p. [1/1/1] *A.T.:* CBD.

5182. Unger, Judith M., & N.R. Marin, 1997. The Flora of North America project: a 21st Century tool for managing plant information. *American Biology Teacher* 59: 338-343. *A.T.:* bd databases; bd catalogs.

5183. Unitarian Service Committee of Canada, 1994. *Sowing Seeds for Change: Sustainable Agriculture, Biodiversity, and Food Security* [workshop proceedings]. Ottawa. 96p. [1/1/1] *A.T.:* Canada.

5184. United Nations, 1993. *Agenda 21: The United Nations Programme of Action From Rio*. New York. 294p. [1/1/1] *A.T.:* international cooperation; UNCED.

5185. ———, 1993. *Report of the United Nations Conference on Environment and Development, Rio de Janeiro, 3-14 June 1992*. New York. 3 vols. [1/1/1].

5186. United States Agency for International Development (USAID), 1989. *Conserving Tropical Forests and Biological Diversity* [1988-1989 report to Congress]. Washington, DC. 44p. [1/1/1].

5187. ———, 1992. *Tropical Forests and Biological Diversity: USAID Report to Congress 1990-1991*. Washington, DC. 56p. [1/1/1] *A.T.:* tropical forestry.

5188. United States Fish and Wildlife Service, 1994. *An Ecosystem Approach to Fish and Wildlife Conservation: An Approach to More Effectively Conserve the Nation's Biodiversity*. Washington, DC. 14p. [1/1/1] *A.T.:* ecosystem management.

5189. United States General Accounting Office, 1997. *U.S. Department of Agriculture Information on the Condition of the National Plant Germplasm System: Report to Congressional Committees*. Washington, DC. 89p. [1/1/-].

5190. United States Man and the Biosphere Program, 1994. *Strategic Plan for the U.S. Biosphere Reserve Program*. Washington, DC. 28p. [1/1/1].

5191. ———, 1995. *Biosphere Reserves in Action: Case Studies of the American Experience*. Washington, DC. 86p. [2/1/1].

5192. Upton, Harold F., May-June 1992. Biodiversity and conservation of the marine environment. *Fisheries* 17(3): 20-25. [1].

5193. Urban, Dean L., R.V. O'Neill, & H.H. Shugart, Jr., 1987. Landscape ecology: a hierarchical perspective can help scientists understand spatial patterns. *Bioscience* 37: 119-127. [2] *A.T.:* vegetation patterns.

5194. Urbanska, Krystyna M., 1995. Biodiversity assessment in ecological restoration above the timberline. *Biodiversity and Conservation* 4: 679-695. [1] *A.T.:* montane environments; Switzerland; Alps; age-state structure.

5195. Urbanska, Krystyna M., N.R. Webb, & P.J. Edwards, eds., 1997. *Restoration Ecology and Sustainable Development* [conference papers]. Cambridge, U.K., and New York: CUP. 397p. [1/-/-].

5196. Urbanski, Mark A., 1995. Chemical prospecting, biodiversity conservation, and the importance of international protection of intellectual property rights in biological materials. *Buffalo J. of International Law* 2: 131-182. [1] *A.T.:* biotechnology.

5197. Uresk, Daniel W., G.L. Schenbeck, & J.T. O'Rourke, eds., 1997. *Conserving Biodiversity on Native Rangelands* [symposium proceedings]. Fort Collins, CO: General Technical Report RM GTR-298, Rocky Mountain Forest and Range Experiment Station. 38p. [1/1/-].

5198. Usher, Michael B., 1986. Invasibility and wildlife conservation: invasive species on nature reserves. *PTRSL B* 314: 695-710. [1] *A.T.:* U.K.; biological invasions.

5199. ———, ed., 1986. *Wildlife Conservation Evaluation*. London and New York: Chapman & Hall. 394p. [1/1/2].

5200. ———, 1988. Biological invasions of nature reserves: a search for generalisations. *Biological Conservation* 44: 119-135. [1].

5201. ———, 1989. Scientific aspects of nature conservation in the United Kingdom. *J. of Applied Ecology* 26: 813-824. [1] *A.T.:* Nature Conservancy Council.

5202. Usher, Michael B., & M. Edwards, 1986. The selection of conservation areas in Antarctica: an example using the arthropod fauna of Antarctic islands. *Environmental Conservation* 13: 115-122. [1].

5203. Usher, Michael B., A.C. Brown, & S.E. Bedford, 1992. Plant species richness in farm woodlands. *Forestry* 65: 1-13. [1] *A.T.:* island biogeography; U.K.

5204. Usher, Michael B., T.J. Crawford, & J.L. Banwell, 1992. An American invasion of Great Britain: the case of the native and alien squirrel (*Sciurus*) species. *CB* 6: 108-115. [1] *A.T.:* introduced species.

5205. Usseglio-Polatera, Philippe, 1994. Theoretical habitat templets, species traits, and species richness: aquatic insects in the Upper Rhône River and its floodplain. *Freshwater Biology* 31: 417-437. [1] *A.T.:* France; river ecology.

5206. Väisänen, R., & K. Heliövaara, 1994. Hot-spots of insect diversity in Northern Europe. *Annales Zoologici Fennici* 31: 71-81. [1] *A.T.:* Scandinavia; Finland; Russia.

5207. Valdéz, Manuel, & Gerardo Ceballos, 1997. Conservation of endemic mammals of Mexico: the Perote ground squirrel (*Spermophilus perotensis*). *J. of Mammalogy* 78: 74-82. *A.T.:* habitat fragmentation; wildlife conservation.

5208. Valencia, R., H. Balslev, & C.G. Paz y Miño, 1994. High tree alpha-diversity in Amazonian Ecuador. *Biodiversity and Conservation* 3: 21-28. [2] *A.T.:* tropical rainforests; species richness.

5209. Valentine, James W., 1968. Climatic regulation of species diversification and extinction. *Geological Society of America Bull.* 79: 273-275. [1] *A.T.:* climatic change; marine environments.

5210. ———, 1970. How many marine invertebrate fossil species?: a new approximation. *J. of Paleontology* 44: 410-415. [1] *A.T.:* species numbers; Phanerozoic.

5211. ———, 1971. Plate tectonics and shallow marine diversity and endemism, an actualistic model. *Systematic Zoology* 20: 253-264. [1] *A.T.:* provinciality; biogeography.

5212. ———, 1980. Determinants of diversity in higher taxonomic categories. *Paleobiology* 6: 444-450. [1] *A.T.:* diversification.

5213. ———, ed., 1985. *Phanerozoic Diversity Patterns: Profiles in Macroevolution* [conference papers]. Princeton, NJ: Princeton University Press, and San Francisco: Pacific Division, AAAS. 441p. [1/1/1] *A.T.:* evolution; diversification.

5214. ———, 1995. Why no new phyla after the Cambrian? Genome and ecospace hypotheses revisited. *Palaios* 10: 190-194. [1] *A.T.:* evolution; diversification.

5215. Valentine, James W., & David Jablonski, 1983. Larval adaptations and patterns of brachiopod diversity in space and time. *Evolution* 37: 1052-1061. [1].

5216. ———, 1991. Biotic effects of sea level change: the Pleistocene test. *J. of Geophysical Research* 96(B4): 6873-6878. [1] *A.T.:* California; molluscs.

5217. Valentine, James W., T.C. Foin, & D. Peart, 1978. A provincial model of Phanerozoic marine diversity. *Paleobiology* 4: 55-66. [1] *A.T.:* provinciality.

5218. Valentine, P.S., 1993. Ecotourism and nature conservation: a definition with some recent developments in Micronesia. *Tourism Management* 14: 107-115. [1] *A.T.:* Marshall Islands.

5219. Valle, Silvio, 1994. Enabling biodiversity [editorial]. *Bio/technology* 12: 1040. [1].

5220. Valone, T.J., & J.H. Brown, 1995. Effects of competition, colonization, and extinction on rodent species diversity. *Science* 267: 880-883. [1] *A.T.:* Chihuahuan Desert; competitive exclusion.

5221. Van der Kamp, Garth, 1995. The hydrogeology of springs in relation to the biodiversity of spring fauna: a review. *J. of the Kansas Entomological Society* 68(2), Suppl.: 4-17. [1] *A.T.:* aquatic insects.

5222. Van der Oost, John, W.M. De Vos, & Garo Antranikian, 1996. Extremophiles. *Trends in Biotechnology* 14: 415-417. [1] Reviews a related conference.

5223. Van der Ryn, Sim, & Stuart Cowan, 1995. *Ecological Design.* Washington, DC: Island Press. 201p. [2/1/1] *A.T.:* environmental policy; human ecology.

5224. Van Dorp, D., & P.F.M. Opdam, 1987. Effects of patch size, isolation and regional abundance on forest bird communities. *Landscape Ecology* 1: 59-73. [2] *A.T.:* Netherlands; woodlots.

5225. Van Dover, Cindy L., 1990. Biogeography of hydrothermal vent communities along seafloor spreading centers. *TREE* 5: 242-246. [1].

5226. Van Emden, H.F., & Z.T. Dabrowski, 1994. Biodiversity and habitat modification in pest management. *Insect Science and Its Application* 15: 605-620. [1].

5227. Van Hook, Tonya, 1997. Insect coloration and implications for conservation. *Florida Entomologist* 80: 193-210. Describes why large conspicuously colored insects require greater conservation attention.

5228. Van Jaarsveld, Albert S., 1995. Where to with reserve selection and conservation planning in South Africa? *South African J. of Zoology* 30: 164-168. [1].

5229. Van Jaarsveld, Albert S., & S.L. Chown, 1996. Strategies and timeframes for implementing the Convention on Biological Diversity: biological requirements. *South African J. of Science* 92: 459-464. [1] *A.T.:* South Africa.

5230. Van Jaarsveld, Albert S., J.W.H. Fergusson, & G.J. Bredenkamp, 1998. The Groenvaly grassland fragmentation experiment: design and initiation. *Agriculture, Ecosystems & Environment* 68: 139-150. *A.T.:* South Africa.

5231. Van Jaarsveld, Albert S., Stefanie Freitag, et al., 1998. Biodiversity assessment and conservation strategies. *Science* 279: 2106-2108. *A.T.:* South Africa; higher taxa.

5232. Van Jaarsveld, Albert S., K.J. Gaston, et al., 1998. Commentary: throwing biodiversity out with the binary data? [editorial] *South African J. of Science* 94: 210-214. *A.T.:* South Africa.

5233. Van Valen, Leigh M., 1973. Body size and numbers of plants and animals. *Evolution* 27: 27-35. [2].

5234. ———, 1973. A new evolutionary law. *Evolutionary Theory* 1: 1-30. [3]. *A.T.:* Red Queen hypothesis.

5235. Van Vuren, Dirk, & P.W. Hedrick, 1989. Genetic conservation in feral populations of livestock. *CB* 3: 312-317. [1] *A.T.:* California; sheep.

5236. Van Welzen, P.C., 1992. Interpretation of historical biogeographic results. *Acta Botanica Neerlandica* 41: 75-87. [1] *A.T.:* cladistics.

5237. Van Wilgen, Brian W., R.M. Cowling, & C.J. Burgers, 1996. Valuation of ecosystem services: a case study from South African fynbos ecosystems. *Bioscience* 46: 184-189. [1] *A.T.:* water supply.

5238. Van Wilgen, Brian W., D.M. Richardson, & A.H.W. Seydack, 1994. Managing fynbos for biodiversity: constraints and options in a fire-prone environment. *South African J. of Science* 90: 322-328. [1] *A.T.:* South Africa.

5239. Vandermeer, John, 1996. *Reconstructing Biology: Genetics and Ecology in the New World Order.* New York: Wiley. 478p. [2/1/1] *A.T.:* human ecology.

5240. Vandermeer, John, & Ivette Perfecto, 1995. *Breakfast of Biodiversity: The Truth About Rain Forest Destruction.* Oakland, CA: Institute for Food and Development Policy. 185p. [2/1/1] *A.T.:* rainforest conservation; tropics; agricultural ecology.

5241. ———, 1997. The agroecosystem: a need for the conservation biologist's lens [editorial]. *CB* 11: 591-592. *A.T.:* sustainable agriculture.

5242. Vandermeer, John, Douglas Boucher, et al., 1996. A theory of disturbance and species diversity: evidence from Nicaragua after Hurricane Joan. *Biotropica* 28: 600-613. [1].

5243. Vandermeer, John, Meine van Noordwijk, et al., 1998. Global change and multi-species agroecosystems: concepts and issues. *Agriculture, Ecosystems & Environment* 67: 1-22. *A.T.:* ecosystem management.

5244. Vane-Wright, Richard I., 1996. Systematics and the conservation of biological diversity. *Annals of the Missouri Botanical Garden* 83: 47-57. [1].

5245. Vane-Wright, Richard I., C.J. Humphries, & P.H. Williams, 1991. What to protect?—systematics and the agony of choice. *Biological Conservation* 55: 235-254. [3] Discusses means of prioritizing areas for conservation. *A.T.:* complementarity; bd measures.

5246. Vaney, Neil, 1995. Biodiversity and beauty. *Pacifica* (Melbourne) 8: 335-345. [1] *A.T.:* environmental ethics.

5247. Vanzolini, Paulo E., & W.R. Heyer, eds., 1988. *Proceedings of a Workshop on Neotropical Distribution Patterns.* Rio de Janeiro: Academia Brasileira de Ciências. 488p. [1/1/1] *A.T.:* biogeography.

5248. Varty, Nigel, 1991. The status and conservation of Jamaica's threatened and endemic forest avifauna and their habitats following Hurricane Gilbert. *Bird Conservation International* 1: 135-151. [1].

5249. Vaughan, Duncan A., & Te-Tzu Chang, 1992. *In situ* conservation of rice genetic resources. *Economic Botany* 46: 368-383. [1].

5250. Vaughan, Ray, 1994. *Endangered Species Act Handbook.* Rockville, MD: Government Institutes. 165p. [1/1/1].

5251. Vaughn, Gerald F., & C.D. Meine, 1996. Acting on principle: Aldo

Leopold and biodiversity conservation. *Choices: The Magazine of Food, Farm and Resource Issues* (3): 32-35. [1].

5252. Vázquez, D., & J.L. Gittleman, 1998. Biodiversity conservation: does phylogeny matter? *Current Biology* 8: R379-R381.

5253. Vecchione, Michael, & B.B. Collette, Nov. 1994. National Systematics Laboratory: the NSL works toward healthy marine biodiversity. *Fisheries* 19(11): 26. [1] Profiles the NSL's work.

5254. ———, 1996. Fisheries agencies and marine biodiversity. *Annals of the Missouri Botanical Garden* 83: 29-36. [1] *A.T.:* systematics.

5255. Vellvé, Renée, 1993. The decline of diversity in European agriculture. *Ecologist* 23: 64-69. [1] Examines the apparent decline in number of fruit and vegetable varieties available to consumers.

5256. Verghese, Abraham, S. Sridhar, & A.K. Chakravarthy, eds., 1995. *Bird Diversity and Conservation: Thrusts for the Nineties and Beyond* [seminar recommendations]. Bangalore, India: Ornithological Society of India. 120p. [1/1/1].

5257. Veríssimo, A., P. Barreto, et al., 1995. Extraction of a high-value natural resource in Amazonia: the case of mahogany. *Forest Ecology and Management* 72: 39-60. [2] *A.T.:* logging; deforestation; Brazil; forest management.

5258. Vermeij, Geerat J., 1978. *Biogeography and Adaptation: Patterns of Marine Life*. Cambridge, MA: Harvard University Press. 322p. [2/1/3].

5259. ———, 1987. The dispersal barrier in the tropical Pacific: implications for molluscan speciation and extinction. *Evolution* 41: 1046-1058. [1].

5260. ———, 1989. Geographical restriction as a guide to the causes of extinction: the case of the cold northern oceans during the Neogene. *Paleobiology* 15: 335-356. [1] *A.T.:* refugia; Tertiary.

5261. ———, 1991. When biotas meet: understanding biotic interchange. *Science* 253: 1099-1104. [2] *A.T.:* Tertiary; paleobiogeography; marine environments.

5262. ———, 1991. Anatomy of an invasion: the trans-Arctic interchange. *Paleobiology* 17: 281-307. [2] *A.T.:* biological invasions; Bering Strait; molluscs; Cenozoic.

5263. ———, 1993. Biogeography of recently extinct marine species: implications for conservation. *CB* 7: 391-397. [1] *A.T.:* refugia; North Pacific; Northeast Atlantic.

5264. Vermeij, Geerat J., & G. Rosenberg, 1993. Giving and receiving: the tropical Atlantic as donor and recipient region for invading species. *American Malacological Bull.* 10: 181-194. [1] *A.T.:* molluscs.

5265. Verner, Jared, 1992. Data needs for avian conservation biology: have we avoided critical research? *Condor* 94: 301-303. [1] *A.T.:* research programs.

5266. Verner, Jared, M.L. Morrison, & C.J. Ralph, eds., 1986. *Wildlife 2000: Modeling Habitat Relationships of Terrestrial Vertebrates* [symposium papers]. Madison: University of Wisconsin Press. 470p. [2/1/1] *A.T.:* vertebrate ecology.

5267. Vernon, P., Guy Vannier, & Paul Tréhen, 1998. A comparative approach to the entomological diversity of polar regions. *Acta Oecologica* 19: 303-308. *A.T.:* Arctic; Antarctic.

5268. Veron, J.E.N., 1992. Conservation of biodiversity: a critical time for the hermatypic corals of Japan. *Coral Reefs* 11: 13-21. [1] *A.T.:* rare species; seastar predation; Northwest Pacific.

5269. ———, 1994. Biodiversity of coral reefs: is there a problem in the Indo-Pacific centre of diversity? In Robert N. Ginsburg, ed., *Proceedings of the Colloquium on Global Aspects of Coral Reefs, Health, Hazards and History* (Miami: Rosenstiel School of Marine and Atmospheric Science, University of Miami): 365-370. [1].

5270. ———, 1995. *Corals in Space and Time: Biogeography and Evolution of the Scleractinia.* Ithaca, NY: Comstock/Cornell. 321p. [1/1/2].

5271. Vetaas, Ole R., 1997. The effect of canopy disturbance on species richness in a central Himalayan oak forest. *Plant Ecology* 132: 29-38. *A.T.:* Nepal; intermediate disturbance hypothesis.

5272. Vice, Daniel, 1997. Implementation of biodiversity treaties: monitoring, fact-finding, and dispute resolution. *New York University J. of International Law & Politics* 29: 577-639. *A.T.:* international treaties; endangered species.

5273. Vidal, Gonzalo, & Malgorzata Moczydlowska-Vidal, 1997. Biodiversity, speciation, and extinction trends of Proterozoic and Cambrian phytoplankton. *Paleobiology* 23: 230-246. *A.T.:* Cambrian explosion.

5274. Villaseñor, José Luis, Guillermo Ibarra, & Daniel Ocaña, 1998. Strategies for the conservation of Asteraceae in Mexico. *CB* 12: 1066-1075.

5275. Villora-Moreno, S., 1997. Environmental heterogeneity and the biodiversity of interstitial Polychaeta. *Bull. of Marine Science* 60: 494-501. *A.T.:* community composition.

5276. Vincent, A., & A. Clarke, 1995. Diversity in the marine environment [editorial]. *TREE* 10: 55-56. [1].

5277. Vincent, W.F., & M.R. James, 1996. Biodiversity in extreme aquatic environments: lakes, ponds and streams of the Ross Sea sector, Antarctica. *Biodiversity and Conservation* 5: 1451-1471. [1] *A.T.:* microorganisms.

5278. Vinson, Hilary, May-June 1996. Outreach and beyond! *Endangered Species Bull.* 21(3): 10-11. [1] *A.T.:* public outreach; North Carolina; South Carolina.

5279. Vinson, Mark R., & C.P. Hawkins, 1996. Effects of sampling area and subsampling procedure on comparisons of taxa richness among streams. *J. of the North American Benthological Society* 15: 392-399. [1] *A.T.:*

invertebrates; bd monitoring.

5280. ———, 1998. Biodiversity of stream insects: variation at local, basin, and regional scales. *Annual Review of Entomology* 43: 271-293. *A.T.:* habitat complexity; spatial scale.

5281. Virk, Parminder S., B.V. Ford-Lloyd, et al., 1995. Use of RAPD for the study of diversity within plant germplasm collections. *Heredity* 74: 170-179. [2] *A.T.:* rice germplasm.

5282. Virk, Parminder S., B.V. Ford-Lloyd, et al., 1996. Predicting quantitative variation within rice germplasm using molecular markers. *Heredity* 76: 296-304. [1] *A.T.:* RAPD; stress tolerance.

5283. Virkkala, R., 1991. Population trends of forest birds in a Finnish Lapland landscape of large habitat blocks: consequences of stochastic environmental variation or regional habitat alteration? *Biological Conservation* 56: 223-240. [1] *A.T.:* boreal forest; population fluctuations.

5284. Visscher, Henk, Henk Brinkhuis, et al., 1996. The terminal Paleozoic fungal event: evidence of terrestrial ecosystem destabilization and collapse. *PNAS* 93: 2155-2158. [2] *A.T.:* mass extinctions.

5285. Vitousek, Peter M., 1990. Biological invasions and ecosystem processes: towards an integration of population biology and ecosystem studies. *Oikos* 57: 7-13. [2] *A.T.:* Hawaii; nutrient cycling.

5286. ———, 1994. Beyond global warming: ecology and global change. *Ecology* 75: 1861-1876. [3] *A.T.:* nitrogen cycle; carbon dioxide enrichment; human disturbance.

5287. Vitousek, Peter M., L.L. Loope, & Henning Adsersen, eds., 1995. *Islands: Biological Diversity and Ecosystem Function.* New York: Springer. 238p. [1/1/1] *A.T.:* island ecology.

5288. Vitousek, Peter M., C.M. D'Antonio, et al., 1996. Biological invasions as global environmental change. *American Scientist* 84: 468-478. [2].

5289. Vitousek, Peter M., H.A. Mooney, et al., 1997. Human domination of Earth's ecosystems. *Science* 277: 494-499. [2] *A.T.:* human disturbance.

5290. Vitt, Laurie J., T.C.S. Avila-Pires, et al., 1998. The impact of individual tree harvesting on thermal environments of lizards in Amazonian rain forest. *CB* 12: 654-664. *A.T.:* single-tree forestry; canopy gaps.

5291. Vogel, Gretchen, 1997. The Pentagon steps up the battle to save biodiversity. *Science* 275: 20. *A.T.:* Dept. of Defense; protected areas; wildlife management; military bases.

5292. Vogel, Joseph H., 1994. *Genes for Sale: Privatization as a Conservation Policy.* New York: OUP. 155p. [1/1/1] *A.T.:* germplasm resources.

5293. Vogelmann, James E., 1995. Assessment of forest fragmentation in southern New England using remote sensing and geographic information systems technology. *CB* 9: 439-449. [1] *A.T.:* Massachusetts; New Hampshire.

5294. Vogler, A.P., & R. DeSalle, 1994. Diagnosing units of conservation management. *CB* 8: 354-363. [2] *A.T.:* tiger beetles; evolutionarily significant units; cladistics.

5295. Volk, Tyler, 1998. *Gaia's Body: Toward a Physiology of Earth.* New York: Copernicus. 269p. [2/-/-] *A.T.:* Gaia hypothesis.

5296. Volkman, John M., May 1992. Making room in the ark: the Endangered Species Act and the Columbia River Basin. *Environment* 34(4): 18-20, 37-43. [1] *A.T.:* fisheries management.

5297. Vora, Robin S., 1994. Integrating old-growth forest into managed landscapes: a northern Great Lakes perspective. *Natural Areas J.* 14: 113-123. [1] *A.T.:* forest management.

5298. Voss, Robert S., & L.H. Emmons, 1996. *Mammalian Diversity in Neotropical Lowland Rainforests: A Preliminary Assessment.* New York: Bull. of the American Museum of Natural History No. 230. 115p. [1/1/1].

5299. Vries, D.J. de, & M.R. Hall, 1994. Marine biodiversity as a source of chemical diversity. *Drug Development Research* 33: 161-173. [1].

5300. Vrijenhoek, Robert C., 1998. Animal clones and diversity: are natural clones generalists or specialists? *Bioscience* 48: 617-628. *A.T.:* genetic variation.

5301. Vuilleumier, François, & Maximina Monasterio, eds., 1986. *High Altitude Tropical Biogeography.* New York: OUP. 649p. [1/1/1] *A.T.:* montane environments.

5302. Vuilleumier, François, & D.S. Simberloff, 1980. Ecology versus history as determinants of patchy and insular distributions in high Andean birds. *Evolutionary Biology* 12: 235-379. [1].

5303. Vuori, K.-M., & I. Joensuu, 1996. Impact of forest drainage on the macroinvertebrates of a small boreal headwater stream: do buffer zones protect lotic biodiversity? *Biological Conservation* 77: 87-95. [1] *A.T.:* Finland; aquatic moss.

5304. Wade, Michael J., & D.E. McCauley, 1988. Extinction and recolonization: their effects on the genetic differentiation of local populations. *Evolution* 42: 995-1005. [3] *A.T.:* colonization.

5305. Wagner, Peter J., 1996. Contrasting the underlying patterns of active trends in morphologic evolution. *Evolution* 50: 990-1007. [1] *A.T.:* gastropods; Paleozoic.

5306. Wagner, Warren H., Jr., 1990. Biological diversity: underlying concepts. *Michigan Academician* 22: 311-317. [1].

5307. Wagner, Warren L., & V.A. Funk, eds., 1995. *Hawaiian Biogeography: Evolution on a Hot Spot Archipelago* [symposium papers]. Washington, DC: Smithsonian Institution Press. 467p. [1/1/2] *A.T.:* island ecology.

5308. Waite, Anya M., spring-summer 1996. Phytoplankton biodiversity. *Oceanus* 39(1): 2-5. [1].

5309. Wake, David B., 1991. Declining amphibian populations. *Science* 253: 860. [2] Reports on the conclusions of a workshop.

5310. Waldman, John R., Sept. 1995. Sturgeons and paddlefishes: a convergence of biology, politics, and greed. *Fisheries* 20(9): 20-21, 49. [1] Overview of a bdc-oriented conference.

5311. Waldren, S., J. Florence, & A.J. Chepstow-Lusty, 1995. Rare and endemic vascular plants of the Pitcairn Islands, south-central Pacific Ocean: a conservation appraisal. *Biological Conservation* 74: 83-98. [1] *A.T.:* endangered plants.

5312. Walker, Brian H., 1992. Biodiversity and ecological redundancy. *CB* 6: 18-23. [2] Argues that conservation efforts should focus on maintaining ecosystem functioning.

5313. ———, 1993. Rangeland ecology: understanding and managing change. *Ambio* 22: 80-87. [1].

5314. ———, 1995. Conserving biological diversity through ecosystem resilience. *CB* 9: 747-752. [2] *A.T.:* ecological redundancy.

5315. Walker, Brian H., & Henry Nix, 1993. Managing Australia's biological diversity. *Search* 24: 173-178. [1] *A.T.:* regional bd modelling.

5316. Walker, Paul A., & D.P. Faith, 1994. DIVERSITY-PD: procedures for conservation evaluation based on phylogenetic diversity. *Biodiversity Letters* 2: 132-139. [1] *A.T.:* computer programs; diversity measures.

5317. Walker, Timothy D., & J.W. Valentine, 1984. Equilibrium models of evolutionary species diversity and the number of empty niches. *American Naturalist* 124: 887-899. [1] *A.T.:* diversification; mathematical models.

5318. Wallace, Alfred Russel, 1860. On the zoological geography of the Malay Archipelago. *J. of Proc. of the Linnean Society, Zoology* 4: 172-184. [1] Original description of the "Wallace Line" faunal discontinuity.

5319. ———, 1869. *The Malay Archipelago.* London: Macmillan. 2 vols. [3/-/2] A celebrated scientific travel book. *A.T.:* natural history.

5320. ———, 1876. *The Geographical Distribution of Animals.* London: Macmillan. 2 vols. [2/-/2] A cornerstone in the development of the field of zoogeography.

5321. ———, 1878. *Tropical Nature, And Other Essays.* London: Macmillan. 356p. [1/-/2] Includes the first mention of the subject of latitudinal diversity gradients. *A.T.:* natural history.

5322. ———, 1880. *Island Life.* London: Macmillan. 526p. [2/-/1] The first major glacial and natural selection theories-based treatment of the evolution and distribution of island biotas. *A.T.:* glacial theory, island biotas.

5323. Wallace, David R., winter 1989. Of buccaneers and biodiversity. *Wilderness* 53(187): 38-51. [1] Concerning bd loss in the Caribbean and Florida Keys.

5324. ———, winter 1992. The Klamath surprise: forestry meets biodiversity on the West Coast. *Wilderness* 56(199): 10-21, 31-33. [1] Describes differences of opinion as to how to manage resources in the Klamath Valley region.

5325. Waller, Geoffrey, ed., 1996. *SeaLife: A Complete Guide to the Ma-*

rine Environment. Washington, DC: Smithsonian Institution Press. 504p. [2/1/1] *A.T.:* marine life.

5326. Wallis, G.P., 1994. Population genetics and conservation in New Zealand: a hierarchical synthesis and recommendations for the 1990s. *J. of the Royal Society of New Zealand* 24: 143-160. [1] *A.T.:* population bottlenecks; inbreeding.

5327. Wallis de Vries, M.F., 1995. Large herbivores and the design of large-scale nature reserves in Western Europe. *CB* 9: 25-33. [1] *A.T.:* keystone species; umbrella species; habitat fragmentation.

5328. Walliser, Otto H., ed., 1986. *Global Bio-events: A Critical Approach* [conference proceedings]. Berlin and New York: Springer. 442p. [1/1/1] *A.T.:* paleobiogeography.

5329. ———, ed., 1996. *Global Events and Event Stratigraphy in the Phanerozoic* [conference papers]. Berlin and New York: Springer. 333p. [1/1/1] *A.T.:* paleobiogeography.

5330. Wallner, William, 15 March 1996. Invasion of the tree snatchers. *American Nurseryman* 183(6): 28-31. [1] *A.T.:* insect pests; Asian gypsy moth.

5331. Walls, Susan C., A.R. Blaustein, & J.J. Beatty, 1992. Amphibian biodiversity of the Pacific Northwest with special reference to old-growth stands. *Northwest Environmental J.* 8: 53-69. [1] *A.T.:* amphibian declines.

5332. Wallwork, John A., 1976. *The Distribution and Diversity of Soil Fauna*. London and New York: Academic Press. 355p. [2/1/2].

5333. Walter, D.E., & H.C. Proctor, 1998. Predatory mites in tropical Australia: local species richness and complementarity. *Biotropica* 30: 72-81. *A.T.:* speciose taxa.

5334. Walter, Heinrich, 1983. *Vegetation of the Earth and Ecological Systems of the Geobiosphere* (3rd ed.). Berlin and New York: Springer. 318p. [3/-/3] *A.T.:* plant ecology; biogeography.

5335. Walter, John, 15 March 1994. How to harness biodiversity: biological pest control can be a byproduct of conservation practices with diverse plantings. *Successful Farming* 92(5): 38-39. [1] *A.T.:* farming practices.

5336. Walters, Jeffrey R., 1991. Application of ecological principles to the management of an endangered species: the case of the red-cockaded woodpecker. *ARES* 22: 505-523. [2].

5337. Walther, B.A., & S. Morand, 1998. Comparative performance of species richness estimation methods. *Parasitology* 116: 395-405. *A.T.:* sampling; jackknife procedure; bootstrap procedure.

5338. Walther, B.A., P. Cotgreave, et al., 1995. Sampling effort and parasite species richness. *Parasitology Today* 11: 306-310. [2].

5339. Waples, Robin S., 1990. Conservation genetics of Pacific salmon. II. Effective population size and the rate of loss of genetic variability. *J. of Heredity* 81: 267-276. [1] *A.T.:* population genetics.

5340. Waples, Robin S., 1991. Pacific salmon, *Oncorhynchus* spp., and the definition of "species" under the Endangered Species Act. *Marine Fisheries Review* 53(3): 11-22. [2].

5341. Wapner, Paul, 1994. On the global dimension of environmental challenges. *Politics and the Life Sciences* 13: 173-181. [1] Argues against attempting to treat all environmental issues at a global level. *A.T.:* environmentalism; globalization.

5342. Ward, David M., C.M. Santegoeds, et al., 1997. Biodiversity within hot spring microbial mat communities: molecular monitoring of enrichment cultures. *Antonie van Leeuwenhoek* 71: 143-150. *A.T.:* 16S rRNA sequences; microorganisms.

5343. Ward, J.P., & S.H. Anderson, 1988. Influences of cliffs on wildlife communities in southcentral Wyoming USA. *J. of Wildlife Management* 52: 673-678. [1] *A.T.:* topographic heterogeneity.

5344. Ward, J.V., 1998. Riverine landscapes: biodiversity patterns, disturbance regimes, and aquatic conservation. *Biological Conservation* 83: 269-278. *A.T.:* ecosystem management.

5345. Ward, Peter D., 1992. *On Methuselah's Trail: Living Fossils and the Great Extinctions.* New York: W.H. Freeman. 212p. [3/3/1].

5346. ———, 1994. *The End of Evolution: On Mass Extinctions and the Preservation of Biodiversity.* New York: Bantam Books. 301p. [2/3/1] *A.T.:* endangered species.

5347. ———, 1995. After the fall: lessons and directions from the K/T debate. *Palaios* 10: 530-538. [1] *A.T.:* asteroid impacts; mass extinctions.

5348. ———, 1997. *The Call of Distant Mammoths: Why the Ice Age Mammals Disappeared.* New York: Copernicus. 241p. [2/2/-] *A.T.:* Pleistocene extinctions.

5349. Ward, Philip S., April 1987. Distribution of the introduced Argentine ant (*Iridomyrmex humilis*) in natural habitats of the Lower Sacramento Valley and its effects on the indigenous ant fauna. *Hilgardia* 55(2): 1-16. [2] *A.T.:* biological invasions.

5350. Ward, Trevor J., 1998. *Marine BioRap Guidelines: Rapid Assessment of Marine Biological Diversity.* Perth, Western Australia: CSIRO. *A.T.:* bd measurement.

5351. Wardell-Johnson, Grant, & P. Horwitz, 1996. Conserving biodiversity and the recognition of heterogeneity in ancient landscapes: a case study from south-western Australia. *Forest Ecology and Management* 85: 219-238. [1] *A.T.:* microhabitats; endemic species.

5352. Wardle, D.A., K.I. Bonner, & K.S. Nicholson, 1997. Biodiversity and plant litter: experimental evidence which does not support the view that enhanced species richness improves ecosystem function. *Oikos* 79: 247-258. [2].

5353. Wardle, D.A., G.W. Yeates, et al., 1995. The detritus food-web and the diversity of soil fauna as indicators of disturbance regimes in agroecosystems. *Plant and Soil* 170: 35-43. [1] *A.T.:* plant litter; decomposition.

5354. Ware, Jane, Aug. 1992. Where the wild things are. The Center for Plant Conservation tracks down rare species to save them from extinction. *American Horticulturist* 71(8): 19-25. [1].

5355. Warner, L.R., 1998. Australian helminths in Australian rodents: an issue of biodiversity. *International J. for Parasitology* 28: 839-846. Describes the evolutionary and biogeographic relations involved.

5356. Warren, Melvin L., Jr., & B.M. Burr, Jan. 1994. Status of freshwater fishes of the United States: overview of an imperiled fauna. *Fisheries* 19(1): 6-18. [1] *A.T.:* endangered species; bd education.

5357. Warren, M.S., 1993. A review of butterfly conservation in central southern Britain (Parts I and II). *Biological Conservation* 64: 25-49. [1] *A.T.:* endangered species; bd management.

5358. Warwick, R.M., & K.R. Clarke, 1994. Relearning the ABC: taxonomic changes and abundance/biomass relationships in disturbed benthic communities. *Marine Biology* 118: 739-744. [1] *A.T.:* body size.

5359. ———, 1995. New "biodiversity" measures reveal a decrease in taxonomic distinctness with increasing stress. *Marine Ecology Progress Series* 129: 301-305. [2] *A.T.:* disturbance; environmental indicators; North Sea.

5360. ———, 1996. Relationships between body-size, species abundance and diversity in marine benthic assemblages: facts or artefacts? *J. of Experimental Marine Biology and Ecology* 202: 63-71. [1] *A.T.:* Northeast Atlantic; sampling.

5361. Watling, Roy, 1995. Assessment of fungal diversity: macromycetes, the problems. *Canadian J. of Botany* 73, Suppl. 1: S15-S24. [1] *A.T.:* bd assessment.

5362. ———, 1997. Pulling the threads together: habitat diversity. *Biodiversity and Conservation* 6: 753-763. *A.T.:* fungi.

5363. Watson, Reginald A., 1995. *Marine Biodiversity Management.* Halifax, NS: School for Resource and Environmental Studies, Dalhousie University. 127p. [1/1/1] *A.T.:* Indonesia.

5364. Watts, Bryan D., 1996. Landscape configuration and diversity hotspots in wintering sparrows. *Oecologia* 108: 512-517. [1] *A.T.:* Georgia; patch use.

5365. Watve, Milind G., & R. Sukumar, 1995. Parasite abundance and diversity in mammals: correlates with host ecology. *PNAS* 92: 8945-8949. [1] *A.T.:* India.

5366. Waugh, John, 1996. The global policy outlook for marine biodiversity conservation. *Global Biodiversity* 6(1): 23-30. [1].

5367. Way, M.J., & K.L. Heong, 1994. The role of biodiversity in the dynamics and management of insect pests of tropical irrigated rice—a review.

Bull. of Entomological Research 84: 567-587. [2] *A.T.:* biological control.

5368. Wayne, P.M., & F.A. Bazzaz, 1991. Assessing diversity in plant communities: the importance of within-species variation. *TREE* 6: 400-404. [1] *A.T.:* ecological diversity.

5369. Weaver, J.C., 1995. Indicator species and scale of observation. *CB* 9: 939-942. [1] *A.T.:* Missouri; arthropods; species-area relation.

5370. Webb, D.W., M.J. Wetzel, et al., 1995. Aquatic biodiversity in Illinois springs. *J. of the Kansas Entomological Society* 68(2), Suppl.: 93-107. [1].

5371. Webb, Gregory E., W.J. Sando, & Anne Raymond, 1997. Mississippian coral latitudinal diversity gradients (Western Interior United States): testing the limits of high resolution diversity data. *J. of Paleontology* 71: 780-791. *A.T.:* sampling; spatial scale.

5372. Webb, N.R., & A.H. Vermaat, 1990. Changes in vegetational diversity on remnant heathland fragments. *Biological Conservation* 53: 253-264. [1] *A.T.:* England.

5373. Webb, S. David, 1991. Ecogeography and the Great American Interchange. *Paleobiology* 17: 266-280. [1] *A.T.:* biotic interchange; New World.

5374. Webb, S. David, & A.D. Barnosky, 1989. Faunal dynamics of Pleistocene mammals. *Annual Review of Earth and Planetary Sciences* 17: 413-438. [1] *A.T.:* phyletic gradualism; punctuated equilibria; North America.

5375. Webb, Thompson, III, 1987. The appearance and disappearance of major vegetational assemblages: long-term vegetational dynamics in eastern North America. *Vegetatio* 69: 177-187. [2] *A.T.:* vegetation change.

5376. Webb, Thompson, III, & P.J. Bartlein, 1992. Global changes during the last 3 million years: climatic controls and biotic responses. *ARES* 23: 141-173. [2] *A.T.:* climatic change; Cenozoic; glacial period.

5377. Weber, Marcel, & Bernard Schmid, 1995. Reductionism, holism, and integrated approaches in biodiversity research. *Interdisciplinary Science Reviews* 20: 49-60. [1].

5378. Weber, Michael L., & Judith Gradwohl, 1995. *The Wealth of Oceans* [exhibition companion book]. New York: Norton. 256p. [2/2/1] *A.T.:* marine conservation.

5379. Weber, Peter K., July-Aug. 1993. Saving the coral reefs. *Futurist* 27(4): 28-33. [1].

5380. Weber, Peter K., Anne Platt, & Ed Ayres, 1993. *Abandoned Seas: Reversing the Decline of the Oceans.* Washington, DC: Worldwatch Paper No. 116. 66p. [2/1/1] *A.T.:* environmental degradation; environmental pollution.

5381. Webster, Jim, 1994. *Conserving Biodiversity in Africa: A Review of the USAID Africa Bureau's Biodiversity Program.* Washington, DC: Biodiversity Support Program. 111p. [1/1/1] *A.T.:* Agency for International

Development.

5382. Wedin, David A., 1992. Biodiversity conservation in Europe and North America. I. Grasslands: a common challenge. *Restoration & Management Notes* 10: 137-143. [1] *A.T.:* landscape conservation; disturbance.

5383. Wedin, David A., & David Tilman, 1996. Influence of nitrogen loading and species composition on the carbon balance of grasslands. *Science* 274: 1720-1723. [2] *A.T.:* carbon cycle; Minnesota.

5384. Weeks, Paul J.D., & K.J. Gaston, 1997. Image analysis, neural networks, and the taxonomic impediment to biodiversity studies. *Biodiversity and Conservation* 6: 263-274. *A.T.:* computers.

5385. Wege, David C., & A.J. Long, 1995. *Key Areas for Threatened Birds in the Neotropics.* Cambridge, U.K.: BirdLife International. 311p. [1/1/1] *A.T.:* bird conservation; Latin America; endangered and rare species.

5386. Weiher, Evan, & C.W. Boylen, 1994. Patterns and prediction of alpha and beta diversity of aquatic plants in Adirondack (New York) lakes. *Canadian J. of Botany* 72: 1797-1804. [1].

5387. Weimann, C., & M. Heinrich, 1997. Indigenous medicinal plants in Mexico: the example of the Nahua (Sierra de Zongolica). *Botanica Acta* 110: 62-72. *A.T.:* ethnobotany.

5388. Weiner, Jonathan, 1990. *The Next One Hundred Years: Shaping the Fate of Our Living Earth.* New York: Bantam. 312p. [3/3/1].

5389. Weir, A., & P.M. Hammond, 1997. Laboulbeniales on beetles: host utilization patterns and species richness of the parasites. *Biodiversity and Conservation* 6: 701-719. *A.T.:* fungi.

5390. Weis, Julie A., 1997. Eliminating the National Forest Management Act's diversity requirement as a substantive standard. *Environmental Law* 27: 641-662. *A.T.:* Forest Service; ecosystem management.

5391. Weisberg, Stephen B., J.A. Ranasinghe, et al., 1997. An estuarine benthic index of biotic integrity (B-IBI) for Chesapeake Bay. *Estuaries* 20: 149-158. *A.T.:* environmental monitoring.

5392. Weitzman, Martin L., 1992. On diversity. *Quarterly J. of Economics* 107: 363-405. [2] *A.T.:* diversity value.

5393. ———, 1993. What to preserve? An application of diversity theory to crane preservation. *Quarterly J. of Economics* 108: 157-183. [1] *A.T.:* conservation policy.

5394. Weizsacker, Ernst U. von, 1994. *Earth Politics.* London and Atlantic Highlands, NJ: Zed Books. 234p. [1/1/2] *A.T.:* environmental policy.

5395. Wellburn, Alan R., 1994. *Air Pollution and Climate Change: The Biological Impact* (2nd ed.). New York: Wiley. 268p. [2/1/2].

5396. Weller, Milton W., 1994. *Freshwater Marshes: Ecology and Wildlife Management* (3rd ed.). Minneapolis: University of Minnesota Press. 154p. [3/1/1].

5397. Wells, Michael P., 1992. Biodiversity conservation, affluence and poverty: mismatched costs and benefits and efforts to remedy them. *Ambio* 21: 237-243. [1] Distinguishes protected area benefits and costs at local, national and global scales. *A.T.:* developing countries.

5398. ———, 1993. Neglect of biological riches: the economics of nature tourism in Nepal. *Biodiversity and Conservation* 2: 445-464. [1] *A.T.:* ecotourism.

5399. ———, 1994. The Global Environment Facility and prospects for biodiversity conservation. *International Environmental Affairs* 6: 69-97. [1] *A.T.:* bd grants.

5400. ———, 1995. Community-based forestry and biodiversity projects have promised more than they have delivered. Why is this and what can be done? In Øyvind Sandbukt, ed., *Management of Tropical Forests: Towards an Integrated Perspective* (Oslo: SUM Occasional Paper No. 95-1): 269-286. [1].

5401. Wells, Michael P., & K.E. Brandon, 1993. The principles and practice of buffer zones and local participation in biodiversity conservation. *Ambio* 22: 157-162. [1] *A.T.:* developing countries; park-people relations.

5402. Wells, Michael P., Katrina Brandon, & Lee Jay Hannah, 1992. *People and Parks: Linking Protected Area Management With Local Communities.* Washington, DC: World Bank, WWF, and USAID. 99p. [1/1/2] *A.T.:* ICDPs.

5403. Wells, Susan M., 30 Oct. 1986. A future for coral reefs. *New Scientist* 112(1532): 46-50. [1].

5404. ———, ed., 1988. *Coral Reefs of the World.* Nairobi: UNEP, and Gland, Switzerland: IUCN. 3 vols. [1/2/1].

5405. Wells, William A., 1997. Seeking extremophiles: Recombinant Biocatalysis, Inc. *Chemistry & Biology* 4: 401-402. *A.T.:* hydrothermal vents; thermophilic bacteria.

5406. Welsh, D.A., & L.A. Venier, 1996. Binoculars and satellites: developing a conservation framework for boreal forest wildlife at varying scales. *Forest Ecology and Management* 85: 53-65. [1].

5407. Welsh, H.H., Jr., 1990. Relictual amphibians and old-growth forests. *CB* 4: 309-319. [1].

5408. Wemmer, Chris, Rasanayagam Rudran, et al., 1993. Training developing-country nationals is the critical ingredient to conserving global biodiversity. *Bioscience* 43: 762-767. [1] *A.T.:* bd education; bdc organizations.

5409. West, J.G., 1998. Floristics and biodiversity research in Australia: the 21st century. *Australian Systematic Botany* 11: 161-174.

5410. West, Neil E., 1993. Biodiversity of rangelands. *J. of Range Management* 46: 2-13. [2] *A.T.:* grazing.

5411. ———, 1994. Biodiversity and land use. *General Technical Report RM* GTR-247: 21-26. [1].

5412. ——, ed., 1995. *Biodiversity on Rangelands* [symposium papers]. Logan, UT: College of Natural Resources, Utah State University. 114p. [1/1/1] *A.T.:* range management.

5413. West, Patrick C., & S.R. Brechin, eds., 1991. *Resident Peoples and National Parks: Social Dilemmas and Strategies in International Conservation.* Tucson: University of Arizona Press. 443p. [1/2/2].

5414. West-Eberhard, Mary Jane, 1989. Phenotypic plasticity and the origins of diversity. *ARES* 20: 249-278. [3] *A.T.:* evolution.

5415. Wester, Lyndon, & Sekson Yongvanit, 1995. Biological diversity and community lore in northeastern Thailand. *J. of Ethnobiology* 15: 71-87. [1].

5416. Western, David R., 1992. The biodiversity crisis: a challenge for biology. *Oikos* 63: 29-38. [1] *A.T.:* research programs.

5417. ——, March-April 1993. The balance of nature. *Wildlife Conservation* 96(2): 52-55. [1] Concerning the place of elephants in African habitats.

5418. Western, David R., & M.C. Pearl, eds., 1989. *Conservation for the Twenty-first Century* [conference papers]. Oxford, U.K., and New York: OUP. 365p. [2/2/2] *A.T.:* wildlife conservation.

5419. Western, David R., R.M. Wright, & S.C. Strum, eds., 1994. *Natural Connections: Perspectives in Community-based Conservation* [workshop proceedings]. Washington, DC: Island Press. 581p. [1/1/2] *A.T.:* local participation.

5420. Westing, Arthur H., 1993. Biodiversity and the challenge of national borders [comment]. *Environmental Conservation* 20: 5-6. [1] *A.T.:* international cooperation; international law.

5421. ——, ed., 1994. *Biodiversity in the Context of Science and Society.* Middletown Springs, VT: Occasional Paper No. 27, Vermont Academy of Arts and Sciences. 62p. [1/1/1].

5422. Westman, Walter E., 1981. Diversity relations and succession in Californian coastal sage scrub. *Ecology* 62: 170-184. [1] *A.T.:* chaparral; fire.

5423. ——, 1990. Managing for biodiversity. *Bioscience* 40: 26-33. [1] *A.T.:* environmental policy; environmental management.

5424. ——, 1990. Park management of exotic plant species: problems and issues. *CB* 4: 251-260. [1] *A.T.:* introduced species; environmental policy.

5425. Weston, Anthony, 1987. Forms of gaian ethics. *Environmental Ethics* 9: 217-230. [1] *A.T.:* environmental philosophy; environmental ethics.

5426. Westrop, S.R., 1996. Temporal persistence and stability of Cambrian biofacies: Sunwaptan (Upper Cambrian) trilobite faunas of North America. *Palaeogeography, Palaeoclimatology, Palaeoecology* 127: 33-46. [2] *A.T.:* faunal turnover; extinction; coordinated stasis.

5427. Wethey, David S., 1985. Catastrophe, extinction, and species diversity: a rocky intertidal example. *Ecology* 66: 445-456. [1] *A.T.:* barnacles; sea ice; New England.

5428. Wheater, R., 1995. World Zoo Conservation Strategy: a blueprint for zoo development. *Biodiversity and Conservation* 4: 544-552. [1] *A.T.:* wildlife conservation; bd education.

5429. Wheeler, B.D., 1988. Species richness, species rarity and conservation evaluation of rich-fen vegetation in lowland England and Wales. *J. of Applied Ecology* 25: 331-352. [1].

5430. Wheeler, Quentin D., 1990. Insect diversity and cladistic constraints. *Annals of the Entomological Society of America* 83: 1031-1047. [1] *A.T.:* inventories; systematics.

5431. ———, 1995. Systematics and biodiversity. *Bioscience* 45, Suppl.: S21-S28. [1].

5432. ———, 1995. Systematics, the scientific basis for inventories of biodiversity. *Biodiversity and Conservation* 4: 476-489. [1].

5433. Whelan, Tensie, ed., 1991. *Nature Tourism: Managing for the Environment*. Washington, DC: Island Press. 223p. [2/1/2] *A.T.:* ecotourism.

5434. Whitcomb, Robert F., A.L. Hicks, et al., 1994. Biogeography of leafhopper specialists of the shortgrass prairie: evidence for the roles of phenology and phylogeny in determination of biological diversity. *American Entomologist* 40: 19-36. [1].

5435. White, Denis, P.G. Minotti, et al., 1997. Assessing risks to biodiversity from future landscape change. *CB* 11: 349-360. *A.T.:* Pennsylvania; land use planning; Monte Carlo simulations.

5436. White, Douglas W., & E.W. Stiles, 1992. Bird dispersal of fruits of species introduced into eastern North America. *Canadian J. of Botany* 70: 1689-1696. [1] *A.T.:* seed dispersal; ornithochory.

5437. White, G.A., J.C. Gardner, & C.G. Cook, 1994. Biodiversity for industrial crop development in the United States. *Industrial Crops and Products* 2: 259-272. [1] *A.T.:* crop germplasm resources.

5438. White, Lee J.T., 1994. Biomass of rain forest mammals in the Lopé Reserve, Gabon. *J. of Animal Ecology* 63: 499-512. [2] *A.T.:* logging.

5439. White, Lynn, Jr., 1967. The historical roots of our ecologic crisis. *Science* 155: 1203-1207. [2] Traces related research in theology, science and technology.

5440. White, Peter S., 1995. Conserving biodiversity: lessons from the Smokies. *Forum for Applied Research and Public Policy* 10(2): 116-120. [1] *A.T.:* Great Smoky Mountains National Park; environmental management; national parks.

5441. White, Peter S., & J.L. Walker, 1997. Approximating nature's variation: selecting and using reference information in restoration ecology. *Restoration Ecology* 5: 338-349.

5442. White, Peter S., R.I. Miller, & G.S. Ramseur, 1984. The species-area relationship of the Southern Appalachian high peaks: vascular plant richness and rare plant distributions. *Castanea* 49(2): 47-61. [1] *A.T.:* montane environments.

5443. White, Peter T., Jan. 1983. Tropical rain forests: nature's dwindling treasures. *National Geographic* 163(1): 2-47. [1] An overview of worldwide trends and problems. *A.T.:* Alwyn Gentry.

5444. Whitehead, M., 1995. Saying it with genes, species and habitats: biodiversity education and the role of zoos. *Biodiversity and Conservation* 4: 664-670. [1] *A.T.:* Agenda 21.

5445. Whitford, Walter G., 1996. The importance of the biodiversity of soil biota in arid ecosystems. *Biodiversity and Conservation* 5: 185-195. [1] *A.T.:* desert soils.

5446. ———, 1997. Desertification and animal biodiversity in the desert grasslands of North America. *J. of Arid Environments* 37: 709-720. *A.T.:* Arizona; breeding birds; small mammals.

5447. Whitham, Thomas G., P.A. Morrow, & B.M. Potts, 1991. Conservation of hybrid plants. *Science* 254: 779-780. [2] *A.T.:* plant hybrid zones.

5448. ———, 1994. Plant hybrid zones as centers of biodiversity: the herbivore community of two endemic Tasmanian eucalypts. *Oecologia* 97: 481-490. [2] *A.T.:* refugia.

5449. Whitlock, Cathy, 1992. Vegetational and climatic history of the Pacific Northwest during the last 20,000 years: implications for understanding present-day biodiversity. *Northwest Environmental J.* 8: 5-28. [1] *A.T.:* climatic change; paleoecology.

5450. Whitmore, Timothy C., 1984. *Tropical Rain Forests of the Far East* (2nd ed.). Oxford, U.K.: Clarendon Press. 352p. [2/1/3] A standard, frequently cited source. *A.T.:* Southeast Asia.

5451. ———, 1998. *An Introduction to Tropical Rain Forests* (2nd ed.). Oxford, U.K., and New York: OUP. 282p. [2/1/2] *A.T.:* rainforest ecology.

5452. Whitmore, Timothy C., & G.T. Prance, eds., 1987. *Biogeography and Quaternary History in Tropical America*. Oxford, U.K.: Clarendon Press. 214p. [1/1/2] *A.T.:* paleobiogeography.

5453. Whitmore, Timothy C., & J.A. Sayer, eds., 1992. *Tropical Deforestation and Species Extinction* [workshop papers]. London and New York: Chapman & Hall. 153p. [1/1/1].

5454. Whittaker, Robert H., 1962. Classification of natural communities. *Botanical Review* 28: 1-239. [1] Includes a detailed historical perspective and extensive bibliography of sources. *A.T.:* community ecology.

5455. ———, 1965. Dominance and diversity in land plant communities: numerical relations of species express the importance of competition in community function and evolution. *Science* 147: 250-260. [2] *A.T.:* dominance-diversity curves; diversity measures; plant communities.

5456. ———, 1970. *Communities and Ecosystems*. New York: Macmillan. 162p. [3/1/3] *A.T.:* community ecology.

5457. ———, 1972. Evolution and measurement of species diversity. *Taxon* 21: 213-251. [3] *A.T.:* alpha, beta and gamma diversity; community diversity.

5458. ———, 1977. Evolution of species diversity in land communities. *Evolutionary Biology* 10: 1-67. [2].

5459. Whittaker, Robert J., 1998. *Island Biogeography: Ecology, Evolution, and Conservation.* Oxford, U.K., and New York: OUP. 285p.

5460. Whittaker, Robert J., M.B. Bush, et al., 1992. Ecological aspects of plant colonisation of the Krakatau Islands. *GeoJournal* 28: 201-211. [1] A summary of plant colonization and succession since the 1883 eruption. *A.T.:* volcanoes.

5461. Whitten, Anthony J., 1987. Indonesia's transmigration program and its role in the loss of tropical rain forests. *CB* 1: 239-246. [1].

5462. Whittier, Thomas R., D.B. Halliwell, & S.G. Paulsen, 1997. Cyprinid distributions in Northeast U.S.A. lakes: evidence of regional scale minnow biodiversity losses. *Canadian J. of Fisheries and Aquatic Sciences* 54: 1593-1607. *A.T.:* introduced species.

5463. Whitworth, William R., & Alison Hill, 1997. *Applicability of Land Condition Trend Analysis Data for Biological Diversity Assessment in the Southeastern United States* [report]. Champaign, IL: USACERL Technical Report 97-67, Construction Engineering Research Laboratories, U.S. Army Corps of Engineers. 82p. [1/1/-].

5464. Wicken, Jeffrey S., 1981. Evolutionary self-organization and the entropy principle: teleology and mechanism. *Nature and System* 3: 129-141. [1].

5465. Wickham, James D., Jianguo Wu, & D.F. Bradford, 1997. A conceptual framework for selecting and analyzing stressor data to study species richness at large spatial scales. *Environmental Management* 21: 247-257. *A.T.:* environmental stress; spatial scale.

5466. Wiegmann, Brian M., Charles Mitter, & Brian Farrell, 1993. Diversification of carnivorous parasitic insects: extraordinary radiation or specialized dead end? *American Naturalist* 142: 737-754. [1] *A.T.:* carnivory.

5467. Wiens, John A., 1976. Population responses to patchy environments. *ARES* 7: 81-120. [2] *A.T.:* population biology; patch structure.

5468. ———, 1989. *The Ecology of Bird Communities.* Cambridge, U.K., and New York: CUP. 2 vols. [2/2/3].

5469. ———, 1989. Spatial scaling in ecology. *Functional Ecology* 3: 385-397. [3].

5470. ———, 1995. Habitat fragmentation: island vs. landscape perspectives on bird conservation. *Ibis* 137, Suppl. 1: S97-S104. [2] *A.T.:* landscape mosaics.

5471. Wieringa, Mark J., & A.G. Morton, 1996. Hydropower, adaptive management, and biodiversity. *Environmental Management* 20: 831-840. [1] *A.T.:* environmental planning; dams; Colorado River.

5472. Wiese, Robert J., & Michael Hutchins, 1994. *Species Survival Plans: Strategies for Wildlife Conservation.* Bethesda, MD: American Zoo and Aquarium Association. 64p. [1/1/1] *A.T.:* endangered species; zoos.

5473. Wigley, T. Bentley, & T.H. Roberts, 1997. Landscape level effects of forest management on faunal diversity in bottomland hardwoods. *Forest Ecology and Management* 90: 141-154. *A.T.:* habitat fragmentation.

5474. Wignall, Paul B., & R.J. Twitchett, 1996. Oceanic anoxia and the end Permian mass extinction. *Science* 272: 1155-1158.

5475. Wijnstekers, Willem, 1995. *The Evolution of CITES: A Reference to the Convention on International Trade in Endangered Species of Wild Fauna and Flora* (4th ed.). Châtelaine-Geneva, Switzerland: CITES Secretariat. 519p. [1/1/1] *A.T.:* wild animal trade.

5476. Wikramanayake, Eric D., Eric Dinerstein, et al., 1998. An ecology-based method for defining priorities for large mammal conservation: the tiger as case study. *CB* 12: 865-878. *A.T.:* wildlife conservation; prioritization.

5477. Wilcove, David S., 1988. Changes in the avifauna of the Great Smoky Mountains, 1947-1983. *Wilson Bull.* 100: 256-271. [1].

5478. ———, 1989. Protecting biodiversity in multiple-use lands: lessons from the US Forest Service. *TREE* 4: 385-388. [1] *A.T.:* forest policy; forest management.

5479. ———, 1993. Getting ahead of the extinction curve. *Ecological Applications* 3: 218-220. [1] Makes suggestions for reducing rate of species loss. *A.T.:* environmental policy; endangered species.

5480. Wilcove, David S., & M.J. Bean, eds., 1994. *The Big Kill: Declining Biodiversity in America's Lakes and Rivers* [case studies]. Washington, DC: Environmental Defense Fund. 275p. [1/1/1].

5481. Wilcove, David S., M.J. Bean, & P.C. Lee, 1992. Fisheries management and biological diversity: problems and opportunities. *Trans. of the 57th North American Wildlife and Natural Resources Conference*: 373-383. [1].

5482. Wilcove, David S., M.J. Bean, et al., 1996. *Rebuilding the Ark: Toward a More Effective Endangered Species Act for Private Land.* Washington, DC: Environmental Defense Fund. 20p. [1/1/1].

5483. Wilcove, David S., David Rothstein, et al., 1998. Quantifying threats to imperiled species in the United States. *Bioscience* 48: 607-615. Summarizes the main threats and ranks them.

5484. Wilcox, Bruce A., 1978. Supersaturated island faunas: a species-age relationship for lizards on post-Pleistocene land-bridge islands. *Science* 199: 996-998. [1] *A.T.:* Baja California; island biogeography.

5485. ———, 1995. Tropical forest resources and biodiversity: the risks of forest loss and degradation. *Unasylva,* no. 181: 43-49. [1] *A.T.:* deforestation; risk analysis.

5486. Wilcox, Bruce A., & K.N. Duin, winter 1994. Indigenous cultural and biological diversity: overlapping values of Latin American ecoregions. *Cultural Survival Quarterly* 18(4): 49-53. [1].

5487. Wilcox, Bruce A., & D.D. Murphy, 1985. Conservation strategy: the effects of fragmentation on extinction. *American Naturalist* 125: 879-887. [2]

5488. Wildt, David E., W.F. Rall, et al., 1997. Genome resource banks: living collections for biodiversity conservation. *Bioscience* 47: 689-698. *A.T.:* endangered species; cryopreservation.

5489. Wiley, E.O., 1981. *Phylogenetics: The Theory and Practice of Phylogenetic Systematics*. New York: Wiley. 439p. [2/1/3] *A.T.:* cladistics.

5490. ———, 1988. Vicariance biogeography. *ARES* 19: 513-542. [2] *A.T.:* phylogenetic systematics.

5491. Wiley, J.W., & J.M. Wunderle, Jr., 1993. The effects of hurricanes on birds, with special reference to Caribbean islands. *Bird Conservation International* 3: 319-349. [1] *A.T.:* habitat destruction.

5492. Wilkes, H.G., 1987. Plant genetic resources: why privatize a public good? [letter] *Bioscience* 37: 215-217. [1].

5493. Wilkinson, Clive R., & R.W. Buddemeier, 1994. *Global Climate Change and Coral Reefs: Implications for People and Reefs* [report]. Gland, Switzerland: IUCN. 124p. [1/1/1].

5494. Willby, Nigel J., & J.W. Eaton, 1996. Backwater habitats and their role in nature conservation on navigable waterways. *Hydrobiologia* 340: 333-338. [1] *A.T.:* U.K.; canals.

5495. Wille, Chris, 1992. Riches from the rainforest. In *Nature Conservancy Annual Report 1992* (Arlington, VA: Nature Conservancy): 10-17. [1].

5496. Willers, W.B. "Bill," ed., 1991. *Learning to Listen to the Land* [essays]. Washington, DC: Island Press. 282p. [2/2/1] Writings on environmental ethics and human ecology.

5497. Williams, Byron K., 1996. Assessment of accuracy in the mapping of vertebrate biodiversity. *J. of Environmental Management* 47: 269-282. [1] *A.T.:* gap analysis.

5498. Williams, B.L., & B.G. Marcot, 1991. Use of biodiversity indicators for analyzing and managing forest landscapes. *Trans. of the 56th North American Wildlife and Natural Resources Conference*: 613-627. [1] *A.T.:* landscape ecology.

5499. Williams, Cindy Deacon, Jan. 1994. Aquatic resources and the Endangered Species Act. *Fisheries* 19(1): 19-21. [1] *A.T.:* wildlife management.

5500. Williams, C.B., 1960. The range and pattern of insect abundance. *American Naturalist* 94: 137-159. [1].

5501. Williams, David F., ed., 1994. *Exotic Ants: Biology, Impact, and Control of Introduced Species*. Boulder, CO: Westview Press. 332p. [1/1/1].

5502. Williams, David M., & T.M. Embley, 1996. Microbial diversity: domains and kingdoms. *ARES* 27: 569-595. *A.T.:* microbial taxonomy.

5503. Williams, Geoff, & Terry Evans, 1993. *Hidden Rainforests: Subtropical Rainforests and Their Invertebrate Biodiversity.* Kensington, NSW: NSWU Press. 188p. [1/1/1] *A.T.:* New South Wales; rainforest ecology.

5504. Williams, Jack E., & R.R. Miller, 1990. Conservation status of the North American fish fauna in fresh water. *J. of Fish Biology* 37, Suppl. A: 79-85. [1] *A.T.:* Endangered Species Act.

5505. Williams, Jack E., & J.N. Rinne, May-June 1992. Biodiversity management on multiple-use federal lands: an opportunity whose time has come. *Fisheries* 17(3): 4-5. [1] *A.T.:* environmental degradation.

5506. Williams, Jack E., J.E. Johnson, et al., Nov.-Dec. 1989. Fishes of North America endangered, threatened, or of special concern. *Fisheries* 14(6): 2-20. [2].

5507. Williams, James D., M.L. Warren, Jr., et al., Sept. 1993. Conservation status of freshwater mussels of the United States and Canada. *Fisheries* 18(9): 6-22. [2].

5508. Williams, J.T., 1991. Plant genetic resources: some new directions. *Advances in Agronomy* 45: 61-91. [1] *A.T.:* crop germplasm resources.

5509. ———, 1992. International aspects of biodiversity. *Forestry Chronicle* 68: 454-458. [1] *A.T.:* forest management.

5510. Williams, Michael, 1989. Deforestation: past and present. *Progress in Human Geography* 13: 176-208. [1].

5511. ———, 1991. *Wetlands: A Threatened Landscape.* Oxford, U.K., and Cambridge, MA: Basil Blackwell. 419p. [2/2/1].

5512. Williams, Nancy M., & Graham Baines, eds., 1993. *Traditional Ecological Knowledge: Wisdom for Sustainable Development* [workshop papers]. Canberra: Centre for Resource and Environmental Studies, Australian National University. 184p. [1/1/1].

5513. Williams, Nigel, 1995. Slow start for Europe's new habitat protection plan. *Science* 269: 320-322. [1] *A.T.:* environmental policy; birds.

5514. Williams, Paul H., 1996. Mapping variations in the strength and breadth of biogeographic transition zones using species turnover. *PRSL B* 263: 579-588. [1] *A.T.:* bumble bees.

5515. Williams, Paul H., & K.J. Gaston, 1994. Measuring more of biodiversity: can higher-taxon richness predict wholesale species richness? *Biological Conservation* 67: 211-217. [2] *A.T.:* top-down approaches.

5516. Williams, Paul H., K.J. Gaston, & C.J. Humphries, 1994. Do conservationists and molecular biologists value differences between organisms in the same way? *Biodiversity Letters* 2: 67-78. [2] *A.T.:* species distinctiveness; diversity measures.

5517. ———, 1997. Mapping biodiversity value worldwide: combining higher-taxon richness from different groups. *PRSL B* 264: 141-148. *A.T.:*

hotspots.

5518. Williams, Paul H., C.J. Humphries, & K.J. Gaston, 1994. Centres of seed-plant diversity: the family way. *PRSL B* 256: 67-70. [2] *A.T.:* higher taxa; endemism.

5519. Williams, Paul H., C.J. Humphries, & R.I. Vane-Wright, 1991. Measuring biodiversity: taxonomic relatedness for conservation priorities. *Australian Systematic Botany* 4: 665-679. [2] *A.T.:* bd measures; bumble bees; species distinctiveness.

5520. Williams, Paul H., K.J. Gaston, et al., 1997. Biodiversity, biospecifics, and ecological services: reply [letter]. *TREE* 12: 66-67.

5521. Williams, Paul H., David Gibbons, et al., 1996. A comparison of richness hotspots, rarity hotspots, and complementary areas for conserving diversity of British birds. *CB* 10: 155-174. [1].

5522. Williams, Paul H., C.J. Humphries, et al., 1996. Value in biodiversity, ecological services and consensus [letter]. *TREE* 11: 385. [1].

5523. Williams, Paul H., G.T. Prance, et al., 1996. Promise and problems in applying quantitative complementary areas for representing the diversity of some Neotropical plants. *Biological J. of the Linnean Society* 58: 125-157. [1].

5524. Williams, S.E., & J.-M. Hero, 1998. Rainforest frogs of the Australian wet tropics: guild classification and the ecological similarity of declining species. *PRSL B* 265: 597-602. *A.T.:* functional groups; amphibian declines; Queensland.

5525. Williams, S.E., & R.G. Pearson, 1997. Historical rainforest contractions, localized extinctions and patterns of vertebrate endemism in the rain forests of Australia's wet tropics. *PRSL B* 264: 709-716. *A.T.:* Queensland; habitat fragmentation; patch shape; patch size.

5526. Williamson, Mark, 1981. *Island Populations.* Oxford, U.K., and New York: OUP. 286p. [2/1/2] *A.T.:* population biology.

5527. ———, 1989. Natural extinction on islands. *PTRSL B* 325: 457-468. [1] *A.T.:* island biogeography; species turnover; rarity.

5528. ———, 1989. The MacArthur and Wilson theory today: true but trivial [editorial]. *J. of Biogeography* 16: 3-4. [1] *A.T.:* island biogeography; species-area relation.

5529. ———, 1993. Invaders, weeds and the risk from genetically manipulated organisms. *Experientia* 49: 219-224. [1].

5530. ———, 1996. *Biological Invasions.* London and New York: Chapman & Hall. 244p. [1/1/2].

5531. Williamson, Mark, & Alastair Fitter, 1996. The varying success of invaders. *Ecology* 77: 1661-1666. [2] *A.T.:* biological invasions.

5532. Willig, Michael R., & S.K. Lyons, 1998. An analytical model of latitudinal gradients of species richness with an empirical test for marsupials and bats in the New World. *Oikos* 81: 93-98. *A.T.:* geographical range size.

5533. Willig, Michael R., & K.W. Selcer, 1989. Bat species density gradients in the New World: a statistical assessment. *J. of Biogeography* 16: 189-195. [1] *A.T.:* latitudinal diversity gradients.

5534. Willig, Michael R., D.L. Moorhead, et al., 1996. Functional diversity of soil bacterial communities in the Tabonuco Forest: interaction of anthropogenic and natural disturbance. *Biotropica* 28: 471-483. [1] *A.T.:* Puerto Rico; hurricanes.

5535. Willis, John C., 1922. *Age and Area: A Study in the Geographical Distribution and Origin of Species.* Cambridge, U.K.: CUP. 259p. [1/-/2] *A.T.:* biogeography; evolution.

5536. Willis, Kenneth G., & J.T. Corkindale, eds., 1995. *Environmental Valuation: New Perspectives.* Wallingford, Oxon, U.K.: CAB International. 249p. [1/1/1] *A.T.:* environmental policy; environmental economics.

5537. Willis, Kenneth G., Guy Garrod, & Peter Shepherd, 1996. *Towards a Methdology for Costing Biodiversity Conservation in the UK* [report]. London: HMSO. 197p. [1/1/1].

5538. Wills, Christopher, 23 March 1996. Safety in diversity: the richness of species in a rainforest may rely on the presence of parasites and diseases. *New Scientist* 149(2022): 38-42. [1].

5539. Wills, Christopher, Richard Condit, et al., 1997. Strong density- and diversity-related effects help to maintain tree species diversity in a Neotropical forest. *PNAS* 94: 1252-1257. *A.T.:* Panama.

5540. Willson, Mary F., & T.A. Comet, 1996. Bird communities of northern forests: patterns of diversity and abundance. *Condor* 98: 337-349. [1] *A.T.:* southeast Alaska.

5541. Willson, Mary F., & K.C. Halupka, 1995. Anadromous fish as keystone species in vertebrate communities. *CB* 9: 489-497. [2].

5542. Wilson, Cathleen J., R.S. Reid, et al., 1997. Effects of land use and tsetse fly control on bird species richness in southwestern Ethiopia. *CB* 11: 435-447.

5543. Wilson, David S., 1992. Complex interactions in metacommunities, with implications for biodiversity and higher levels of selection. *Ecology* 72: 1984-2000. [2] *A.T.:* simulation studies; community structure.

5544. Wilson, Don E., & Abelardo Sandoval, eds., 1997. *Manu: The Biodiversity of Southeastern Peru.* Washington, DC: Smithsonian Institution Press. 679p. [1/-/-].

5545. Wilson, Don E., F.R. Cole, et al., eds., 1996. *Measuring and Monitoring Biological Diversity: Standard Methods for Mammals.* Washington, DC, and London: Smithsonian Institution Press. 409p. [1/1/1].

5546. Wilson, Edward O., 1961. The nature of the taxon cycle in the Melanesian ant fauna. *American Naturalist* 95: 169-193. [2] *A.T.:* saturation curves; island biogeography.

5547. ———, summer 1984. Million-year histories: species diversity as an ethical goal. *Wilderness* 48(165): 12-17. [1].

5548. ———, 1984. *Biophilia.* Cambridge, MA: Harvard University Press. 157p. [2/3/2] *A.T.:* nature philosophy.

5549. ———, 1985. The biological diversity crisis: despite unprecedented extinction rates, the extent of biological diversity remains unmeasured. *Bioscience* 35: 700-706. [1].

5550. ———, 1987. The little things that run the world: the importance and conservation of invertebrates. *CB* 1: 344-346. [1].

5551. ———, ed., 1988. *Biodiversity* [forum papers]. Washington, DC: National Academy Press. 521p. [3/2/3] *A.T.:* bdc.

5552. ———, Sept. 1989. Threats to biodiversity. *Scientific American* 261(3): 108-116. [2] *A.T.:* extinction rates; deforestation; habitat destruction.

5553. ———, Nov.-Dec. 1992. Toward renewed reverence for life [editorial]. *Technology Review* 95(8): 72-73. [1] *A.T.:* bd loss; rainforests.

5554. ———, 1992. *The Diversity of Life.* Cambridge, MA: Belknap Press of Harvard University Press. 424p. [3/3/3] *A.T.:* bd; bdc.

5555. ———, 1992. The effects of complex social life on evolution and biodiversity. *Oikos* 63: 13-18. [1] *A.T.:* social organization; population genetics.

5556. ———, 30 May 1993. Is humanity suicidal? *New York Times Magazine* 142(49347): 24-29. [1] Concerning the juggernaut theory of human nature.

5557. ———, 1994. Biodiversity: challenge, science, opportunity. *American Zoologist* 34: 5-11. [1] A plea for efforts to "salvage and restore" bd.

5558. ———, 1996. *In Search of Nature.* Washington, DC: Island Press. 214p. [2/3/1] *A.T.:* nature philosophy; human ecology.

5559. Wilson, John W., III, 1974. Analytical zoogeography of North American mammals. *Evolution* 28: 124-140. [1] *A.T.:* latitudinal diversity gradients; areography.

5560. Wilson, J. Bastow, H. Gitay, et al., 1998. Relative abundance distributions in plant communities: effects of species richness and of spatial scale. *J. of Vegetation Science* 9: 213-220.

5561. Wilson, J. Bastow, T.C.E. Wells, et al., 1996. Are there assembly rules for plant species abundance? An investigation in relation to soil resources and successional trends. *J. of Ecology* 84: 527-538. [1] *A.T.:* community structure; U.K.; relative abundance distributions.

5562. Wilson, R.T., 1997. Animal genetic resources and domestic animal diversity in Nepal. *Biodiversity and Conservation* 6: 233-251. *A.T.:* genetic diversity.

5563. Winchester, N.N., & R.A. Ring, 1996. Northern temperate coastal Sitka spruce forests with special emphasis on canopies: studying arthropods in an unexplored frontier. *Northwest Science* 70, Suppl.: 94-103. [1] *A.T.:* British Columbia.

5564. Winckler, Suzanne, Jan. 1992. Stopgap measures. *Atlantic Monthly* 269(1): 74-81. [1] Discusses problems in the administration of the Endangered Species Act.

5565. Windsor, Donald A., 1997. Equal rights for parasites. *Perspectives in Biology and Medicine* 40: 222-229. Considers how parasites may help stabilize ecosystem interactions. *A.T.:* animal behavior.

5566. Winemiller, Kirk O., 1989. Must connectance decrease with species richness? *American Naturalist* 134: 960-968. [2] *A.T.:* Venezuela; Costa Rica; food webs.

5567. ———, 1996. Chronicling changing biological diversity. *Environmental Biology of Fishes* 45: 211-213. [1] *A.T.:* IUCN.

5568. Winfield, Ian J., & J.S. Nelson, eds., 1991. *Cyprinid Fishes: Systematics, Biology, and Exploitation.* London and New York: Chapman & Hall. 667p. [1/1/1].

5569. Winker, Kevin, 1996. The crumbling infrastructure of biodiversity: the avian example. *CB* 10: 703-707. [1] *A.T.:* zoological specimens.

5570. Winston, Judith E., & K.L. Metzger, 1998. Trends in taxonomy revealed by the published literature. *Bioscience* 48: 125-128. *A.T.:* systematics.

5571. Winston, M.R., & P.L. Angermeier, 1995. Assessing conservation value using centers of population density. *CB* 9: 1518-1527. [1] *A.T.:* Virginia; Tennessee; fish.

5572. Winter, Brian D., & R.M. Hughes, Jan. 1997. Biodiversity. *Fisheries* 22(1): 22-29. [1] American Fisheries Society position statement. *A.T.:* environmental policy; fisheries management.

5573. Wirgin, Isaac I., J.E. Stabile, & J.R. Waldman, 1997. Molecular analysis in the conservation of sturgeons and paddlefish. *Environmental Biology of Fishes* 48: 385-398. *A.T.:* genetic diversity.

5574. Wise, K.P., & T.J.M. Schopf, 1981. Was marine faunal diversity in the Pleistocene affected by changes in sea level? *Paleobiology* 7: 394-399. [1].

5575. Wisheu, I.C., & P.A. Keddy, 1996. Three competing models for predicting the size of species pools: a test using eastern North American wetlands. *Oikos* 76: 253-258. [1].

5576. Witcombe, J.R., A. Joshi, et al., 1996. Farmer participatory crop improvement. I. Varietal selection and breeding methods and their impact on biodiversity. *Experimental Agriculture* 32: 445-460. [1] *A.T.:* plant breeding.

5577. Witte, Frans, Tijs Goldschmidt, et al., 1992. Species extinction and concomitant ecological changes in Lake Victoria. *Netherlands J. of Zoology* 42: 214-232. [1] *A.T.:* cichlids; introduced fish; biological invasions; endangered species.

5578. Witter, Richard L., L.F. Lee, & J.M. Sharma, 1990. Biological diversity among serotype 2 Marek's disease viruses. *Avian Diseases* 34: 944-957. [1].

5579. Witting, Lars, & Volker Loeschcke, 1993. Biodiversity conservation: reserve optimization or loss minimization? [letter] *TREE* 8: 417. [1] *A.T.:* in situ conservation.

5580. ———, 1995. The optimization of biodiversity conservation. *Biological Conservation* 71: 205-207. [1] *A.T.:* in situ conservation; multispecies risk analysis; loss minimization.

5581. Wittmann, Helmut, 1993. In Central Europe. *Naturopa*, no. 73: 19-20. [1] A short review of nature conservation strategies.

5582. Wivagg, Dan, 1994. Out of touch [editorial]. *American Biology Teacher* 56: 131. [1] Encourages teachers to get their students outdoors and in touch with the natural world.

5583. Woelkerling, W.J., 1997. The biodiversity of Corallinales (Rhodophyta) in southern Australia: 1976 vs. 1996 with implications for generating a world biodiversity database. *Cryptogamie: Algologie* 18: 225-261. *A.T.:* coralline red algae.

5584. Wohlgemuth, Thomas, 1998. Modelling floristic species richness on a regional scale: a case study in Switzerland. *Biodiversity and Conservation* 7: 159-177.

5585. Woinarski, J.C.Z., 1992. Biogeography and conservation of reptiles, mammals and birds across north-western Australia: an inventory and base for planning an ecological reserve system. *Wildlife Research* 19: 665-705. [1].

5586. Woinarski, J.C.Z., O. Price, & D.P. Faith, 1996. Application of a taxon priority system for conservation planning by selecting areas which are most distinct from environments already reserved. *Biological Conservation* 76: 147-159. [1] *A.T.:* phylogenetic diversity; Northern Territory; reserve selection.

5587. Woinarski, J.C.Z., P.J. Whitehead, et al., 1992. Conservation of mobile species in a variable environment: the problem of reserve design in the Northern Territory, Australia. *Global Ecology and Biogeography Letters* 2: 1-10. [1].

5588. Woiwode, Anne M., 1994. *A New Way of Thinking: Biological Diversity and Forestry Policy in the Northwoods of the Great Lakes States.* Lansing, MI: Sierra Club. 100p. [1/1/1].

5589. ———, spring 1998. Pittman Robertson: an old law opens new possibilities for biodiversity restoration. *Wild Earth* 8(1): 66-71. *A.T.:* Michigan; environmental legislation.

5590. Wolbach, Wendy S., Iain Gilmour, et al., 1988. Global fire at Cretaceous-Tertiary boundary. *Nature* 334: 665-669. [1] *A.T.:* mass extinctions; asteroid impacts.

5591. Wolda, Henk, 1978. Fluctuations in abundance of tropical insects. *American Naturalist* 112: 1017-1045. [2].

5592. ———, 1992. Trends in abundance of tropical forest insects. *Oecologia* 89: 47-52. [1] *A.T.:* Panama.

5593. Wolf, Edward C., 1985. Challenges and priorities in conserving biological diversity. *Interciencia* 10: 236-242. [1].

5594. ———, 1987. *On the Brink of Extinction: Conserving the Diversity of Life.* Washington, DC: Worldwatch Paper No. 78. 54p. [2/1/1] *A.T.:* endangered species.

5595. Wolff, Torben, 1977. Diversity and faunal composition of the deep-sea benthos. *Nature* 267: 780-785. [1] *A.T.:* marine invertebrates.

5596. Wollenburg, Jutta E., & Andreas Mackensen, 1998. Living benthic foraminifers from the central Arctic Ocean: faunal composition, standing stock and diversity. *Marine Micropaleontology* 34: 153-185.

5597. Wolseley, P.A., C. Moncrieff, & B. Aguirre-Hudson, 1994. Lichens as indicators of environmental stability and change in the tropical forests of Thailand. *Global Ecology and Biogeography Letters* 4: 116-123. [1] *A.T.:* biological indicators; fire.

5598. Wood, Christopher A., A.P. Martin, & J.E. Williams, Nov.-Dec. 1993. *Bring Back the Natives*: restoring native aquatic species on public lands. *Endangered Species Technical Bull.* 18(4): 5-10. [1] *A.T.:* species reintroductions; Forest Service; Bureau of Land Management.

5599. Wood, David, 18 Jan. 1992. A matter of good breeding [opinion]. *New Scientist* 133(1804): 8. [1] Argues that bdc should concentrate on the preservation of "useful" organisms.

5600. ———, 1993. *Agrobiodiversity in Global Conservation Policy.* Nairobi: African Centre for Technology Studies. 35p. [1/1/1] *A.T.:* international cooperation.

5601. ———, 1995. Conserved to death: are tropical forests being overprotected from people? *Land Use Policy* 12: 115-135. [1] *A.T.:* park-people relations; forest conservation; developing countries.

5602. Wood, David, & J.M. Lenne, 1997. The conservation of agrobiodiversity on-farm: questioning the emerging paradigm. *Biodiversity and Conservation* 6: 109-129. *A.T.:* crop genetic resources; farming practices; in situ conservation.

5603. Wood, Don A., B.A. Millsap, & P.M. Rose, July-Aug. 1992. Florida's nongame and endangered species programs. *Endangered Species Update* 9(9-10): 8, 10-12. [1] *A.T.:* wildlife conservation.

5604. Wood, Henry, Melissa McDaniel, & Katherine Warner, eds., 1995. *Community Development and Conservation of Forest Biodiversity Through Community Forestry: Proceedings of an International Seminar.* Bangkok: Regional Community Forestry Training Center, Kasetsart University. 346p. [1/1/1] *A.T.:* citizen participation.

5605. Wood, Paul M., 1997. Biodiversity as the source of biological resources: a new look at biodiversity values. *Environmental Values* 6: 251-268.

5606. Wood, Thomas K., & K.L. Olmstead, 1984. Latitudinal effects on treehopper species richness (Homoptera: Membracidae). *Ecological*

Entomology 9: 109-115. [1] *A.T.:* Western Hemisphere.

5607. Woodley, Stephen J., James Kay, & George Francis, eds., 1993. *Ecological Integrity and the Management of Ecosystems* [conference proceedings]. Delray Beach, FL: St. Lucie Press. 220p. [2/1/2] *A.T.:* Canada.

5608. Woodroffe, Colin D., 1990. The impact of sea-level rise on mangrove shoreline. *Progress in Physical Geography* 14: 483-502. [1] *A.T.:* Quaternary.

5609. Woodruff, D.S., & G.A.E. Gall, 1992. Genetics and conservation. *Agriculture, Ecosystems & Environment* 42: 53-73. [1] *A.T.:* genetic engineering; conservation prioritization.

5610. Woodward, F.I., 1987. *Climate and Plant Distribution.* Cambridge, U.K., and New York: CUP. 174p. [2/2/3] *A.T.:* plant geography.

5611. Woodwell, George M., ed., 1990. *The Earth in Transition: Patterns and Processes of Biotic Impoverishment* [conference papers]. Cambridge, U.K., and New York: CUP. 530p. [2/2/1] *A.T.:* human ecology; environmental degradation; bd loss.

5612. ———, 1995. A habitat for all life [letter]. *Bioscience* 45: 67-68. [1] Argues that all local habitats are equally deserving of bdc attention.

5613. Woodwell, George M., & H.H. Smith, eds., 1969. *Diversity and Stability in Ecological Systems* [symposium report]. Upton, NY: Brookhaven Symposia in Biology No. 22, Brookhaven National Laboratory. 264p. [1/1/1].

5614. Woolf, Norma Bennett, 1986. New hope for exotic species. *Bioscience* 36: 594-597. [1] *A.T.:* cryopreservation.

5615. Woolhouse, M.E.J., 1987. On species richness and nature reserve design: an empirical study of UK woodland avifauna. *Biological Conservation* 40: 167-178. [1].

5616. Woomer, Paul L., & M.J. Swift, eds., 1994. *The Biological Management of Tropical Soil Fertility.* Chichester, U.K., and New York: Wiley. 243p. [1/1/1] *A.T.:* soil ecology.

5617. World Bank, 1995. *Mainstreaming Biodiversity in Development: A World Bank Assistance Strategy for Implementing the Convention on Biological Diversity.* Washington, DC: Environment Dept. Paper (Biodiversity Series) No. 29. 45p. [1/1/1].

5618. ———, 1998. *Biodiversity in World Bank Projects: A Portfolio Review.* Washington, DC: Environment Dept. Paper (Natural Habitats and Ecosystems Management Series) No. 59. 57p.

5619. World Commission on Environment and Development (WCED), 1987. *Our Common Future.* Oxford, U.K., and New York: OUP. 383p. [2/2/3] *A.T.:* environmental policy; human ecology.

5620. World Conservation Monitoring Centre (WCMC), 1991-1992. *Protected Areas of the World: A Review of National Systems.* Gland, Switzerland: IUCN. 4 vols. [1/1/1].

5621. World Health Organization (WHO), IUCN, & WWF, 1993. *Guidelines on the Conservation of Medicinal Plants.* Gland, Switzerland. 58p. [1/1/1].

5622. World Resources Institute (WRI), 1984. *Recommendations for a United States Strategy to Conserve Biological Diversity in Developing Countries.* Washington, DC. 134p. [1/1/1].

5623. ———, 1985. *Tropical Forests: A Call for Action.* Washington, DC. 3 vols. [2/1/2] *A.T.:* deforestation.

5624. World Resources Institute (WRI), IUCN, & UNDP, 1992. *Global Biodiversity Strategy: Guidelines for Action to Save, Study, and Use Earth's Biotic Wealth Sustainably and Equitably.* Washington, DC. 244p. [2/1/2].

5625. World Wildlife Fund (WWF), 1988. *World Wildlife Fund's Tropical Andes Program: Protecting a Global Center of Biological Diversity.* Washington, DC. 27p. [1/1/1].

5626. Worthington, E.B., & R. Lowe-McConnell, 1994. African lakes reviewed: creation and destruction of biodiversity. *Environmental Conservation* 21: 199-213. [1] *A.T.:* human disturbance; biological invasions; fish.

5627. Wright, David H., 1983. Species-energy theory: an extension of species-area theory. *Oikos* 41: 496-506. [2] *A.T.:* island biogeography.

5628. ———, 1990. Human impacts on energy flow through natural ecosystems, and implications for species endangerment. *Ambio* 19: 189-194. [1] *A.T.:* species-energy curves.

5629. ———, 1991. Correlations between incidence and abundance are expected by chance. *J. of Biogeography* 18: 463-466. [2] *A.T.:* null hypotheses; spatial patterns.

5630. Wright, D.D., J.H. Jessen, et al., 1997. Tree and liana enumeration and diversity on a one-hectare plot in Papua New Guinea. *Biotropica* 29: 250-260.

5631. Wright, Martin, 11 Feb. 1995. Death by a thousand cuts. *New Scientist* 145(1964): 36-40. [1] Describes the destruction of old-growth forests in Finland. *A.T.:* extinction; environmental policy.

5632. Wright, M.G., & M.J. Samways, 1998. Insect species richness tracking plant species richness in a diverse flora: gall-insects in the Cape Floristic Region, South Africa. *Oecologia* 115: 427-433. *A.T.:* fynbos; endemism.

5633. Wright, R. Gerald, & John Lemons, eds., 1996. *National Parks and Protected Areas: Their Role in Environmental Protection.* Oxford, U.K., and Cambridge, MA: Blackwell Science. 470p. [1/1/1] *A.T.:* ecosystem management; Canada; United States.

5634. Wright, R.Gerald, & P.D. Tanimoto, 1998. Using GIS to prioritize land conservation actions: integrating factors of habitat diversity, land ownership, and development risk. *Natural Areas J.* 18: 38-44. *A.T.:* Washington.

5635. Wright, S. Joseph, 1981. Intra-archipelago vertebrate distributions:

the slope of the species-area relation. *American Naturalist* 118: 726-748. [1] *A.T.:* biogeography.

5636. ———, 1985. How isolation affects rates of turnover of species on islands. *Oikos* 44: 331-340. [1] *A.T.:* Panama; population fluctuations; birds.

5637. ———, 1988. Patterns of abundance and the form of the species-area relation. *American Naturalist* 131: 401-411. [1] *A.T.:* Panama; West Indies; species-abundance relation; island biogeography.

5638. Wright, S. Joseph, & C.C. Biehl, 1982. Island biogeographic distributions: testing for random, regular, and aggregated patterns of species occurrence. *American Naturalist* 119: 345-357. [1] *A.T.:* colonization.

5639. Wright, S. Joseph, & S.P. Hubbell, 1983. Stochastic extinction and reserve size: a focal species approach. *Oikos* 41: 466-476. [1] *A.T.:* population fluctuations.

5640. Wu, Jianguo, & O.L. Loucks, 1995. From balance of nature to hierarchical patch dynamics: a paradigm shift in ecology. *Quarterly Review of Biology* 70: 439-466. [2] *A.T.:* ecological theory.

5641. Wu, Ning, 1997. Indigenous knowledge and sustainable approaches for the maintenance of biodiversity in nomadic society: experiences from the eastern Tibetan Plateau. *Die Erde* 128: 67-80.

5642. Wuerthner, George, May-June 1992. No home for snails. *Defenders* 67(3): 8-14. [1] *A.T.:* endangered species; Idaho; hotsprings.

5643. Wuketits, F.M., 1997. The status of biology and the meaning of biodiversity. *Naturwissenschaften* 84: 473-479. *A.T.:* bd education; bioethics; bdc; systematics.

5644. Wyatt, A.R., 1993. Phanerozoic shallow water diversity driven by changes in sea-level. *Geologische Rundschau* 82: 203-211. [1] *A.T.:* marine habitats.

5645. Wylie, John L., & D.J. Currie, 1993. Species-energy theory and patterns of species richness (in two parts). *Biological Conservation* 63: 137-148. [1] *A.T.:* island biogeography; Australia; birds; mammals; angiosperms; protected areas.

5646. Wylynko, David, winter 1995. The Rio challenge. *Nature Canada* 24(1): 21-25. [1] *A.T.:* Canada; action plans.

5647. Wyman, Richard L., 1990. What's happening to the amphibians? [letter] *CB* 4: 350-352. [2] *A.T.:* amphibian declines.

5648. ———, ed., 1991. *Global Climate Change and Life on Earth* [conference papers]. New York: Routledge. 282p. [2/2/1] *A.T.:* greenhouse effect; global warming.

5649. Wynne, Graham, 1994. *Biodiversity Challenge: An Agenda for Conservation Action in the UK* (2nd ed.). Sandy, Beds., U.K.: Royal Society for the Protection of Birds. 285p. [1/1/1].

5650. Xu, Fu-Liu, 1997. Exergy and structural exergy as ecological indicators for the development state of the Lake Chaohu ecosystem. *Ecological*

Modelling 99: 41-49. *A.T.:* China.

5651. Yablokov, Aleksei V., & S.A. Ostroumov, 1991. *Conservation of Living Nature and Resources: Problems, Trends, and Prospects.* Berlin and New York: Springer. 271p. [1/1/1] *A.T.:* wildlife conservation.

5652. Yaffee, Steven Lewis, 1994. *The Wisdom of the Spotted Owl: Policy Lessons for a New Century.* Washington, DC: Island Press. 430p. [2/1/2] *A.T.:* environmental policy; forest management.

5653. Yahnke, C.J., I.G. de Fox, & F. Colman, 1998. Mammalian species richness in Paraguay: the effectiveness of national parks in preserving biodiversity. *Biological Conservation* 84: 263-268. *A.T.:* wildlife management.

5654. Yamin, Farhana, 1995. Biodiversity, ethics and international law. *International Affairs* 71: 529-546. [1] *A.T.:* CBD.

5655. Yang, S.-L., & A.W. Meerow, 1996. The *Cycas pectinata* (Cycadaceae) complex: genetic structure and gene flow. *International J. of Plant Sciences* 157: 468-483. [1] *A.T.:* cycads; Asia; population genetics.

5656. Yap, Son-Kheong, & Su Win Lee, eds., 1992. *In Harmony With Nature: Proceedings of the International Conference on Conservation of Tropical Biodiversity.* Kuala Lumpur: Malayan Nature Society. 656p. [1/1/1].

5657. Yap, W.H., X. Li, et al., 1996. Quantitative comparisons of *in situ* microbial biodiversity by signature biomarker analysis. *J. of Industrial Microbiology & Biotechnology* 17: 179-184. [1] *A.T.:* soil microorganisms; DNA analysis.

5658. Yasuno, M., & M.M. Watanabe, eds., 1994. *Biodiversity: Its Complexity and Role* [symposium papers]. Tokyo: Global Environmental Forum. 273p. [1/1/1].

5659. Yates, C.J., & R.J. Hobbs, 1997. Woodland restoration in the western Australian wheatbelt: a conceptual framework using a state and transition model. *Restoration Ecology* 5: 28-35. *A.T.:* environmental degradation.

5660. Yates, Steve, July 1984. On the cutting edge of extinction. *Audubon* 86(4): 62-85. [1] Profiles the endemic bird species of Hawaii.

5661. Yeager, Kurt E., 1996. The exuberant planet: a global look at the role of utilities in protecting biodiversity. *Environmental Management* 20: 967-971. [1].

5662. Yeatman, Christopher W., D.L. Kafton, & Garrison Wilkes, eds., 1984. *Plant Genetic Resources: A Conservation Imperative* [symposium papers]. Boulder, CO: Westview Press. 164p. [1/1/1] *A.T.:* plant germplasm resources.

5663. Yen, A.L., & R.J. Butcher, 1992. Practical conservation of non-marine invertebrates. *Search* 23: 103-105. [1] *A.T.:* Australia.

5664. Yipp, May W., C.H. Hau, & G. Walthew, 1995. Conservation evaluation of nine Hong Kong mangals. *Hydrobiologia* 295: 323-333. [1] *A.T.:* mangroves; bd value.

5665. Yonzon, Pralad B., & M.L. Hunter, Jr., 1991. Cheese, tourists, and red pandas in the Nepal Himalayas. *CB* 5: 196-202. [1] *A.T.:* human disturbance; endangered species.

5666. Yoon, Carol K., 1993. Counting creatures great and small. *Science* 260: 620-622. [1] Reports on a conference that explored how to carry out an All Taxa Biodiversity Inventory.

5667. York, A., 1994. The long-term effects of fire on forest and communities: management implications for the conservation of biodiversity. *Memoirs of the Queensland Museum* 36: 231-239. [1] *A.T.:* New South Wales; ants.

5668. Young, Allen M., March-April 1985. Nature's riotous variety. *Américas* 37(2): 24-31. [1] Describes a tropical rainforest in Costa Rica.

5669. Young, C.M., M.A. Sewell, et al., 1997. Biogeographic and bathymetric ranges of Atlantic deep-sea echinoderms and ascidians: the role of larval dispersal. *Biodiversity and Conservation* 6: 1507-1522.

5670. Young, Kenneth R., 1996. Threats to biological diversity caused by *coca*/cocaine deforestation in Peru. *Environmental Conservation* 23: 7-15. [1] *A.T.:* illegal drug trade; montane forests.

5671. Young, Mike, & Bruce Howard, 1996. Can Australia afford a representative reserve network by 2000? *Search* 27: 22-26. [1] *A.T.:* protected areas; environmental management.

5672. Young, Patrick, 1996. The "new science" of wetland restoration. *Environmental Science & Technology* 30: 292A-296A. [1] *A.T.:* Everglades; California; habitat loss.

5673. Young, Stephen, 15 March 1997. All life is here. *New Scientist* 153(2073): 24-26. Focuses on the microhabitat of tree holes.

5674. Young, Truman P., 1994. Natural die-offs of large mammals: implications for conservation. *CB* 8: 410-418. [2] *A.T.:* population viability; literature reviews.

5675. Yurtsev, B.A., 1997. Effect of climate change on biodiversity of arctic plants. In W.C. Oechel, ed., *Global Change and Arctic Terrestrial Ecosystems* (New York: Springer): 229-244. [1].

5676. Zainal Abidin, A.H., & Akbar Zubaid, eds., 1997. *Conservation and Faunal Biodiversity in Malaysia* [collection of papers]. Bangi: Penerbit Universiti Kebangsaan Malaysia. 163p. [1/1/-].

5677. Zaitsev, I.P., & V.O. Mamaev, 1997. *Marine Biological Diversity in the Black Sea: A Study of Change and Decline.* New York: United Nations Publications. 208p. [1/1/-].

5678. Zak, John C., 1993. The enigma of desert ecosystems: the importance of interactions among the soil biota to fungal biodiversity. In *Aspects of Tropical Mycology* (Cambridge, U.K., and New York: CUP): 59-71. [1].

5679. Zak, John C., & S. Visser, 1996. An appraisal of soil fungal biodiversity: the crossroads between taxonomic and functional biodiversity.

Biodiversity and Conservation 5: 169-183. [1].
  5680. Zak, John C., M.R. Willig, et al., 1994. Functional diversity of microbial communities: a quantitative approach. *Soil Biology & Biochemistry* 26: 1101-1108. [3] *A.T.:* Biolog; bacteria; substrate use.
  5681. Zakaria-Ismail, M., 1994. Zoogeography and biodiversity of the freshwater fishes of Southeast Asia. *Hydrobiologia* 285: 41-48. [1].
  5682. Zakri, A. Hamid, ed., 1991. *Conservation of Plant Genetic Resources Through In Vitro Methods* [workshop proceedings]. Kuala Lumpur: Forest Research Institute Malaysia. 270p. [1/1/1] *A.T.:* plant germplasm resources.
  5683. Zapparoli, Marzio, 1997. Urban development and insect biodiversity of the Rome area, Italy. *Landscape and Urban Planning* 38: 77-86. *A.T.:* urban bd.
  5684. Zaret, Thomas M., & R.T. Paine, 1973. Species introduction in a tropical lake. *Science* 182: 449-455. [2] *A.T.:* fish.
  5685. Zavala, M.A., & J.A. Oria, 1995. Preserving biological diversity in managed forests: a meeting point for ecology and forestry. *Landscape and Urban Planning* 31: 363-378. [1] *A.T.:* Pacific Northwest; Mediterranean region.
  5686. Zebich-Knös, Michele, 1997. Preserving biodiversity in Costa Rica: the case of the Merck-INBio agreement. *J. of Environment & Development* 6: 180-186. *A.T.:* pharmaceutical industry; chemical prospecting.
  5687. Zedaker, Sheppard M., 1991. Biodiversity: boom or bust for forest vegetation management? *Proc., Southern Weed Science Society* 44: 222-232. [1].
  5688. Zedler, Joy B., 1991. The challenge of protecting endangered species habitat along the Southern California coast. *Coastal Management* 19: 35-53. [1] *A.T.:* coastal zone management; environmental policy; wetlands.
  5689. Zeide, Boris, 1997. Assessing biodiversity. *Environmental Monitoring and Assessment* 48: 249-260. *A.T.:* ecosystem health; environmental assessment.
  5690. Zenetos, A., 1997. Diversity of marine Bivalvia in Greek waters: effects of geography and environment. *J. of the Marine Biological Association* 77: 463-472.
  5691. Zentilli, Bernardo, 1992. Forests, trees and people. *Environmental Science & Technology* 26: 1096-1099. [1] *A.T.:* forest management; deforestation.
  5692. Zhang, Yongzu, 1994. *Protected Areas in China, With Emphasis on Conserving Mountain Biodiversity*. Honolulu: East-West Center. 32p. [1/1/1].
  5693. Zholdasova, I.M., 1997. Sturgeons and the Aral Sea ecological catastrophe. *Environmental Biology of Fishes* 48: 373-380. *A.T.:* environmental degradation; local extinction.

5694. Zhou, Z.H., & W.S. Pan, 1997. Analysis of the viability of a giant panda population. *J. of Applied Ecology* 34: 363-374. *A.T.:* China; endangered species.

5695. Zhu, J., M.D. Gale, et al., 1998. AFLP markers for the study of rice biodiversity. *Theoretical and Applied Genetics* 96: 602-611. *A.T.:* DNA fingerprinting; plant germplasm.

5696. Zimmer, Carl, 1994. More productive, less diverse. *Discover* 15: 24-26. [1] Reports on ecosystem productivity research by Michael Rosenzweig and David Tilman.

5697. Zimmerer, Karl S., 1991. Managing diversity in potato and maize fields of the Peruvian Andes. *J. of Ethnobiology* 11: 23-49. [1] *A.T.:* landscape perspective; mixed-cultivar fields.

5698. ———, 1992. Biological diversity and local development: "popping beans" in the central Andes. *Mountain Research and Development* 12: 47-61. [1] *A.T.:* traditional agriculture.

5699. ———, 1992. The loss and maintenance of native crops in mountain agriculture. *GeoJournal* 27: 61-72. [1] *A.T.:* Peru; Andes.

5700. ———, 1993. Agricultural biodiversity and peasant rights to subsistence in the central Andes during Inca rule. *J. of Historical Geography* 19: 15-32. [1].

5701. ———, 1996. *Changing Fortunes: Biodiversity and Peasant Livelihood in the Peruvian Andes*. Berkeley: University of California Press. 308p. [1/2/2] *A.T.:* human ecology.

5702. Zimmerer, Karl S., & D.S. Douches, 1991. Geographical approaches to crop conservation: the partitioning of genetic diversity in Andean potatoes. *Economic Botany* 45: 176-189. [1].

5703. Zimmerer, Karl S., & K.R. Young, eds., 1998. *Nature's Geography: New Lessons for Conservation in Developing Countries*. Madison, WI: University of Wisconsin Press. 351p.

5704. Zimmerman, Barbara L., & R.O. Bierregaard, 1986. Relevance of the equilibrium theory of island biogeography and species-area relations to conservation with a case from Amazonia. *J. of Biogeography* 13: 133-143. [2] *A.T.:* frogs.

5705. Zimmerman, John L., 1992. Density-independent factors affecting the avian diversity of the tallgrass prairie community. *Wilson Bull.* 104: 85-94. [1] *A.T.:* Kansas.

5706. Zink, Robert M., 1996. Bird species diversity [letter]. *Nature* 381: 566. [1] *A.T.:* phylogenetic analysis.

5707. Zink, Robert M., & M.C. McKitrick, 1995. The debate over species concepts and its implications for ornithology. *Auk* 112: 701-719. [2] *A.T.:* phylogenetic analysis.

5708. Ziswiler, Vinzenz, 1967. *Extinct and Vanishing Animals: A Biology of Extinction and Survival*. New York: Springer. 133p. [2/1/1].

5709. Zobel, Martin, 1997. The relative role of species pools in determining plant species richness: an alternative explanation of species coexistence? *TREE* 12: 266-269. [2].

5710. Zonneveld, Isaak S., & R.T.T. Forman, eds., 1990. *Changing Landscapes: An Ecological Perspective* [contributed papers]. New York: Springer. 286p. [1/1/1] *A.T.:* landscape ecology.

5711. Zulka, K.P., N. Milasowszky, & C. Lethmayer, 1997. Spider biodiversity potential of an ungrazed and a grazed inland salt meadow in the national park "Neusiedler See Seewinkel" (Austria): implications for management (Arachnida: Araneae). *Biodiversity and Conservation* 6: 75-88.

5712. Zurick, David N., 1995. Preserving paradise. *Geographical Review* 85: 157-172. [1] Concerning environmental degradation in the South Pacific region.

5713. Zweifel, Helen, fall 1997. The gendered nature of biodiversity conservation (in two parts). *NWSA J.* 9(3): 107-123. *A.T.:* ecofeminism.

5714. Zwick, Peter, 1992. Stream habitat fragmentation—a threat to biodiversity. *Biodiversity and Conservation* 1: 80-97. [1] *A.T.:* Europe; aquatic insects.

Bibliography II: Special Issues

5715. *Africa* 66(1): "The Social Shaping of Biodiversity: Perspectives on the Management of Biological Variety in Africa" (1996).

5716. *Agriculture, Ecosystems & Environment* 40(1-4): "Biotic Diversity in Agroecosystems" (May 1992).

5717. ——— 42(1-2): "Integrating Conservation Biology and Agricultural Production" (Oct. 1992).

5718. ——— 62(2-3): "Biodiversity in Agriculture for a Sustainable Future" (Feb. 1997).

5719. *Ambio* 20(2): "Environmental Economics" (April 1991).

5720. ——— 21(3): "The Economics of Biodiversity Loss" (May 1992).

5721. ——— 22(2-3): "Biodiversity: Ecology, Economics, Policy" (May 1993).

5722. *American Zoologist* 34(1): "Science as a Way of Knowing: Biodiversity" (1994).

5723. ——— 34(3): "Contribution of Long-term Ecological Research to Current Issues in the Conservation of Biological Diversity" (1994).

5724. *Amicus Journal* 15(1): "Biotechnology & Ecology" (spring 1993).

5725. *Anais da Academia Brasileira de Ciências* 66, Supl. 1: "Ecology and Biodiversity" (June 1995).

5726. *Annales Zoologici Fennici* 31(1): "Biodiversity in the Fennoscandian Boreal Forests: Natural Variation and Its Management" (1994).

5727. *Annals of the Missouri Botanical Garden* 70(3-4): "Biogeographical Relationships between Temperate Eastern Asia and Temperate Eastern North America" (1983).

5728. ——— 77(1): "Conserving Biological Diversity: Prospects for the 21st Century" and "A Symposium on the Biological Diversity and Evolution of the Tarweeds" (winter 1990).

5729. ——— 85(1): "New Tools for Investigating Biodiversity" (winter 1998).

5730. *Antonie van Leeuwenhoek* 72(1): "Biodiversity and Evolution of Micro-organisms" (July 1997).

5731. *Applied Soil Ecology* 6(1): "Agricultural Intensification, Soil Biodiversity and Agroecosystem Function in the Tropics" (July 1997).

5732. ——— 10(3): "Functional Aspects of Animal Diversity in Soil" (Nov. 1998).

5733. *Australasian Biotechnology* 5(4): "Biodiversity and Biotechnology" (Aug. 1995).

5734. *Australian Biologist* 5(1): "Australia's Biota and the National Interest: The Role of Biological Collections" (March 1992).

5735. ——— 11(1): "Evolutionary Biology at High Southern Latitudes" (1998).

5736. *Australian Systematic Botany* 4(1): "Austral Biogeography" (1991).

5737. ——— 6(5): "Southern Temperate Ecosystems—Origin and Diversification" (1993).

5738. ——— 9(2): "Origin and Evolution of the Flora of the Monsoon Tropics" (1996).

5739. *Australian Zoologist* 28(1-4): "Communication Skills and the Debate on Conserving Biodiversity" (Dec. 1992).

5740. *Biodiversity and Conservation* 2(3): "Global Biodiversity and Conservation of Insects" (June 1993).

5741. ——— 2(5): "Peatlands and People" (Oct. 1993).

5742. ——— 3(9): "Biodiversity and Conservation in the Gulf of Guinea Islands" (Dec. 1994).

5743. ——— 4(5): "Systematics Agenda 2000" (July 1995).

5744. ——— 4(6): "Captive Propagation and Effective Conservation" (Aug. 1995).

5745. ——— 4(8): "Ecologists and Ethical Judgments" (Nov. 1995).

5746. ——— 5(3): "Ecotourism, Biodiversity and Local Development" (1996).

5747. ——— 5(5): "Biodiversity and Conservation on Table Mountain and the Cape Peninsula" (1996).

5748. ——— 5(9): "Environmental Discontinuities and Synergisms" (1996).

5749. ——— 5(11): "Antarctic Microbial Diversity" (1996).

5750. ——— 6(3): "Biodiversity and Environmental Stability" (March 1997).

5751. ——— 6(5): "Fungal Biodiversity" (May 1997).

5752. ——— 6(11): "Deep-sea Biodiversity" (Nov. 1997).

5753. ——— 7(2): "Ecological Aspects of Biodiversity in Forests" (Feb. 1998).

5754. ——— 7(4): "The Biological Diversity of Namibia" (April 1998).

5755. *Biodiversity Letters* 1(3-4): "New Caledonia: A Case Study in Biodiversity" (May-July 1993).

5756. *Biological Conservation* 73(2): "Applications of Population Viability Analysis to Biodiversity Conservation" (1995).

5757. ——— 83(3): "Conservation Biology and Biodiversity Strategies" (1998).

5758. *Biological Journal of the Linnean Society* 28(1-2): "Island Biogeography of Mammals" (1986).

5759. ——— 56, Suppl. A: "The National Trust and Nature Conservation 100 Years On" (Dec. 1995).

5760. *Biology and Fertility of Soils* 19(2-3): "Microbial and Faunal Biomass in Soils: Part I" (Feb. 1995).

5761. *Bioscience* 45, Suppl.: "Science and Biodiversity Policy" (1995).

5762. *Boletim do Museu Municipal do Funchal* (Madeira), Supl. No. 4A: "Fauna and Flora of the Atlantic Islands" (1995).

5763. *BOS Nieuwsletter* 15(2): "The Guyana Shield: Recent Developments and Alternatives for Sustainable Development" (1996).

5764. *Buffalo Journal of International Law* 1(1): "Biodiversity and Biotechnology" (spring 1994).

5765. *Bulletin of Marine Science* 47(1): "Marine Biogeography and Evolution in the Pacific" (July 1990).

5766. ——— 57(1): "Indian River Lagoon Biodiversity" (July 1995).

5767. *Bulletin of the National Research Institute of Aquaculture*, Suppl. No. 3: "Biodiversity and Aquaculture for Sustainable Development" (Sept. 1997).

5768. *California Agriculture* 49(6): "Biological Diversity: What Is It and Why Do We Care?" (Nov.-Dec. 1995).

5769. *Canadian Journal of Forest Research* 22(11): "Emerging Issues in Northern Hardwood Management: Air Pollution, Climate Change, and Biodiversity" (Nov. 1992).

5770. *Canadian Journal of Plant Science* 75(1): "Symposium on Plant Gene Resources" (Jan. 1995).

5771. *Climatic Change* 27(1): "Climate Change: Significance for Agriculture and Forestry: Systems Approaches Arising from an IPCC Meeting" (May 1994).

5772. *Colorado Journal of International Environmental Law and Policy* 5(1): "Endangered Peoples: Indigenous Rights and the Environment" (winter 1994).

5773. ——— 7(1): "Biodiversity and Biotechnology" (winter 1996).

5774. *Commonwealth Forestry Review* 74(1): "Forestry and Nature Conservation" (March 1995).

5775. ——— 74(4): "Tropical Silviculture" (June 1995).

5776. *Cultural Survival Quarterly* 15(3): "Intellectual Property Rights: The Politics of Ownership" (1991).

5777. *Deep-Sea Research, Part II*, 45(1-3): "Deep-Sea Biodiversity: A Compilation of Recent Advances in Honor of Robert R. Hessler" (1998).

5778. *Defenders* 67(5): "Biodiversity after Rio" (Sept. 1992).

5779. *Different Drummer* 1(3): "Incentives for Protecting North American Biodiversity" (summer 1994).

5780. *Diversity* 12(4): "Convention on Biological Diversity" (1996).

5781. *Drug Development Research* 33(2): "Biological and Chemical Diversity" (Oct. 1994).

5782. *Ecodecision,* no. 18: "Feeding Humanity" (autumn 1995).

5783. *Ecological Applications* 2(2): "NAPAP and Biodiversity: Orchestrating Environmental Research and Assessment" (May 1992).

5784. ——— 5(4): "Integrated Conservation and Development: Applying Ecological Research at Integrated Conservation and Development Projects" and "Plant Diversity in Managed Forests: Impacts of Forest Management on Plant Diversity" (Nov. 1995).

5785. *Ecology* 70(4): "Periodic Extinctions in Earth History" (Aug. 1989).

5786. *Endangered Species Update* 14(7-8): "Habitat Conservation Planning" (July-Aug. 1997).

5787. *Environmental & Resource Economics* 4(1): "Biodiversity Loss: Measurement and Policy" (Feb. 1994).

5788. *Environmental Biology of Fishes* 48(1-4): "Sturgeon Biodiversity and Conservation" (1997).

5789. *Environmental Law* 24(2): "Endangered Species Act at Twenty-one: Issues of Reauthorization" (April 1994).

5790. *Environmental Management* 20(6): "Managing for Biodiversity: Emerging Ideas for the Electric Utility Industry" (Nov.-Dec. 1996).

5791. *Environmental Monitoring and Assessment* 49(2-3): "Atmospheric Change and Biodiversity: Formulating a Canadian Science Agenda" (Feb. 1998).

5792. *Environmental Professional* 13(3): "Restoration of Ecosystems" (1991).

5793. *Eurasian Soil Science* 29(1): "Soils and Biological Diversity" (1996).

5794. *Fisheries* 17(3): "Biodiversity" (May-June 1992).

5795. ——— 19(1): "Endangered Species" (Jan. 1994).

5796. *Forest Ecology and Management* 35(1-2): "Conservation of Diversity in Forest Ecosystems" (15 June 1990).

5797. ——— 85(1-3): "Conservation of Biological Diversity in Temperate and Boreal Forest Ecosystems" (Sept. 1996).

5798. *Forest Science* 41(2), Suppl.: "Invasion by Exotic Forest Pests: A Threat to Forest Ecosystems" (May 1995).

5799. *Forestry on the Hill,* Special Issue No. 3: "Biodiversity and Monocultures" (1992).

5800. *Galaxea* 13(1): "Biodiversity and Adaptive Strategies of Coral Reef

Organisms" (1996).

5801. *GeoJournal* 35(3): "Feeding 4 Billion People: The Challenge for Rice Research in the 21st Century" (March 1995).

5802. *Global Change Biology* 2(6): "Coral Reefs and Global Change" (1996).

5803. *Global Ecology and Biogeography Letters* 3(4-6): "The Political Ecology of Southeast Asian Forests: Transdisciplinary Discourses" (July-Nov. 1993).

5804. ——— 7(1): "Biodiversity and Function of Mangrove Ecosystems" (Jan. 1998).

5805. *Great Lakes Research Review* 3(1): "Great Lakes Exotic Species" (April 1997).

5806. *Growth and Change* 26(3): "Wilderness Areas" (summer 1995).

5807. *Historical Biology* 2(1): "Rare Events, Mass Extinction and Evolution" (March 1989).

5808. ——— 5(2-4): "Innovations and Revolutions in the Biosphere" (Dec. 1991).

5809. *Hydrobiologia* 287(1): "Biogeography of Subterranean Crustaceans: The Effects of Different Scales" (15 July 1994).

5810. *IDS Bulletin* 28(4): "Community-based Sustainable Development: Consensus or Conflict?" (Oct. 1997).

5811. *Impact of Science on Society,* no. 158: "Conserving and Managing Our Genetic Resources" (1990).

5812. *International Legal Materials* 31(4): "United Nations Conference on Environment and Development" (July 1992).

5813. *Israel Environment Bulletin* 20(1): "Biodiversity and Nature Conservation" (winter 1997).

5814. *Israel Journal of Zoology* 42(4): "Faunal Biodiversity and Variability" (1996).

5815. *Journal of Biogeography* 17(4-5): "Savanna Ecology and Management: Australian Perspectives and Intercontinental Comparisons" (July-Sept. 1990).

5816. ——— 22(2-3): "Terrestrial Ecosystem Interactions with Global Change" (March-May 1995).

5817. ——— 23(4): "Fig Trees and Their Associated Animals" (July 1996).

5818. *Journal of Biotechnology* 64(1): "Genome Analysis and the Changing Face of Biotechnology" (1998).

5819. *Journal of Ethnopharmacology* 51(1-3): "Intellectual Property Rights, Naturally Derived Bioactive Compounds and Resource Conservation" (April 1996).

5820. *Journal of Industrial Microbiology & Biotechnology* 17(3-4): "Microbial Diversity" (Oct. 1996).

5821. ——— 19(3): "Biological Control" (Sept. 1997).

5822. *Journal of Mammalogy* 75(2): "Biodiversity" (May 1994).

5823. *Journal of Parasitology* 78(4): "Parasite Biogeography and Coevolution" (Aug. 1992).

5824. *Journal of Sustainable Forestry* 4(3-4): "Sustainable Forests: Global Challenges and Local Solutions" (1997).

5825. *Journal of the Iowa Academy of Science* 105(2): "Perspectives on the Declining Flora and Fauna of Iowa" (June 1998).

5826. *Journal of the Marine Biological Association* 76(1): "Marine Biodiversity: Causes and Consequences" (Feb. 1996).

5827. *Journal of Vegetation Science* 7(1): "Community Ecology and Conservation Biology" (Feb. 1996).

5828. *Landscape and Urban Planning* 28(1): "Wildlife Habitat Conservation: Its Relationship to Biological Diversity and Landscape Sustainability" (Feb. 1994).

5829. ——— 37(1-2): "The Future of Our Landscapes" (June 1997).

5830. ——— 38(3-4): "Wildlife Habitats in Human-dominated Landscapes" (1997).

5831. ——— 40(1-3): "Ecosystem Management" (1998).

5832. *Madroño* 42(2): "The Future of California Floristics and Systematics: Research, Education, Conservation (April-June 1995).

5833. *Marine Technology Society Journal* 25(3): "Global Change I" (fall 1991).

5834. ——— 25(4): "Global Change II" (winter 1991-1992).

5835. *Memoirs of the Queensland Museum* 36(1): "Invertebrate Biodiversity and Conservation" (1994).

5836. *Michigan Academician* 22(4): "Biodiversity" (fall 1990).

5837. *Molecular Ecology* 4(5): "Molecular Microbial Ecology" (1995).

5838. *Molecular Screening News*, no. 11: "Molecular Tools for Biodiversity" (1996).

5839. *National Geographic* 195(2): "Biodiversity: The Fragile Web" (Feb. 1999).

5840. *Natural Resources and Environment* 8(1): "Endangered Species Protection" (summer 1993).

5841. [not used]

5842. [not used]

5843. *Nature and Resources* 29(1-4): "Biosphere Reserves: The Theory and the Practice" (1993).

5844. *Nature Conservancy* 44(1): "Biodiversity" (Jan.-Feb. 1994).

5845. *Naturopa*, no. 73: "Biodiversity" (1993).

5846. *New Zealand Journal of Zoology* 16(4): "Panbiogeography of New

Zealand" (1989).

5847. *Northwest Environmental Journal* 8(1): "Biological Diversity: Focus on Northwestern North America" (spring 1992).

5848. *Northwest Science* 70, Special Issue: "Northwest Forest Canopies" (1996).

5849. *Ocean & Coastal Management* 29(1-3): "Earth Summit Implementation: Progress Achieved on Oceans and Coasts" (1995).

5850. *Oceanus* 38(2): "Marine Biodiversity" (fall-winter 1995).

5851. *Oxford Economic Papers* 46, Suppl.: "Environmental Economics" (Oct. 1994).

5852. *Pacific Science* 49(1): "The Ecology and Evolutionary Biology of Islands with Implications for Conservation Research" (Jan. 1995).

5853. *Palaeogeography, Palaeoclimatology, Palaeoecology* 127(1-4): "New Perspectives on Faunal Stability in the Fossil Record" (1996).

5854. *Philosophical Transactions of the Royal Society of London B* 325(1228): "Evolution and Extinction" (1989).

5855. ——— 335(1275): "Tropical Rain Forest: Disturbance and Recovery" (1992).

5856. ——— 344(1308): "Estimating Extinction Rates" (1994).

5857. ——— 345(1311): "Biodiversity: Measurement and Estimation" (1994).

5858. ——— 353(1366): "Evolution of Biological Diversity: From Population Differentiation to Speciation" (1998).

5859. *Plant and Soil* 159(1): "Management of Mycorrhizas in Agriculture, Horticulture and Forestry" (Feb. 1994).

5860. ——— 170(1): "The Significance and Regulation of Soil Biodiversity" (March 1995).

5861. *Proceedings of the Academy of Environmental Biology* 3(1-2): "Toxicology, Conservation and Biodiversity" (1994).

5862. *Progress in Oceanography* 34(2-3): "Ecology of the Deep Sea Floor and Pelagic Biogeography" (Sept. 1994).

5863. *Researches on Population Ecology* 40(2): "Speciation, Its Ecological and Biogeographical Consequences" (1998).

5864. *Review of European Community & International Environmental Law* 6(3): "Focus on Biodiversity" (Nov. 1997).

5865. *Revista de Biologia Tropical* 44, Suppl. 4: "Fungi of Costa Rica: Selected Studies on Biodiversity and Ecology" (1996).

5866. *Science* 253(5021): "Perspectives on Biodiversity" (16 Aug. 1991).

5867. *Sea Wind* 11(3): "Global Freshwater Biodiversity: Striving for the Integrity of Freshwater Ecosystems" (July-Sept. 1997).

5868. *Sierra* 81(3): "Wetlands" (May-June 1996).

5869. *Silva Fennica* 30(2-3): "Climate Change, Biodiversity, and Boreal

Forest Ecosystems" (1996).

5870. *South African Journal of Science* 90(6): "Biotic Diversity and Function in Fynbos Ecosystems" (June 1994).

5871. *Systematic Zoology* 37(3, 4): "Vicariance Biogeography" (1988).

5872. *Texas Journal of Science* 49(3), Suppl.: "Biodiversity and Ecology of the West Gulf Coastal Plain Landscape" (1996).

5873. *Touro Journal of Transnational Law* 4: "New Diplomacy for the Biodiversity Trade: Biodiversity, Biotechnology, and Intellectual Property in the Convention on Biological Diversity" (spring 1993).

5874. *Trends in Ecology & Evolution* 7(6): "Arctic Marine Biogeography" (June 1992).

5875. *Tulane Environmental Law Journal* 8(1): "Biodiversity" (winter 1994).

5876. *Vanderbilt Journal of Transnational Law* 28(4): "Biodiversity: Opportunities and Obligations" (Oct. 1995).

5877. *Verhandlungen der Internationalen Vereinigung für Theoretische und Angewandte Limnologie* 25(4): "Biodiversity of Lake Tanganyika, With Special Reference to Inshore-fishes" (1994).

5878. *Vie et Milieu* 47(4): "Biodiversity in Dispersive Environments" (Dec. 1994).

5879. *Weed Technology* 10(2): "Role of Forest and Rangeland Vegetation Management in Conservation Biology" (April-June 1996).

5880. *Wild Earth*, Special Issue: "The Wildlands Project: Plotting a North American Wilderness Recovery Strategy" (1992).

The Indexes

The Indexes: Introduction

There are three indexes to this work. Index I covers all general subjects and is by far the longest of the three. It is alphabetically arranged by subject, and continuous in organization. Index II, on geographical subjects, is divided into nine sublists mirroring regional affinities (note that the classification provided does not implicitly distinguish ecological conditions; thus, the "Ocean Regions" sublist includes both oceans and oceanic islands, and the "California" entry may refer either to the mainland or the waters nearby). Index III, on organism-centered subjects, also is arranged into nine sublists, this time according to taxonomic affinities.

The indexing style features many cross-referrals; I decided to set things up in this fashion not only for the sake of completeness, but because many or most of the significant concepts in the developing field of biodiversity are associational. I believe therefore it should be of some reference value to have all two- or three-word terms involving, for example, the concept "conservation" listed in that part of the index, even if this means that many of the entries there are merely referrals to other parts of the list. Along the same lines, closely related concepts or modifiers are grouped in many of the entries, separated from the feature term or modifier by brackets and slashes [/—/—], but indexed in that entry along with them.

The other major feature of the indexes derives from the bibliometric approach taken in Bibliography I. Item numbers in the Indexes that are underlined represent works in Bibliography I which were rated with the higher scores. More specifically, index item numbers are underlined if (1) they correspond to an article/essay that scored a '2' or '3' in the (citation impact) ratings; *or* (2) they correspond to a monograph that scored at least two '2s' out of the three ratings; *or* (3) they correspond to a monograph that scored at least a '2' in the citation impact field ratings. This feature may make it easier to quickly prioritize readings (for example, in assigning required materials in related course work); however, one should recognize that this system has the weakness of ignoring the most recent literature, to which ratings could not be applied.

Note that Index I contains a few special entries itemizing monographic publications in Bibliography I by some of the more prominent nonprofit organizations (e.g., Island Press, the WWF, and the IUCN), and an entry summarizing early works (*see* "pre-1950 studies"). Note also that all cross-referrals are to items appearing in that same index *unless indicated otherwise.*

The Indexes: Table of Contents

Index I: General Subjects — 405

Index II: Geographical Subjects
- Regional Combinations — 443
- Canada and Greenland — 443
- United States — 443
- Latin America — 445
- Europe, Russia, and Southwest and Central Asia — 446
- South, Southeast, and East Asia — 446
- Africa — 447
- Australia, New Guinea, and New Zealand — 448
- Ocean Regions and Antarctica — 448

Index III: Organism-Centered Subjects
- Multi-Category Groupings — 451
- Mammals — 452
- Birds — 453
- Reptiles and Amphibians — 454
- Fishes — 454
- Arthropods — 455
- Lower (Complex) Animals — 456
- Plants — 456
- Algae, Fungi, Microorganisms, Viruses, etc. — 459

Index I: General Subjects

abundance, animal, 135, 203, 465, 1139, 2814, 4560, 4561
abundance, bird. *See* 'bird abundance' in Index III
abundance, insect. *See* 'insect diversity' in Index III
abundance, species. *See* species abundance
abundance-body size relations, 465, 2349, 2841, 3593, 4654, 5358, 5360
abundance-distribution relations, 634, 1469, 1771, 2110, 3184, 4357, 5046
abundance-diversity relations, 936, 4654, 5360, 5365
abundance-energy use relations, 1139, 4357
abundance-range size relations, 489, 2068, 2842
abundance-species relations, 2204, 2712, 3360, 3809, 5637
acidification, 2178, 2276, 2588, 3388, 3698, 4455, 4496, 4497, 5783
action [/recovery/survival] plans and programs, 151, 256, 294, 338, 399, 527, 603, 732, 779, 782, 917, 918, 1056, 1197, 1254, 1263, 1548, 1702, 1863, 2054, 2160, 2342, 2375, 2477, 3010, 3277, 3353, 3407, 3604, 3682, 3693, 3804, 3943, 4216, 4324, 4387, 4477, 4608, 4623, 4626, 4944, 4964, 5015, 5472, 5624, 5646, 5649, 5880
adaptation, 255, 363, 396, 940, 1068, 1533, 1589, 1877, 1890, 1909, 2396, 2878, 3849, 3853, 4313, 5215, 5258, 5800
afforestation [/reforestation], 68, 184, 738, 1041, 1230, 2922, 3349
AFLP markers, 5134, 5695
African Centre for Technology Studies (ACTS), publications by, 692, 1390, 2640, 3435, 3480, 4449, 5600
age and area, 3338, 5535
age-state structure, 5194
Agenda 21, 2579, 4324, 4719, 5106, 5184, 5444
agreements. *See* treaties and agreements
agricultural development [/intensification], 97, 3111, 3176, 3279, 3469, 3811, 4848, 5731
agricultural ecology, 99, 100, 219, 600, 815, 1842, 1843, 2638, 2733, 3254, 3811, 4734, 4848, 4849, 5063, 5240
agricultural economics, 455, 1010, 3752, 4198
agricultural landscapes, 705, 1153, 1369, 1493, 2208, 2216, 2762, 3380, 3759
agricultural [/farming] practices, 380, 609, 1155, 1624, 1796, 1802, 3254, 3836, 4315, 4864, 5080, 5335, 5602
agricultural productivity [/production], 380, 391, 691, 1732, 1741, 1855, 2336, 2338, 2409, 2580, 2863, 3347, 3378, 3741, 3907, 4396, 4417, 4434, 4758, 5717
agriculture, alternative, 432, 435, 3846, 4315
agriculture, sustainable. *See* sustainable agriculture
agriculture, traditional [/local/indigenous]. *See* traditional agriculture
agrobiodiversity, 2087, 2437, 2733, 4600, 4636, 4734, 4953, 5076, 5600, 5602, 5700
agroecosystems, 96, 99, 695, 896, 987, 1298, 1624, 2019, 2615, 2896, 3114, 3495, 3735, 3836, 3837, 3908, 3909, 3911, 4000, 4297, 4611, 4890, 5241, 5243, 5353, 5716, 5731
agroforestry, 1734, 3775, 3847
AIBS, 153
All-Taxa Biodiversity Inventory, 2524, 5022, 5666
Allen's rule, 76, 4187
allometry, 465, 1139, 3255
allopatric distribution, 559
alluvial environments (*see also* floodplains), 574, 3474, 4506, 5136
alpha diversity, 2101, 3971, 4579, 4638, 5130, 5208, 5386, 5457
alternative agriculture, 432, 435, 3846, 4315
altitudinal [/elevational] effects, 1243, 1264, 1277, 1539, 2294, 2403, 2424, 2438, 2806, 3797, 3798, 4131, 4132, 4556, 4710, 4881,

405

5027, 5343
altitudinal gradients, 1264, 1277, 2424, 4132, 4556, 4710, 4881, 5027
American Fisheries Society, 147, 5572
amphibians. *See* listings in Index III
ancient ecosystems [/environments], 614, 951, 1018, 2076, 3153, 3658, 5351
animal rights and welfare, 763, 1584, 1657, 2463, 3329, 3331, 3663, 3670, 4026, 4218, 4219, 4328, 4408, 4706, 5014
animals. *See* listings in Index III
anoxia, 67, 3697, 5474
anthropic principle, 822, 2887
anthropochory, 855
anthropological perspectives, 207, 668, 743, 1850, 1922, 2395, 3779, 4843, 4852
antibiotics, 608, 829, 1241
antitropical distribution, 2926
aquaculture, 226, 352, 433, 1062, 1870, 3805, 3896, 3906, 5767
aquaria and aquarium fishes, 1062, 2382, 2649, 3624, 5472
aquatic ecosystems: misc. subjects, 352, 419, 484, 520, 773, 1080, 1090, 1110, 1681, 1833, 1991, 2880, 3024, 3063, 3204, 3239, 3388, 3478, 3568, 3700, 3745, 3817, 3871, 3941, 4004, 4044, 4295, 4618, 4705, 5056, 5131, 5221, 5277, 5344, 5370, 5386, 5499, 5598, 5714
aquatic insects, 3092, 4004, 4925, 5205, 5221, 5280, 5714
aquatic plants, 351, 1833, 2190, 2880, 3000, 5386
Arber, Werner, 1739
archaeological studies, 743, 1279, 3025, 4852, 4866
architecture [/design], bd-conscious, 1407, 1926, 3241
arctic environs. *See* boreal environs, tundra (*see also* geographical listings in Index II)
area-species relations. *See* species-area relations
AREM, 42
areography, 1023, 1752, 4150, 5559
arid environs. *See* desert environs
artificial intelligence, 1187
assemblage fidelity, 3744
assembly theory, 640, 1352, 2837, 2949, 3476, 5561
assessment, bd. *See* biodiversity assessment
assessment, rapid. *See* rapid assessment
assessment, risk. *See* risk assessment

asteroid [/comet] collisions, 103, 105, 436, 2020, 2422, 2589, 3253, 3483, 4091, 4168, 4609, 4663, 5347, 5590
asymptotic diversity, 1504
atlas data-based studies, 68, 1459, 2157, 3785
atlases, 980, 983, 2081, 2322, 4483
atmosphere, earth's, 424, 1577, 1681, 1902, 2149, 2461, 2909, 2957, 3053, 3127, 4025, 4051, 4178, 5115, 5791
atolls, 1735, 4893
Australian Tree Seed Centre, 185
avalanche index, 1737
background extinction, 546, 2387, 2677
backyard bd, 2271, 2940, 3131, 3132
Backyard Wildlife Habitat Program, 2940
"bad genes" theory of extinction, 4166, 4787
ballast transport and dumping, 798, 803, 2253, 3369
banks, gene. *See* gene banks
banks, seed. *See* seed banks
barrier islands, 3843, 4191
Bayesian methods, 2913, 4793
behavior, animal, 1142, 1234, 1235, 1590, 1877, 2577, 3178, 4919, 5565
behavioral diversity, 1094, 1877, 4919
benthic environs, freshwater, 787, 951, 1219, 2014, 3092, 4223, 5391
benthic environs, marine. *See* marine benthic environs
Bergmann's rule, 420, 464, 972, 2413, 3262, 4187
Bern Convention, 1057, 5174
beta diversity, 2101, 3971, 4579, 4638, 5386, 5457
binary data, 2290, 3383, 5232
biochemical approaches, 1213, 1529, 3019, 3739, 4116, 5039, 5680
biodiversity and climate. *See* climate and bd
biodiversity and forestry. *See* forestry and bd
biodiversity and women. *See* women and bd
biodiversity assessment, 228, 291, 430, 499, 762, 1161, 1174, 1222, 1239, 1292, 1333, 1470, 1516, 1517, 1519, 1587, 1605, 1685, 1748, 1809, 1830, 1950, 2015, 2080, 2117, 2183, 2197, 2277, 2420, 2546, 2811, 2903, 2962, 2971, 3005, 3044, 3186, 3195, 3213, 3556, 3578, 3601, 3688, 3717, 3867, 3889, 3900, 3943, 3961, 4052, 4072, 4116, 4372, 4499, 4834, 4903, 5074, 5194, 5231, 5293, 5298, 5361, 5368, 5435, 5463, 5571, 5689,

5783
biodiversity, backyard, 2271, 2940, 3131, 3132
biodiversity, canopy. *See* canopy bd studies
biodiversity [/biogeographic] congruence, 259, 1083, 1086, 1095, 1761, 2224, 3808, 3986
biodiversity conservation: misc. overviews, 34, 186, 187, 227, 343, 485, 562, 831, 949, 1245, 1312, 1550, 1594, 1654, 1663, 1664, 1782, 1917, 1965, 1983, 2034, 2441, 2738, 2745, 2747, 2809, 2814, 2985, 3001, 3285, 3286, 3353, 3504, 3540, 3659, 3662, 3664, 3666, 3684, 3771, 3901, 3912, 3919, 4230, 4232, 4403, 4404, 4426, 4535, 4630, 5070, 5112, 5557, 5612, 5624, 5757, 5813
biodiversity, definitions of, 646, 699, 1238, 1782, 2134, 2312, 2861, 3631, 3899, 4031, 4182, 4426, 4796, 4991, 5643
biodiversity, economic valuation of (*see also* contingent valuation), 223, 280, 281, 290, 401, 458, 552, 644, 647, 670, 928, 1132, 1226, 1281, 1744, 1850, 1898, 2050, 2052, 2053, 2138, 2139, 2412, 2473, 2720, 2727, 2744, 2928, 3084, 3211, 3397, 3438, 3439, 3487, 3572, 3712, 3754, 3777, 3876, 3879, 3929, 3934, 3967, 4001, 4020, 4274, 4437, 4700, 4701, 4830, 4883, 4884, 5237, 5536
biodiversity, esthetic and practical value of, 180, 222, 252, 384, 401, 458, 645, 671, 724, 768, 865, 889, 928, 1048, 1123, 1167, 1226, 1270, 1411, 1414, 1427, 1513, 1550, 1747, 1849, 1850, 1873, 2000, 2139, 2309, 2593, 2595, 3211, 3213, 3286, 3288, 3304, 3454, 3530, 3532, 3631, 3664, 3666, 3712, 3731, 3734, 3879, 3880, 3914, 3967, 4049, 4147, 4183, 4228, 4352, 4535, 4660, 4698, 4699, 4733, 4961, 5108, 5299, 5392, 5522, 5605
biodiversity, floristic. *See* floristic diversity
biodiversity, functional. *See* functional diversity
biodiversity [/environmental] indicators, 8, 58, 200, 248, 366, 368, 430, 649, 748, 817, 905, 915, 923, 1052, 1370, 1515, 1516, 1608, 1624, 1928, 1952, 2043, 2281, 2363, 2607, 2678, 2740, 2750, 2751, 2753, 2754, 2804, 2844, 3217, 3246, 3333, 3643, 3680, 3833, 3850, 3889, 3890, 3891, 3939, 4061, 4237, 4332, 4372, 4411, 4414, 4894, 5060, 5353, 5359, 5369, 5498, 5597, 5650
biodiversity integrity. *See* biological integrity
biodiversity law. *See* environmental law
biodiversity [/species] loss, 12, 144, 258, 622, 744, 856, 862, 875, 895, 949, 995, 1125, 1235, 1318, 1419, 1422, 1431, 1638, 1643, 1647, 1709, 1759, 1868, 2112, 2223, 2297, 2490, 2493, 2564, 2646, 2693, 2725, 2762, 2851, 2933, 3010, 3189, 3269, 3277, 3303, 3400, 3455, 3477, 3501, 3511, 3535, 3631, 3656, 3687, 3877, 3909, 3911, 3917, 3918, 3969, 3996, 4045, 4060, 4108, 4146, 4183, 4225, 4256, 4320, 4403, 4590, 4656, 4676, 4742, 4808, 4811, 4888, 4913, 4959, 4975, 4993, 5152, 5154, 5156, 5166, 5323, 5462, 5479, 5553, 5611, 5720, 5787
biodiversity loss [/decline] rates, 2852, 4808, 5051, 5339, 5479
biodiversity, marine. *See* marine bd
biodiversity measurement [/estimation], 118, 123, 263, 381, 532, 696, 787, 990, 1066, 1282, 1319, 1379, 1459, 1611, 1714, 1737, 1757, 2006, 2018, 2044, 2079, 2094, 2133, 2166, 2175, 2204, 2233, 2308, 2423, 2578, 2637, 2660, 2742, 2847, 2913, 2914, 2946, 2955, 3080, 3085, 3553, 3588, 3589, 3608, 3627, 3696, 3697, 3714, 3751, 3752, 3819, 3821, 3834, 3859, 3899, 3959, 3960, 3961, 4000, 4034, 4158, 4487, 4489, 4560, 4561, 4660, 4694, 4741, 4750, 4793, 4794, 4796, 4798, 4799, 4968, 4990, 5245, 5316, 5337, 5350, 5359, 5368, 5455, 5457, 5516, 5519, 5549, 5787, 5857
biodiversity: misc. overviews, 3, 12, 90, 166, 173, 186, 227, 354, 762, 802, 847, 852, 994, 1238, 1420, 1427, 1444, 1445, 1446, 1757, 1759, 1760, 1775, 1917, 1949, 1950, 1968, 1983, 2006, 2008, 2119, 2197, 2226, 2244, 2286, 2297, 2337, 2346, 2441, 2502, 2595, 2710, 2717, 2732, 2747, 2809, 2814, 2861, 2985, 2989, 3001, 3086, 3126, 3208, 3269, 3288, 3429, 3467, 3522, 3528, 3533, 3548, 3631, 3657, 3679, 3684, 3782, 3877, 3879, 3918, 4047, 4181, 4182, 4199, 4222, 4226, 4306, 4530, 4590, 4634, 4714, 4777, 4784, 4890, 4964, 4991, 5070, 5112, 5116, 5165, 5240, 5306, 5346, 5416, 5421, 5480, 5551, 5552, 5557, 5624, 5643, 5658, 5720, 5721, 5722, 5729, 5761, 5768, 5794, 5836, 5839, 5844, 5845, 5850, 5866
biodiversity [/ecological] monitoring, 89, 238, 341, 454, 514, 554, 556, 649, 748, 773, 890, 993, 996, 1011, 1033, 1092, 1163, 1164, 1165, 1166, 1267, 1356, 1391, 1720, 1862, 2014, 2193, 2365, 2551, 2654, 2749, 2750, 2752, 2753, 2754, 2806, 2944, 3117, 3120, 3168, 3212, 3217, 3310, 3444, 3517, 3537, 3648, 3680, 3755, 3818, 3939,

4008, 4263, 4289, 4332, 4402, 4670, 4683, 4684, 4829, 4833, 4905, 4924, 5060, 5279, 5342, 5391, 5545, 5620

biodiversity, nutritional, 3160

biodiversity prospecting (*see also* chemical prospecting, drug discovery), 117, 195, 222, 281, 958, 1545, 1747, 2313, 2386, 2504, 2554, 3083, 3209, 3302, 3841, 4105, 4227, 4236, 4385, 4698, 4700, 4701, 4776, 5001

biodiversity visualization, 2037, 4057

biodiversity-conscious architecture [/design], 1407, 1926, 3241

bioethics (*see also* environmental ethics), 734, 961, 1014, 1135, 3670, 4030, 4218, 4328, 4631, 4706, 5643, 5654

biogeochemical cycles, 67, 346, 356, 2957, 4178

biogeography, 108, 125, 139, 381, 403, 410, 445, 515, 582, 589, 595, 636, 639, 642, 703, 741, 804, 929, 970, 1049, 1075, 1084, 1088, 1090, 1095, 1099, 1105, 1106, 1140, 1152, 1275, 1373, 1465, 1469, 1502, 1560, 1562, 1568, 1605, 1617, 1648, 1712, 1735, 1770, 1826, 1858, 1936, 1938, 1960, 1985, 2012, 2023, 2033, 2143, 2148, 2154, 2155, 2169, 2173, 2175, 2196, 2208, 2224, 2251, 2291, 2306, 2320, 2333, 2387, 2389, 2416, 2435, 2438, 2480, 2481, 2507, 2513, 2525, 2574, 2575, 2749, 2772, 2850, 2923, 2966, 2967, 2969, 2997, 3014, 3032, 3037, 3067, 3105, 3139, 3186, 3231, 3237, 3293, 3328, 3421, 3457, 3459, 3508, 3509, 3599, 3600, 3699, 3703, 3705, 3706, 3791, 3797, 3808, 3831, 3850, 3962, 3984, 3986, 3988, 3989, 4016, 4050, 4052, 4115, 4121, 4150, 4287, 4342, 4356, 4361, 4362, 4368, 4419, 4473, 4474, 4619, 4648, 4653, 4737, 4893, 4989, 5013, 5068, 5144, 5149, 5150, 5169, 5225, 5236, 5247, 5258, 5263, 5270, 5301, 5307, 5321, 5322, 5334, 5355, 5434, 5452, 5459, 5490, 5514, 5535, 5585, 5629, 5635, 5669, 5736, 5765, 5823, 5862, 5863, 5871, 5874

biogeography, bird. *See* 'bird biogeography' in Index III

biogeography, fish. *See* 'fish biogeography' in Index III

biogeography, historical. *See* historical biogeography

biogeography, insect. *See* 'insect biogeography' in Index III

biogeography, island. *See* island biogeography

biogeography, mammal. *See* 'mammal biogeography' in Index III

biogeography, plant. *See* 'plant biogeography' in Index III

biological control, 242, 922, 1024, 1298, 2125, 2130, 2263, 2762, 2789, 2840, 3114, 3344, 3510, 4447, 4943, 5367, 5821

biological [/bd/biotic/ecological/ecosystem] integrity, 110, 145, 1698, 2297, 2551, 2660, 3027, 3085, 3185, 3387, 4438, 5391, 5607, 5867

biological invasions (*see also* introduced species), 128, 129, 216, 264, 292, 324, 371, 387, 445, 502, 508, 618, 799, 801, 803, 845, 855, 941, 945, 969, 1037, 1104, 1109, 1124, 1180, 1260, 1266, 1353, 1460, 1602, 1609, 1730, 1880, 1923, 1954, 1959, 1960, 2170, 2172, 2220, 2240, 2253, 2259, 2304, 2468, 2539, 2669, 2789, 2953, 2973, 2975, 2981, 2984, 3007, 3040, 3041, 3042, 3052, 3095, 3182, 3252, 3324, 3345, 3369, 3370, 3427, 3428, 3476, 3985, 4019, 4078, 4109, 4122, 4137, 4241, 4268, 4269, 4276, 4278, 4330, 4384, 4390, 4618, 4906, 4908, 4934, 5023, 5198, 5200, 5204, 5262, 5264, 5285, 5288, 5330, 5349, 5529, 5530, 5531, 5577, 5626

biology, island. *See* island biology

biology, population. *See* population biology

biomass, 29, 645, 653, 654, 894, 1020, 1154, 2352, 2402, 2447, 2560, 2830, 2870, 2909, 2910, 3730, 3809, 4136, 4153, 4620, 4710, 4827, 5358, 5438, 5760

BioNET International, 1022, 2492

biophilia, 2596, 2598, 5548

biopiracy, 723, 925, 2554, 4632

BioRap, 213, 5350

bioregionalism, 66, 3354, 4768, 5128

bioremediation, 2270, 4951

biosphere, 406, 551, 919, 1067, 1107, 1362, 1583, 2568, 2924, 2993, 2994, 2995, 3002, 3350, 3430, 3602, 3715, 3803, 4423, 4520, 4585, 4977, 5151, 5334, 5388, 5808

biosphere reserves (*see also* nature reserves, protected areas), 151, 294, 333, 335, 828, 1184, 1327, 1932, 2307, 2534, 2612, 2800, 3384, 3595, 4079, 4191, 4683, 4845, 4870, 5190, 5191, 5843

Biosphere 2, 955, 3602

biotas, 137, 387, 627, 673, 674, 795, 940, 1083, 1084, 1129, 1146, 1403, 1485, 1600, 1815, 1856, 2104, 2150, 2319, 2324, 2576, 2836, 2889, 2926, 2953, 3064, 3065, 3106, 3136, 3339, 3369, 3388, 3519, 3520, 3606,

Index I 409

4000, 4178, 4247, 4272, 4378, 4379, 4464, 4665, 4666, 4691, 4750, 4860, 4862, 4872, 4896, 4906, 5216, 5261, 5322, 5373, 5376, 5445, 5611, 5624, 5678, 5734
biotechnology, 24, 35, 36, 218, 246, 521, 656, 693, 694, 713, 772, 780, 1067, 1111, 1150, 1520, 1699, 1751, 1860, 1884, 2088, 2295, 2409, 2511, 2544, 2582, 2640, 2648, 2696, 2763, 2768, 2807, 2891, 3108, 3109, 3110, 3256, 3431, 3715, 3739, 3825, 4046, 4082, 4101, 4105, 4125, 4144, 4563, 4626, 4627, 4629, 4643, 4846, 4853, 4951, 5196, 5724, 5733, 5764, 5773, 5818, 5873
biotechnology and plants. *See* 'plants and biotechnology' in Index III
Biotic Exploration Fund, 1437
biotic integrity. *See* biological integrity
biotic interchange, 2889, 2926, 3146, 3147, 4872, 5261, 5262, 5264, 5373
biotopes, 915, 4402
birds. *See* listings in Index III
blitzkreig hypothesis. *See* Pleistocene overkill
body mass trends. *See* body size trends
body shape [/morphology] trends, 1068, 3425, 4619
body size, animal, 420, 465, 1139, 5233
body size, bird, 464, 1934, 2413, 3593, 5127
body size gradients, latitudinal, 300, 464, 2126, 2129, 3262, 4619
body size, insect, 2590, 4512, 4654
body size, mammal, 76, 641, 825, 3442
body size [/mass] trends, 76, 300, 420, 464, 465, 641, 825, 972, 1065, 1139, 1144, 1269, 1286, 1552, 1581, 1752, 1767, 1823, 1934, 1976, 2126, 2129, 2349, 2413, 2841, 2889, 2950, 3068, 3194, 3262, 3442, 3590, 3593, 3809, 3902, 4032, 4187, 4258, 4512, 4619, 4654, 5127, 5233, 5358, 5360
body size-abundance relations, 465, 2349, 2841, 3593, 4654, 5358, 5360
body size-diversity relations, 1581, 1823, 1934, 4654, 5360
bogs. *See* peatlands and bogs
bootstrap procedure, 3383, 4741, 5337
boreal environs, 505, 632, 1171, 1463, 1613, 1920, 2004, 2069, 2073, 2074, 2075, 2776, 2928, 3154, 3184, 3632, 3671, 3726, 4494, 4897, 5066, 5120, 5283, 5303, 5406, 5726, 5797, 5869
borealization, 1463
botanical gardens, 510, 567, 1134, 2184, 2194, 2555, 3394, 4681

bottlenecks, population, 177, 237, 1450, 2156, 2755, 3597, 3707, 3709, 3710, 4161, 5326
bottom-up and top-down regulation, 2318, 2869, 2897, 5515
breeding birds. *See* 'birds, breeding' in Index III
breeding, captive. *See* captive breeding
breeding, pedigree, 5024
buffer zones, 437, 645, 715, 1010, 1184, 1499, 2361, 2534, 3505, 3603, 4480, 4574, 4812, 5141, 5303, 5401
Bureau of Land Management, 1032, 1033, 1189, 5598
bushmeat trade (*see also* hunting), 111, 1503
caatinga, 2560
California Endangered Species Act, 1394
Cambrian, 2661, 5273, 5426
Cambrian diversification "explosion," 794, 2256, 2661, 3259, 5273
canopy bd studies, 326, 327, 461, 911, 1557, 2743, 3419, 3543, 3925, 4204, 4294, 4831, 4916, 4921, 5271
canopy disturbance, 3162, 5271
canopy gaps. *See* treefall gaps
capital, natural, 1044, 1048, 1896, 3605, 3915
captive breeding (*see also* ex situ conservation), 136, 395, 603, 1401, 1547, 1803, 1819, 1820, 2382, 3343, 3746, 3942, 4130, 4407, 4545, 4769, 5744
carbon dioxide [/carbon cycle], 345, 346, 347, 351, 503, 504, 714, 752, 1257, 1709, 1903, 1977, 2279, 2461, 2468, 2616, 2670, 2703, 2867, 3817, 4025, 4326, 4533, 4625, 4891, 5286
Caring for the Earth, 4320
carrying capacity, human, 189, 954, 1158, 1418
case studies, 53, 111, 178, 192, 221, 290, 322, 426, 487, 511, 552, 571, 705, 715, 782, 800, 918, 928, 951, 1096, 1120, 1166, 1218, 1230, 1537, 1569, 1588, 1627, 1686, 1713, 1718, 1742, 1778, 2157, 2202, 2223, 2282, 2331, 2405, 2496, 2744, 2751, 2915, 2925, 3021, 3218, 3247, 3377, 3434, 3489, 3604, 3686, 3763, 3766, 3822, 4000, 4027, 4041, 4042, 4057, 4099, 4152, 4253, 4352, 4436, 4504, 4711, 4727, 4889, 4890, 5078, 5104, 5132, 5191, 5237, 5351, 5476, 5480, 5584, 5755
catastrophic natural events, 60, 103, 104, 105, 436, 524, 536, 942, 1123, 1838, 2024,

2589, 3097, 3483, 4091, 4143, 4405, 4580, 4609, 4663, 5347, 5427, 5590
caves, 181, 307, 1128, 1129, 2091, 2092, 2103, 2283, 4464
cells, origin of, 837, 2108
Cenozoic, 1622, 4860, 5262, 5376
Center for Plant Conservation, 158, 1692, 5354
cerrado, 790, 1152, 4081, 4156
CGIAR, 1705, 4195
Champion International, 3795
change, climatic. See climatic change
change, environmental. See environmental change
change, sea-level. See sea-level change
change, vegetation. See vegetation change
chaos, 77, 4787
chaparral, 501, 4809, 4814, 5422
character richness, 833
checklists. See inventories and checklists
chemical diversity, 1255, 1529, 3219, 4354, 5299, 5781
chemical prospecting (*see also* bd prospecting, drug discovery), 1436, 1437, 4309, 5196, 5686
CITES, 1540, 2167, 5475
cladistics, 175, 1083, 1095, 1508, 1640, 2302, 2306, 2621, 3458, 3459, 3600, 3808, 3986, 4362, 4581, 5236, 5294, 5430, 5489
classification, 627, 696, 838, 1403, 1491, 2670, 2834, 3043, 3753, 3963, 4651, 4977, 5008, 5169, 5454, 5524
clearcutting (*see also* logging), 2314, 3317, 3937, 4389, 4968, 5141
climate and bd, 3133, 3268, 3850, 3932, 4025, 4233, 4234, 4533
climate and biology, 1780, 5376, 5395, 5648
climate and distribution, 1036, 3845, 4727, 5610
climate and diversity, 1585, 3237, 3858, 4213, 4336, 5157, 5209
climate and extinction, 921, 1115, 5209
climate and faunas, 314, 1027, 1574, 1634, 2850, 4860, 5209
climate and forests [/deforestation], 346, 506, 908, 1161, 1672, 1903, 1977, 2149, 2650, 3154, 3518, 3521, 4051, 4602, 4645, 5771, 5869
climate and plant diversity, 1072, 1786, 3502, 3705, 3706, 4326, 4419, 4533
climate and plants, 316, 351, 505, 1634, 2670, 3315, 5675

climate and species richness, 938, 3705, 3706, 4419
climate and vegetation, 316, 505, 701, 1233, 2461, 2670, 4051, 4064, 4326, 4831, 5449
climate: misc. subjects, 89, 203, 348, 351, 728, 785, 811, 863, 888, 1101, 1143, 1314, 1473, 1574, 1601, 1613, 1634, 2316, 2323, 2401, 2413, 2850, 2909, 2910, 2992, 3094, 3095, 3154, 3502, 3845, 3903, 3923, 3924, 4039, 4059, 4365, 4727, 4860, 5209, 5395, 5493, 5610, 5675, 5771
climatic change (*see also* environmental change, global warming, greenhouse effect), 196, 203, 345, 346, 351, 365, 503, 504, 505, 506, 508, 555, 701, 728, 811, 888, 908, 921, 1073, 1101, 1115, 1143, 1161, 1233, 1314, 1473, 1574, 1585, 1613, 1634, 1780, 1903, 1919, 1952, 1977, 2149, 2277, 2279, 2316, 2401, 2518, 2541, 2650, 2670, 2749, 2910, 2933, 2992, 3095, 3133, 3154, 3268, 3471, 3502, 3518, 3525, 3644, 3850, 3923, 3924, 3932, 4025, 4064, 4233, 4234, 4307, 4326, 4365, 4419, 4533, 4602, 4645, 4746, 4860, 4995, 5209, 5376, 5395, 5449, 5493, 5648, 5675, 5769, 5771, 5869
climax, ecological. See ecological climax
clines. See gradients
clones, natural, 2832, 5300
cloud forests, 2035, 2556, 2557, 4037, 4041
coarse filter approach, 2117, 2317, 3835
coarse woody debris (*see also* dead and dying trees), 1683, 1975, 3261, 4443
coastal environs (*see also* coastal zone management, estuaries, lagoons), 132, 301, 348, 434, 575, 798, 804, 856, 1045, 1125, 1172, 1209, 1797, 1900, 1917, 1940, 1953, 1963, 2014, 2603, 2730, 2787, 3113, 3139, 3313, 3352, 3449, 3508, 3856, 4188, 4189, 4191, 4192, 4233, 4234, 4340, 4375, 4565, 4599, 4868, 4936, 5054, 5175, 5422, 5563, 5872
coastal zone management, 357, 401, 430, 443, 902, 1045, 1209, 1210, 1211, 1223, 1375, 1797, 1879, 1917, 2014, 2206, 2227, 2698, 2760, 3461, 3506, 3653, 3793, 3945, 4079, 4191, 4234, 4620, 4770, 5078, 5688, 5849
cocaine, 5670
coevolution, 71, 1533, 1534, 1535, 1740, 2395, 3465, 3495, 5823
collections (of specimens, etc.; *see also* core collections, gene banks, germplasm collections, seed banks), 39, 40, 59, 330, 467, 624, 777, 1051, 1297, 1365, 2231, 2343, 2488, 2673, 2674, 2723, 2893, 3363, 3959,

Index I   411

4001, 4024, 4245, 4262, 4593, 4774, 4877, 4933, 5069, 5281, 5488, 5734
collections, core, 624, 1297, 2231
collections, germplasm, 330, 1297, 2488, 2893, 4774, 5281
collections, plant, 777, 1365, 2231, 4933, 5069
colonization [/recolonization], 369, 410, 477, 730, 819, 1025, 1505, 2148, 3136, 3390, 3615, 3902, 4764, 4826, 5050, 5071, 5086, 5220, 5304, 5460, 5638
colonization [/immigration/recruitment] rates, 812, 1841, 2148, 3902
combinatorial chemistry, 1255
comet collisions. *See* asteroid collisions
commensalism. *See* symbiosis
Commission on Sustainable Development, 902
commodities. *See* markets
commons, 169, 607, 735, 904, 2083, 2084, 3218, 3790, 4635, 4967, 4971
communities, arthropod. *See* 'arthropod communities' in Index III
communities, bird. *See* 'bird communities' in Index III
communities, fish, 2018, 2269, 2300, 2393, 2447, 2448, 3220, 4262, 4428, 4646, 5541
communities, freshwater. *See* freshwater communities
communities, grassland. *See* grassland communities
communities, hydrothermal vent, 1648, 1911, 3022, 4653, 5225, 5405
communities, intertidal. *See* marine communities
communities, marine. *See* marine communities
communities, microbial, 1707, 2079, 3019, 3834, 5123, 5342, 5534, 5680
communities, parasite, 727, 1749, 2617, 2618, 2619, 4032, 4033, 4034, 4035, 4930
communities, plant. *See* 'plant communities' in Index III
communities, soil. *See* soil communities
communities, wildlife. *See* wildlife ecology
community diversity, 604, 629, 1219, 1474, 1677, 1694, 1708, 1788, 1816, 2414, 2447, 2472, 2619, 2806, 2869, 2884, 3220, 3858, 3861, 4285, 4540, 5041, 5043, 5368, 5457, 5574
community ecology, 336, 627, 684, 2390, 2828, 2952, 3062, 3112, 3157, 3158, 3381, 3968, 3985, 4006, 4083, 4211, 4244, 4668,

4793, 4817, 4837, 5118, 5448, 5454, 5456, 5458, 5541, 5667, 5827
community forestry, 581, 2440, 5400, 5604
community structure [/organization], 274, 275, 394, 429, 627, 629, 633, 708, 773, 924, 934, 1039, 1177, 1305, 1352, 1737, 1904, 1990, 2010, 2049, 2318, 2550, 2552, 2712, 2837, 2869, 2870, 2967, 2982, 3139, 3178, 3193, 3250, 3265, 3549, 3751, 3814, 4033, 4221, 4285, 4287, 4442, 4499, 4792, 4937, 4975, 5041, 5088, 5089, 5114, 5136, 5458, 5543, 5561
community succession. *See* ecological succession
community-related conservation participation. *See* local conservation participation
competition, 18, 19, 774, 2010, 2347, 3000, 3033, 3307, 3472, 4019, 4673, 5087, 5092, 5220, 5445
competitive exclusion, 774, 2347, 5220
competitive replacement, 4019
complementarity, 990, 1379, 1515, 2281, 2678, 5245, 5333, 5521, 5523
complex threat, 1830
computers, 1, 453, 499, 780, 781, 1011, 1410, 2037, 4504, 4999, 5316, 5384
congruence, bd [/biogeographic]. *See* biodiversity congruence
connectivity, 2002, 2293, 3179, 3321, 3686, 5012
Connell, Joseph H., 717, 4531
conservation, animal, 4459, 4518, 5144
conservation, bd. *See* biodiversity conservation
conservation biology, 106, 171, 181, 257, 676, 711, 807, 835, 836, 873, 916, 945, 951, 1304, 1315, 1412, 1570, 1732, 1746, 1945, 2006, 2404, 2533, 2571, 2573, 2584, 2606, 2864, 2974, 3096, 3097, 3098, 3177, 3204, 3301, 3321, 3411, 3453, 3499, 3610, 3701, 3769, 3837, 3957, 4083, 4084, 4085, 4282, 4445, 4804, 4805, 4806, 4810, 4813, 4835, 4907, 4944, 4997, 5061, 5146, 5241, 5265, 5717, 5757, 5827, 5879
conservation, bird. *See* 'bird conservation' in Index III
conservation, crop germplasm, 663, 664, 666, 2709, 3538, 3996
conservation easements, 2558
conservation, environmental, 247, 820, 1019, 1051, 1344, 3878, 4274, 4290, 4528, 5100
conservation evaluation. *See* conservation

value
conservation, ex situ. *See* ex situ conservation
conservation, fish. *See* 'fish conservation' in Index III
conservation, forest. *See* forest conservation
conservation, gene resources. *See* gene resources conservation
conservation, genetic diversity, 1524, 2630, 3190, 3431, 3452, 3853, 4027
conservation genetics, 177, <u>220</u>, <u>221</u>, 330, 395, <u>599</u>, <u>806</u>, <u>1665</u>, <u>2156</u>, 2741, 2956, 3300, <u>3366</u>, 3707, <u>4668</u>, <u>4984</u>, 5035, <u>5134</u>, 5339
conservation, habitat. *See* habitat conservation
conservation, in situ. *See* in situ conservation
conservation, insect. *See* 'insect conservation' in Index III
Conservation International, 1013, 2171, 2939, 4831
conservation, invertebrate. *See* 'invertebrate conservation' in Index III
conservation, mammal. *See* 'mammal conservation' in Index III
conservation, marine. *See* marine conservation
conservation, microorganism, 1582, 4261, 4358, 4851, 4854, 4942
conservation, nature. *See* nature conservation
conservation participation, local (community-related). *See* local conservation participation
conservation, plant. *See* 'plant conservation' in Index III
conservation prioritization, 580, 739, <u>1116</u>, 1688, 1689, 2471, <u>2652</u>, 3497, 3536, <u>5245</u>, 5476, 5609, 5634
Conservation Reserve Program, 1383, 4982
conservation, species. *See* species conservation
conservation value [/evaluation], 148, 458, <u>1507</u>, 1639, 1689, 1928, 3586, 3587, 4834, <u>5199</u>, 5316, 5429, 5517, 5571, 5664
conservation, wildlife. *See* wildlife conservation
consumer behavior, <u>1387</u>, 2510, 3577, <u>5016</u>, 5255
continental drift. *See* plate tectonics
continental slope and rise, <u>1914</u>, <u>2589</u>, 3137, <u>4014</u>

contingent valuation, 928, <u>1132</u>, <u>1281</u>, 1744, <u>2050</u>, 2053, 2412, 2473, <u>3397</u>, 3438, 3439, <u>4020</u>, <u>4830</u>, <u>4884</u>
Convention on Biological Diversity, 156, 167, 192, 195, 229, 273, 278, 335, 383, 447, 521, 544, 564, 565, 602, 656, 713, 797, 848, 849, 897, 1054, 1057, 1087, 1112, 1209, 1249, 1337, 1347, 1495, 1596, 1598, 1700, 1701, 1846, 1847, 1879, 1881, <u>1961</u>, 2088, 2102, 2111, 2263, 2329, 2367, 2385, 2444, 2658, 2673, 2731, 2773, 2890, 2891, 2892, 3222, 3230, 3284, 3346, 3353, 3480, 3492, 3507, 3832, 3848, 3964, 4046, 4104, 4176, 4177, 4195, 4385, 4456, 4659, 4864, 4876, 4913, 4958, 4964, 4967, 5005, 5085, 5098, 5099, 5174, 5177, 5181, 5229, 5617, 5654, 5780, 5873
Convention on Climate Change, 192, 3471
Convention on Nature Protection and Wild Life Preservation, 4341
convergent evolution, <u>948</u>, <u>4029</u>, 4442
conversion, forest. *See* forest conversion
coordinated stasis, <u>583</u>, 2383, 3251, <u>5426</u>
coral bleaching, 626, 697, <u>1848</u>, <u>1883</u>
coral reefs, 211, 451, 525, 621, 626, 697, 850, 855, 974, <u>1003</u>, <u>1038</u>, 1190, <u>1328</u>, <u>1333</u>, 1675, 1735, <u>1848</u>, <u>1883</u>, 2029, 2206, 2207, 2266, <u>2299</u>, 2309, <u>2334</u>, <u>2341</u>, 2448, 2480, 2698, 2920, 3258, 3436, 3984, 4106, <u>4397</u>, <u>4428</u>, <u>4429</u>, 4559, 4646, <u>4755</u>, 4893, 4917, 4918, <u>5002</u>, 5268, 5269, <u>5270</u>, 5379, 5403, 5404, 5493, 5800, 5802
core collections, <u>624</u>, 1297, 2231
core-satellite hypothesis, <u>984</u>, 985, <u>2059</u>, <u>2067</u>
corporations [/private sector], 487, 540, 791, 813, 871, 1055, 2605, 2794, 3110, 4915
correlations [/correlates], <u>182</u>, 235, <u>489</u>, 927, 1370, <u>1471</u>, <u>1514</u>, 1576, 1601, <u>2826</u>, 3164, 3713, 3982, <u>4143</u>, 4956, 5365, <u>5629</u>
corridors, 32, <u>199</u>, 340, <u>377</u>, <u>397</u>, 398, 704, <u>1159</u>, 1810, 1821, 1923, <u>2099</u>, 2174, <u>2188</u>, <u>2215</u>, 2706, 2708, <u>2931</u>, 3062, 3103, 3322, 3505, <u>3554</u>, <u>3677</u>, 4363, <u>4467</u>, <u>4674</u>, <u>4678</u>, <u>4812</u>, 4828, 4871
corridors, wildlife. *See* wildlife habitat
cost-benefit analysis, 552, <u>1302</u>, 2727, 2928, 3422, 4649, 5397
cover, land, 2903, <u>3326</u>, 5084
Cretaceous and Cretaceous-Tertiary boundary, <u>103</u>, <u>176</u>, 591, <u>946</u>, <u>1028</u>, 1063, 1091, 1454, 1838, 2020, 2280, 2328, <u>2387</u>, <u>2589</u>,

2590, 2677, 3064, 3065, 3722, 4091, 4265, 4405, 5590
crop diversity, 667, 669, 943, 2120, 4747, 4758, 4838
crop genetic diversity, 37, 933, 1243, 1392, 2367, 5702
crop germplasm conservation, 663, 664, 666, 2709, 3538, 3996
crop germplasm [/genetic] resources, 101, 379, 670, 1662, 1700, 2090, 2111, 2120, 2241, 2242, 2288, 2437, 2709, 3561, 3997, 4195, 4465, 4733, 4838, 4953, 5437, 5508, 5602
crop rotation, 435, 3019, 894
crop varieties, 937, 1660, 3056, 5140
crops, food. *See* food crops
crops: general and misc. subjects, 85, 102, 242, 435, 667, 1522, 1709, 1750, 1751, 1884, 2090, 2120, 2288, 2338, 2580, 2582, 3019, 3563, 3822, 4601, 4749, 5576, 5699, 5702
cross-boundary management, 1184, 4516, 4517
cryopreservation, 392, 1726, 5488, 5614
cryptic species, 922, 3408
cryptogenic species, 801
cultural diversity, 165, 172, 253, 578, 865, 1390, 1717, 2034, 2273, 2355, 2636, 2786, 2788, 3276, 3786, 4194, 4738, 4768, 5486
cultural ecology, 3541
currents, ocean, 4336, 4427
cycle, carbon. *See* carbon dioxide
cycle, nitrogen. *See* nitrogen
cycles, biogeochemical, 67, 346, 356, 2957, 4178
dams, 2259, 2998, 3829, 4044, 5471
databases, 53, 88, 445, 453, 499, 1087, 1192, 1216, 1289, 1491, 1881, 2641, 2873, 3120, 3460, 4369, 4504, 5182, 5583
dead and dying trees (*see also* coarse woody debris), 1683, 1975, 3844, 4443
debt-for-nature swaps, 1327, 2058, 2691, 3232, 3437, 3862, 4386, 5101
decentralization and bdc, 3021
decline, amphibian. *See* 'amphibian decline' in Index III
decline, bird. *See* 'bird decline' in Index III
decline, fish. *See* 'fish extinction' in Index III
decline, mammal. *See* 'mammal extinction' in Index III
decline rates, bd. *See* biodiversity loss rates

decomposers and decomposition, 2835, 2951, 2952, 3641, 3844, 5353
deep ecology, 877, 953, 1259, 1653, 1866, 3329, 3550, 3551, 4591, 4972
deep-sea environs, 108, 526, 737, 805, 946, 1100, 1494, 1648, 1706, 1724, 1725, 1871, 1911, 1912, 1913, 1914, 2092, 2189, 2442, 2620, 2729, 2906, 3022, 3408, 4257, 4258, 4259, 4305, 4625, 4653, 4762, 4763, 4764, 4765, 5225, 5405, 5595, 5669, 5752, 5777, 5862
defaunation, tropical forest. *See* tropical forest defaunation
Defenders of Wildlife, 1386, 2516
deforestation, 54, 64, 115, 141, 157, 258, 268, 284, 304, 458, 518, 615, 616, 652, 654, 701, 744, 753, 908, 967, 1019, 1214, 1257, 1337, 1343, 1348, 1388, 1397, 1542, 1709, 1874, 1901, 1905, 1986, 2254, 2278, 2331, 2350, 2496, 2642, 2716, 2829, 2895, 2933, 2987, 3010, 3081, 3357, 3390, 3513, 3517, 3518, 3521, 3634, 3757, 3840, 3888, 3970, 3972, 4051, 4179, 4252, 4253, 4273, 4306, 4388, 4482, 4645, 4686, 4716, 4725, 4822, 5009, 5031, 5065, 5171, 5257, 5453, 5485, 5510, 5552, 5623, 5670, 5691
deforestation and climate. *See* climate and forests
deforestation rates, 1542, 3521
degradation, habitat [environmental/ecological]. *See* habitat degradation
density, fish. *See* 'fish diversity' in Index III
density, population. *See* species density
density, species. *See* species density
desert [/arid/dryland] environs, 8, 21, 416, 428, 446, 630, 896, 940, 1016, 1032, 1072, 1074, 1654, 1659, 1801, 2238, 2359, 2399, 2513, 2850, 2975, 3139, 3300, 3378, 3463, 3464, 3469, 3474, 3538, 3542, 3606, 3623, 3791, 3952, 3953, 4005, 4006, 4028, 4048, 4599, 4618, 4845, 5220, 5445, 5446, 5678
desertification, 446, 5446
developing countries, 11, 13, 14, 41, 94, 97, 98, 101, 223, 313, 320, 577, 609, 651, 680, 689, 723, 879, 1014, 1029, 1178, 1206, 1302, 1303, 1467, 1620, 1719, 1872, 1922, 1981, 2210, 2492, 2519, 2520, 2521, 2591, 2653, 2690, 3056, 3263, 3279, 3482, 3488, 3827, 3996, 4066, 4197, 4227, 4252, 4253, 4274, 4301, 4498, 4538, 4607, 4629, 4823, 4853, 5397, 5401, 5601, 5622, 5703
Devonian, 67, 404, 2622, 3236, 4193
dieback, forest, 2205, 3479

disease [/pathogens], 586, 830, 1105, 1298, 1314, 1472, 1473, 1500, 1522, 1684, 1743, 1811, 2090, 2188, 2258, 2395, 2831, 2849, 3203, 3215, 3348, 3561, 3701, 4146, 4200, 4844, 5538, 5578
dispersal, 22, 340, 582, 704, 741, 798, 804, 1084, 1160, 1204, 1371, 1911, 2091, 2103, 2155, 2670, 3054, 3175, 3869, 3988, 4021, 4088, 4115, 4474, 5259, 5436, 5669, 5878
dispersal barriers, 582, 789, 1997, 3508, 4722, 5259
dispersal, plant, 3175, 4021, 4088
dispersal, seed. *See* seed dispersal
distinctiveness, genetic [/taxonomic]. *See* genetic distinctiveness
distribution, allotopic, 559
distribution and climate, 1036, 3845, 4727, 5610
distribution, animal. *See* 'animal distribution' in Index III
distribution, antitropical, 2926
distribution, bird. *See* 'bird biogeography' in Index III
distribution, fish. *See* 'fish biogeography' in Index III
distribution, geographical. *See* geographical distribution
distribution, insect. *See* 'insect biogeography' in Index III
distribution, mammal. *See* 'mammal biogeography' in Index III
distribution, plant. *See* 'plant biogeography' in Index III
distribution-abundance relations, 634, 1469, 1771, 2110, 3184, 4357, 5046
disturbance, canopy, 3162, 5271
disturbance ecology, 537, 869, 2190, 2362, 4292, 5108
disturbance, ecosystem [/ecological]. *See* ecosystem disturbance
disturbance, forest. *See* forest disturbance
disturbance: general and misc. subjects, 49, 183, 649, 1003, 1250, 1677, 1742, 1841, 1844, 2031, 2048, 2057, 2219, 2259, 2299, 2396, 2402, 2730, 3390, 3436, 3916, 3936, 4122, 4430, 4816, 4817, 4841, 5077, 5358, 5359, 5382
disturbance, habitat. *See* habitat disturbance
disturbance, human. *See* human disturbance
disturbance, landscape, 4394, 5159, 5382
disturbance-diversity relations, 1003, 2402, 3936, 4800, 5242, 5353

diversification, 303, 308, 396, 406, 409, 640, 660, 794, 929, 1064, 1082, 1083, 1086, 1091, 1373, 1395, 1483, 1534, 1535, 1628, 1861, 1877, 1906, 1907, 2174, 2235, 2359, 2384, 2407, 2621, 2622, 2704, 2765, 2780, 2878, 3023, 3164, 3165, 3259, 3341, 3636, 3838, 3866, 4050, 4337, 4368, 4478, 4582, 4587, 4611, 4732, 5004, 5010, 5209, 5212, 5213, 5214, 5317, 5466, 5737
diversification, plant. *See* 'plant evolution' in Index III
diversification rates. *See* evolutionary rates
diversity, alpha. *See* alpha diversity
diversity and climate. *See* climate and diversity
diversity, animal. *See* 'animal diversity' in Index III
diversity, asymptotic, 1504
diversity, behavioral, 1094, 1877, 4919
diversity, beta. *See* beta diversity
diversity, bird. *See* 'bird diversity' in Index III
diversity, chemical. *See* chemical diversity
diversity, community. *See* community diversity
diversity, crop. *See* crop diversity
diversity, crop genetic, 37, 933, 1243, 1392, 2367, 5702
diversity, cultural. *See* cultural diversity
diversity, ecological. *See* ecological diversity
diversity, ecosystem. *See* ecosystem diversity
diversity, extratropical, 462, 466
diversity, faunal. *See* faunal diversity
diversity, fish. *See* 'fish diversity' in Index III
diversity, floristic. *See* floristic diversity
diversity, forest. *See* forest diversity
diversity, functional. *See* functional diversity
diversity, gamma, 4579, 4638, 5130, 5457
diversity: general and misc. subjects, 23, 159, 168, 302, 303, 325, 513, 605, 633, 883, 926, 1003, 1018, 1123, 1414, 1508, 1527, 1885, 1964, 1982, 1994, 2013, 2096, 2168, 2204, 2265, 2269, 2327, 2332, 2632, 2728, 2911, 3203, 3205, 3207, 3359, 3424, 3446, 3581, 3795, 3859, 3861, 4015, 4175, 4225, 4228, 4310, 4433, 4434, 4491, 4566, 4579, 4657, 4685, 4731, 4873, 4898, 4916, 4949, 4974, 4998, 5026, 5032, 5033, 5076, 5157, 5212, 5213, 5300, 5390, 5392, 5393, 5414, 5422, 5523, 5539, 5554, 5594, 5697, 5716
diversity, genetic. *See* genetic diversity
diversity, global. *See* global diversity
diversity gradients, 19, 1277, 1675, 2208,

Index I   415

3220, 3307, 4367
diversity gradients, latitudinal. *See* latitudinal diversity gradients
diversity, habitat. *See* habitat diversity
diversity, human. *See* human evolution
diversity indices [/measures], 375, 378, 773, 1066, 1340, 1737, 1748, 2018, 2161, 2204, 2247, 2308, 2414, 2847, 2914, 2946, 2972, 3078, 3648, 3751, 3752, 3859, 3899, 3959, 3960, 4450, 4487, 4489, 4694, 4792, 4968, 5118, 5129, 5316, 5455, 5516, 5523
diversity, insect. *See* 'insect diversity' in Index III
diversity, intraspecific. *See* intraspecific diversity
diversity, invertebrate. *See* 'invertebrate diversity' in Index III
diversity, landscape. *See* landscape diversity
diversity, mammal.
diversity, marine. *See* marine bd
diversity measures. *See* diversity indices
diversity, microorganism. *See* 'microorganism diversity' in Index III
diversity, morphological. *See* morphological diversity
diversity, Phanerozoic, 1618, 3340, 4575, 4576, 4577, 4578, 4584, 5213, 5217
diversity, phylogenetic, 1507, 1510, 1511, 1556, 1707, 2741, 3452, 5316, 5586
diversity, plant. *See* 'plant diversity' in Index III
diversity, population, 1423, 2296, 3415
diversity, regional. *See* regional diversity
diversity, seasonal. *See* seasonal patterns
diversity, species. *See* species diversity
diversity, taxonomic. *See* taxonomic diversity
diversity, tree, 1008, 2143, 2681, 2858, 2872, 3237, 3958, 4038, 4531, 4612, 5630
diversity, vegetation. *See* vegetation diversity
diversity vs. richness, 1693
diversity-abundance relations, 936, 4654, 5360, 5365
diversity-body size relations, 1581, 1823, 1934, 4654, 5360
diversity-disturbance relations, 1003, 2402, 3936, 4800, 5242, 5353
diversity-productivity relations, 18, 19, 21, 29, 1006, 2537, 2870, 4396, 5094
diversity-stability relations, 127, 1006, 1103, 1305, 1831, 1875, 2462, 2659, 2870, 2982, 3265, 3495, 5041, 5088, 5090, 5092, 5613

DNA, 33, 35, 36, 109, 220, 237, 603, 1395, 1450, 1695, 2156, 2289, 2501, 2546, 2547, 2742, 2822, 2873, 3161, 3446, 3450, 3866, 4643, 4984, 5122, 5124, 5657, 5695
domains, 1462, 3126, 5502
domestic animals, 598, 4493, 4844, 5562
dominance-diversity curves, 5455
drainage basins [/watersheds], 426, 574, 3170, 3171, 3179, 3389, 3804, 4855, 5280, 5296
drought, 1358, 1475, 2611, 3293, 5090, 5191
drug cultivation, illegal, 1874, 5670
drug discovery (*see also* bd prospecting, chemical prospecting), 194, 222, 240, 251, 788, 851, 1077, 2664, 3083, 5299
dryland environs. *See* desert environs
dumping, waste, 805, 1415, 2682, 4625, 4763
dunes. *See* sand and dunes
echelons, 3537
ecocolonialism, 1078, 3011
ecofeminism, 877, 1273, 1653, 1716, 2082, 3330, 4197, 4327, 4417, 5713
ecological climax (*see also* ecological succession), 524, 934
ecological degradation. *See* habitat degradation
ecological disturbance. *See* ecosystem disturbance
ecological diversity, 1021, 1069, 2004, 3080, 5147, 5368
ecological [/environmental] economics, 126, 189, 222, 225, 279, 280, 283, 284, 296, 362, 418, 455, 540, 646, 648, 683, 848, 849, 997, 1043, 1047, 1167, 1302, 1411, 1440, 1537, 1543, 1745, 1872, 1895, 1899, 2052, 2058, 2102, 2147, 2720, 2727, 2758, 2877, 3266, 3316, 3320, 3422, 3423, 3441, 3488, 3514, 3605, 3675, 3751, 3752, 3763, 3765, 3881, 3882, 3883, 3885, 3915, 3916, 3918, 4002, 4003, 4353, 4360, 4393, 4687, 4698, 4743, 4822, 4830, 4853, 4955, 4959, 4962, 4963, 4966, 4985, 4988, 5065, 5082, 5100, 5101, 5109, 5110, 5536, 5719, 5720, 5851
ecological heterogeneity. *See* environmental heterogeneity
ecological integrity. *See* biological integrity
ecological monitoring. *See* biodiversity monitoring
ecological productivity. *See* primary productivity
ecological research, long-term, 1163, 2848,

3076, 3077, 3335, 3399, 3622, 5723
ecological restoration, 249, 344, 417, 541, 673, 685, 722, 750, 941, 1059, 1200, 1315, 1403, 1451, 1463, 1526, 1527, 1727, 2049, 2248, 2284, 2285, 2399, 2460, 2498, 2499, 2503, 2581, 2922, 3072, 3210, 3568, 3582, 3682, 3684, 3804, 3961, 3990, 4335, 4455, 4469, 4887, 5194, 5195, 5441, 5589, 5659, 5672, 5792
ecological [/community] succession, 119, 730, 1250, 1844, 2031, 2137, 2982, 3244, 3615, 3632, 3958, 4141, 4409, 4430, 4494, 4532, 5002, 5003, 5120, 5135, 5422, 5460, 5561
ecology, agricultural. *See* agricultural ecology
ecology, bird. *See* 'bird ecology' in Index III
ecology, community. *See* community ecology
ecology, cultural. *See* cultural ecology
ecology, disturbance. *See* disturbance ecology
ecology, evolutionary, 466, 727, 826, 947, 1469, 2872, 3178, 3442, 4366, 4489, 5067
ecology, functional. *See* functional diversity
ecology, human. *See* human ecology
ecology, insect, 1153, 1590, 1613, 2095, 3351, 4928
ecology, island. *See* island biology
ecology, landscape. *See* landscape ecology
ecology, mammal, 134, 894, 1171, 1684, 1908, 2567, 2921, 3297, 4048, 4885
ecology: misc. general subjects, 2, 3, 32, 80, 81, 110, 135, 213, 517, 523, 633, 636, 750, 766, 770, 806, 820, 947, 1006, 1007, 1021, 1259, 1282, 1289, 1304, 1333, 1412, 1444, 1460, 1465, 1571, 1589, 1641, 1674, 1812, 1849, 1862, 1867, 1875, 1959, 2074, 2116, 2243, 2244, 2245, 2250, 2344, 2345, 2347, 2348, 2349, 2423, 2467, 2568, 2721, 2752, 2804, 2871, 2907, 2931, 2996, 3002, 3013, 3032, 3037, 3076, 3080, 3202, 3335, 3428, 3479, 3550, 3551, 3769, 3947, 3955, 3956, 3957, 3968, 4006, 4137, 4203, 4287, 4354, 4418, 4426, 4429, 4461, 4591, 4667, 4803, 4813, 4824, 4825, 4833, 4834, 4962, 5003, 5081, 5149, 5163, 5175, 5223, 5286, 5287, 5302, 5307, 5336, 5388, 5451, 5456, 5468, 5469, 5476, 5503, 5512, 5607, 5613, 5640, 5723, 5724, 5725, 5753, 5784, 5815, 5827, 5837, 5852, 5862
ecology, plant, 80, 550, 1233, 2031, 2766, 3296, 3360, 4128, 4354, 4409, 5334, 5460
ecology, restoration. *See* ecological restoration
ecology, wildlife. *See* wildlife ecology
economic incentives, 540, 909, 1721, 1865, 2287, 2558, 2937, 3266, 3270, 3277, 3763, 3765, 3766, 3915, 4958, 5102, 5110
economics, agricultural, 455, 1010, 3752, 4198
economics, ecological [/environmental]. *See* ecological economics
ecoregions, 232, 233, 1292, 3747, 5486
ecosystem concept, 1867, 5003
ecosystem [/ecological/environmental] disturbance, 5, 247, 248, 386, 617, 725, 1188, 1880, 2381, 2496, 2607, 4221, 5344
ecosystem diversity, 1127, 2811, 2860, 3024, 4322, 5796, 5870
ecosystem engineers, 2485, 2843
ecosystem fragmentation. *See* habitat fragmentation
ecosystem health, 389, 430, 767, 768, 1046, 1127, 1635, 2660, 2839, 3242, 3387, 4152, 5689
ecosystem integrity. *See* biological integrity
ecosystem management, 171, 389, 793, 973, 1498, 1698, 1731, 1864, 1964, 1966, 1970, 1973, 2117, 2176, 2220, 2275, 2585, 2635, 2699, 2812, 2829, 2839, 3009, 3396, 3745, 4200, 4296, 4434, 4435, 4436, 4440, 4615, 4660, 4821, 4895, 5066, 5095, 5188, 5243, 5344, 5390, 5633, 5831
ecosystem services, 1048, 1411, 1430, 1543, 1625, 1626, 5237
ecosystems, ancient. *See* ancient ecosystems
ecosystems, aquatic. *See* aquatic ecosystems
ecosystems, Mediterranean-climate. *See* Mediterranean-climate ecosystems
ecosystems: misc. general subjects, 90, 120, 130, 233, 245, 323, 345, 386, 481, 503, 504, 535, 555, 561, 578, 674, 714, 767, 768, 830, 860, 861, 862, 864, 1104, 1157, 1432, 1472, 1589, 1635, 1670, 1731, 1867, 1946, 1965, 2055, 2139, 2244, 2339, 2462, 2484, 2486, 2572, 2585, 2660, 2784, 2799, 2814, 2848, 2923, 2979, 2996, 3176, 3189, 3215, 3239, 3260, 3351, 3417, 3430, 3544, 3545, 3547, 3548, 3549, 3671, 3673, 3687, 3688, 3773, 3776, 3784, 3817, 3857, 3904, 3921, 3923, 3957, 4045, 4152, 4210, 4221, 4277, 4282, 4299, 4310, 4322, 4469, 4470, 4530, 4621, 4664, 4766, 4869, 4957, 4975, 5003, 5088, 5093, 5095, 5155, 5163, 5168, 5188, 5285, 5287, 5312, 5314, 5352, 5456, 5607, 5628, 5737, 5792, 5796, 5797, 5816,

5867, 5869
ecosystems, threatened. *See* endangered ecosystems and habitats
ecotones [/edges], 868, 1153, 1242, 1286, 1650, 1999, 2056, 2248, 2769, 2781, 3094, 3202, 3494, 3852, 4298, 4571, 4675, 4757
ecotourism, 511, 682, 844, 928, 1301, 1415, 1797, 1878, 2171, 2406, 2446, 2668, 2927, 3084, 3438, 3439, 3675, 4624, 5103, 5218, 5398, 5433, 5665, 5746
ecotoxicology. *See* toxicology
Ecotron, 880, 2537, 2538, 2843, 3547
edges. *See* ecotones
education [/teaching] (*see also* environmental perception), 230, 312, 573, 619, 675, 807, 1001, 1002, 1112, 1224, 1591, 1714, 2165, 2246, 2271, 2404, 2405, 2625, 3299, 3662, 3783, 3807, 4238, 4294, 4323, 4550, 4595, 4610, 4729, 4767, 5128, 5182, 5278, 5356, 5408, 5428, 5444, 5582, 5643, 5832
effective population size, 3696, 4266, 5339
Ehrlich, Paul, 354
El Niño, 4039
elevational effects. *See* altitudinal effects
emergy, 5173
emphasis-use, 1498, 1499
endangered animals, 25, 236, 412, 580, 4459, 5708
endangered birds. *See* 'birds, endangered' in Index III
endangered [/threatened] ecosystems and habitats, 54, 120, 458, 1524, 2305, 2825, 3024, 3687, 3688, 4309, 4440, 4618, 4688
endangered fishes. *See* 'fishes, endangered' in Index III
endangered [/threatened] insects, 2072, 2825, 3498, 5357, 5642
endangered mammals. *See* 'mammals, endangered' in Index III
endangered [/threatened] peoples and cultures, 120, 3276, 4194, 4843, 5772
endangered plants. *See* 'plants, endangered' in Index III
endangered [/threatened] species, 25, 56, 147, 187, 236, 358, 399, 412, 414, 415, 455, 541, 597, 782, 897, 918, 962, 964, 998, 999, 1151, 1172, 1244, 1290, 1309, 1317, 1318, 1338, 1382, 1393, 1400, 1405, 1425, 1455, 1526, 1540, 1571, 1595, 1607, 1756, 1803, 1921, 2255, 2261, 2373, 2412, 2475, 2491, 2565, 2591, 2718, 2770, 2813, 2814, 2928, 2941, 3017, 3043, 3044, 3104,

3145, 3215, 3318, 3319, 3320, 3411, 3511, 3512, 3519, 3520, 3707, 3708, 3762, 3795, 3860, 3972, 4002, 4069, 4097, 4098, 4110, 4124, 4198, 4270, 4333, 4349, 4377, 4383, 4437, 4459, 4543, 4553, 4623, 4688, 4704, 4769, 4832, 4959, 4967, 5142, 5272, 5346, 5475, 5479, 5483, 5488, 5594, 5603, 5688, 5795, 5840
Endangered Species Act 1973, 147, 163, 186, 187, 296, 349, 426, 1290, 1338, 1360, 1400, 1571, 2162, 2264, 2326, 2491, 2516, 2718, 2856, 3018, 3088, 3101, 3145, 3575, 3689, 3733, 3772, 3860, 4346, 4347, 4500, 4623, 4740, 5015, 5250, 5296, 5340, 5482, 5499, 5504, 5564, 5789
endangered [/threatened] wildlife, 25, 552, 580, 707, 1295, 1597, 3028, 3687, 4096, 4586, 5040, 5472, 5708
endemic birds. *See* 'birds, endemic' in Index III
endemic fishes. *See* 'fishes, endemic' in Index III
endemic invertebrates, 969, 1230, 1246, 1993, 2252, 5632
endemic [/rare] mammals, 182, 841, 843, 1502, 2148, 2531, 2633, 5132, 5207
endemic plants. *See* 'plants, endemic' in Index III
endemism and endemic species: misc. subjects, 122, 258, 749, 756, 1081, 1103, 1174, 1601, 1688, 1791, 2627, 3522, 4098, 5211, 5351, 5518
endophytes, 1798, 4573, 4916
enemies hypothesis, 2896, 4611
energy use-abundance relations, 1139, 4357
energy-richness relations, 1658, 1675, 2238, 2628, 2821
energy-species relations. *See* species-energy relations
entropy, 613, 5464
environmental change, 761, 1180, 1473, 1478, 2106, 2324, 2383, 2541, 2589, 2590, 2616, 2961, 3064, 3108, 3420, 3430, 3490, 3502, 3574, 3644, 3851, 3878, 4370, 4715, 4755, 4784, 4789, 4852, 5151, 5243, 5286, 5288, 5816, 5833, 5834
environmental conservation, 247, 820, 1019, 1051, 1344, 3878, 4274, 4290, 4528, 5100
environmental degradation. *See* habitat degradation
environmental disturbance. *See* ecosystem disturbance
environmental economics. *See* ecological

418   Index I

economics
environmental ethics (*see also* bioethics), 66, 208, 299, 339, 517, 763, 764, 765, 766, 768, 877, 1030, 1053, 1153, 1167, 1362, 1451, 1653, 1828, 1866, 1971, 1979, 2034, 2463, 2861, 2888, 2948, 3011, 3331, 3337, 3380, 3555, 3558, 3663, 3670, 3731, 4026, 4030, 4031, 4349, 4350, 4351, 4352, 4353, 4585, 4644, 4909, 4973, 5011, 5063, 5246, 5425, 5496, 5547, 5643, 5654, 5745
environmental gradients, 21, 1788, 2552, 2751, 4372, 4831, 5120
environmental [/ecological/habitat/landscape] heterogeneity, 18, 568, 587, 719, 758, 1891, 2093, 2314, 2318, 2363, 2439, 2628, 2721, 2735, 3607, 3628, 3632, 3957, 5043, 5148, 5150, 5159, 5161, 5275, 5343, 5351
environmental impact analysis, 350, 683, 1061, 2180, 2998
environmental indicators. *See* biodiversity indicators
environmental [/bd] law and legislation (*see also* law review articles), 20, 169, 209, 322, 350, 371, 389, 426, 442, 479, 481, 560, 564, 578, 628, 796, 824, 846, 904, 909, 913, 914, 1147, 1338, 1347, 1394, 1406, 1536, 1587, 1794, 1865, 1980, 2191, 2273, 2381, 2453, 2584, 2636, 2686, 2688, 2718, 2856, 3075, 3104, 3112, 3145, 3209, 3327, 3346, 3507, 3806, 3860, 4105, 4124, 4148, 4249, 4254, 4346, 4373, 4377, 4385, 4392, 4425, 4449, 4501, 4524, 4557, 4631, 4760, 4761, 4910, 4913, 5005, 5056, 5062, 5098, 5178, 5390, 5589, 5772, 5789, 5864, 5875
environmental perception and nature attitudes, 807, 886, 1825, 1869, 1971, 2000, 2246, 2460, 2591, 2592, 2593, 2594, 2597, 2649, 3233, 3557, 3807, 3994, 4190, 4535, 4559, 4830, 5278
environmental planning, 1951, 2335, 2471, 2943, 3274, 3280, 3461, 3894, 4070, 4092, 4435, 4678, 4809, 5471
environmental policy, 61, 215, 276, 277, 290, 295, 315, 349, 402, 443, 483, 484, 485, 512, 523, 539, 558, 628, 651, 679, 756, 779, 782, 840, 846, 871, 909, 916, 917, 919, 1112, 1148, 1155, 1189, 1192, 1302, 1318, 1331, 1345, 1349, 1394, 1424, 1433, 1536, 1554, 1588, 1625, 1626, 1699, 1728, 1796, 1846, 1882, 1895, 1927, 1951, 1966, 1980, 1988, 2003, 2084, 2134, 2261, 2311, 2331, 2365, 2369, 2374, 2375, 2449, 2491, 2515, 2586, 2624, 2639, 2691, 2976, 2991, 3000, 3075, 3081, 3174, 3270, 3277, 3282, 3319, 3323, 3336, 3440, 3528, 3653, 3667, 3735, 3806, 3883, 3885, 4017, 4069, 4176, 4181, 4225, 4227, 4237, 4252, 4253, 4324, 4341, 4347, 4364, 4392, 4425, 4606, 4623, 4719, 4759, 4825, 4847, 4910, 4960, 4967, 4982, 4985, 5007, 5037, 5079, 5082, 5107, 5112, 5176, 5223, 5394, 5422, 5424, 5479, 5513, 5536, 5572, 5619, 5631, 5652, 5688
environmental protection, 523, 748, 1587, 1882, 1988, 2186, 2494, 2740, 3256, 3482, 3657, 4353, 4524, 4935, 5633
environmental stress, 507, 626, 2205, 2258, 3849, 3850, 3851, 3852, 3853, 3923, 4170, 4621, 4737, 5359, 5465
environmentalism, 325, 577, 1429, 1866, 1972, 2917, 3107, 3667, 3668, 4418, 5341
enzymes, 887, 1213, 2289, 3161, 4043, 4846
Eocene, 4095, 5083
epifauna, 4400
epiphytes, 2199, 2776, 3416, 3543, 3598, 4372, 5113
equilibrium models, 49, 406, 1274, 2819, 2911, 3036, 3240, 4086, 4667, 5071, 5317, 5704
Erwin, Terry, 1753, 5049
estimation, bd. *See* biodiversity measurement
estuaries, 119, 210, 1045, 1210, 2014, 3138, 5054, 5391
ethics, environmental. *See* environmental ethics
ethnobotany, 250, 251, 254, 373, 400, 851, 1076, 1077, 1604, 1829, 2753, 3148, 3992, 4056, 4385, 4528, 4529, 5387
ethnopharmacology, 239, 851, 1449, 2386, 2664, 3083, 3209, 3302, 4776, 5819
European Ecological Network, 402, 4871
eutrophication, 2402, 2564, 3138, 4564
evaluation, conservation. *See* conservation evaluation
evapotranspiration, 1658, 2238, 2821, 3432, 4891
evenness, 1050, 1539, 2161, 2204
event-driven systems, 3916
evolution, 60, 67, 78, 173, 174, 270, 303, 363, 396, 529, 530, 545, 551, 583, 590, 612, 613, 640, 814, 816, 822, 839, 854, 885, 929, 942, 947, 948, 1018, 1088, 1094, 1099, 1101, 1128, 1182, 1320, 1371, 1399, 1445, 1466, 1486, 1488, 1500, 1538, 1553, 1631, 1633, 1663, 1733, 1739, 1834, 1861, 1885, 1888, 1889, 1902, 1907, 1910, 2066,

Index I

2108, 2121, 2192, 2226, 2256, 2324, 2330, 2368, 2384, 2389, 2391, 2394, 2397, 2416, 2526, 2576, 2587, 2622, 2662, 2666, 2701, 2702, 2704, 2832, 2854, 2878, 2911, 2934, 2982, 3128, 3147, 3203, 3204, 3206, 3207, 3250, 3251, 3259, 3264, 3425, 3465, 3533, 3534, 3559, 3591, 3607, 3633, 3654, 3655, 3725, 3792, 3851, 4015, 4029, 4031, 4033, 4161, 4169, 4171, 4175, 4178, 4248, 4337, 4344, 4399, 4442, 4478, 4587, 4651, 4692, 4693, 4737, 4742, 4786, 4813, 4857, 4868, 5004, 5010, 5133, 5213, 5214, 5234, 5270, 5302, 5304, 5305, 5307, 5317, 5322, 5355, 5414, 5457, 5458, 5464, 5535, 5555, 5728, 5730, 5738, 5765, 5807, 5852

evolution [/systematics], bird, 660, 885, 1049, 1086, 1996, 142, 4651, 4652

evolution, convergent, 948, 4029, 4442

evolution, human. *See* human evolution

evolution [/systematics], insect, 816, 1907, 2780, 5430, 5466

evolution, molecular, 548, 1885, 2662, 4651

evolution, plant. *See* 'plant evolution' in Index III

evolutionarily significant units (ESU), 3204, 3451, 5294

evolutionary ecology, 466, 727, 826, 947, 1469, 2872, 3178, 3442, 4366, 4489, 5067

evolutionary [/speciation/diversification] rates, 529, 530, 1088, 2148, 2765, 3627, 3838, 4478, 4583

ex situ conservation, 9, 957, 1409, 2036, 2087, 3995, 4942

exergy, 3138, 5650

existence value, 2720, 2727, 4883, 4884

exotic species. *See* biological invasions, introduced species

Expert Center for Taxonomic Identification, 1491

extinct [/fossil] animals, 118, 720, 3484, 4581, 5574, 5708

(recently) extinct animals, 75, 572, 1434, 1930, 5263, 5708

extinct invertebrates. *See* 'invertebrates, fossil' in Index III

extinct mammals. *See* 'mammals, extinct' in Index III

extinct [/fossil] plants, 118, 1091, 1679, 4193, 4255, 5083

extinction and climate, 921, 1115, 5209

extinction, animal, 720, 721, 2212, 4580

extinction, background, 546, 2387, 2677

extinction, bird. *See* 'bird extinction' in Index III

extinction curves, 5071, 5479

extinction, fish. *See* 'fish extinction' in Index III

extinction: general and misc. subjects (*see also* mass extinction), 60, 77, 144, 158, 257, 262, 319, 409, 493, 501, 508, 528, 547, 572, 583, 590, 595, 615, 616, 637, 687, 710, 716, 720, 800, 854, 862, 878, 917, 921, 965, 1115, 1125, 1126, 1145, 1185, 1244, 1278, 1280, 1401, 1418, 1422, 1424, 1425, 1430, 1431, 1436, 1438, 1443, 1451, 1452, 1492, 1532, 1537, 1595, 1617, 1622, 1668, 1763, 1813, 1939, 2020, 2114, 2142, 2192, 2212, 2221, 2286, 2296, 2388, 2389, 2398, 2407, 2422, 2434, 2523, 2563, 2565, 2701, 2724, 2765, 2815, 2854, 2855, 2888, 2889, 2905, 2936, 2948, 2967, 2987, 2992, 3026, 3044, 3069, 3087, 3100, 3249, 3334, 3345, 3358, 3372, 3379, 3399, 3515, 3516, 3531, 3591, 3618, 3619, 3645, 3655, 3692, 3851, 3869, 3930, 3954, 3970, 4054, 4058, 4095, 4110, 4118, 4163, 4165, 4166, 4168, 4169, 4170, 4183, 4202, 4225, 4243, 4267, 4268, 4318, 4476, 4482, 4511, 4581, 4583, 4598, 4641, 4673, 4718, 4722, 4732, 4735, 4787, 4791, 4802, 4861, 4866, 4959, 4960, 4967, 5004, 5016, 5019, 5091, 5096, 5156, 5165, 5209, 5220, 5259, 5260, 5273, 5354, 5426, 5427, 5453, 5487, 5527, 5577, 5631, 5639, 5660, 5708, 5854

extinction, invertebrate. *See* 'invertebrate extinction' in Index III

extinction, local. *See* local extinction

extinction, mammal. *See* 'mammal extinction' in Index III

extinction, marine. *See* marine extinction

extinction, mass. *See* mass extinction

extinction, periodic. *See* periodic extinction

extinction [/loss], plant, 473, 1532, 2724, 5154, 5156

extinction rates, 369, 529, 1278, 1841, 2032, 2148, 2765, 2845, 3043, 3627, 3690, 3902, 3976, 3977, 4281, 4513, 4744, 4745, 5058, 5549, 5552, 5856

extinction risk, 267, 473, 477, 549, 1697, 3043, 3249, 3559, 3974, 4167, 4552, 5008, 5127

extinction, stepwise, 2020, 2340, 2713

extinction-immigration relations, 637, 984, 1841, 2967, 5071

extinction-origination dynamics. *See* origination-extinction dynamics

extinction-prone species, 144, 2826, 5026
extraction [/extractivism], 513, 620, 645, 1026, 1176, 2015, 5257
extrapolation methods, 118, 990, 2729, 3819, 3821, 4441, 4990
extraterrestrial intelligence, 4399
extratropical diversity, 462, 466
extremophiles, 884, 2934, 5222, 5405
FAO, 160, 1696
farming and farmers: general and misc. subjects (*see also* agriculture/agricultural), 98, 219, 388, 435, 664, 666, 879, 943, 994, 1392, 1541, 1624, 1676, 1796, 1802, 2437, 2443, 3056, 3424, 3774, 3812, 4028, 4388, 4600, 4628, 4803, 4952, 4953, 5080, 5335, 5576, 5602
farming practices. *See* agricultural practices
farming, sustainable. *See* sustainable agriculture
farming, traditional [/local/indigenous]. *See* traditional agriculture
faunal diversity, 46, 118, 743, 1227, 1801, 1904, 2189, 2439, 2561, 2694, 2999, 3473, 4011, 4037, 4095, 4544, 4556, 4581, 4765, 5332, 5353, 5473, 5574
faunas and climate. *See* climate and faunas
faunas, bird. *See* 'bird faunas' in Index III
faunas, fish. *See* 'fish faunas' in Index III
faunas, insect. *See* 'insect biogeography' in Index III
faunas, mammal. *See* 'mammal biogeography' in Index III
faunas: misc. subjects, 47, 113, 118, 132, 149, 307, 314, 326, 327, 367, 391, 423, 546, 629, 698, 702, 718, 743, 834, 874, 969, 1126, 1145, 1162, 1227, 1229, 1247, 1279, 1308, 1321, 1341, 1358, 1371, 1539, 1574, 1575, 1634, 1724, 1783, 1801, 1816, 1920, 2103, 2170, 2179, 2189, 2259, 2268, 2294, 2439, 2566, 2694, 2705, 2850, 2938, 2971, 3016, 3028, 3149, 3287, 3436, 3615, 3716, 3797, 3800, 3868, 4058, 4095, 4096, 4202, 4400, 4473, 4474, 4491, 4556, 4577, 4578, 4580, 4581, 4586, 4765, 4844, 5202, 5221, 5318, 5332, 5349, 5353, 5426, 5473, 5484, 5546, 5574, 5595, 5596, 5676, 5762, 5814, 5825, 5853
feral animals and plants, 414, 944, 2730, 4938, 5235
fidelity, assemblage, 3744
fire, 243, 244, 604, 698, 834, 958, 1069, 1073, 1180, 1191, 1240, 1283, 1814, 1815, 1856, 2468, 2560, 2600, 2909, 2910, 3292, 3298, 3916, 3953, 4359, 4540, 4851, 5238, 5422, 5590, 5597, 5667

Fish and Wildlife Service, 964, 2176, 4315
fisheries, 147, 152, 160, 292, 348, 496, 735, 812, 950, 1375, 1748, 2041, 2326, 2352, 2418, 2491, 2569, 2791, 2812, 3063, 3238, 3258, 3493, 3646, 3647, 3700, 3712, 3724, 3778, 3871, 3872, 3941, 3981, 4126, 4249, 4304, 4412, 4452, 4460, 4503, 4738, 4751, 5073, 5074, 5077, 5254, 5296, 5481, 5499, 5572
fishes. *See* listings in Index III
fjords, 1940, 2439
flagship species, 3612, 4670
floodplains (*see also* alluvial environments), 344, 520, 534, 1329, 1565, 1892, 1900, 2282, 2483, 3704, 3926, 4507, 5205
floods and flooding, 978, 1565, 1892, 2282, 2908, 4009, 4845, 5135
Flora of North America project, 4504, 5182
floras, 118, 629, 777, 855, 929, 1091, 1102, 1354, 1556, 1634, 1679, 1712, 1786, 1833, 1859, 2268, 2456, 2513, 2530, 2776, 2850, 2868, 3324, 3360, 3443, 3474, 3705, 4128, 4424, 4504, 5017, 5138, 5153, 5182, 5409, 5475, 5584, 5632, 5738, 5762, 5825, 5832
floristic biodiversity. *See* floristic diversity
floristic diversity [/bd], 550, 1091, 1786, 1833, 1859, 2479, 2513, 4128, 5584, 5832
flow regulation. *See* river flow regulation
flowering plants, 706, 1466, 2573, 2642, 4478, 4867
focal species [/taxa], 1201, 4414, 5639
food crops, 217, 726, 1014, 1029, 1169, 1647, 3997, 4633, 4953
food patches, 4764, 4765
food plants, 102, 726, 1136, 2120
food sources [/supply/security], 85, 217, 218, 376, 726, 809, 1014, 1029, 1040, 1136, 1213, 1435, 1467, 1564, 1647, 1927, 2443, 2786, 3524, 3997, 4067, 4068, 4195, 4374, 4417, 4493, 4600, 4633, 4783, 4878, 4953, 4954, 5075, 5076, 5183, 5240, 5782, 5801
food webs, 2124, 2349, 2862, 2869, 3159, 3178, 3193, 3194, 3195, 3813, 3814, 3975, 4005, 4007, 4044, 5131, 5353, 5566
forest birds. *See* 'birds, forest' in Index III
forest communities. *See* 'plant communities' in Index III
forest conservation [/preservation], 27, 245, 482, 539, 569, 779, 813, 832, 967, 1078, 1183, 1388, 1408, 1670, 1901, 2035, 2558,

Index I

2599, 2610, 2744, 3154, 3246, 3357, 3658, 3661, 3840, 3983, 4274, 4393, 4480, 4606, 4822, 5032, 5240, 5601
forest conversion, 572, 1348, 1709, 1873, 2468, 3513, 3577
forest dieback, 2205, 3479
forest disturbance, 486, 731, 2033, 2362, 3410, 3927, 4205, 4567, 5047, 5855
forest diversity, 502, 649, 910, 1011, 1264, 1294, 1874, 2143, 2144, 2274, 2501, 2548, 2860, 3293, 3723, 3933, 4036, 4430, 5796
forest floor and understorey, 1036, 1539, 2817, 3162, 3175, 3632, 3982, 4371, 4411, 4556, 5086, 5120, 5352, 5353
forest mammals. *See* 'mammals, forest' in Index III
forest management, 106, 116, 142, 228, 245, 280, 288, 342, 434, 460, 513, 553, 698, 699, 752, 1041, 1147, 1240, 1242, 1364, 1567, 1682, 1731, 1745, 1849, 2004, 2116, 2145, 2247, 2314, 2315, 2459, 2493, 2495, 2527, 2528, 2529, 2558, 2610, 2644, 2660, 2667, 2777, 2779, 2823, 2937, 3226, 3242, 3272, 3552, 3555, 3598, 3630, 3741, 3752, 3770, 3775, 3795, 3806, 3827, 3898, 3933, 3999, 4310, 4490, 4606, 4660, 4822, 4957, 4968, 4978, 5257, 5297, 5390, 5473, 5478, 5498, 5509, 5652, 5691, 5784
forest [/woodland] plants, 412, 1708, 3928, 4036, 4310, 5028, 5154, 5203
forest preservation. *See* forest conservation
forest products (*see also* natural products), 513, 1850, 2015, 2478, 3827, 3929, 3993, 4749
Forest Service (Canada), 779, 3246
Forest Service (U.S.), 116, 502, 931, 1967, 2635, 2681, 3242, 3414, 4432, 5390, 5478, 5598
forestry and bd, 73, 142, 155, 185, 553, 554, 556, 699, 724, 1041, 1368, 1519, 1554, 1567, 1682, 1849, 2282, 2312, 2315, 2373, 2440, 2644, 2660, 2763, 2777, 2937, 3356, 3422, 3555, 3740, 3966, 4094, 4660, 5187, 5324, 5400, 5509, 5588, 5604, 5685, 5799
forestry, single-tree, 5290
forestry, sustainable. *See* sustainable forestry
forests and climate. *See* climate and forests
forests, cloud, 2035, 2556, 2557, 4037, 4041
forests, old-growth. *See* old-growth forests
forests, tropical. *See* tropical forests, tropical rainforests
fossil animals. *See* extinct animals
fossil invertebrates. *See* 'invertebrates, fossil' in Index III
fossil plants. *See* extinct plants
fossil record, 78, 407, 1101, 1481, 1631, 2780, 2815, 4170, 4172, 4379, 4583, 4584, 4657, 4658, 5853
Founder effect, 427
fractal landscapes, 3375, 3820
fragmentation, habitat [/ecosystem/landscape]. *See* habitat fragmentation
fragments, habitat. *See* habitat fragments
freshwater benthic environs, 787, 951, 1219, 2014, 3092, 4223, 5391
freshwater communities, 148, 574, 773, 787, 1052, 1219, 1694, 1991, 2269, 2393, 2806, 3698, 3858, 3861, 4262, 4372, 5342
freshwater environs, 15, 16, 69, 144, 190, 235, 385, 493, 715, 811, 977, 1109, 1514, 1580, 1655, 1853, 1870, 1976, 2096, 2110, 2178, 2228, 2259, 2564, 2618, 2669, 2737, 2819, 2897, 2898, 2954, 3170, 3231, 3413, 3475, 3476, 3624, 3625, 3700, 3817, 3861, 3939, 4240, 4268, 4453, 4496, 4605, 4724, 4888, 5167, 5356, 5396, 5504, 5507, 5681, 5867
freshwater fishes. *See* 'fishes, freshwater' in Index III
freshwater invertebrates, 783, 787, 977, 1219, 1514, 1694, 2403, 2806, 5125, 5279, 5303
freshwater liming, 2178, 3698, 4455
frugivores, 1261, 1615, 5436
functional diversity [/bd/ecology], 1072, 1194, 1285, 1488, 1590, 1625, 1755, 2348, 2932, 3365, 3429, 3621, 3834, 3857, 4192, 4396, 4804, 4869, 4891, 4920, 5095, 5469, 5524, 5534, 5679, 5680, 5732
fynbos, 1069, 1196, 1212, 1332, 2932, 3854, 4540, 4682, 5138, 5139, 5237, 5238, 5632, 5870
Gaia hypothesis, 317, 1330, 1733, 1941, 2803, 2879, 2993, 2994, 2995, 3127, 3337, 4423, 4505, 4585, 4604, 5295, 5425
galling insects. *See* parasitic insects
game theory, 455, 4198, 4248
gamma diversity, 4579, 4638, 5130, 5457
GAP and gap analysis, 79, 498, 747, 827, 853, 1012, 1034, 1035, 1097, 1193, 1215, 1413, 1544, 1608, 2367, 2445, 2536, 2652, 3047, 3048, 3245, 3560, 4092, 4142, 4546, 4547, 4549, 4551, 4640, 4903, 4904, 4929, 5497
gaps, treefall [/canopy]. *See* treefall gaps
gardening, 2940, 3054, 3132
GATT, 24, 1122, 2102, 2890, 3230, 4101

gender. *See* ecofeminism
gene banks (*see also* germplasm collections), 9, 35, 379, 956, 957, 1169, 1661, 1662, 2231, 3485, 3996, 3997, 4465, 4878, 5488
gene flow, 289, 882, 1458, 2211, 2764, 3070, 4267, 4757, 5655
gene [/genetic/germplasm] resources conservation, 35, 36, 63, 71, 662, 666, 857, 859, 1014, 1243, 1604, 1660, 1705, 1937, 1984, 2268, 2400, 2533, 2610, 2711, 3191, 3450, 3468, 3538, 3591, 3734, 3778, 3995, 4112, 4144, 4196, 4226, 4293, 4465, 4518, 5235, 5811
gene sequences, 33, 1204, 1336, 1706, 2873, 3895, 3897, 4478, 4589, 5342
generalized tracks, 1095, 1099, 1936
generic age, 4873
genes, homeobox, 816
genes, homeotic, 1258, 2359
genes, hox, 2256
genetic [/taxonomic] distinctiveness, 1116, 1440, 1688, 2018, 3201, 4799, 5359, 5519, 5516
genetic diversity, 662, 882, 887, 922, 1021, 1116, 1117, 1243, 1268, 1392, 1510, 1647, 1666, 1667, 1895, 2001, 2211, 2420, 2681, 2859, 3049, 3408, 3452, 3596, 3606, 3607, 3710, 3855, 3942, 3996, 4412, 4612, 4662, 4684, 4751, 5021, 5024, 5124, 5562, 5573
genetic diversity conservation, 1524, 2630, 3190, 3431, 3452, 3853, 4027
genetic diversity, crop, 37, 933, 1243, 1392, 2367, 5702
genetic diversity, plant, 1523, 1974, 2047, 2983, 3468, 4068
genetic drift, 1458, 2359, 2785, 2801
genetic engineering (*see also* transgenic organisms), 272, 1750, 1751, 1781, 1984, 2582, 3426, 3428, 3492, 3715, 4217, 4951, 5081, 5529, 5609
genetic engineering and plants. *See* 'plants and biotechnology' in Index III
genetic fingerprinting methods, 37, 109, 603, 1450, 2156, 5695
genetic markers, 330, 3366, 4043, 5134, 5282, 5657, 5695
genetic resources. *See* germplasm resources
genetic resources, animal. *See* 'animal germplasm resources' in Index III
genetic resources, crop. *See* crop germplasm resources
genetic resources, plant. *See* 'plant germplasm resources' in Index III

genetic variation [/variability], 71, 109, 599, 1471, 1636, 1665, 1666, 2089, 2547, 2755, 2785, 2802, 3597, 3700, 4116, 4266, 4856, 5059, 5134, 5300, 5339
genetically modified organisms. *See* genetic engineering, transgenic organisms
genetics, conservation. *See* conservation genetics
genetics, fish, 3408, 3700, 5073, 5074, 5339, 5573
genetics: misc. subjects, 393, 554, 624, 662, 700, 1111, 1204, 1399, 1525, 1635, 2107, 2156, 2511, 2587, 2894, 3191, 3561, 3709, 4235, 4334, 4812, 4853
genetics, population. *See* population genetics
genomes, 734, 1947, 2873, 5214, 5488, 5818
Gentry, Alwyn, 5443
geographical distribution: misc. subjects (*see also* geographical range), 95, 121, 135, 203, 255, 355, 412, 559, 632, 634, 635, 826, 881, 968, 977, 984, 985, 1036, 1113, 1151, 1179, 1181, 1228, 1242, 1317, 1363, 1378, 1382, 1469, 1514, 1566, 1623, 1649, 1677, 1690, 1763, 1766, 1771, 1791, 1804, 1840, 1911, 2059, 2068, 2110, 2146, 2151, 2152, 2175, 2209, 2263, 2276, 2356, 2392, 2478, 2567, 2577, 2697, 2713, 2814, 2880, 2926, 2969, 2970, 2972, 3015, 3032, 3115, 3184, 3314, 3315, 3360, 3361, 3362, 3449, 3497, 3509, 3559, 3732, 3810, 3845, 3895, 3928, 4011, 4077, 4088, 4096, 4139, 4333, 4356, 4357, 4397, 4410, 4652, 4723, 4724, 4727, 4752, 4828, 4885, 4945, 5046, 5158, 5247, 5302, 5318, 5320, 5322, 5332, 5349, 5442, 5462, 5535, 5610, 5635, 5638
geographical information systems (GIS), 53, 56, 79, 203, 461, 498, 715, 827, 853, 910, 1034, 1035, 1097, 1193, 1215, 1216, 1521, 1544, 1669, 2009, 2488, 2962, 2965, 3245, 3312, 3362, 3410, 3552, 3556, 3588, 3628, 3998, 4263, 4436, 4477, 4552, 4555, 4640, 4727, 4736, 4831, 4832, 4903, 4904, 4999, 5293, 5634
geographical range (*see also* geographical distribution), 95, 121, 129, 314, 489, 582, 634, 1469, 1655, 1711, 1752, 1754, 1767, 1954, 2068, 2713, 2842, 2846, 2965, 2966, 2969, 3068, 3338, 3810, 3845, 4446, 4880, 4881, 5144, 5532
geography, animal. *See* 'animal distribution' in Index III
geography, plant. *See* 'plant biogeography' in Index III
geomorphological agents, 719, 1082, 3628,

4044, 4430, 4992
germplasm collections, 330, 1297, 2488, 2893, 4774, 5281
germplasm conservation, crop, 663, 664, 666, 2709, 3538, 3996
germplasm [/genetic] resources, 37, 38, 40, 914, 956, 957, 1029, 1206, 1252, 1402, 1594, 1660, 1692, 2030, 2090, 2213, 2242, 2532, 2533, 2640, 2674, 2890, 2892, 2956, 3191, 3480, 3734, 3777, 3997, 4027, 4067, 4236, 4312, 4339, 4449, 4518, 4563, 4626, 4699, 4751, 4853, 4877, 4912, 4935, 4953, 4954, 5074, 5140, 5292
germplasm resources, animal. *See* 'animal germplasm resources' in Index III
germplasm resources, crop. *See* crop germplasm resources
germplasm resources, plant. *See* 'plant germplasm resources' in Index III
germplasm, rice, 859, 2400, 5249, 5281, 5282, 5695
glacial extinction hypothesis, 30
glacial periods, 30, 1027, 2689, 3237, 3433, 4021, 4342, 4862, 5322, 5348, 5376
Global Biodiversity Assessment, 762, 2197, 4230
Global Biodiversity Forum, 2377, 2378, 2379, 2380
Global Biodiversity Strategy, 1206
global cooling, 876, 4862
global diversity, 406, 592, 841, 1578, 1581, 2028, 2729, 3339, 4489
Global Environment Facility, 159, 538, 1120, 2210, 2494, 3169, 3332, 3402, 3403, 3471, 3487, 4607, 4876, 5079, 5176, 5399
Global Positioning System (GPS), 1521
Global Taxonomic Facility, 4541
global warming, 10, 443, 505, 538, 1309, 1574, 1903, 2003, 2106, 2616, 2620, 2867, 2895, 2918, 3095, 3180, 3268, 3845, 3849, 3930, 3931, 3973, 4336, 4486, 4722, 4746, 5286, 5648
Global 2000, 4687
global-scale studies and effects (*see also* climatic change, environmental change, global warming), 168, 192, 234, 293, 317, 346, 347, 406, 453, 475, 477, 505, 555, 563, 567, 577, 590, 592, 595, 628, 697, 701, 726, 746, 749, 756, 803, 811, 841, 844, 871, 872, 891, 903, 957, 1019, 1048, 1070, 1074, 1085, 1139, 1180, 1257, 1350, 1351, 1353, 1359, 1367, 1416, 1421, 1438, 1578, 1581, 1623, 1658, 1754, 1758, 1767, 1770,

1783, 1797, 1825, 1918, 1930, 1932, 1949, 1983, 2017, 2025, 2028, 2055, 2191, 2197, 2213, 2243, 2260, 2278, 2316, 2369, 2375, 2376, 2377, 2378, 2379, 2380, 2382, 2476, 2494, 2541, 2562, 2586, 2616, 2693, 2697, 2724, 2726, 2729, 2779, 2821, 2909, 2910, 2961, 2969, 2970, 2984, 2987, 3108, 3159, 3174, 3221, 3238, 3263, 3269, 3275, 3326, 3339, 3348, 3353, 3418, 3429, 3440, 3441, 3479, 3490, 3574, 3650, 3659, 3702, 3747, 3787, 3871, 3878, 3884, 3890, 3997, 4030, 4051, 4101, 4118, 4119, 4135, 4179, 4180, 4259, 4273, 4370, 4374, 4418, 4422, 4462, 4469, 4489, 4593, 4601, 4606, 4609, 4622, 4715, 4718, 4770, 4781, 4783, 4785, 4789, 4824, 4876, 4886, 4910, 4954, 4962, 4964, 5009, 5101, 5109, 5133, 5151, 5169, 5178, 5179, 5180, 5243, 5269, 5288, 5328, 5329, 5341, 5366, 5376, 5388, 5397, 5428, 5443, 5517, 5590, 5600, 5620, 5624, 5661, 5740, 5802, 5816, 5824, 5833, 5834, 5867
globalization, 168, 1963, 3651, 5341
Gloger's rule, 4149
goal-oriented planning [/management], 153, 328, 699, 1046, 1155, 3102, 4432, 4535, 4730, 4811, 5061, 5547
gradients, altitudinal. *See* altitudinal gradients
gradients, diversity. *See* diversity gradients
gradients, environmental. *See* environmental gradients
gradients, latitudinal. *See* latitudinal gradients
gradients, latitudinal body size, 300, 464, 2126, 2129, 3262, 4619
gradients, latitudinal diversity. *See* latitudinal diversity gradients
gradients, latitudinal species richness. *See* latitudinal species richness gradients
gradients, species richness, 991, 1560, 4131
grasslands (*see also* prairie), 58, 184, 198, 248, 428, 645, 673, 714, 984, 985, 986, 1868, 1928, 2182, 2284, 2461, 2468, 2503, 2769, 2825, 2833, 2834, 2876, 2886, 3342, 3577, 4136, 4396, 4522, 4654, 4891, 5086, 5087, 5089, 5090, 5091, 5094, 5135, 5230, 5382, 5383, 5434, 5446, 5705
grassland [/prairie] communities, 2042, 2182, 5605
grazing and overgrazing (*see also* rangelands), 119, 986, 1307, 1419, 1610, 1651, 3393, 4396, 4397, 4415, 5410, 5711
Great American Interchange, 2889, 3146,

3147, 5373, 5374
Great Chain of Being, 4426
Green movement, 26, 41, 667, 1311, 1347, 1362, 1428, 2554, 2917, 3883, 3885, 4973
greenhouse effect, 10, 284, 365, 503, 960, 1149, 1316, 1941, 3248, 3521, 3761, 3849, 3923, 3930, 3931, 4307, 4365, 4419, 4562, 4602, 4746, 4899, 4995, 5133, 5648
greenways, 2942, 4739
groundwater, 1807, 2260, 3136, 3895, 5115, 5809
guilds, 3824, 5524
habitat conservation [/habitat conservation planning], 48, 349, 358, 681, 1017, 1249, 1702, 2623, 3018, 3145, 3500, 3689, 4266, 4623, 5612, 5786
habitat [/environmental/ecological] degradation, 16, 70, 158, 170, 268, 458, 562, 575, 856, 949, 1156, 1177, 1248, 1315, 1397, 1419, 1610, 1643, 1719, 2029, 2106, 2206, 2207, 2219, 2943, 2973, 3176, 3258, 3269, 3378, 3390, 3490, 3688, 3738, 3826, 3847, 3947, 4228, 4332, 4415, 5097, 5380, 5485, 5505, 5611, 5659, 5693, 5712
habitat disturbance, 1821, 2033, 2458, 2844, 3686, 4802, 5171
habitat diversity [/complexity], 1341, 2719, 2955, 3240, 3642, 4201, 5280, 5362, 5634
habitat [/ecosystem/landscape] fragmentation, 65, 134, 199, 211, 310, 311, 319, 397, 440, 501, 531, 618, 708, 709, 716, 870, 895, 1145, 1146, 1162, 1177, 1284, 1285, 1286, 1396, 1493, 1810, 1814, 1826, 1908, 1977, 1999, 2045, 2063, 2072, 2073, 2077, 2098, 2163, 2181, 2182, 2199, 2209, 2222, 2258, 2293, 2360, 2557, 2599, 2716, 2762, 2825, 2828, 2829, 2830, 2851, 2872, 2883, 2978, 3012, 3118, 3122, 3225, 3234, 3294, 3322, 3494, 3640, 3716, 3758, 3760, 4080, 4099, 4113, 4154, 4318, 4319, 4325, 4469, 4470, 4536, 4537, 4668, 4669, 4672, 4725, 4814, 4850, 4917, 5032, 5053, 5096, 5152, 5207, 5230, 5293, 5327, 5372, 5470, 5473, 5487, 5525, 5714
habitat fragments [/remnants] (*see also* habitat fragmentation, landscape mosaics, patches), 65, 202, 311, 397, 440, 501, 531, 1349, 1493, 1505, 1650, 1810, 1977, 2181, 2199, 2530, 2600, 2602, 2825, 2828, 2830, 2851, 2883, 2981, 3322, 4471, 4537, 5152, 5154, 5372
habitat heterogeneity. *See* environmental heterogeneity
habitat loss [/destruction], 69, 258, 276, 744, 839, 856, 949, 1422, 1431, 1557, 1832, 2623, 2725, 2857, 2912, 3334, 3630, 3687, 3688, 3738, 3970, 4703, 4917, 4918, 5096, 5461, 5485, 5491, 5552, 5611, 5672
habitat selection, 2552, 3038, 4366
habitat, wildlife. *See* wildlife habitat
habitats, threatened. *See* endangered ecosystems and habitats
hatcheries, 3300
Hawksworth, David, 3197
hedgerows, 3210, 3839
herbivores [/herbivory], 127, 369, 416, 1037, 1171, 1684, 2303, 2425, 3000, 3799, 3802, 3857, 4081, 4397, 4930, 5327, 5448
Hessler, Robert, 5777
heterogeneity, environmental. *See* environmental heterogeneity
heterozygosity, 599, 2107, 3596, 3597, 3607, 3853, 4856, 5134
hierarchical richness index (HRI), 1693
higher taxon approaches, 113, 260, 261, 404, 1117, 1556, 1764, 1769, 1776, 1779, 1813, 2234, 3200, 3338, 3633, 4052, 4061, 4174, 4382, 4576, 4577, 4578, 4857, 5212, 5231, 5515, 5517, 5518
higher-taxon richness, 113, 260, 261, 1117, 1776, 1779, 5515, 5517
historical biogeography: overviews, 125, 2154, 3457, 3988, 4362, 5236
historical factors and effects, 1177, 1373, 1416, 1445, 1465, 1675, 1985, 2299, 2513, 2543, 2876, 3175, 3237, 3703, 3928, 4287, 5302
historical perspectives, 572, 870, 972, 1026, 1071, 1177, 1381, 1465, 1489, 1603, 1675, 1821, 1867, 1970, 2065, 2513, 2543, 2567, 2848, 2876, 2936, 2970, 2976, 3006, 3134, 3175, 3222, 3237, 3334, 3369, 3558, 3867, 4426, 4456, 4457, 4567, 4591, 4717, 4818, 4866, 5048, 5136, 5154, 5439, 5454, 5525
holistic approaches, 2139, 3583, 4890, 5377
Holocene, 1920, 4852
homeobox genes, 816
homeotic genes, 1258, 2359
host mammals, 1740, 3442, 5365
host plants, 1299, 2739, 4879
host species, 71, 327, 1037, 1299, 1500, 1569, 1740, 1976, 2124, 2125, 2225, 2739, 2951, 4032, 4035, 4248, 4679, 4879, 5055, 5365, 5389
hotspots, 210, 872, 1071, 1137, 1174, 1291, 1317, 1382, 1515, 1601, 1607, 1769, 1779,

1883, 2011, 2157, 2410, 2436, 2939, 2960, 3295, 3401, 3405, 3406, 3433, 3437, 3519, 3520, 3522, 3598, 3616, 3823, 3972, 4010, 4060, 4062, 4063, 4231, 4333, 4724, 5206, 5364, 5517, 5521
hotsprings. *See* springs and hotsprings
hox genes, 2256
human carrying capacity, 189, 954, 1158, 1418
human disturbance (*see also* human impacts), 862, 1176, 1457, 1876, 2055, 2206, 2676, 3099, 3325, 3869, 4295, 4590, 4866, 5077, 5286, 5289, 5626, 5665
human ecology, 111, 325, 551, 667, 671, 807, 1060, 1157, 1446, 1603, 1625, 1626, 1718, 1719, 2562, 2595, 2596, 2598, 2609, 3093, 3218, 3241, 3330, 3601, 4017, 4767, 4843, 5151, 5223, 5239, 5496, 5558, 5611, 5619, 5701
human evolution [/diversity], 33, 289, 548, 551, 734, 1947, 2394, 2419, 2596, 2854, 3134, 3167, 3576, 4457
Human Genome Diversity Project, 734, 1947
human [/public] health, 254, 1310, 1472, 1473, 1944, 4146
human impacts (*see also* human disturbance, recreation activity impacts), 62, 68, 683, 707, 820, 862, 864, 1176, 1188, 1503, 1605, 1876, 1896, 1956, 2015, 2055, 2223, 2406, 2775, 2859, 2895, 3144, 3254, 3325, 3455, 3456, 3470, 3724, 3774, 3788, 3892, 3893, 3969, 4048, 4058, 4156, 4279, 4466, 4936, 5048, 5074, 5077, 5290, 5303, 5628
human population growth, 157, 720, 875, 1158, 1426, 1435, 2083, 4181, 4718
human trampling, 617, 1868, 2920
hunting [/hunters], 111, 237, 491, 492, 809, 1464, 1503, 1649, 1809, 3135, 3334, 4205
Hurlbert's index, 378
hurricanes, 524, 606, 2051, 2362, 3436, 5242, 5248, 5491, 5534
Huston, Michael, 49
hybrids and hybridization, 473, 894, 1395, 2764, 2905, 3906, 4267, 4747, 5447, 5448
hydropower industry. *See* utility industry
hydrothermal vent communities, 1648, 1911, 3022, 4653, 5225, 5405
hyperoxia, 1902
hypersea, 3260
ICDPs, 94, 288, 570, 651, 2752, 2753, 2754, 3216, 5784
illegal drug cultivation, 1874, 5670

immigration, 637, 984, 1810, 1841, 2060, 2964, 2967, 5071
immigration rates. *See* colonization rates
immigration-extinction relations, 637, 984, 1841, 2967, 5071
impacts, human. *See* human impacts
imperialism, 1106, 2520, 2521, 3082
in situ conservation, 101, 415, 663, 664, 666, 669, 672, 1169, 1467, 2087, 2122, 2367, 2437, 2983, 3190, 3507, 3735, 3995, 4062, 4293, 4484, 5249, 5579, 5580, 5602
in vitro methods, 35, 1323, 4196, 5682
INBio, 487, 1054, 1545, 1736, 1851, 2313, 2430, 2504, 3323, 3841, 4309, 4721, 5022, 5686
inbreeding, 603, 882, 1458, 1665, 1667, 1668, 2156, 2785, 2801, 2802, 3372, 3709, 5326
incentives, economic. *See* economic incentives
incremental cost, 848, 849, 2494
index, avalanche, 1737
index, hierarchical richness, 1693
index, Hurlbert's, 378
index of biotic integrity, 3027, 3387, 5391
index, Pielou's, 378
index, Shannon. *See* Shannon index
index, Simpson, 2078, 4694
indicators, bd [/environmental]. *See* biodiversity indicators
indices, diversity. *See* diversity indices
indigenous agriculture [/farming]. *See* traditional agriculture
indigenous knowledge. *See* traditional knowledge
indigenous medicine. *See* traditional medicine
indigenous peoples, 170, 400, 450, 609, 643, 668, 671, 898, 925, 966, 1078, 1133, 1218, 1723, 1915, 1922, 1981, 2171, 2272, 2273, 2355, 2369, 2370, 2609, 2664, 2788, 2793, 3075, 3148, 3779, 4013, 4022, 4023, 4194, 4206, 4207, 4209, 4264, 4373, 4635, 4712, 4713, 4843, 5164, 5387, 5486, 5641, 5772
insects. *See* listings in Index III
integrity, biological [/bd/biotic/ecological/ecosystem]. *See* biological integrity
integrity, index of biotic, 3027, 3387, 5391
intellectual property, 24, 229, 239, 383, 666, 668, 671, 723, 898, 1118, 1402, 1700, 1701, 1722, 1860, 1884, 1922, 2102, 2213, 2273, 2329, 2386, 2521, 2640, 2664, 2891,

3125, 3230, 3256, 4046, 4101, 4301, 4312, 4563, 4630, 4631, 4635, 4659, 4912, 4935, 4961, 5005, 5196, 5776, 5819, 5873
interchange, biotic. *See* biotic interchange
interdisciplinary approaches, 312, 1803, 2587, 2753, 3917, 4609, 4961, 5065, 5377, 5803
intermediate disturbance hypothesis, 4540, 4800, 4801, 5125, 5271
International Cocoa Gene Bank, 379
international cooperation, 92, 313, 334, 371, 402, 521, 1194, 1209, 1884, 1988, 2003, 2385, 2444, 2492, 2494, 2517, 2688, 2691, 2933, 3074, 3075, 3108, 3174, 3336, 3353, 3364, 3437, 3480, 3793, 3964, 4104, 4176, 4225, 4229, 4338, 4364, 4449, 4456, 4538, 4753, 4839, 4853, 4927, 4958, 4960, 4964, 4983, 5001, 5075, 5076, 5109, 5184, 5420, 5600
International Development Research Centre, 2793
international law, 229, 371, 383, 442, 452, 490, 521, 544, 563, 1122, 1318, 2191, 2313, 2381, 2444, 3074, 3075, 3125, 3140, 3832, 4046, 4341, 5085, 5098, 5099, 5196, 5272, 5420, 5654, 5764, 5812
International Livestock Center, 4220
international trade, 282, 723, 1118, 1347, 1540, 1597, 2167, 2890, 4101, 4104, 5005, 5475, 5873
intertidal communities. *See* marine communities
intertidal [/littoral/nearshore] environs, 314, 429, 575, 617, 1742, 2091, 2092, 2269, 2964, 3000, 3308, 3387, 3813, 4400, 4523, 4816, 5427, 5608
interviews, 165, 1739, 2810, 3466, 3770, 3771
intraspecific diversity [/variation], 1909, 2791, 4376, 4412, 4588
introduced animals, 2259, 2901, 3427, 3625, 4515, 4906
introduced birds, 2900, 2949, 2970, 3472, 4673
introduced fish. *See* 'fish introductions' in Index III
introduced insects, 845, 903, 1037, 4122, 5330
introduced mammals, 1391, 1623, 1684, 2899
introduced plants. *See* 'plant introductions' in Index III
introduced species (*see also* biological invasions), 128, 190, 212, 264, 387, 559, 749, 775, 826, 845, 866, 868, 903, 945, 1024, 1037, 1062, 1109, 1280, 1358, 1391, 1559, 1602, 1623, 1684, 1846, 1954, 2259, 2508, 2619, 2630, 2663, 2686, 2868, 2886, 2899, 2900, 2901, 2902, 2919, 2949, 2953, 2954, 2970, 2999, 3040, 3042, 3052, 3334, 3345, 3369, 3426, 3427, 3472, 3477, 3503, 3625, 3724, 3981, 4330, 4384, 4390, 4476, 4514, 4515, 4673, 4705, 4735, 4906, 4919, 4934, 4938, 4943, 5081, 5204, 5349, 5424, 5436, 5462, 5501, 5577, 5684
invasions, biological. *See* biological invasions
invasions, fish. *See* 'fish introductions' in Index III
invasions, plant. *See* 'plant introductions' in Index III
inventories and checklists (*see also* species lists), 59, 236, 260, 499, 759, 777, 790, 975, 1033, 1163, 1267, 1356, 1369, 1504, 1720, 1986, 2016, 2055, 2142, 2257, 2271, 2432, 2487, 2524, 2654, 2723, 2751, 2844, 2962, 2971, 3005, 3013, 3221, 3448, 3583, 3586, 3587, 3678, 3856, 4244, 4493, 4640, 4728, 4896, 4924, 5430, 5432, 5585, 5666
invertebrates. *See* listings in Index III
IPCC, 2277, 5771
IPGRI, 1696, 2371
island biogeography, 224, 631, 632, 635, 637, 708, 709, 825, 971, 1025, 1068, 1103, 1247, 1256, 1274, 2098, 2300, 2465, 2466, 2470, 2522, 2604, 2623, 2719, 2819, 2964, 2968, 3010, 3036, 3037, 3070, 3123, 3214, 3294, 3615, 3620, 3642, 3954, 4110, 4113, 4409, 4511, 4667, 4668, 4671, 4672, 4675, 4676, 4677, 4722, 5053, 5071, 5203, 5211, 5459, 5484, 5527, 5528, 5546, 5627, 5635, 5637, 5638, 5645, 5704, 5758
island biology [/ecology], 427, 795, 1910, 2974, 3182, 5287, 5307, 5459, 5852
island birds. *See* 'birds, island' in Index III
island mammals, 703, 2964, 5758
island plants, 1102, 1256, 1609, 2039, 3182, 3642, 4409, 4515, 4637, 5311
Island Press, publications by, 10, 106, 296, 325, 343, 417, 571, 671, 684, 918, 1046, 1047, 1157, 1197, 1302, 1429, 1527, 1687, 1872, 1901, 1944, 1968, 1969, 2167, 2265, 2293, 2311, 2595, 2596, 2598, 2700, 2718, 2758, 3117, 3166, 3227, 3274, 3658, 3659, 3684, 3689, 3828, 3894, 3993, 4018, 4436, 4440, 4490, 4519, 4761, 4803, 4810, 4966, 5016, 5070, 5146, 5223, 5419, 5433, 5496, 5558, 5652

island settings, 2, 46, 297, 369, 370, 376, 410, 566, 616, 703, 721, 825, 826, 944, 1025, 1068, 1079, 1102, 1103, 1235, 1247, 1256, 1274, 1609, 1623, 1667, 1668, 1909, 1910, 2038, 2039, 2142, 2223, 2356, 2357, 2365, 2416, 2439, 2465, 2466, 2475, 2476, 2517, 2522, 2576, 2604, 2719, 2783, 2872, 2949, 2964, 2974, 3010, 3113, 3182, 3231, 3257, 3334, 3400, 3434, 3503, 3604, 3615, 3642, 3780, 3843, 3869, 3902, 4007, 4058, 4097, 4110, 4114, 4191, 4215, 4281, 4409, 4476, 4484, 4511, 4515, 4592, 4637, 4647, 4665, 4676, 4677, 4866, 4931, 4986, 5046, 5071, 5131, 5202, 5218, 5287, 5307, 5311, 5318, 5319, 5322, 5459, 5460, 5484, 5491, 5526, 5636, 5637, 5742, 5758, 5762, 5852
island-like settings, 224, 631, 632, 635, 637, 708, 709, 971, 1025, 1220, 1275, 2098, 2300, 2428, 2470, 2623, 2819, 2968, 3010, 3012, 3070, 3123, 3294, 3618, 3620, 3954, 4113, 4211, 4490, 4594, 4668, 4671, 4672, 4675, 4722, 4826, 5053, 5203, 5302, 5470, 5645, 5704
isolation effects, 46, 1247, 2522, 2530, 2872, 2964, 2965, 3012, 3070, 3759, 3869, 4113, 4215, 4222, 4696, 5154, 5224, 5636
Itza Maya people, 207, 873
IUBS, 4780
IUCN, 1455, 2080, 3044, 4119, 5567
IUCN, publications by, 236, 482, 844, 976, 1199, 1209, 1376, 1595, 1847, 2019, 2194, 2196, 2288, 2375, 2376, 2377, 2378, 2379, 2380, 2382, 2571, 2688, 2747, 2915, 3060, 3061, 3172, 3266, 3271, 3283, 3285, 3407, 3489, 3693, 4079, 4216, 4384, 4480, 4915, 4931, 5169, 5404, 5493, 5620, 5621, 5624
IUPAC, 137
ivory trade, 285
jackknife procedure, 2166, 3383, 4741, 5337
Jakarta Mandate, 1879
Janzen, Dan, 717, 2810, 3466, 4531
Jurassic, 3291
Kayapó Indians, 4023, 5164
Keystone Dialogue, 3056
keystone species, 356, 864, 1079, 1710, 2485, 2869, 3215, 3308, 3342, 3373, 3399, 3543, 3815, 4045, 4670, 4920, 4941, 5002, 5327, 5541
kill curves, 4167, 4168
kingdoms, 838, 1324, 1336, 1462, 1573, 1893, 3126, 3130, 5133, 5502
Kothari, Smitu, 165
Lack, David, 4215

lagoons, 301, 525, 892, 1211, 3508, 4375, 4948, 5766
lakes (*see also* ponds), 70, 231, 286, 292, 559, 575, 614, 866, 888, 899, 950, 951, 952, 1042, 1052, 1325, 1326, 1379, 1861, 1880, 1943, 2151, 2152, 2269, 2297, 2393, 2402, 2508, 2564, 2588, 2604, 2819, 2839, 2897, 2999, 3077, 3095, 3153, 3345, 3369, 3370, 3386, 3387, 3388, 3399, 3493, 3581, 3697, 3698, 3724, 3858, 3981, 4108, 4126, 4129, 4239, 4409, 4455, 4460, 4564, 4800, 4845, 4855, 5119, 5131, 5277, 5386, 5462, 5480, 5577, 5626, 5650, 5684, 5805, 5877
land cover, 2903, 3326, 5084
land ethic, 763, 764, 912, 2699, 2882, 3072
land plants, 67, 2368, 2384, 2621, 2622, 3636, 3854, 4193
Land Use History of North America, 4717
land use: misc. subjects, 50, 322, 353, 399, 470, 569, 751, 1605, 1708, 1897, 2208, 2278, 2848, 3111, 3326, 3390, 3788, 3963, 4557, 4717, 4809, 4986, 5105, 5411, 5435, 5542, 5601
landraces, 664, 669, 879, 1243, 4983
landscape disturbance, 4394, 5159, 5382
landscape diversity, 1058, 2955, 3676, 4315, 4507
landscape ecology [/-level approaches], 31, 32, 58, 134, 199, 200, 243, 244, 249, 274, 310, 328, 397, 398, 457, 695, 705, 709, 775, 881, 910, 1173, 1384, 1606, 1641, 1642, 1731, 1891, 2009, 2054, 2056, 2073, 2077, 2078, 2163, 2217, 2218, 2222, 2248, 2267, 2291, 2293, 2325, 2337, 2460, 2540, 2657, 2735, 2772, 2811, 2921, 3089, 3234, 3268, 3312, 3321, 3410, 3421, 3464, 3582, 3583, 3584, 3628, 3676, 3678, 3740, 3755, 3760, 3820, 3999, 4154, 4155, 4188, 4189, 4272, 4315, 4355, 4394, 4502, 4507, 4547, 4549, 4697, 4890, 4894, 4895, 4979, 4992, 5012, 5126, 5159, 5160, 5161, 5162, 5163, 5193, 5224, 5283, 5344, 5364, 5382, 5435, 5470, 5473, 5498, 5511, 5685, 5710, 5828, 5829, 5830
landscape fragmentation. *See* habitat fragmentation
landscape heterogeneity. *See* environmental heterogeneity
landscape mosaics, 22, 65, 910, 1641, 2073, 2078, 5470
landscape planning, 32, 4355, 4890
landscapes, agricultural, 705, 1153, 1369, 1493, 2208, 2216, 2762, 3380, 3759

landscapes, fractal, 3375, 3820
large mammals, 367, 718, 1126, 1279, 1363, 1684, 3149, 3620, 3799, 3800, 5476, 5674
Latin American Maize Project, 4983
latitudinal body size gradients, 300, 464, 2126, 2129, 3262, 4619
latitudinal diversity gradients, 87, 140, 462, 463, 466, 891, 920, 1049, 1070, 1088, 1091, 1110, 1144, 1186, 1277, 1585, 1614, 1617, 1724, 1779, 1917, 2389, 2426, 2561, 2651, 2739, 2821, 3068, 3224, 3810, 3951, 4120, 4121, 4223, 4259, 4343, 4344, 4367, 4381, 4395, 4619, 4874, 5130, 5158, 5321, 5371, 5533, 5559
latitudinal gradients, 1762, 1765, 2694, 3068, 3705, 3903, 4880, 4881
latitudinal species richness gradients, 7, 466, 1655, 2127, 2739, 4345, 4679, 5532, 5606
law and legislation, bd. *See* environmental law
law and legislation, environmental. *See* environmental law
law, international. *See* international law
law reviews, articles in, 169, 195, 229, 350, 371, 383, 479, 480, 481, 490, 521, 560, 563, 564, 578, 628, 712, 796, 897, 909, 1054, 1122, 1150, 1290, 1318, 1338, 1347, 1360, 1406, 1442, 1536, 1587, 1794, 1846, 1860, 1865, 1879, 1980, 2088, 2191, 2262, 2273, 2275, 2313, 2381, 2444, 2506, 2516, 2519, 2520, 2542, 2584, 2636, 2658, 2684, 2687, 2773, 3074, 3075, 3112, 3125, 3140, 3145, 3242, 3256, 3327, 3507, 3806, 3832, 3860, 4046, 4074, 4105, 4148, 4249, 4254, 4301, 4341, 4377, 4385, 4392, 4398, 4501, 4557, 4563, 4740, 4760, 4796, 4864, 4912, 4913, 5005, 5006, 5007, 5056, 5085, 5098, 5099, 5145, 5196, 5272, 5390, 5764, 5772, 5789, 5812, 5864, 5873, 5875, 5876
laws and regulations, state. *See* state laws
Leopold, Aldo, 763, 764, 765, 912, 1849, 2699, 4030, 5251
levels-of-organization concept, 80, 312, 3066
life history studies, 623, 888, 911, 2396, 3164, 3468, 4337, 4926
life, origin of. *See* origin of life
life zones, 2245, 3314, 3315
liming, freshwater, 2178, 3698, 4455
lists, species. *See* species lists
litigation, 2516, 3414
litter, plant, 1539, 4556, 5086, 5352, 5353
littoral environs. *See* intertidal environs

local (community-related) conservation participation, 82, 388, 422, 578, 732, 1170, 1385, 1713, 1723, 2282, 2514, 2529, 2675, 2916, 2958, 3056, 3135, 3148, 3603, 3830, 3965, 4069, 4634, 4649, 4650, 4712, 4713, 4892, 4985, 4986, 5006, 5102, 5106, 5401, 5402, 5746, 5824
local extinction, 413, 477, 491, 549, 819, 1492, 1505, 1841, 2059, 2100, 2557, 3026, 3215, 3620, 4421, 4669, 5050, 5152, 5525, 5693
local knowledge. *See* traditional knowledge
local [/regional] richness, 1039, 1943, 2618, 3179, 4132, 5333, 5584
local-scale studies and effects, 77, 261, 413, 477, 489, 491, 549, 580, 760, 819, 1038, 1039, 1234, 1370, 1469, 1492, 1505, 1506, 1744, 1771, 1805, 1841, 1914, 1943, 2059, 2068, 2091, 2100, 2316, 2543, 2557, 2608, 2618, 2729, 3026, 3135, 3159, 3171, 3215, 3606, 3620, 4004, 4081, 4088, 4285, 4421, 4669, 5050, 5119, 5152, 5280, 5304, 5333, 5397, 5525, 5693, 5698, 5746, 5824
logging, 342, 539, 597, 652, 1176, 1177, 1561, 1703, 1905, 2203, 2272, 2457, 2459, 3142, 3289, 3317, 3806, 3937, 4254, 4603, 4987, 5044, 5047, 5257, 5438
longlines, 3646
long-term ecological research, 1163, 2848, 3076, 3077, 3335, 3399, 3622, 5723
loss, bd. *See* biodiversity loss
loss, fish. *See* 'fish extinction' in Index III
loss, habitat. *See* habitat loss
loss minimization, 5579, 5580
loss, plant. *See* 'extinction, plant'
loss rates, bd. *See* biodiversity loss rates
loss, species. *See* biodiversity loss
Lovejoy, Thomas, 4939
Lujan v. Defenders of Wildlife, 2516
MacArthur, Robert, 635, 3036, 5528
macroecology, 636, 1763, 1767, 3179, 3590
macroevolution, 533, 611, 636, 1082, 1631, 2235, 2383, 2387, 2878, 3339, 3340, 3341, 3425, 4786, 4787, 4857, 4858, 5213
magnetic reversals, 4162
mainstreaming, 3811, 5617
mammals. *See* listings in Index III
Man and the Biosphere, 151, 294, 334, 335, 993, 1359, 4870
management, coastal zone. *See* coastal zone management
management, cross-boundary, 1184, 4516, 4517

management, ecosystem. *See* ecosystem management
management, forest. *See* forest management
management, goal-oriented. *See* goal-oriented planning
management, natural resource. *See* natural resource management
management, wildlife. *See* wildlife conservation
Maori people, 1828
maps and mapping (*see also* atlases), 353, <u>439</u>, 739, <u>1012</u>, 1215, 1386, 1413, 1521, 1651, 1669, <u>1764</u>, <u>1776</u>, 1812, 2055, 2146, <u>2322</u>, 2363, 2436, 2500, 2536, 2698, <u>2766</u>, 2767, 2833, 2969, 3015, 3047, 3090, 3359, 3401, 3648, <u>3792</u>, 4013, 4075, 4263, 4402, 4546, 4547, 4832, 4903, 4904, 4905, 4999, 5051, 5497, 5514, 5517
maps and mapping, vegetation, 1413, 2363, <u>2766</u>, 2833, 3015, <u>3792</u>
marine animals, 655, 737, <u>1725</u>, 1783, 2103, <u>2189</u>, 4491, <u>4580</u>, <u>4581</u>, <u>4765</u>, 5574
marine bd [/diversity]: general subjects, 140, 160, <u>409</u>, <u>588</u>, 594, 655, 736, 802, 1125, 1207, 1209, 1398, 1439, 1574, 1917, <u>2028</u>, 2372, 2506, <u>2798</u>, 2812, 2855, 2992, 3071, 3087, 3113, 3199, 3200, 3249, 3291, 3336, 3339, 3499, 3573, 3647, <u>3659</u>, 3660, 3781, 3782, 3935, 4190, 4192, <u>4450</u>, 4452, 4491, <u>4576</u>, <u>4584</u>, 4621, 4762, 4763, <u>4764</u>, <u>4765</u>, 4795, 4815, 4869, <u>5070</u>, 5175, 5192, 5211, 5217, 5253, <u>5258</u>, <u>5261</u>, 5276, 5325, 5363, 5366, <u>5378</u>, 5380, 5574, 5595, 5644, 5677, 5690, 5765, 5826, 5849, 5850, 5862, 5874
marine benthic environs, 6, <u>265</u>, 584, 741, 805, 924, 929, <u>946</u>, 1100, <u>1494</u>, 1724, 1871, 1913, <u>1914</u>, 2014, 2472, <u>2523</u>, 2535, <u>2614</u>, <u>2620</u>, <u>2798</u>, <u>2906</u>, 3137, 3699, <u>3842</u>, <u>4014</u>, 4257, <u>4259</u>, <u>4450</u>, 4662, 4763, <u>4764</u>, 4766, 4860, 5077, 5358, 5360, 5391, 5595, 5596, 5862
marine [/intertidal] communities, 44, <u>314</u>, 429, 617, 850, 924, <u>1038</u>, 1207, 1208, 1492, 1648, 1656, <u>1707</u>, 1742, <u>1912</u>, <u>1929</u>, 1991, 2018, <u>2299</u>, 2447, 2448, 2472, 2566, <u>2614</u>, <u>2906</u>, <u>3000</u>, <u>3022</u>, 3220, <u>4014</u>, <u>4428</u>, 4646, 4653, 4936, 5225, 5275, 5358
marine conservation, 50, 51, 86, 357, 602, <u>799</u>, 949, 1398, 1917, 2123, 2227, 2381, 2418, 2612, 3461, 3499, 3646, <u>3659</u>, 3793, 3945, 4189, 4249, 4770, 4771, <u>5070</u>, 5192, 5366, <u>5378</u>
marine diversity. *See* marine bd

marine extinction, <u>409</u>, 800, 876, 921, 1125, 1813, 2021, 2032, <u>2387</u>, 2422, <u>2589</u>, <u>2590</u>, <u>2620</u>, <u>2677</u>, 2701, 2855, 2992, 3087, 3249, 3339, 3655, <u>3868</u>, <u>4167</u>, 4168, <u>4172</u>, <u>4173</u>, 4521, <u>4578</u>, <u>4580</u>, 4666, 4859, 4860, 4862, 5209, 5263, 5474
marine fishes. *See* 'fishes, marine' in Index III
marine invertebrates. *See* 'invertebrates, marine' in Index III
marine microorganisms, 305, <u>839</u>, 1239, <u>1707</u>, <u>3842</u>
marine reserves, 50, 51, 86, 162, 357, <u>496</u>, 823, 1301, <u>1303</u>, 2227, 2448, <u>2586</u>, 2612, 2613, 2760, 3257, 3287, 3368, 3945, 4079, 4304, 4503, 4770, 4771, 5078
markers, genetic. *See* genetic markers
markets [/commodities], <u>124</u>, 390, 1414, 1537, 1687, 1865, 1897, 1898, 1899, 2147, 2937, 3316, 3712, 4360
Markov processes, 4731
Mars, <u>3243</u>
marshes. *See* swamps and marshes
mass, body. *See* body size
mass effects, 2769
mass extinction, 87, <u>103</u>, <u>104</u>, 105, <u>176</u>, 367, 404, 405, 406, 407, 591, 720, 876, 932, <u>946</u>, <u>1028</u>, 1063, 1064, 1115, 1126, 1278, 1279, <u>1334</u>, 1454, 1476, <u>1477</u>, <u>1478</u>, 1479, 1480, 1482, <u>1538</u>, 1645, 1838, <u>1839</u>, 1904, 1919, 1920, 2020, 2021, 2022, 2024, 2104, 2234, 2236, 2237, 2280, 2328, <u>2340</u>, <u>2387</u>, 2408, 2559, <u>2589</u>, <u>2590</u>, <u>2620</u>, 2629, <u>2677</u>, <u>2703</u>, 2874, 3064, 3149, 3150, <u>3151</u>, <u>3152</u>, 3155, <u>3236</u>, <u>3253</u>, 3339, 3341, 3501, 3523, 3527, 3534, 3591, 3617, <u>3654</u>, <u>3721</u>, <u>3722</u>, 3749, 3799, 3800, 3864, <u>3868</u>, <u>4161</u>, 4162, 4165, <u>4167</u>, <u>4170</u>, <u>4172</u>, 4193, 4265, 4379, 4405, 4521, <u>4578</u>, 4609, 4666, <u>4744</u>, 4859, 4862, <u>5146</u>, <u>5284</u>, <u>5304</u>, <u>5345</u>, <u>5346</u>, 5347, <u>5348</u>, 5474, 5590, 5807
mathematical models, 118, 814, 1011, 1566, 1845, 2086, 2722, 2725, 2837, 2870, 3255, 3588, 3617, <u>3819</u>, <u>4337</u>, <u>4560</u>, 4772, 4787, 4792, 4918, <u>5093</u>, <u>5096</u>, 5317
matrix models, 881, <u>5002</u>
maximal covering problem, 191
meadows, 1307, 1738, 2364, 5711
measurement. *See* biodiversity measurement
measures, diversity. *See* diversity indices
measures, similarity, 787, 2290, <u>2392</u>
medicinal plants. *See* 'plants, medicinal' in Index III

medicine, traditional [/indigenous], 170, 252, 254, 643, 5387
Mediterranean-climate ecosystems (*see also* chaparral, fynbos), 948, 1041, 1073, 1195, 1266, 1332, 1939, 1960, 2214, 2420, 2525, 2876, 3042, 3295, 3469, 3475, 3727, 3824, 4277, 4378, 4394, 5147, 5685
megafauna, 367, 1126, 1279, 3149, 3799, 3800
Merck, 487, 1054, 1545, 1799, 2313, 2504, 3841, 4309, 5686
Mesozoic, 885, 1483
metapopulations, 319, 1818, 2061, 2062, 2063, 2064, 2065, 2066, 2067, 2070, 2071, 2072, 2100, 2113, 2114, 2188, 2790, 3202, 3203, 3226, 3227, 3758, 3760, 4325, 4511, 4669, 5050
Michaelis-Menten model, 2578
microbial communities, 1707, 2079, 3019, 3834, 5123, 5342, 5534, 5680
microclimate, 3606, 3909
microcosms, 130, 880, 1674, 2537, 2538, 3545, 3547, 3549, 3602
microorganisms. *See* listings in Index III
migration, animal. *See* 'animal migration' in Index III
migratory birds. *See* 'birds, migratory' in Index III
military lands, 497, 1290, 1357, 2655, 4296, 5121, 5291
mineralogy, 305, 4089
minimum viable population size (*see also* extinction risk), 2071, 2801, 3695, 4243, 4596, 4598, 4812
Miocene, 95
Mississippian, 2526, 5371
Mittermeier, Russell, 2692
model, Michaelis-Menten, 2578
models, equilibrium. *See* equilibrium models
models, mathematical. *See* mathematical models
models, matrix, 881, 5002
models, nonequilibrium. *See* nonequilibrium models
models, rule-based, 4723, 4929
models, spatially explicit. *See* spatially explicit models
models, stochastic. *See* stochastic models
molecular approaches, 107, 109, 149, 177, 241, 518, 519, 548, 660, 808, 1028, 1204, 1255, 1380, 1399, 1462, 1529, 1573, 1885, 1893, 1992, 2155, 2295, 2501, 2545, 2546, 2547, 2662, 2765, 3126, 3208, 3366, 3445, 3450, 3452, 3708, 3803, 3855, 3895, 4043, 4331, 4651, 4756, 5021, 5134, 5282, 5342, 5516, 5573, 5837, 5838
molecular evolution, 548, 1885, 2662, 4651
molecular phylogenies, 1573, 1893, 2765, 3803, 4651
monitoring, bd [/ecological]. *See* biodiversity monitoring
monoculture, 1741, 4629, 5799
montane environs, 184, 231, 296, 631, 860, 900, 1220, 1264, 1453, 1463, 1473, 1528, 1600, 1821, 1874, 1920, 1985, 2012, 2035, 2163, 2294, 2424, 2438, 2470, 2556, 2557, 2597, 2650, 2827, 2831, 2883, 2938, 2967, 2968, 3420, 3831, 3900, 4037, 4041, 4136, 4211, 4372, 4411, 4495, 4526, 4603, 4710, 4722, 4841, 4892, 4894, 4895, 5052, 5064, 5194, 5271, 5301, 5440, 5442, 5477, 5625, 5665, 5670, 5692, 5699
montane mammals, 631, 1920, 2967, 2968, 4722
Monte Carlo methods, 3588, 3589, 5435
morphological diversity, 655, 940, 1072, 1628, 1629, 1630, 1632, 1633, 1877, 2911, 3633, 3635, 3654, 4305, 4313, 4380, 4619, 4637
mortality studies, 759, 1008, 2125, 5055
mosaics, landscape. *See* landscape mosaics
MRP, 4451
multiple-use, 2057, 2937, 3210, 3685, 3744, 5478, 5505
multivariate techniques, 119, 382, 1471, 1514, 1830, 4752
museums, 59, 230, 241, 1198, 1355, 1365, 1715, 2723, 3629, 4024, 4078, 4245, 4262, 4541, 4593
mutualism. *See* symbiosis
Naeem, Shahid, 130, 2537
NAPAP, 4497, 5783
National Audubon Society, 1619
National Environmental Policy Act, 350, 1061, 4821
National Forest Management Act, 3806, 5390
National Park Service, 1339, 1967
national parks and forests, 82, 116, 20, 306, 316, 557, 645, 833, 867, 915, 1191, 1218, 1221, 1245, 1339, 1528, 1565, 1651, 1800, 1932, 1964, 1967, 2406, 2512, 2778, 2934, 2973, 2988, 3047, 3242, 3271, 3278, 3505, 3603, 3619, 3661, 3806, 3830, 4093, 4433, 4517, 4624, 4649, 4894, 4895, 4896, 5390,

5413, 5440, 5633, 5653, 5711
National Plant Germplasm System, 2421, 3566, 5189
National Research Council (NRC), 802
National Systematics Laboratory (NSL), 5253
natural capital, 1044, 1048, 1896, 3605, 3915
natural clones, 2832, 5300
natural fluctuations, 5, 394, 531, 923, 1027, 1786, 2028, 2569, 3892, 3893, 4108, 4370, 4446, 4513, 5283, 5587, 5591, 5636, 5639, 5814
Natural Heritage Program, 1223, 1958, 4728, 4898
natural history, 362, 1051, 1198, 1355, 1603, 1715, 1725, 1925, 2427, 2666, 2882, 3057, 3228, 3351, 3481, 4055, 4078, 4541, 5038, 5319, 5321
natural products (*see also* forest products), 137, 194, 240, 246, 269, 281, 390, 788, 821, 1135, 1241, 1437, 1729, 2450, 2478, 2510, 2753, 2807, 3219, 4569, 5819
natural resource management, 171, 282, 434, 522, 751, 884, 1216, 1218, 1237, 1521, 1828, 1982, 2335, 2551, 2624, 2630, 2635, 2675, 3005, 3242, 3579, 4094, 4102, 4821, 4890, 4971, 5061, 5066, 5162, 5257
natural selection (*see also* sexual selection), 257, 774, 1182, 2086, 2764, 2879, 4222, 4331, 5322, 5543
nature attitudes. *See* environmental perception
Nature Conservancy, 1957, 1958, 2265, 3678, 4282, 4479
Nature Conservancy Council (U.K.), 5201
nature conservation, 399, 458, 576, 688, 1288, 1289, 1389, 1414, 1570, 1744, 2074, 2077, 2222, 2490, 2555, 3119, 3141, 4031, 4078, 4090, 4203, 4458, 4467, 4468, 4469, 4471, 4688, 4819, 4834, 5037, 5106, 5201, 5218, 5494, 5581, 5759, 5774, 5813
nature [/ecological/environmental] philosophy, 81, 746, 764, 771, 1296, 1330, 2595, 2596, 2598, 2887, 3550, 3551, 3667, 3668, 3956, 4351, 4352, 4972, 4991, 5425, 5548, 5558
nature reserves (*see also* biosphere reserves, national parks, protected areas), 162, 243, 716, 988, 1175, 1418, 1490, 1878, 2455, 2460, 2534, 2668, 2671, 2808, 2963, 2975, 3040, 3042, 3123, 3930, 3954, 4318, 4414, 4420, 4516, 4594, 4812, 4929, 5121, 5198,
5200, 5327, 5615
nearshore environs. *See* intertidal environs
NEPA, 796
nestedness, 501, 1025, 1145, 1146, 1646, 2875, 4677
neural networks, 3171, 5384
New Forestry, 1817, 3770, 4957
new species, 2351, 4491
niche relationships, 214, 376, 1580, 1582, 1962, 2345, 2526, 2563, 2694, 3033, 5006, 5113, 5317
nitrogen [/nitrogen cycle], 39, 161, 2580, 3418, 4734, 5115, 5286, 5383
nocturnality, 1160, 1373, 2567
nonequilibrium models, 406, 631, 2174, 2332, 2552, 4086, 4221, 4816, 5034
nongovernmental organizations (NGOs), 151, 192, 1988, 3323, 4773, 5408
Noosphere, 1733, 4585
North American Breeding Bird Survey, 2415
North American Wetlands Conservation Act, 1921
North-South relations, 298, 848, 1131, 2329, 2773, 2893, 3082, 3279, 3778, 3827, 3884, 4839, 4963, 5085, 5109
null hypotheses, 4731, 5629
numbers, insect. *See* 'insect diversity' in Index III
nutrient upwelling, 4336, 4427
nutrients and nutrient cycling, 391, 938, 1036, 1069, 1659, 1673, 2029, 2303, 2318, 2333, 2468, 2560, 2863, 2979, 3298, 3857, 4204, 4336, 4664, 4716, 4788, 4800, 5093, 5094, 5136, 5285
nutritional biodiversity, 3160
ocean currents, 4336, 4427
Office of Technology Assessment, 1447, 4617
old fields, 183, 3496, 4522, 5135
old-growth forests, 597, 618, 731, 1154, 1197, 1364, 1682, 1683, 1691, 1894, 2076, 2743, 2776, 2817, 2824, 3410, 3658, 3726, 3728, 3770, 3806, 4495, 5062, 5297, 5331, 5407, 5631
Oligocene, 4095
OPOSA decision, 4254
optimization procedures, 310, 329, 434, 1845, 3320, 5580
orbital forcing, 1100, 2323
Ordovician, 47, 876, 1628, 3338, 3340
Organization for Tropical Studies, 4996

origin of cells, 837, 2108
origin of life, 901, 3128, 3129, 3243, 3767, 3803
origination-extinction dynamics, 547, 1813, 2389, 3069
outbreeding depression, 2156
overgrazing. *See* grazing
overkill hypothesis. *See* Pleistocene overkill
oxygen, 67, 424, 1553, 1885, 1902, 2906, 2957, 3697, 5474
ozone, 161, 196, 757, 1203, 4308, 4754
paleobiogeography and paleobiology: misc. subjects, 78, 95, 118, 175, 265, 367, 381, 408, 546, 547, 589, 590, 593, 595, 794, 814, 854, 876, 921, 1469, 1616, 1617, 1618, 1628, 1629, 1630, 1632, 1679, 1813, 1858, 1889, 2022, 2023, 2224, 2390, 2391, 2398, 2407, 2422, 2523, 2704, 2713, 2889, 2926, 3146, 3147, 3264, 3338, 3340, 3341, 3627, 3654, 3749, 3789, 3799, 4059, 4095, 4158, 4159, 4160, 4167, 4168, 4170, 4171, 4193, 4255, 4368, 4379, 4575, 4576, 4577, 4578, 4579, 4581, 4587, 4696, 4872, 5004, 5010, 5210, 5212, 5217, 5260, 5261, 5262, 5273, 5328, 5329, 5373, 5426, 5452, 5574, 5807, 5853
Paleocene, 175, 1679, 2620
Paleozoic, 1483, 1629, 4581, 4862, 5305
panbiogeography, 1098, 1936, 5846
parasite communities, 727, 1749, 2617, 2618, 2619, 4032, 4033, 4034, 4035, 4930
parasites and parasitism, 45, 71, 368, 384, 385, 530, 727, 992, 1037, 1299, 1306, 1488, 1500, 1560, 1578, 1740, 1935, 1976, 1992, 2110, 2124, 2125, 2128, 2130, 2224, 2225, 2258, 2330, 2469, 2617, 2618, 2619, 2739, 2832, 2896, 3215, 3235, 3442, 3694, 4015, 4032, 4033, 4034, 4035, 4133, 4248, 4325, 4463, 4679, 4844, 4930, 5054, 5055, 5337, 5338, 5355, 5365, 5389, 5466, 5538, 5565, 5632
parasitic [/galling] insects, 1037, 1560, 2128, 3235, 4879, 5466, 5632
park-people relations, 1651, 2714, 3603, 3830, 4415, 4649, 5401, 5601
patch size, 2735, 2796, 4495, 5224, 5525
patches and patch dynamics (*see also* habitat fragmentation, habitat fragments, landscape mosaics, patch size), 243, 472, 719, 1004, 1159, 1505, 1506, 1810, 2062, 2063, 2064, 2067, 2100, 2113, 2188, 2483, 2540, 2735, 2796, 3035, 3184, 3317, 3375, 3685, 3759, 3954, 3955, 4363, 4490, 4495, 4639,

4814, 5224, 5302, 5364, 5467, 5525, 5640
patents and patenting, 24, 421, 723, 1118, 1150, 2088, 2329, 2519, 2520, 2521, 3263, 4301, 4631, 4632
pathogens. *See* disease
peatlands and bogs, 1833, 2857, 3633, 4290, 4291, 5741
pedigree breeding, 5024
peninsular effects, 2838, 3376, 4473, 4989, 5013
periodic extinction, 3864, 4143, 4173, 4174, 4580, 5785
Permian and Permo-Triassic boundary, 87, 1231, 1476, 1477, 1478, 1479, 2022, 2703, 3155, 3749, 4161, 4255, 4521, 4666, 5284, 5474
persistence, 413, 477, 1007, 1805, 1813, 2026, 2072, 3096, 3097, 4802, 4850, 5050, 5426
pesticides, 247, 550, 820, 1577, 1624, 2637, 2782, 4421
pests and pest control, 96, 98, 99, 530, 749, 922, 1298, 1806, 2789, 3114, 3344, 3495, 4297, 4702, 4930, 5226, 5330, 5335, 5367, 5798
pests, insect, 96, 99, 922, 1613, 3344, 4920, 5330, 5367
pH, 1694, 4129
Phanerozoic, 87, 265, 409, 424, 533, 546, 1064, 1115, 1618, 2023, 2235, 3339, 3340, 3465, 3636, 4157, 4159, 4160, 4167, 4168, 4170, 4575, 4576, 4577, 4578, 4584, 4657, 4658, 5210, 5213, 5214, 5329, 5644
Phanerozoic diversity, 1618, 3340, 4575, 4576, 4577, 4578, 4584, 5213, 5217
pharmaceutical industry, 788, 1729, 2519, 3841, 4701, 5686
phenotypic diversity, 608, 5124, 5414
philosophy, nature. *See* nature philosophy
phyla, 838, 1462, 2632, 3130, 5214
phyletic gradualism, 1889, 3253, 5374
phylogenetic diversity, 1507, 1510, 1511, 1556, 1707, 2741, 3452, 5316, 5586
phylogenetics, 175, 241, 611, 1081, 1090, 1116, 1117, 1258, 1507, 1510, 1511, 1556, 1563, 1573, 1631, 1707, 1906, 1907, 2177, 2192, 2295, 2741, 2742, 3023, 3425, 3452, 3707, 3803, 4451, 4582, 4731, 5316, 5489, 5490, 5586, 5706, 5707
phylogenies, molecular, 1573, 1893, 2765, 3803, 4651
phylogeny, 612, 885, 1049, 1117, 2302, 2741, 3425, 3692, 3838, 3866, 4032, 4035,

4175, 5252, 5434
phylogeography, 3984, 4331, 4984, 3452
physiology, 351, 543, 921, 1902, 2394, 3014, 4313
Pielou's index, 378
Pittman Robertson Act, 5589
planning, environmental. *See* environmental planning
planning, goal-oriented. *See* goal-oriented planning
planning, habitat conservation. *See* habitat conservation
planning, landscape, 32, 4355, 4890
plantations, 894, 1041, 1734, 1814, 1924, 3847, 3908, 3909, 3910, 4155, 4371, 4603, 4836, 4968, 4969, 5000
plants. *See* listings in Index III
plate tectonics and continental drift, 589, 794, 1616, 1648, 2103, 4361, 4521, 5211, 5225
playas, 2118
Pleistocene, 367, 720, 1278, 1786, 1904, 1919, 2689, 3149, 3152, 3293, 3799, 3800, 4059, 5216, 5348, 5374, 5484, 5574
Pleistocene overkill hypothesis, 1919, 2570, 3149, 3150, 3151, 3152
Pliocene, 1100, 2398
poaching, 610, 2778, 3757, 4902, 5111
polar wandering, 4875
political ecology, 646, 680, 4327, 5803
political economy, 124, 1337, 2394, 2696, 4701
political factors, 124, 192, 205, 441, 521, 571, 646, 680, 871, 873, 961, 1056, 1147, 1311, 1337, 1647, 1800, 2331, 2394, 2440, 2679, 2692, 2696, 2705, 2763, 3238, 4101, 4176, 4181, 4327, 4415, 4418, 4633, 4647, 4701, 5101, 5112, 5310, 5341, 5394, 5776, 5803
pollen, 1204, 2322
pollination [/pollinators], 85, 684, 1079, 2573, 2583, 2637, 2940, 5025
pollution [/pollutants], 16, 161, 295, 773, 820, 890, 905, 923, 949, 952, 1577, 1635, 2178, 2207, 2381, 2518, 2759, 3217, 3269, 3418, 3490, 3698, 3738, 4424, 4455, 5060, 5137, 5380, 5395, 5769
polynyas, 657, 1656
ponds (*see also* lakes), 374, 1321, 1580, 1694, 2026, 2151, 2819, 4574, 5277
population biology (*see also* metapopulations, population dynamics), 492, 710, 1421, 1506, 2026, 2066, 2172, 2261, 2630, 2972, 3026, 3186, 3250, 3857, 4243, 4318, 4560, 4639, 5285, 5467, 5526
population bottlenecks, 177, 237, 1450, 2156, 2755, 3597, 3707, 3709, 3710, 4161, 5326
population density. *See* species density
population diversity, 1423, 2296, 3415
population dynamics (*see also* metapopulations, population biology), 1008, 1384, 2064, 2841, 3193, 4099
population genetics, 1665, 1666, 1667, 2546, 3695, 3708, 3984, 5326, 5555, 5655
population growth, human, 157, 720, 875, 1158, 1426, 1435, 2083, 4181, 4718
population size, 413, 1158, 1458, 1506, 1666, 1767, 2071, 2209, 2801, 3695, 3696, 4266, 4513, 4596, 4598, 5339
population size, minimum viable. *See* minimum viable population size
population viability analysis, 56, 549, 2071, 2785, 2796, 2801, 2930, 3309, 3695, 4243, 4391, 4597, 4812, 5008, 5021, 5674, 5694, 5756
populations, bird. *See* 'bird populations' in Index III
populations, mammal, 1908, 2785, 4956
populations, small. *See* small populations
populations, unisexual, 2748
power plants, 2998, 3745, 5471
prairie (*see also* grasslands), 984, 985, 986, 1727, 1841, 2042, 2109, 2284, 2538, 2730, 2851, 2884, 3072, 3342, 4439, 4440, 4522, 4675, 5434, 5705
prairie communities. *See* grassland communities
pre-1950 studies, 75, 76, 188, 331, 420, 934, 1182, 1282, 1593, 1836, 1837, 2344, 2392, 2881, 2882, 3144, 3314, 3315, 3481, 3978, 4076, 4692, 5003, 5318, 5319, 5320, 5321, 5322, 5535
precautionary principle, 897, 3528, 5099
precipitation [/rainfall] effects, 21, 2178, 3797, 3890, 4956, 21, 938, 3623
predation [/predators and prey], 974, 1406, 1993, 2152, 2318, 2447, 2663, 2764, 2869, 3114, 3178, 3307, 3399, 3813, 4007, 4514, 4704, 4840, 5029, 5131, 5268, 5333
preservation, forest. *See* forest conservation
preservation, species. *See* species conservation
primary [/ecological] productivity, 18, 19, 21, 29, 49, 604, 1006, 1416, 1741, 1857, 2462, 2537, 2870, 2924, 3378, 3546, 3796,

3798, 3949, 4009, 4396, 4710, 4831, 4975, 5086, 5090, 5093, 5094, 5696
prioritization, conservation. *See* conservation prioritization
private lands and landowners, 48, 399, 619, 964, 1536, 1958, 2310, 3101, 3711, 3729, 4002, 4557, 4558, 5482
private sector. *See* corporations
privatization, 1537, 5292, 5492
productivity, agricultural. *See* agricultural productivity
productivity, primary [/ecological]. *See* primary productivity
productivity-diversity relations, 18, 19, 21, 29, 1006, 2537, 2870, 4396, 5094
prospecting, bd. *See* biodiversity prospecting
prospecting, chemical. *See* chemical prospecting
protected areas (*see also* biosphere reserves, national parks, nature reserves), 50, 51, 238, 288, 450, 558, 570, 571, 576, 585, 810, 828, 844, 966, 1218, 1237, 1291, 1301, 1302, 1331, 1357, 1385, 1461, 1518, 1713, 1718, 1800, 2019, 2159, 2210, 2305, 2358, 2361, 2370, 2374, 2376, 2410, 2448, 2512, 2586, 2609, 2613, 2633, 2641, 2668, 2678, 2714, 2738, 2745, 2760, 2853, 2915, 2962, 3021, 3058, 3060, 3061, 3106, 3202, 3271, 3273, 3274, 3278, 3283, 3438, 3439, 3491, 3580, 3603, 3618, 3637, 3780, 3830, 3933, 3945, 3965, 4071, 4083, 4185, 4415, 4480, 4481, 4594, 4598, 4770, 4773, 4819, 4986, 4987, 5078, 5101, 5291, 5397, 5402, 5620, 5633, 5645, 5671, 5692
Proterozoic, 2957, 4520, 5273
provinces and provinciality, 4381, 5169, 5211, 5217
public awareness. *See* environmental perception
public health. *See* human health
public lands, 497, 1032, 1419, 2262, 2542, 2584, 2639, 3106, 3665, 5598
punctuated equilibria, 1889, 5374
quadrat methods, 123, 1000, 4153
Quaternary (*see also* individual epochs), 403, 1027, 1233, 1279, 2320, 2397, 3151, 3433, 5452, 5608
rainfall. *See* precipitation effects
rainforest mammals. *See* 'mammals, forest' in Index III
rainforests, temperate [/subtropical], 980, 1650, 2530, 2603, 4519, 5503
rainforests, tropical. *See* tropical rainforests

ranching. *See* rangelands and ranching
range boundaries (*see also* ecotones), 1650, 3559, 4356, 4571, 5514
range collapse, 2965, 2966
range, geographical. *See* geographical range
range size, 121, 489, 582, 1655, 1752, 1754, 1763, 1767, 2068, 2842, 2846, 2969, 4880, 5532
range size-abundance relations, 489, 2068, 2842
rangelands and ranching, 1307, 1419, 1610, 2325, 2611, 2805, 2914, 3311, 3378, 3649, 3916, 4090, 5057, 5197, 5313, 5410, 5412, 5879
RAPD, 5281, 5282
rapid assessment, 360, 784, 1096, 1159, 1369, 2939, 3611, 3742, 3743, 4037, 4526, 4527, 4736, 4894, 5350
Rapoport's rule, 991, 1655, 4132, 4345, 4381, 4880, 4881, 4882
rare birds. *See* 'birds, endemic' in Index III
rare fishes. *See* 'fishes, endemic' in Index III
rare mammals. *See* endemic mammals
rare plants. *See* 'plants, rare' in Index III
rare species, 25, 138, 179, 182, 332, 473, 833, 999, 1246, 1361, 1448, 1458, 1514, 1524, 1525, 1756, 1774, 1852, 1955, 1979, 2097, 2157, 2316, 2531, 2556, 2633, 2666, 2761, 2834, 2905, 3045, 3168, 3212, 3251, 3296, 3300, 3310, 3361, 3362, 3379, 3396, 3434, 3460, 3818, 4062, 4069, 4076, 4081, 4115, 4492, 4534, 4616, 4728, 4802, 4840, 4948, 4976, 5025, 5050, 5052, 5068, 5132, 5268, 5311, 5354, 5385, 5429, 5442, 5521
rarefaction analysis, 1629, 2414, 4158, 4450
rarity, 182, 805, 833, 1246, 1361, 1756, 1979, 2060, 2316, 2527, 2556, 2633, 2770, 3859, 4076, 4077, 4948, 5521, 5527
rates, bd loss [/decline], 2852, 4808, 5051, 5339, 5479
rates, colonization. *See* colonization rates
rates, deforestation, 1542, 3521
rates, diversification. *See* evolutionary rates
rates, evolutionary. *See* evolutionary rates
rates, extinction. *See* extinction rates
rates, immigration. *See* colonization rates
rates, recruitment. *See* colonization rates
rates, speciation. *See* evolutionary rates
rates, turnover, 637, 3948, 4510, 5636
recolonization. *See* colonization
recovery plans. *See* action plans
recreation activity impacts, 617, 1415, 1868,

Index I 435

2309, 2700, 2775, 2920
recruitment, 119, 812, 1008, 2425, 2428, 3732, 3948, 5089
recruitment rates. *See* colonization rates
Red Queen hypothesis, 4288, 5234
redundancy, 924, 1432, 1822, 3544, 4354, 5312, 5314
reforestation. *See* afforestation
refugia, 403, 728, 729, 812, 868, 1113, 1375, 1465, 1600, 1996, 1997, 2040, 2125, 2130, 2225, 2501, 3293, 3294, 3910, 4342, 4567, 5125, 5260, 5263, 5448
regional diversity, 760, 1179, 1770, 1914, 4314
regional richness. *See* local richness
regional-scale studies and effects, 66, 202, 204, 260, 335, 449, 575, 714, 760, 881, 984, 1038, 1039, 1074, 1179, 1358, 1512, 1519, 1528, 1605, 1688, 1755, 1770, 1771, 1833, 1840, 1914, 1943, 2059, 2146, 2209, 2435, 2507, 2543, 2608, 2618, 3179, 3220, 3283, 3354, 3360, 3361, 3537, 3538, 3554, 3579, 3676, 3793, 4072, 4132, 4285, 4314, 4325, 4469, 4472, 4736, 4768, 5119, 5128, 5151, 5224, 5280, 5283, 5315, 5462, 5584
reintroduction, 83, 84, 561, 1450, 1526, 1527, 1702, 1820, 2001, 2683, 2966, 3039, 3183, 3399, 4586, 4642, 5598
relaxation times, 1274
relicts, 1102, 5407
remnants, habitat. *See* habitat fragments
remote sensing, 53, 89, 461, 514, 747, 870, 1215, 1521, 1572, 1601, 2009, 2363, 2500, 2698, 2833, 2834, 3090, 3312, 3517, 3553, 3556, 3648, 3998, 4013, 4057, 4436, 4549, 4725, 4905, 4929, 5084, 5148, 5293, 5406
renewable resources, 907
representativeness. *See* reserve representativeness
reproduction, 65, 1142, 2818, 4115, 4566, 4702, 4934
reptiles. *See* listings in Index III
research needs, programs and agenda, 11, 13, 14, 93, 131, 241, 275, 315, 414, 680, 736, 797, 1019, 1085, 1192, 1355, 1374, 1433, 1549, 1674, 2030, 2165, 2218, 2679, 2793, 3002, 3051, 3076, 3169, 3335, 3411, 3527, 3562, 3567, 3573, 3737, 3888, 3917, 4282, 4310, 4407, 4475, 4707, 4733, 4759, 4778, 4779, 4780, 4807, 4810, 4868, 4905, 4997, 5265, 5409, 5416, 5783, 5832
reserve design and selection, 191, 224, 243, 370, 372, 382, 729, 1012, 1121, 1146, 1189, 1275, 1518, 1649, 1689, 1690, 1840, 2099, 2174, 2227, 2281, 2358, 2460, 2612, 2627, 2633, 2634, 2671, 2672, 2772, 2796, 2884, 2960, 2963, 3123, 3124, 3247, 3361, 3500, 3835, 3954, 4042, 4070, 4071, 4073, 4080, 4113, 4231, 4318, 4414, 4420, 4516, 4598, 4672, 4812, 4929, 5202, 5228, 5327, 5585, 5586, 5587, 5615
reserve networks, 372, 402, 1498, 1608, 1690, 2045, 2641, 2963, 3124, 4871, 5671
reserve representativeness, 372, 1264, 2586, 3247, 3637, 3747, 3933, 4070, 4897, 5671
reserve selection. *See* reserve design
reserve size (*see also* reserve design, SLOSS), 1966, 2174, 2428, 4595, 5042, 5639
reserves, biosphere. *See* biosphere reserves
reserves, marine. *See* marine reserves
reserves, nature. *See* nature reserves
resilience, 1177, 2244, 2250, 3399, 3916, 4742, 5314
resource allocation, 2852, 4003, 4876
Resource Management Act, 522
restoration, ecological. *See* ecological restoration
restoration ecology. *See* ecological restoration
Return a Gift to Wildlife, 164
richness, character, 833
richness, higher-taxon, 113, 260, 261, 1117, 1776, 1779, 5515, 5517
richness index, hierarchical, 1693
richness, insect. *See* 'insect diversity' in Index III
richness, invertebrate. *See* 'invertebrate diversity' in Index III
richness, local. *See* local richness
richness, regional. *See* local richness
richness, species. *See* species richness
richness vs. diversity, 1693
richness-energy relations, 1658, 1675, 2238, 2628, 2821
riparian habitats [/zones], 42, 704, 725, 1307, 1588, 1610, 1891, 2048, 2667, 2706, 2708, 2781, 3089, 3293, 3294, 3322, 3554, 3638, 3639, 3985, 4009, 4957, 4980, 5141
riparian plants, 1821, 1891, 3638, 3639, 4009, 4828, 4980
risk assessment, 711, 1333, 2650, 5435, 5485, 5580
river flow regulation, 978, 1396, 2908, 4044, 4855

rivers (*see also* streams), 69, 331, 344, 426, 520, 534, 574, 787, 789, 890, 1329, 1396, 1821, 1891, 1892, 2300, 2311, 2452, 2483, 2806, 3027, 3170, 3171, 3638, 3639, 3702, 3703, 3704, 3804, 3861, 3926, 4044, 4246, 4430, 4455, 4855, 5136, 5205, 5296, 5344, 5471, 5480, 5766
rivet-popper hypothesis, 1432
RNA, 107, 1573, 1706, 1893, 3767, 3895, 3897, 4029, 5342
roads and transportation corridors, 72, 291, 759, 1531, 1576, 1923, 2180, 3070, 3290, 4448, 4820, 4940
Rosenzweig, Michael, 5696
rule, Allen's, 76, 4187
rule, Bergmann's. *See* Bergmann's rule
rule, Gloger's, 4149
rule, Rapoport's. *See* Rapoport's rule
rule-based models, 4723, 4929
rural environs, 17, 58, 101, 311, 432, 936, 1118, 1543, 1713, 1809, 2440, 2456, 3788, 3830, 4600
sacred groves, 858, 1225, 4711, 5108
safe minimum standard, 426, 455, 456, 4147, 4198, 4398, 5102
sampling, 330, 379, 501, 760, 784, 790, 818, 910, 1011, 1319, 1459, 1504, 1593, 1632, 1804, 1912, 1935, 2233, 2393, 2535, 2712, 2713, 2749, 2875, 2903, 2971, 2980, 3077, 3078, 3212, 3310, 3340, 3588, 3589, 3635, 3950, 3959, 4153, 4244, 4441, 4613, 4614, 4762, 4772, 4829, 4933, 4980, 5279, 5337, 5338, 5360, 5371
sampling bias, 87, 381, 991, 1459, 4063, 4193
sampling intensity, 1319, 3310, 4980
sand and dunes, 132, 630, 1321, 1953, 2452
savannas, 970, 2835, 3040, 3293, 3294, 3298, 4757, 4785, 5815
scale: general and misc. subjects (*see also* spatial scale), 80, 139, 328, 457, 580, 719, 758, 881, 1045, 1074, 1140, 1233, 1369, 1764, 1766, 1767, 1771, 1779, 1891, 2157, 2390, 2706, 2903, 2907, 2967, 3159, 3237, 3632, 3685, 4259, 4370, 4378, 4379, 4391, 4399, 4535, 4640, 4869, 5163, 5369, 5397, 5406, 5809
scale, spatial. *See* spatial scale
SCOPE, 323, 1194, 4780
Scottish Natural Heritage, 2045
sea ice, 543, 2566, 5427
sea-level change, 301, 430, 590, 591, 876, 1172, 2021, 2022, 3291, 3843, 4234, 4521, 5216, 5574, 5608
seasonal patterns [/diversity], 112, 2871, 4127, 5157
seed banks, 428, 733, 1340, 1692, 2036, 3539, 3585, 4522
seed dispersal, 3175, 4021, 4088, 5436
seedling survival, 3205
seeds, 40, 185, 388, 428, 733, 1340, 1692, 2036, 2511, 2695, 2696, 2983, 3054, 3056, 3175, 3539, 3585, 4021, 4088, 4522, 4633, 4636, 4864, 4886, 5183, 5436, 5518
Seeds of Hope, 3056, 4838
self-organizing systems, 2113, 3922, 5464
sequences, gene. *See* gene sequences
set-aside lands, 1676
sexual selection (*see also* natural selection), 3398, 4288, 4564
Shannon index, 378, 2972, 3078, 3960, 5129
shifting cultivation (swidden agriculture), 141, 1283, 1708, 1709, 2493, 2496, 2560, 3055
sighting data, 3135, 4791
Silurian, 2622, 4193
silviculture, 1242, 3741, 4085, 5775
similarity measures, 787, 2290, 2392
Simpson index, 3078, 4694
simulations, 316, 330, 340, 377, 1714, 2188, 2563, 2925, 3184, 3420, 3588, 3589, 3820, 4441, 5435, 5543
single-species approach, 1571, 2060, 2160, 4670
single-tree forestry, 5290
slash and burn. *See* shifting cultivation
SLOSS, 1256, 2433, 2963, 3287, 4589, 4672, 4812
small mammals, 397, 1225, 3139, 3395, 3649, 4571, 4981, 5446
small populations, 109, 256, 835, 1458, 2060, 2211, 2755, 2801, 2885, 3026, 4243, 4850, 4856, 5035
social networks, 1267, 1784, 2820, 4119, 4468, 4689
Society of American Foresters, 155, 4775
socioeconomic factors, 1802, 1870, 3047, 3245, 3333, 3651
sociology, 2594, 3046, 3079
soil, 132, 149, 266, 356, 391, 423, 486, 494, 518, 519, 608, 629, 673, 674, 733, 938, 979, 1108, 1230, 1308, 1383, 1575, 1659, 1673, 1677, 1678, 1785, 1816, 1859, 1923, 2013, 2079, 2112, 2115, 2170, 2276, 2333, 2336, 2468, 2548, 2615, 2835, 2836, 2862,

Index I 437

2863, 2868, 2977, 2979, 3019, 3141, 3801, 3833, 3834, 3846, 3903, 3924, 3927, 4000, 4028, 4283, 4335, 4424, 4656, 4734, 4788, 4851, 5093, 5094, 5122, 5123, 5124, 5137, 5332, 5353, 5445, 5534, 5561, 5616, 5657, 5678, 5679, 5680, 5731, 5732, 5760, 5793, 5860

soil communities, 494, 608, 629, 1677, 1816, 2079, 2862, 3019, 3834, 5123, 5534

soil fauna, 132, 149, 391, 423, 629, 1108, 1308, 1575, 1816, 2170, 2977, 4656, 5332, 5353, 5678, 5732, 5760

soil fertility, 423, 2977, 3141, 4788, 5616, 5760

soil microorganisms. *See* 'microorganisms, soil' in Index III

spatial patterns, 44, 463, 464, 817, 1384, 1471, 1758, 2110, 2143, 3410, 4697, 5160, 5193, 5629

spatial scale, 139, 471, 495, 641, 760, 818, 1138, 1172, 1233, 1841, 2064, 2144, 2543, 2735, 2978, 3159, 3375, 3669, 3801, 3823, 3834, 3922, 4370, 4531, 4904, 5051, 5280, 5371, 5465, 5469, 5560

spatially explicit models, 319, 1384, 2070, 4099, 5162

speciation (*see also* evolution, diversification), 262, 307, 308, 583, 614, 728, 1068, 1081, 1086, 1395, 1481, 1530, 1646, 1763, 1909, 1996, 1997, 2148, 2192, 2224, 2689, 2764, 3408, 3627, 3792, 3808, 3866, 4331, 4583, 4588, 4691, 4732, 5144, 5259, 5273, 5858, 5863

speciation rates. *See* evolutionary rates

species abundance, 135, 183, 203, 465, 471, 489, 634, 642, 733, 826, 936, 984, 985, 1050, 1139, 1229, 1242, 1283, 1469, 1514, 1539, 1771, 1804, 1816, 1933, 2038, 2039, 2110, 2175, 2298, 2414, 2637, 2841, 2842, 2869, 3029, 3030, 3142, 3184, 3187, 3463, 3592, 3593, 3623, 4011, 4076, 4077, 4081, 4292, 4357, 4397, 4448, 4494, 4556, 4560, 4561, 4571, 4654, 4799, 4827, 4937, 4945, 5046, 5113, 5114, 5132, 5224, 5358, 5360, 5365, 5500, 5540, 5560, 5561, 5591, 5592, 5629, 5637

species abundance, plant, 1283, 2039, 2814, 5560, 5561

species accumulation function, 1605, 4772

species boundaries, 45, 4720

species concepts, 906, 1081, 2142, 3091, 3204, 3816, 4348, 4720, 5340, 5707

species conservation [/preservation], 56, 395, 710, 916, 1080, 1393, 1571, 1782, 1942, 2813, 2960, 3074, 3088, 3423, 3454, 3832, 3968, 4002, 4406, 4570, 4596

species, cryptic, 922, 3408

species, cryptogenic, 801

species [/population] density, 123, 148, 634, 1023, 1293, 1908, 2175, 2651, 2667, 3187, 3442, 3809, 4485, 4695, 5533, 5539, 5571

species density, bird. *See* 'birds, species density of' in Index III

species density, mammal. *See* 'mammals, species density of' in Index III

species diversity, 18, 29, 87, 286, 337, 385, 504, 592, 594, 630, 731, 740, 741, 785, 786, 890, 968, 986, 1006, 1049, 1066, 1068, 1072, 1084, 1094, 1100, 1110, 1137, 1154, 1250, 1251, 1261, 1277, 1283, 1340, 1372, 1423, 1453, 1494, 1551, 1580, 1659, 1673, 1749, 1787, 1804, 1835, 1844, 1857, 1871, 1912, 1917, 1934, 1990, 2031, 2093, 2124, 2137, 2168, 2284, 2327, 2332, 2334, 2360, 2402, 2452, 2464, 2467, 2611, 2719, 2798, 2811, 2817, 2833, 2838, 2876, 2914, 2944, 2945, 3000, 3008, 3012, 3031, 3034, 3038, 3188, 3244, 3293, 3307, 3322, 3543, 3549, 3685, 3732, 3789, 3801, 3806, 3810, 3813, 3854, 3899, 3936, 3952, 3953, 3958, 3960, 4052, 4085, 4136, 4159, 4160, 4201, 4284, 4287, 4318, 4368, 4395, 4575, 4619, 4638, 4654, 4661, 4816, 4940, 4968, 4969, 5013, 5027, 5028, 5044, 5129, 5135, 5170, 5220, 5242, 5317, 5427, 5457, 5458, 5539, 5547, 5706

species diversity, bird, 337, 1049, 1084, 3034, 4121, 4201, 4284, 5027, 5044, 5129

species diversity, plant. *See* 'plant species diversity' in Index III

species diversity, tree, 29, 717, 731, 786, 1251, 2425, 3322, 3732, 3958, 4395, 5539

species, endangered. *See* endangered species

species, extinction-prone, 144, 2816, 5026

species, flagship, 3612, 4670

species, focal. *See* focal species

species groups, 4950

species, host. *See* host species

species, introduced. *See* introduced species

species, keystone. *See* keystone species

species lists (*see also* inventories and checklists), 236, 975, 976, 1400, 1504, 2142, 2257, 3583, 4244, 4493, 4948

species loss. *See* biodiversity loss

species numbers, 327, 696, 1129, 1487, 1551, 1552, 1583, 1593, 1614, 1753, 1755,

1768, 1770, 2354, 2425, 2465, 2466, 2470, 2712, 2715, 2729, 2783, 2846, 2913, 3192, 3193, 3194, 3195, 3198, 3199, 3200, 3201, 3383, 3484, 3768, 4052, 4416, 4485, 4622, 4793, 4815, 4922, 4923, 5049, 5210, 5233
species numbers, plant, 1755, 3928, 4682, 4867, 5233
species pools, 1474, 5575, 5709
species preservation. *See* species conservation
species richness and climate, 938, 3705, 3706, 4419
species richness, animal, 1138, 1329, 1341, 1379, 1560, 2841
species richness, ants and termites, 7, 113, 582, 1191, 1144
species richness, arthropod, 1319, 1484, 4204, 4441, 4827, 4840, 5333
species richness, bird. *See* 'bird species richness' in Index III
species richness, butterfly, 366, 817, 1246, 1778, 1985, 4772, 5052
species richness covariance, 3398, 4061
species richness, crustacean, 5, 1325, 1326, 1655, 4014, 4133
species richness, fish. *See* 'fish species richness' in Index III
species richness, fungi, 927, 5389
species richness: general and misc. subjects, 140, 148, 260, 261, 263, 356, 462, 530, 532, 756, 760, 990, 991, 1012, 1039, 1065, 1215, 1319, 1486, 1560, 1652, 1675, 1714, 1764, 1766, 1769, 1810, 1816, 1900, 1914, 1976, 2166, 2281, 2527, 2535, 2627, 2792, 2950, 3023, 3121, 3255, 3388, 3389, 3588, 3640, 3730, 3819, 3821, 3890, 3906, 4061, 4113, 4131, 4153, 4285, 4552, 4741, 4904, 5086, 5337, 5338, 5352, 5465, 5538, 5566, 5645
species richness gradients, 991, 1560, 4131
species richness gradients, latitudinal. *See* latitudinal species richness gradients
species richness, helminth, 727, 1935, 2617, 2618, 2619, 4033
species richness, insect. *See* 'insect species richness' in Index III
species richness, invertebrate, 235, 265, 1038, 2543, 3642, 5279
species richness, mammal. *See* 'mammal species richness' in Index III
species richness, parasitoid, 530, 727, 1935, 1976, 2125, 2128, 2225, 2617, 2618, 2619, 3442, 4032, 4033, 4034, 4035, 4133, 4463,

5338, 5389
species richness, plant. *See* 'plant species richness' in Index III
species richness, reptiles and amphibians, 704, 1539, 2151, 2152, 2483, 3797
species richness, tree. *See* 'tree species richness' in Index III
species richness, vertebrate, 1284, 1576, 3225
species, threatened. *See* endangered species
species, umbrella. *See* umbrella species
species-abundance relations, 2204, 2712, 3360, 3809, 5637
species-age relations, 5484
species-area relations, 146, 530, 687, 1000, 1009, 1256, 1341, 1490, 1749, 2068, 2093, 2182, 2208, 2535, 2604, 2712, 2875, 3156, 3240, 3255, 3360, 3642, 3823, 3970, 4077, 4113, 4300, 4367, 4540, 4575, 4638, 4794, 4826, 4879, 5369, 5442, 5528, 5627, 5635, 5637, 5704
species-energy relations, 1065, 1138, 1139, 1140, 1416, 2821, 3237, 5157, 5158, 5627, 5628, 5645
speciesism (*see also* animal rights), 4408
speciosity and speciose groups, 1269, 1556, 2044, 3164, 3165, 5333
springs and hotsprings, 306, 1475, 4618, 5221, 5342, 5370, 5642
stability, 477, 667, 992, 1006, 1007, 1045, 1103, 1177, 1219, 1305, 1472, 1599, 1600, 1601, 1831, 2250, 2383, 2462, 2552, 2784, 2870, 2959, 2982, 3071, 3251, 3417, 3495, 3814, 3924, 4083, 4120, 4743, 4816, 4971, 5029, 5426, 5597, 5750, 5853
stability-diversity relations. *See* diversity-stability relations
stability-time hypothesis, 1100, 4305, 4450
state laws and regulations, 28, 846, 913, 1794, 2180, 2684
statistical analyses, 118, 696, 1011, 1305, 1561, 2328, 2629, 3939, 5092, 5153, 5533
Steadman, David, 2936
stepwise extinction, 2020, 2340, 2713
stochastic models and processes, 211, 1244, 2725, 2875, 4175, 4731, 5092, 5283, 5639
stocks and stocking, 231, 2046, 2326, 4126, 4503, 5074, 5596
stream stones, 1341, 1346
streams (*see also* rivers), 146, 148, 574, 1219, 1341, 1346, 1403, 1652, 1821, 2297, 2403, 2792, 3027, 3092, 3313, 3389, 4262, 4272, 4502, 4616, 4828, 4855, 4925, 5056,

5125, 5277, 5279, 5280, 5303, 5714
stress, environmental. *See* environmental stress
subspecies, 2142, 4406
suburban environs. *See* urban environs
succession, ecological [/community]. *See* ecological succession
supernovas, 1454
surrogate use, 113, 260, 261, 1516, 1519, 1566, 1764
surveys, 777, 1011, 1035, 1696, 1812, 1881, 2005, 2415, 2524, 2566, 2656, 3013, 3077, 3119, 3120, 3135, 3397, 3513, 3569, 3580, 3587, 3611, 3614, 3648, 3693, 3743, 4011, 4024, 4184, 4244, 4270, 4994, 5022
survival plans. *See* action plans
sustainable agriculture [/farming], 26, 97, 98, 100, 432, 435, 937, 956, 1014, 1141, 1435, 1467, 1604, 1842, 1843, 1884, 2131, 2409, 2411, 2511, 2615, 2638, 3176, 3424, 3571, 3825, 3837, 3846, 4028, 4315, 4715, 4743, 4758, 4803, 4848, 4849, 5063, 5075, 5105, 5183, 5241, 5718
Sustainable Biosphere Initiative, 3002
sustainable development [/growth], 17, 41, 110, 225, 226, 315, 320, 321, 322, 400, 418, 422, 438, 447, 448, 456, 522, 564, 628, 650, 671, 752, 902, 919, 997, 1040, 1043, 1044, 1047, 1158, 1217, 1337, 1344, 1345, 1351, 1357, 1512, 1543, 1564, 1854, 1872, 1896, 1949, 1971, 2082, 2254, 2374, 2417, 2446, 2635, 2734, 2758, 2793, 2877, 2933, 2989, 2990, 3004, 3050, 3364, 3391, 3424, 3464, 3470, 3487, 3605, 3652, 3825, 3826, 3876, 3882, 4053, 4090, 4227, 4235, 4236, 4320, 4324, 4340, 4454, 4458, 4568, 4620, 4621, 4624, 4650, 4915, 5104, 5106, 5195, 5512, 5763, 5767, 5810
sustainable forestry, 142, 342, 434, 460, 581, 620, 652, 658, 699, 752, 753, 1499, 2116, 2529, 2644, 2645, 2646, 3671, 3993, 4660
sustainable use, 115, 225, 309, 321, 649, 770, 896, 1112, 1288, 1467, 1687, 1736, 1747, 1829, 1945, 2041, 2191, 2222, 2313, 2417, 2446, 2615, 2635, 2645, 2646, 2732, 2746, 2863, 2892, 3002, 3098, 3289, 3489, 3609, 3673, 3683, 3763, 3766, 4090, 4094, 4200, 4208, 4290, 4291, 4295, 4431, 4454, 4620, 4621, 4624, 4712, 4825, 4931, 5094, 5173, 5180, 5624, 5641, 5824, 5828
swamps and marshes, 2096, 2190, 3726, 4401, 4567, 5396

swidden agriculture. *See* shifting cultivation
symbiosis [/commensalism/mutualism], 783, 839, 1553, 1990, 1992, 2269, 2450, 2469, 2568, 2836, 3126, 4015, 4376
synergisms, 2106, 3516, 3531, 3947, 3969, 4189, 5748
systematics (*see also* cladistics, phylogenetics), 45, 241, 255, 467, 777, 1002, 1022, 1051, 1202, 1324, 1444, 1489, 1508, 1509, 1529, 1549, 1639, 1640, 1925, 1992, 2135, 2177, 2432, 2492, 2625, 2736, 3204, 3229, 3275, 3344, 3363, 3382, 3392, 3600, 3863, 3864, 3979, 4053, 4242, 4406, 4475, 4542, 4685, 4690, 4733, 4807, 4932, 4974, 4977, 5244, 5245, 5253, 5254, 5430, 5431, 5432, 5489, 5490, 5519, 5570, 5643, 5743, 5832
Systematics Agenda 2000, 3275, 1489, 3229, 3350, 4977, 5743
systematics, bird. *See* evolution, bird
systematics, insect. *See* evolution, insect
systematics [/taxonomy], plant, 1893, 2432, 3392, 4733, 4932
taxa, focal. *See* focal species
taxon cycles, 4286, 5546
taxonomic distinctiveness. *See* genetic distinctiveness
taxonomic diversity, 381, 468, 814, 1507, 1509, 1628, 1629, 1630, 3164, 3627, 4157, 4158, 4491, 4576, 4577, 4578, 4665, 4874, 4875
taxonomy, 694, 1092, 1185, 1462, 1772, 1773, 1919, 2416, 2864, 2932, 3196, 3381, 3635, 4652, 5502, 5570
taxonomy, plant. *See* systematics, plant
teaching. *See* education
technology: misc. subjects (*see also* biotechnology), 218, 225, 383, 586, 656, 1131, 1289, 1447, 1802, 1854, 1860, 2409, 2547, 2763, 2773, 2893, 2894, 3362, 3424, 3553, 3568, 3719, 3720, 4013, 4385, 4633, 4659, 4699, 5293, 5439
technology transfer, 383, 1447, 1860, 2773, 2891, 2893, 3424, 4105, 4659, 4699
Teilhard de Chardin, Pierre, 1733
teleology, 1030, 2344, 5464
temperate environs (*see also* geographical listings in Index II), 504, 554, 555, 731, 812, 1367, 1423, 1569, 2074, 2239, 2320, 2402, 2549, 2588, 2603, 2614, 2676, 3077, 3154, 3458, 3477, 3671, 3672, 3673, 3987, 4339, 4488, 4499, 4506, 4512, 4831, 4837, 4936, 5028, 5043, 5175, 5540, 5563, 5727, 5737, 5797

temperate [/subtropical] rainforests, 980, 1650, 2530, 2603, 4519, 5503
temperature effects, 355, 921, 1622, 2403, 3315, 3433, 3797, 3798, 4860
temporal effects, 139, 394, 471, 531, 984, 1233, 1261, 1940, 2239, 2925, 2952, 3307, 3808, 4284, 4370, 4378, 4499, 4513, 4930, 5426
Terborgh, John, 1294, 3349
thermophiles. *See* extremophiles
Third World. *See* developing countries
threatened animals. *See* endangered animals
threatened birds. *See* 'birds, endangered' in Index III
threatened ecosystems and habitats. *See* endangered ecosystems and habitats
threatened fishes. *See* 'fishes, endangered' in Index III
threatened insects. *See* endangered insects
threatened mammals. *See* 'mammals, endangered' in Index III
threatened peoples and cultures. *See* endangered peoples and cultures
threatened plants. *See* 'plants, endangered' in Index III
threatened species. *See* endangered species
threatened wildlife. *See* endangered wildlife
Tilman, David, 21, 1127, 1831, 2538, 3417, 5696
timed species counts, 4011
top-down regulation. *See* bottom-up regulation
toxicology [/ecotoxicology], 748, 773, 2782, 4926, 5861
trade, bushmeat (*see also* hunting), 111, 1503
trade, international. *See* international trade
trade, ivory, 285
trade, wild [/ornamental] animal. *See* wild animal trade
traditional [/local/indigenous] agriculture [/farming], 96, 101, 102, 207, 380, 432, 609, 662, 666, 879, 2019, 2896, 2983, 3585, 3735, 3836, 4758, 4864, 5000, 5045, 5698, 5700
traditional [/local/indigenous] knowledge, 178, 668, 671, 1133, 1225, 1723, 2664, 2793, 4023, 4207, 4301, 4635, 4712, 4713, 4900, 5512, 5641
traditional [/indigenous] medicine, 170, 252, 254, 643, 5387
trampling by humans, 617, 1868, 2920

transboundary studies, 178, 585, 638, 5036, 5420
transfer payments, 3482, 4393
transfer, technology. *See* technology transfer
transferable development rights, 3827
transgenic organisms (*see also* genetic engineering), 149, 1522, 1559, 1750, 1751, 4082, 4951, 5081
translocation, 190, 998, 1322, 1669, 1942, 2229, 3079, 4545, 4647
transportation corridors. *See* roads
treaties and agreements (*see also* Convention on Biological Diversity, environmental law, international cooperation, international law), 229, 487, 723, 1054, 1390, 1596, 1699, 1961, 2385, 2579, 2919, 3346, 3712, 3841, 4105, 4249, 4256, 4312, 4341, 4377, 4628, 4753, 4864, 5005, 5098, 5272, 5686
treefall [/canopy] gaps, 461, 568, 731, 911, 1004, 1251, 3162, 4038, 4395, 5271, 5290
trees. *See* listings in Index III
Triassic (*see also* Permian), 118, 405, 408
TRIPs, 1390, 2890, 5005
tropical birds, 259, 431, 1271, 2350, 2716, 2959, 3768, 4141, 4889, 5044, 5045
tropical environs (*see also* geographical listings in Index II), 65, 91, 207, 245, 254, 259, 320, 327, 342, 346, 354, 462, 482, 513, 539, 542, 554, 556, 569, 596, 658, 717, 744, 755, 777, 889, 908, 993, 1003, 1005, 1056, 1163, 1251, 1257, 1271, 1299, 1313, 1320, 1349, 1397, 1408, 1438, 1447, 1466, 1484, 1485, 1615, 1673, 1703, 1791, 1792, 1793, 1811, 1850, 1856, 1872, 1901, 1905, 1906, 1976, 1994, 2015, 2105, 2251, 2278, 2294, 2353, 2389, 2424, 2425, 2431, 2457, 2459, 2529, 2549, 2552, 2558, 2601, 2610, 2644, 2645, 2745, 2816, 2817, 2828, 2835, 2912, 2922, 2923, 2951, 2952, 2959, 3008, 3009, 3040, 3061, 3093, 3143, 3188, 3189, 3205, 3211, 3293, 3304, 3357, 3406, 3419, 3504, 3513, 3517, 3518, 3519, 3520, 3521, 3525, 3526, 3571, 3634, 3720, 3776, 3828, 3840, 3947, 3948, 3949, 3950, 3973, 3983, 3987, 3993, 4050, 4051, 4083, 4179, 4252, 4283, 4290, 4291, 4311, 4322, 4481, 4482, 4490, 4512, 4531, 4613, 4614, 4664, 4749, 4788, 4880, 4925, 4950, 4996, 5028, 5033, 5060, 5065, 5152, 5186, 5187, 5240, 5301, 5321, 5367, 5400, 5443, 5451, 5453, 5485, 5591, 5601, 5616, 5623, 5656, 5684, 5731, 5738, 5775, 5855
tropical forest defaunation (*see also* bushmeat trade, hunting), 1294, 4205

tropical forests, 65, 82, 254, 320, 342, 482, 539, 556, 569, 620, 653, 654, 755, 786, 889, 983, 993, 1026, 1056, 1163, 1271, 1286, 1348, 1438, 1447, 1703, 1811, 1901, 1994, 2015, 2081, 2202, 2203, 2278, 2362, 2425, 2468, 2529, 2610, 2644, 2645, 2744, 2828, 2844, 2871, 2912, 2922, 2951, 2952, 3090, 3162, 3188, 3189, 3294, 3304, 3357, 3519, 3520, 3521, 3526, 3578, 3720, 3768, 3776, 3828, 3947, 3948, 3950, 4332, 4477, 4481, 4483, 4556, 4613, 4614, 4664, 4749, 4822, 4989, 4990, 5028, 5032, 5043, 5186, 5187, 5400, 5592, 5597, 5601, 5623
tropical insects, 331, 4081, 5591, 5592
tropical plants, 777, 1673, 1708, 4036, 5154, 5738
tropical rainforests, 27, 54, 91, 115, 245, 250, 252, 400, 440, 458, 513, 524, 542, 652, 658, 682, 733, 809, 813, 910, 911, 930, 967, 1003, 1005, 1008, 1026, 1076, 1078, 1089, 1159, 1177, 1232, 1251, 1261, 1313, 1335, 1337, 1377, 1408, 1456, 1644, 1790, 1793, 1908, 2143, 2144, 2145, 2232, 2254, 2272, 2274, 2331, 2365, 2457, 2459, 2495, 2496, 2501, 2505, 2510, 2610, 2680, 2716, 2826, 2827, 2831, 2835, 2923, 2945, 2971, 2973, 2997, 3009, 3055, 3093, 3143, 3205, 3216, 3228, 3293, 3297, 3367, 3390, 3419, 3577, 3634, 3732, 3840, 3925, 3927, 3929, 3949, 3983, 3992, 3993, 4016, 4036, 4051, 4080, 4087, 4141, 4179, 4186, 4194, 4210, 4275, 4289, 4294, 4306, 4317, 4371, 4393, 4480, 4537, 4757, 4788, 4841, 4889, 4921, 4950, 5016, 5033, 5034, 5042, 5044, 5045, 5065, 5208, 5152, 5154, 5171, 5240, 5290, 5298, 5438, 5443, 5450, 5451, 5461, 5495, 5524, 5525, 5668, 5855
tundra, 863, 3067, 3244, 3502
turnover, 46, 175, 637, 730, 1251, 1679, 2101, 2383, 2434, 2438, 3077, 3250, 3744, 3948, 3971, 3973, 4095, 4510, 5426, 5514, 5527, 5636
turnover rates, 637, 3948, 4510, 5636
ultraviolet radiation, 478, 757, 2616, 4754
umbrella species, 416, 2825, 3612, 4670, 5327
UNCED, 156, 205, 318, 383, 389, 480, 602, 871, 902, 1057, 1596, 1854, 1863, 1961, 1989, 2210, 2474, 2477, 2579, 2929, 3082, 3364, 3402, 3832, 3848, 3884, 4338, 4353, 4364, 4418, 4508, 4620, 4719, 5174, 5184, 5185, 5812, 5849
understorey. See forest floor

UNDP, 5176
UNEP, 762, 5176
UNEP, publications by, 781, 1087, 1881, 2375, 2890, 4452, 5176, 5177, 5178, 5179, 5180, 5181, 5404
UNESCO, 2165, 2374, 4780
unisexual populations, 2748
U.S. National Biological Survey, 1035, 2656, 3569, 4994
upwelling, nutrient, 4336, 4427
urban [/suburban] environs, 31, 32, 337, 358, 470, 501, 936, 1284, 1591, 1814, 2211, 2360, 2481, 3005, 3235, 3990, 4212, 4355, 4739, 4809, 4940, 5683
urban wildlife, 31, 32, 3005, 4809
USAID, 93, 778, 4538, 4748, 5186, 5187, 5381
utility [/hydropower] industry, 579, 973, 1704, 2794, 2998, 3181, 3745, 5020, 5471, 5661, 5790
utilization, wildlife. *See* wildlife utilization
valuation, bd. *See* biodiversity, economic valuation of
valuation, contingent. *See* contingent valuation
valuation, wildlife. *See* wildlife utilization
variation, intraspecific. *See* intraspecific diversity
variation [/variability], genetic. *See* genetic variation
vegetation and climate, 316, 505, 701, 1233, 2461, 2670, 4051, 4064, 4326, 4831, 5449
vegetation change, 1233, 1650, 2730, 3792, 4789, 5372, 5375, 5449
vegetation diversity, 127, 287, 1264, 2247, 2303, 2896, 4326, 5064, 5372
vegetation: general and misc. subjects, 58, 287, 560, 606, 747, 777, 1264, 1544, 1650, 1832, 2048, 2833, 2903, 3012, 3015, 3175, 3443, 3469, 3586, 3753, 3961, 3982, 4016, 4141, 4142, 4156, 4335, 4471, 4515, 4599, 4789, 5120, 5135, 5193, 5334, 5879
vegetation maps and mapping, 1413, 2363, 2766, 2833, 3015, 3792
vegetation structure, 1036, 1264, 2181, 2553
vertical effects (marine and forest environs), 1186, 1261, 1494, 4947, 5028
veterinarians, 1468, 2606
vicariance, 125, 660, 728, 1083, 1084, 1086, 1465, 1997, 2480, 2501, 3214, 3600, 3988, 3989, 4361, 4362, 4474, 5490, 5871
visualization, bd, 2037, 4057

volcanoes [/volcanism], 730, 819, 1063, 1482, 1502, 1838, 2589, 3231, 3615, 3722, 4316, 4609, 4680, 5460
Wallace, Alfred Russel, 4110, 4592, 4901
Wallace Line, 4110, 4592, 5318, 5319
Wasichu, 2788
waste dumping, 805, 1415, 2682, 4625, 4763
water supply, 2035, 5237
water transfers, 5056
watersheds. *See* drainage basins
weeds, 216, 1298, 2220, 2840, 3054, 3649, 3822, 4943, 5023, 5529
wetlands, 338, 443, 575, 715, 868, 1249, 1376, 1576, 1654, 1727, 1805, 1824, 1900, 1921, 1953, 2002, 2007, 2118, 2150, 2183, 2774, 3384, 3873, 4009, 4018, 4108, 4605, 4993, 5054, 5511, 5575, 5672, 5688, 5868
wild [/ornamental] animal trade, 124, 136, 285, 707, 1062, 1540, 1597, 2167, 3371, 3624, 4026, 4321, 5111, 5475
wild relatives (of domestic spp.), 598, 1169, 2089, 2090, 2122, 2288, 2488
Wilderness Act, 840, 1148
wilderness [/wildland] areas, 769, 771, 824, 840, 1148, 1184, 1189, 1869, 1873, 1971, 2305, 2431, 2460, 2626, 3102, 3107, 3221, 3406, 3542, 3557, 3681, 3682, 3683, 4214, 4292, 5806, 5880
wildland areas. *See* wilderness areas
Wildlands Project, 2460, 3102, 5880
wildlife communities. See wildlife ecology
wildlife conservation [/management], 4, 32, 58, 150, 164, 184, 256, 398, 399, 454, 488, 492, 500, 509, 562, 580, 596, 689, 707, 791, 832, 867, 918, 1017, 1201, 1227, 1288, 1295, 1296, 1381, 1461, 1497, 1571, 1594, 1649, 1676, 1695, 1703, 1728, 1809, 1937, 1945, 2160, 2176, 2307, 2315, 2375, 2405, 2457, 2509, 2567, 2581, 2594, 2597, 2675, 2700, 2813, 2852, 2853, 2881, 2930, 2931, 2941, 3016, 3025, 3028, 3062, 3098, 3099, 3185, 3227, 3329, 3331, 3374, 3411, 3512, 3646, 3664, 3666, 3670, 3746, 3800, 3865, 3991, 4018, 4185, 4280, 4321, 4477, 4550, 4641, 4642, 4650, 4806, 4809, 4981, 4982, 5014, 5111, 5132, 5166, 5188, 5198, 5199, 5207, 5291, 5396, 5406, 5418, 5428, 5472, 5476, 5499, 5603, 5651, 5653, 5828
wildlife corridors. *See* wildlife habitat
wildlife ecology [/communities], 457, 500, 1363, 1614, 1904, 1908, 1916, 2407, 2967, 3139, 3753, 5266, 5343, 5828
wildlife, endangered. *See* endangered wildlife
wildlife: general and misc. subjects, 164, 203, 282, 552, 596, 610, 846, 928, 1413, 1666, 1685, 1916, 2315, 2411, 2592, 2700, 2714, 3005, 3135, 3213, 3687, 3701, 3887, 3920, 4315, 4966, 5040, 5830
wildlife habitat [/corridors], 457, 2931, 3099, 3103, 3380, 3505, 3865, 3873, 3874, 4018, 4605, 5828, 5830
wildlife, threatened. *See* endangered wildlife
wildlife utilization [/value/valuation], 552, 928, 1226, 1461, 1564, 1686, 1687, 2720, 3140, 3514, 3934, 4065, 4066, 4280, 4321, 4883, 4884
Wilson, Edward O., 354, 535, 635, 5029, 5528
women and bd, 17, 380, 516, 1716, 3004, 4197
woodland birds. *See* 'birds, forest' in Index III
woodland plants. *See* 'forest plants'
woodlands, 311, 461, 1130, 1340, 1557, 1977, 2076, 2185, 2285, 3040, 3928, 4420, 5203, 5615, 5659
World Bank, 159, 539, 1120, 1873, 3886, 4626, 5079, 5177, 5617, 5618
World Bank, publications by, 393, 558, 661, 889, 930, 1120, 1292, 2266, 2586, 2675, 2737, 2744, 2797, 3021, 3081, 3318, 3332, 3489, 3811, 4847, 4848, 4849, 5176, 5177, 5402, 5617, 5618
World Commission on Forests and Sustainable Development, 3530
World Conservation Monitoring Centre, 4263
World Resources Institute, publications by, 11, 13, 14, 277, 516, 3354, 3355, 3391, 3578, 4232, 4233, 4236, 4237, 5075, 5622, 5623, 5624
World Zoo Conservation Strategy, 2382, 5428
WWF, 573, 2008, 3011, 4307, 5625
WWF, publications by, 349, 511, 1135, 1597, 2008, 2167, 2196, 2288, 2310, 2375, 2756, 3059, 3133, 3271, 3371, 3524, 3662, 3748, 5402, 5621, 5625
zooarchaeology, 3025, 4866
zoogeography (*see also* biogeography), 121, 122, 174, 588, 1098, 1181, 2228, 2260, 2283, 3036, 4059, 4696, 4842, 5318, 5320, 5559, 5681
zoos, 230, 1017, 1819, 2342, 2343, 2382, 2606, 3670, 4117, 4130, 4323, 4407, 4484, 5146, 5428, 5444, 5472

Index II: Geographical Subjects

Regional Combinations

Arctic Ocean (and Arctic region), 657, 761, 860, 1656, 1891, 2046, 2677, 2903, 3067, 3213, 4683, 4750, 4928, 4947, 5262, 5267, 5596, 5675, 5874
Asia, 197, 653, 667, 983, 1344, 2555, 2737, 2933, 3371, 5104, 5655
Caribbean Sea (and region), 49, 449, 606, 1457, 2029, 2259, 2475, 2782, 3010, 3748, 3793, 4314, 4361, 4401, 5323, 5491
Holarctic region, 3758
Indo-Pacific region, 381, 1291, 5269
Mediterranean Sea (and region), 62, 387, 1041, 1266, 1730, 1939, 2091, 2092, 2420, 3295, 3824, 4133, 4463, 5685
New World, 75, 443, 463, 466, 976, 1165, 1729, 1764, 1766, 2081, 2553, 2561, 2889, 3146, 3149, 3791, 4223, 4341, 4852, 4872, 5030, 5373, 5532, 5533, 5606
North America, 29, 30, 76, 138, 175, 307, 489, 537, 627, 640, 641, 642, 719, 760, 817, 1023, 1138, 1140, 1179, 1326, 1614, 1655, 1904, 1919, 1943, 2128, 2169, 2228, 2320, 2415, 2426, 2460, 2461, 2597, 2627, 2628, 2651, 2689, 2743, 3007, 3187, 3224, 3252, 3300, 3314, 3342, 3358, 3421, 3427, 3703, 4021, 4065, 4095, 4201, 4268, 4302, 4339, 4342, 4356, 4439, 4440, 4504, 4534, 4619, 4648, 4717, 4981, 5130, 5182, 5374, 5504, 5506, 5507, 5559, 5779, 5880
North America, eastern, 198, 201, 1679, 1833, 2205, 3792, 3970, 4120, 5375, 5436, 5575, 5727
North America, northwestern, 1672, 3770, 5847
North America, western, 818, 348, 1610, 2042, 2109, 2141, 2570, 2967, 3618, 3952
Old World, 464, 1166, 1569
Pacific region, 197, 3904, 4074
Southern Hemisphere, 4278

Canada and Greenland

Alberta, 1442, 3019, 4494
British Columbia, 698, 1240, 1409, 1554, 1555, 2085, 2123, 2149, 2366, 2667, 3999, 4148, 4495, 4532, 4885, 5563
Canada, 209, 385, 447, 454, 553, 575, 657, 779, 886, 888, 914, 936, 1109, 1364, 1434, 1470, 1613, 1681, 1731, 1840, 1927, 2111, 2305, 2352, 2477, 2630, 3067, 3211, 3246, 3380, 3388, 3467, 3653, 3858, 3933, 4369, 4497, 4733, 5066, 5183, 5507, 5607, 5633, 5646, 5770, 5791, 5799
Great Lakes (and region), 70, 899, 1691, 2508, 2608, 2839, 3095, 3369, 3370, 3387, 3409, 3581, 5297, 5588, 5805
Greenland, 657
Maritime Provinces, 1682, 2637, 4571, 4573
Newfoundland, 3468
Northwest Territories, 1891, 2903, 4750
Ontario, 179, 1576, 2151, 2152, 2190, 2393, 2608, 3386, 3697, 4114, 4129
Pacific Northwest. *See under* "United States"
Quebec, 866, 1256, 1356, 2456
Rocky Mountains. *See under* "United States"
Saskatchewan, 1798

United States

Adirondack Mountains, 4211, 4509, 5386
Alabama, 3024
Alaska, 863, 928, 1188, 3213, 3244, 3502, 3780, 3934, 4009, 5540
Appalachians, 161, 731, 3395
Appalachians, Southern, 512, 1453, 3360, 3361, 3937, 4122, 5442
Arizona, 637, 2399, 3542, 4295, 5446
Arkansas, 1090, 3179
California, 1, 120, 287, 314, 338, 353, 377, 470, 501, 559, 804, 948, 1193, 1208, 1394, 1448, 1592, 1697, 1939, 2221, 2304, 2325,

2449, 2467, 2525, 2581, 2624, 2626, 2649, 2730, 2790, 2825, 2876, 3018, 3103, 3112, 3145, 3475, 3477, 3478, 4005, 4069, 4241, 4336, 4414, 4558, 4623, 4768, 4809, 4814, 4818, 4868, 4903, 4943, 5006, 5040, 5056, 5216, 5235, 5349, 5422, 5672, 5688, 5768, 5832
Chesapeake Bay (and region), 4340, 5391
Colorado River (and region), 42, 426, 725, 5296, 5471
Connecticut, 202, 2593
Delaware, 3175
Everglades, 527, 1200, 2908, 5672
Florida, 212, 580, 845, 1210, 1211, 1321, 1923, 2588, 3225, 3310, 3374, 3686, 3843, 4334, 4644, 4948, 5323, 5603, 5766
Georgia, 550, 604, 1319, 2267, 2522, 5364
Grand Canyon, 978
Great Basin, 631, 1145, 1307, 1920, 2470
Great Lakes. See under "Canada"
Great Plains, 95, 714, 1298, 1977, 2707, 2886, 3180
Great Smoky Mountains, 5440, 5477
Hawaii. See under "Ocean Regions"
Idaho, 747, 853, 2117, 2652, 3048, 3389, 4904, 5642
Illinois, 146, 472, 704, 2182, 3005, 3691, 3835, 5370
Indiana, 309, 618, 1588, 1676, 4499
Iowa, 1727, 5825
Kansas, 984, 985, 986, 1841, 2833, 2834, 4522, 5705
Kentucky, 4557, 5141
Louisiana, 1900
Maine, 1242, 1805, 2314, 3088, 3309, 3376, 4389
Massachusetts, 5293
Michigan, 471, 2183, 2811, 2865, 2925, 2976, 3399, 3410, 3496, 3729, 4038, 4272, 4488, 4723, 5589
Minnesota, 146, 243, 244, 3317, 3520, 3633, 4126, 5383
Mississippi, 1900
Missouri, 150, 1090, 5369
Montana, 1221, 2578, 2611, 2884, 3079, 4855
Nevada, 1920
New England, 429, 1229, 4573, 4884, 5293, 5427
New Hampshire, 1229, 5293
New Jersey, 183
New Mexico, 2093, 2118

New York, 775, 1042, 4211, 4728
North Carolina, 3303, 3823, 3937, 5278
Ohio, 310, 494, 4929, 5118
Oregon, 191, 476, 793, 803, 804, 1121, 1215, 1386, 1504, 3234, 3308, 3598, 3753, 3762
Ozarks, 150, 1090
Pacific Northwest, 296, 396, 1670, 1697, 1926, 2031, 2057, 2123, 2796, 2823, 2824, 3423, 3500, 3658, 3870, 4519, 4738, 4916, 4957, 5062, 5331, 5449, 5685, 5848
Pennsylvania, 971, 3175, 3537, 5435
Rhode Island, 3628
Rocky Mountains, 296, 1528, 2163, 2597, 4894, 4895
Sierra Nevada, 559, 1358, 1643, 3944
South Carolina, 715, 1172, 1804, 2522, 3893, 4099, 5278
Tennessee, 148, 3106, 3303, 5571
Texas, 1093, 1844, 2118, 2589, 3649, 3796, 3797, 4486, 5006, 5036, 5872
United States, 20, 186, 232, 390, 481, 484, 597, 628, 677, 846, 909, 913, 927, 1020, 1035, 1062, 1151, 1155, 1295, 1317, 1339, 1381, 1382, 1434, 1524, 1532, 1607, 1824, 1955, 1967, 2326, 2394, 2413, 2421, 2585, 2592, 2639, 2656, 2778, 2809, 2814, 2942, 3560, 3566, 3569, 3647, 3657, 3661, 3687, 3688, 3713, 3717, 3718, 3804, 3818, 4026, 4176, 4233, 4234, 4296, 4333, 4369, 4485, 4602, 4746, 4747, 4887, 4890, 4896, 5023, 5111, 5112, 5188, 5189, 5190, 5191, 5291, 5356, 5437, 5478, 5480, 5483, 5505, 5598, 5633
United States, eastern, 531, 1197, 2126, 4728, 5462
United States, northwestern, 792, 1702, 3771
United States, southeastern, 743, 1220, 1991, 2170, 3157, 3261, 4019, 4968, 5463
United States, southwestern, 1032, 2968, 3180, 3942, 4850, 4980
United States, western, 231, 1403, 1419, 2116, 2706, 3379, 3538, 3539, 4618, 5025, 5057
United States-Mexico border region, 638, 5036
Utah, 1034, 1189, 2093
Vermont, 4411, 4828
Virginia, 144, 148, 3395, 5571
Washington, 172, 364, 476, 793, 827, 5634
Wisconsin, 471, 519, 552, 575, 2851, 2954, 4129, 5119
Wyoming, 3343, 3867, 4292, 5343

Index II 445

Yellowstone Park (and region), 306, 316, 884, 1864, 2934, 4261
Yosemite, 1358, 3481

Latin America

Amazon River, 1892, 2282, 4246, 4569
Amazonia, 115, 331, 400, 440, 491, 492, 518, 649, 679, 701, 728, 729, 933, 1026, 1162, 1286, 1337, 1377, 1456, 1464, 1543, 1565, 1637, 1708, 1789, 1892, 1996, 1997, 2254, 2282, 2495, 2496, 2767, 2829, 2830, 3115, 3163, 3446, 3716, 3875, 3886, 3939, 3992, 4055, 4056, 4209, 4246, 4289, 4430, 4529, 4537, 4568, 4569, 4691, 4716, 4725, 4843, 4939, 5148, 5149, 5150, 5171, 5208, 5290, 5704
American tropics, 1372, 2361, 2397, 2398, 2757, 4012, 5030, 5452
Andes, 665, 667, 672, 900, 1786, 1874, 2556, 2557, 3139, 3563, 3866, 4037, 4116, 4250, 4526, 5027, 5302, 5625, 5697, 5698, 5699, 5700, 5701, 5702
Argentina, 428, 582, 4396, 5135
Atacama Desert, 3139
Baja California (Mexico), 576, 1659, 2467, 2838, 4281, 4473, 5013, 5484
Belize, 252, 253, 823, 2600, 2602, 3293, 3294, 4989, 4990
Bolivia, 620, 1740, 2171, 3289, 4247, 4527
Brazil, 22, 72, 224, 284, 309, 376, 620, 679, 682, 780, 1449, 2150, 2560, 2987, 3306, 3390, 3445, 4053, 4081, 4156, 5032
Brazil, Amazonian, 1456, 1542, 1544, 1709, 3081, 3367, 4023, 4645, 4661, 5164, 5257
Brazil, eastern (Atlantic) areas of, 615, 790, 895, 1627, 1852, 3322, 3449, 3807
Caribbean. *See under* "Regional Combinations"
Central America, 1614, 2012, 2509, 3577, 3775, 4474, 4921
Chile, 948, 1546, 2525, 3139, 3435
Colombia, 1377, 1963, 2488, 2556, 2557
Costa Rica, 82, 281, 487, 557, 568, 733, 738, 911, 1119, 1159, 1539, 1736, 1799, 1851, 2246, 2333, 2406, 2427, 2428, 2429, 2430, 2468, 2524, 2534, 2810, 2938, 2971, 3003, 3228, 3323, 3466, 3520, 3757, 3768, 3841, 3908, 3909, 3911, 3925, 4039, 4041, 4042, 4262, 4309, 4371, 4477, 4556, 4721, 4889, 4998, 5022, 5566, 5668, 5686, 5865
Cuba, 2357, 4458
Dominican Republic, 4525

Ecuador, 431, 515, 1107, 1120, 1261, 2403, 2725, 2939, 3520, 3888, 4037, 4388, 4526, 4727, 5208
French Guiana, 869, 1908, 3313, 5042, 5043, 5044, 5047
Guadeloupe, 2362
Guatemala, 1204, 1393, 1924, 3578
Gulf of California, 825, 1469, 4007, 4414
Guyana (and region), 2478, 2479, 5763
Honduras, 1393, 2186
Jamaica, 5248
Latin America, 96, 97, 182, 204, 366, 449, 511, 571, 686, 789, 813, 900, 938, 1086, 1228, 1292, 1294, 1342, 1348, 1545, 1606, 1644, 1765, 1786, 1787, 1790, 1998, 2350, 2370, 2577, 2782, 2808, 3116, 3185, 3187, 3748, 3825, 3863, 3910, 3983, 4052, 4132, 4151, 4206, 4207, 4302, 4303, 4321, 4661, 4689, 4822, 4829, 4927, 4983, 5247, 5298, 5385, 5486, 5523
Martinique, 2977
Mexico, 181, 388, 498, 499, 742, 753, 842, 843, 1015, 1016, 1220, 1287, 1393, 1469, 1493, 1502, 1530, 1734, 1986, 2012, 2187, 2199, 2896, 2983, 3027, 3055, 3162, 3297, 3735, 4139, 4302, 4802, 5207, 5220, 5387
Montserrat, 4316
Neotropical. *See* Latin America
Nicaragua, 5242
Panama, 146, 1000, 2550, 2866, 2871, 2872, 3735, 5539, 5592, 5636, 5637
Pantanal, 683
Paraguay, 5653
Peru, 431, 492, 662, 1026, 2880, 3139, 3349, 4430, 4526, 5027, 5034, 5148, 5544, 5670, 5697, 5699, 5701
Puerto Rico, 572, 603, 697, 1434, 3847, 5534
South America, 660, 1152, 1371, 1858, 3459, 4696, 4823
Trinidad, 379, 4262
Uruguay, 4457
Venezuela, 52, 2452, 2602, 3294, 3298, 3391, 4332, 4989, 5566
West Indies (*see also* 'Caribbean Sea' under "Regional Combinations"), 2051, 2153, 2154, 2155, 4059, 4286, 4764, 5637
Yucatán, 207, 873

Europe, Russia, and Southwest and Central Asia

Alps, 507, 5194
Arabian Peninsula, 2512
Aral Sea, 5693
Asia, Central, 1696
Asia, West, 2580
Austria, 5711
Baltic Sea, 2041
Belgium, 3020
Black Sea (and region), 2734, 3938, 5677
British Isles, 2466, 5068
Caucasus Mountains, 1696
Corsica, 3296
Cyprus, 4400
Czech Republic, 1463, 3444
Denmark, 893, 2402, 2403, 2819, 4726
England, 374, 461, 732, 924, 1284, 1340, 1501, 2014, 2209, 2537, 2538, 2769, 3015, 3168, 3546, 3928, 5372, 5429
Estonia, 2364
Europe, 30, 402, 537, 878, 1058, 1059, 1106, 1144, 1164, 1246, 1266, 1325, 1368, 1495, 1649, 1769, 1952, 1953, 2003, 2195, 2205, 2208, 2230, 2320, 2322, 2408, 2443, 2503, 2619, 2669, 2686, 3421, 3704, 3812, 4624, 4684, 4984, 5048, 5080, 5255, 5513, 5714
Europe, Central and Eastern, 178, 488, 585, 2650, 3333, 4021, 5581
Europe, Northern, 403, 1450, 2434, 4419, 5206, 5726
Europe, Western, 673, 2175, 2407, 2438, 3703, 4934, 5327
Finland, 2070, 2072, 2157, 2158, 2435, 2481, 2776, 2777, 3632, 3633, 3902, 3982, 5119, 5120, 5206, 5283, 5303, 5631
France, 58, 62, 132, 520, 534, 574, 705, 892, 924, 936, 1230, 1234, 1329, 1749, 2091, 2092, 2483, 2822, 2848, 2883, 3012, 3049, 3171, 3296, 3299, 3926, 4375, 4506, 4507, 4993, 5043, 5136, 5205
Georgia (former USSR), 610
Germany, 2592, 4800, 5121, 5136
Great Britain, 301, 943, 1065, 1247, 1711, 1933, 1934, 2032, 2101, 2201, 2257, 2507, 2536, 2618, 2846, 3039, 3592, 3593, 3887, 4008, 4060, 4115, 4155, 4827, 5064, 5157, 5158, 5204, 5357, 5521
Greece, 4108, 4676, 5142, 5690
Hungary, 248, 1548
Iberian Peninsula, 833, 4689
Iran, 2661
Israel, 1730, 3559, 3606, 4917, 5814
Italy, 55, 289, 430, 507, 905, 1870, 2822, 5683
Jersey, 1863
Kazakhstan, 38
Kuwait, 630
Kyrgyzstan, 38
Lake Baikal, 613, 3153
Lake Constance region, 495
Near East, 896, 1169, 1558
Netherlands, 1293, 1952, 2276, 3299, 3760, 4335, 4655, 5115, 5224
Norway, 2178, 3416, 3728, 4420, 4455, 4897
Oman, 1558
Poland, 1153, 2107, 3738
Pyrenees, 1230, 2883
Romania, 3938
Russia, 133, 142, 661, 988, 1450, 1726, 2756, 3094, 3616, 3903, 4000, 4647, 4773, 5206
Saudi Arabia, 3623, 4599
Scotland, 203, 1612, 2045, 2440
Sinai, 3474
Slovak Republic, 4402, 4870, 4871
Spain, 759, 946, 1738, 1742, 1868, 1985, 2876, 3469, 3727, 5132
Sweden, 142, 393, 412, 425, 460, 2041, 2076, 2077, 2178, 2795, 3092, 3637, 3638, 3639, 3641, 3642, 3643, 3698, 3726, 3939, 4409, 4410, 4567
Switzerland, 1975, 2806, 5194, 5584
Turkey, 411, 1441, 5147
Turkmenistan, 1568
United Kingdom, 394, 1254, 1728, 1744, 1745, 1774, 1863, 2070, 2229, 2623, 3901, 3963, 4063, 4127, 4176, 4608, 4752, 5051, 5053, 5058, 5127, 5201, 5203, 5494, 5561, 5615, 5649
USSR, 4096, 4648
Wales, 5429

South, Southeast, and East Asia

Andaman Islands, 5046
Asia, East, 30, 5727
Asia, South, 269, 654, 2454, 4715, 5038
Asia, Southeast, 616, 654, 1569, 3090, 3258, 3579, 3624, 3826, 4087, 4715, 4971, 5450,

5681, 5803
Borneo, 1089, 1404, 2493, 2945, 4085, 4086, 4317
Brunei, 1089, 1183
Burma. *See* Myanmar
China, 162, 785, 786, 887, 1870, 2455, 2797, 2943, 3028, 3057, 3059, 5102, 5103, 5106, 5641, 5650, 5692, 5694
Himalayas, 1248, 1262, 1263, 1264, 1978, 3830, 3831, 3900, 4136, 4138, 4372, 4603, 4710, 5271, 5665
Hindu Kush, 3900
Hong Kong, 2943, 3461, 5664
India, 778, 857, 858, 890, 1000, 1120, 1175, 1264, 1558, 1654, 1718, 1719, 1720, 1801, 1863, 1982, 2011, 2641, 2642, 2643, 2647, 2717, 2731, 2733, 2738, 2797, 3302, 3384, 3553, 3594, 3595, 3829, 3830, 3856, 4048, 4049, 4057, 4128, 4136, 4141, 4144, 4185, 4415, 4603, 4630, 4633, 4636, 4707, 4708, 4709, 4710, 4711, 4712, 4713, 4714, 4748, 5105, 5108, 5365
Indonesia, 141, 224, 276, 277, 284, 322, 754, 1343, 2033, 2148, 2203, 2274, 3142, 3385, 3751, 3872, 4901, 4914, 4942, 5363, 5461
Japan, 128, 803, 2200, 2360, 2592, 2676, 5268
Java, 2614
Korea, 1357, 2655
Krakatau, 730, 3615, 4680, 5071, 5460
Laos, 1497
Malay Archipelago, 593, 718, 1354, 4592, 5318, 5319
Malaysia, 54, 1000, 2143, 2144, 2514, 3082, 3732, 5153, 5676
Mt. Kinabalu, Sabah, Malaysia, 1404
Myanmar (Burma), 4123
Nepal, 53, 645, 646, 651, 776, 1651, 2159, 2160, 3603, 4372, 4710, 5271, 5398, 5562, 5665
Oriental Realm, 3058, 4303
Pakistan, 1801, 4048
Philippines, 26, 230, 271, 828, 1558, 2148, 3943, 4235, 4254
Sarawak, 401, 4086
Singapore, 3625, 5154, 5156
Sri Lanka, 261, 437, 1562, 3488
Sulawesi, 2232, 2668, 3694
Sumatra, 141, 2274, 5045
Thailand, 1191, 1217, 1218, 2678, 2712, 2958, 3277, 5415, 5597
Thar Desert, 1801, 4048
Vietnam, 1995, 4841
Western Ghats, India, 1173, 1174, 1176, 2642, 3312, 4049, 4142

Africa

Africa, 94, 131, 269, 448, 509, 1114, 1134, 1164, 1600, 1809, 1858, 2127, 2227, 2417, 2490, 2513, 2666, 2675, 2723, 2808, 3073, 3270, 3433, 4010, 4220, 4483, 4757, 4888, 4931, 5097, 5381, 5417, 5715
Africa, Central, 809, 930
Africa, East, 718, 2665, 2997, 4011
Africa, Southern, 739, 1072, 1130, 1363, 1937, 2238, 2319, 2321, 2359, 3041, 3172, 3705, 3706, 3953, 5059
Africa, tropical, 304, 1113, 1335, 1601, 2633, 2634, 2898, 3060, 4012
Africa, West, 930, 1225, 1976, 2300
African Great Lakes, 2897, 2999, 3345, 3724, 3981, 5626
Bioko, 4484
Botswana, 2185
Cameroon, 92, 486, 1232, 2272, 2844, 4294, 4393
Egypt, 3474, 4845
Equatorial Guinea, 1503, 4294, 4484
Ethiopia, 380, 459, 1243, 2653, 5017, 5542
Gabon, 3895, 5438
Ghana, 643, 1120, 1225, 1987, 2202
Ivory Coast, 870, 3927, 4016
Karoo-Namib region, 1074
Kenya, 2511, 3438, 3439, 3497, 3675, 4327, 4650, 4659
Kruger National Park, 867, 915
Lake Malawi, 292, 613, 3493, 4239
Lake Nasser, 4845
Lake Tanganyika, 613, 950, 951, 952, 2269, 3153, 5877
Lake Victoria, 292, 1861, 1880, 2564, 4564, 5577
Lesotho, 3785
Liberia, 2550, 2716
Madagascar, 224, 268, 288, 721, 1126, 1385, 1547, 1876, 2054, 2482, 2744, 2751, 2752, 2753, 2987, 3084, 3400, 3407, 3488, 4111, 4186, 4240, 4649, 4736, 4831
Morocco, 4424, 5052
Mozambique, 3306
Namibia, 2436, 4280, 4427, 5754
Nigeria, 111, 797, 2386, 3788
Rwanda, 4838
Sahara Desert, 2850

São Tomé and Principe, 1391, 3898
Senegal, 1748, 2500, 3648
South Africa, 68, 118, 184, 365, 832, 867, 1069, 1112, 1355, 1520, 1688, 1689, 2714, 2960, 2962, 3235, 3710, 3800, 4078, 4448, 4724, 4827, 5228, 5229, 5230, 5231, 5232
South Africa, Cape region of, 1069, 1071, 1196, 1212, 1332, 1859, 1939, 2932, 3854, 4279, 4540, 4682, 5138, 5139, 5237, 5238, 5632, 5747, 5870
Tanzania, 458, 2998, 3083, 3362, 3448, 3505, 3506, 3620, 4892
Tunisia, 2589
Uganda, 809, 868, 2281, 4036
Zaire, 284, 809, 2646, 4036
Zambia, 4, 1809, 3135, 3507
Zimbabwe, 1557, 3485, 4053, 4460, 4650

Australia, New Guinea, and New Zealand

Australasia, 174, 1164, 1603
Australia, 7, 27, 45, 110, 113, 123, 185, 190, 215, 216, 239, 332, 360, 398, 399, 450, 560, 703, 760, 783, 824, 834, 897, 995, 1083, 1084, 1095, 1137, 1141, 1187, 1201, 1253, 1289, 1345, 1378, 1468, 1531, 1747, 1815, 1832, 1928, 1945, 1980, 2214, 2268, 2301, 2515, 2525, 2574, 2582, 2606, 2636, 2680, 2760, 2805, 2827, 2831, 2930, 3016, 3119, 3122, 3247, 3290, 3453, 3462, 3463, 3464, 3470, 3486, 3672, 3743, 3744, 3854, 3945, 3953, 3994, 4070, 4201, 4202, 4203, 4255, 4467, 4468, 4471, 4485, 4586, 4641, 4642, 4702, 4704, 4735, 4831, 4973, 5037, 5084, 5167, 5175, 5315, 5333, 5355, 5409, 5645, 5671, 5734, 5736, 5835
Australia, northern, 112, 1684, 4359, 5585
Australia, southeastern, 1814, 3587, 3588, 3586, 3589, 3940, 5175
Great Barrier Reef, 855, 2309, 4397, 5002
New Guinea (incl. Papua New Guinea), 61, 1095, 1938, 2576, 3216, 3508, 3786, 4572, 4999, 5630
New South Wales, 119, 213, 372, 562, 1036, 1519, 1650, 2567, 3118, 3124, 4071, 4204, 5503, 5667
New Zealand, 206, 522, 941, 977, 1096, 1185, 1219, 1280, 1828, 1945, 2030, 2146, 2240, 2604, 2663, 3196, 3220, 3231, 3712, 3723, 4097, 4312, 4471, 4586, 4676, 5077, 5326, 5846
Northern Territory, 2363, 4080, 5586, 5587

Papua New Guinea. *See* New Guinea
Queensland, 326, 2501, 2826, 3452, 4275, 5524, 5525
South Australia, 4281
Tasmania, 1283, 2530, 2671, 2672, 5448
Victoria, 397, 1341, 1346, 1514, 2931
Western Australia, 8, 702, 1939, 2048, 2181, 2221, 2222, 2583, 3085, 4090, 4466, 5351, 5659

Ocean Regions and Antarctica

Antarctica (and region), 336, 543, 584, 929, 1236, 1677, 1940, 2439, 2566, 2631, 2919, 4308, 4754, 5202, 5267, 5749
Arctic Ocean (and region). *See under* "Regional Combinations"
Atlantic, East, 2142, 3068
Atlantic, North/Northeast, 924, 139, 1186, 1871, 1914, 3199, 4258, 4763, 4765, 5083, 5263, 5360
Atlantic Ocean, 1707, 2103, 3137, 4259, 4947, 5264, 5669, 5762
Atlantic, Southwest/West, 805, 1213, 2029, 4565
Bahamas, 4215, 4511, 4840
Bermuda, 735
Canary Islands, 566, 2356, 4637
Fiji, 370, 2447, 2698, 3699
Galápagos Islands, 545, 1909, 2039, 2416, 2465, 2974, 3182, 4515
Guam, 974, 1079, 2001, 4330, 4476
Gulf of Guinea, 2489, 2517, 4215, 4484, 5742
Gulf of Mexico, 805, 5872
Hawaii, 158, 297, 622, 819, 969, 998, 1068, 1434, 1993, 2097, 2223, 2755, 2949, 2973, 2974, 3427, 3472, 3646, 3969, 4313, 4331, 4383, 4548, 4673, 4906, 4907, 4908, 5285, 5307
Henderson Island. *See* Pitcairn Islands
Indian Ocean, 46, 1871
Kuril Islands, Russia, 4647
Marshall Islands, 5218
Mauritius, 2981, 4421
New Caledonia, 525, 528, 874, 2252, 2410, 3405, 3443, 3508, 5755
North Sea, 1929, 2614, 4799, 5359
Oceania, 3113, 3453, 4819, 5736
Pacific, central, 1669, 3976
Pacific, eastern, 2906, 4381, 4382

Pacific, North, 348, 5263, 5268
Pacific Ocean, 826, 983, 1707, 2576, 2613, 2933, 3842, 3984, 4882, 5765
Pacific, South, 1079, 1735, 5712
Pacific, Southwest, 1274, 3220, 3699, 5546
Pacific, tropical, 3113, 4866, 4893, 5259
Pacific, West, 4612
Pitcairn Islands, 410, 4058, 5311
Red Sea, 4917
Réunion, 850
St. Helena, 1102, 1103
Samoa, 1076, 1078, 1079
Seychelles, 1609, 2446, 2448
Shetland Islands, 2719
Society Islands, 3503
South Georgia, 5131
Spitsbergen, 2439, 2614
Spratly Islands, South China Sea, 3257
Tahiti/Moorea, 44, 2949, 3324
Tonga, 2936
Vanuatu (New Hebrides), 4985, 4986, 4987

Index III: Organism-Centered Subjects

Multi-Category Groupings

animal abundance, 135, 203, 465, 1139, 2814, 4560, 4561
animal behavior, 1142, 1234, 1235, 1590, 1877, 2577, 3178, 4919, 5565
animal body size, 420, 465, 1139, 5233
animal conservation, 4459, 4518, 5144
animal distribution [/geography], 135, 203, 1181, 2146, 2705, 2814, 3315, 5320, 5332
animal diversity, 118, 743, 1801, 2189, 2632, 2694, 4220, 4581, 4765, 5300, 5332, 5473, 5732
animal extinction, 720, 721, 2212, 4580
animal geography. *See* animal distribution
animal germplasm [/genetic] resources, 63, 221, 939, 2268, 3570, 4493, 4684, 5073, 5562
animal migration, 200, 201, 531, 623, 716, 1228, 1606, 1998, 2323, 2350, 2549, 2577, 3185, 3187, 4041, 4042, 4121, 4151, 4302, 4325, 5000, 5030
animals, domestic, 598, 4493, 4844, 5562
animals, endangered [/threatened], 25, 236, 412, 580, 4459, 5708
animals, extinct [/fossil], 118, 720, 3484, 4581, 5574, 5708
animals, (recently) extinct, 75, 572, 1434, 1930, 5263, 5708
animals, fossil. *See* animals, extinct
animals, introduced, 2259, 2901, 3427, 3625, 4515, 4906
animals, marine, 1725, 1783, 2103, 2189, 4491, 4580, 4581, 4765, 5574
animals: misc. subjects, 282, 412, 414, 421, 610, 623, 874, 1017, 1142, 1150, 1162, 1329, 1356, 1634, 1703, 1724, 1725, 1783, 1801, 1893, 2666, 2757, 2850, 3016, 3028, 3206, 3259, 3315, 3484, 3625, 4809, 4844, 4926, 5144, 5221, 5233, 5300, 5475, 5760, 5814
eukaryotes, 114, 837, 892, 1042, 1992, 2702,
3260
invertebrate conservation, 9, 977, 1059, 2418, 3608, 3610, 3614, 3743, 5073, 5550, 5663, 5835
invertebrate diversity [/richness], 265, 303, 584, 1096, 1219, 1629, 1632, 1694, 2131, 2304, 2354, 2403, 2806, 3340, 3632, 4159, 4575, 4576, 4577, 4578, 4653, 4874, 5210, 5279, 5503, 5550, 5835
invertebrate extinction, 800, 876, 921, 1645, 1813, 2021, 2022, 2032, 2387, 4580, 4666, 5058
invertebrate richness. *See* invertebrate diversity
invertebrates, endemic, 969, 1230, 1246, 1993, 2252, 5632
invertebrates, fossil [/extinct], 265, 876, 1629, 1632, 1645, 1813, 2021, 2022, 2387, 3340, 4159, 4575, 4576, 4577, 4578, 4580, 4666, 5210
invertebrates, freshwater, 783, 787, 977, 1219, 1514, 1694, 2403, 2806, 5125, 5279, 5303
invertebrates: general and misc. subjects, 314, 783, 923, 982, 1092, 1501, 1514, 2131, 2135, 2304, 2416, 2593, 3743, 3744, 4653, 5055, 5303, 5503
invertebrates, marine, 265, 314, 584, 800, 921, 1629, 1632, 1724, 1813, 1911, 2021, 2022, 2387, 2416, 2418, 3340, 4576, 4577, 4578, 4653, 4666, 5210, 5595
lichens, 905, 1894, 2530, 2776, 3003, 3416, 3420, 3598, 3643, 3940, 4573, 5597
mycorrhizae [/rhizobia], 39, 74, 1154, 1268, 2468, 2469, 2580, 3445, 3621, 3801, 5859
phytoplankton, 49, 890, 892, 1052, 2347, 2402, 2701, 3220, 3855, 4754, 4800, 4801, 4815, 5129, 5273, 5308
rhizobia. *See* mycorrhizae
tetrapods, mass extinctions of, 407
tetrapods, Phanerozoic diversity of, 404
vertebrates, 174, 698, 702, 759, 825, 1083, 1121, 1138, 1227, 1284, 1805, 1935, 1997, 2154, 2155, 2652, 2707, 2748, 2804, 3374,

3744, 3749, 4032, 4059, 4202, 4485, 5166, 5266, 5497, 5525, 5541, 5635
vertebrates, endemic, 1083, 5525
zooplankton, 1325, 1326, 1379, 1580, 1940, 2523, 2589, 3698, 3858, 4336, 4375, 4842

Mammals

antelopes, 2633, 2634
bats, 181, 1557, 2771, 4486, 5532, 5533
bears, 928, 1956, 3079, 3213, 3371, 4984
beavers, 1450, 3039
bison, European (*Bison bonasus*), 2107
caribou, 3213
carnivores, 1823, 2597, 3796, 5132
cats, 3693, 4919
cheetahs, 806, 3709, 3710
dolphins and porpoises, 1122, 2943, 3913, 4216
elephants, 4, 285, 1130, 2185, 2852, 5417
ferret, black-footed, 917, 3343
flying foxes (Chiroptera), 1079, 1710
goats, 944, 1749
lemurs, 268, 3407, 4736
livestock, 63, 414, 1307, 1564, 1610, 2016, 2017, 2230, 2531, 2693, 3393, 3570, 4135, 4220, 4493, 4684, 4844, 5235
mammal biogeography [/distribution/faunas], 76, 632, 968, 1036, 1502, 1616, 1840, 1904, 2665, 2964, 2965, 2966, 2967, 2968, 3067, 3115, 3116, 3146, 3147, 3314, 3810, 3863, 4095, 4648, 4696, 4722, 4828, 4885, 5374, 5559, 5585, 5758
mammal body size, 76, 641, 825, 3442
mammal conservation, 241, 395, 968, 1502, 1689, 1840, 2683, 2921, 2966, 3116, 4641, 4702, 4828, 4981, 5014, 5132, 5207, 5476, 5585, 5653, 5674
mammal decline. *See* mammal extinction
mammal distribution. *See* mammal biogeography
mammal diversity, 655, 809, 841, 843, 1616, 1734, 2561, 2838, 3067, 3649, 3810, 4095, 4571, 4619, 4648, 5298
mammal ecology, 134, 894, 1171, 1684, 1908, 2567, 2921, 3297, 4048, 4885
mammal extinction [/decline], 1145, 1904, 1919, 2889, 3149, 3618, 3619, 3620, 4641, 5348
mammal faunas. *See* mammal biogeography
mammal populations, 1908, 2785, 4956
mammal species richness, 703, 1171, 1823, 2628, 2964, 3297, 3798, 4486, 4537, 4571, 4772, 4828, 5532, 5645, 5653
mammals and hunting, 111, 237, 491, 492, 809, 1464, 1503, 1809, 3135
mammals, endangered [/threatened], 75, 655, 841, 843, 917, 1350, 1710, 1820, 2140, 2230, 2683, 2755, 2965, 3079, 3343, 4135, 4334, 4642, 4647, 4702, 5665, 5694
mammals, endemic [/rare], 182, 841, 843, 1502, 2148, 2531, 2633, 5132, 5207
mammals, extinct, 75, 95, 175, 1616, 1904, 1919, 2889, 3146, 3147, 3149, 4095, 4696, 5348, 5374
mammals, forest [/rainforest], 182, 397, 632, 792, 809, 1036, 1464, 1614, 1920, 2826, 2967, 2968, 3395, 4571, 5298, 5438
mammals: general and misc. subjects, 241, 1363, 1399, 1908, 2190, 3863, 4021, 5545
mammals, host, 1740, 3442, 5365
mammals, introduced, 1391, 1623, 1684, 2899
mammals, island, 703, 2964, 5758
mammals, large, 367, 718, 1126, 1279, 1363, 1684, 3149, 3620, 3799, 3800, 5476, 5674
mammals, montane, 631, 1920, 2967, 2968, 4722
mammals, rainforest. *See* mammals, forest
mammals, rare. *See* mammals, endemic
mammals, small, 397, 1225, 3139, 3395, 3649, 4571, 4981, 5446
mammals, species density and numbers of, 123, 1614, 1908, 4695
mammals, threatened. *See* mammals, endangered
marine mammals, 655, 737
marsupials, 22, 1036, 2567, 2931, 4642, 4702, 5532
otter, sea, 1492, 3025
panda, giant, 5694
panda, red, 1651, 5665
pigs, 2531, 2730
pinnipeds, 2224, 2755, 5008
prairie dogs, 3342
primates, 224, 1547, 1823, 2080, 2458, 3143, 3807, 3838, 4213, 4537
puma (cougar and Florida panther), 377, 4334, 4644
rabbits (*Oryctolagus cuniculus*), 1623
rhinoceroses, 415, 416, 2852
rodents, 21, 22, 792, 2407, 3139, 3446, 3796, 3824, 3866, 4735, 4814, 5013, 5220, 5355
sea-cow, Steller's, 1492

sheep, bighorn, 413
sheep, feral, 5235
shrews, 3902
squirrels (*Sciurus*), 5204
tigers, 5476
whales, 172, 237, 2442, 3913, 4216
wolves, 1702, 3409

Birds

bird abundance, 642, 936, 1933, 2038, 3029, 3142, 3592, 3593, 4011, 4292, 4357, 4494, 4945, 5046, 5224
bird biogeography [/distribution], 370, 632, 1049, 1084, 1113, 2209, 2435, 2567, 2575, 2969, 3184, 3497, 4011, 4121, 4356, 4357, 4652, 4677, 4885, 5046, 5071, 5158, 5585
bird body size, 464, 1934, 2413, 3593, 5127
bird communities, 202, 394, 472, 531, 472, 895, 1050, 1177, 2182, 2414, 2550, 2552, 2575, 2866, 2949, 2959, 3421, 3759, 4049, 4120, 4122, 4532, 4993, 5043, 5046, 5224, 5468, 5540
bird conservation, 1, 259, 311, 431, 439, 443, 692, 790, 1152, 1173, 1270, 1271, 1272, 1621, 1669, 1998, 2577, 2663, 3166, 3604, 3760, 3898, 4114, 4127, 4145, 4244, 4245, 4247, 4608, 4865, 4889, 4897, 4927, 5042, 5045, 5256, 5265, 5385, 5393, 5470, 5513, 5585
bird decline, 198, 200, 201, 601, 895, 1621, 1711, 2350, 2935, 3630, 3650, 4151, 4292, 4302, 4421, 5000, 5030, 5031
bird distribution. *See* bird biogeography
bird diversity, 46, 370, 534, 789, 936, 1113, 1202, 1605, 2163, 2185, 2550, 2553, 2667, 2706, 3317, 3497, 3633, 3649, 3713, 3829, 3898, 4011, 4037, 4114, 4544, 5043, 5047, 5130, 5141, 5256, 5521
bird ecology, 2, 134, 199, 501, 531, 894, 1173, 1221, 1767, 1998, 2552, 2575, 2577, 3184, 3185, 3234, 3586, 3622, 4244, 4495, 4885, 5302, 5468
bird evolution [/systematics], 660, 885, 1049, 1086, 1996, 2142, 4651, 4652
bird extinction, 572, 1145, 1307, 1434, 1774, 1930, 2434, 2557, 2936, 3559, 3970, 3976, 4476, 4669, 4866, 5019, 5127
bird faunas, 46, 133, 640, 1028, 1086, 1274, 1280, 1307, 1765, 1766, 2240, 2434, 2435, 2553, 2556, 3559, 4037, 4141, 4286, 4476, 4544, 4548, 4673, 4677, 5248, 5477, 5615
bird populations, 68, 471, 1924, 2209, 2415,
3758, 3760, 4099, 4145, 4669, 4945, 4956, 5636
bird species diversity, 337, 1049, 1084, 3034, 4121, 4201, 4284, 5027, 5044, 5129
bird species richness, 42, 463, 466, 495, 532, 534, 891, 971, 1459, 1764, 1766, 2076, 2208, 2414, 2481, 3214, 3398, 3589, 3713, 3839, 4012, 4127, 4132, 4204, 4246, 4292, 4389, 4494, 4828, 5521, 5542, 5645
bird systematics. *See* bird evolution
birds, breeding, 150, 337, 531, 971, 1774, 2481, 3234, 3559, 3758, 3839, 4420, 4532, 5446
birds, endangered [/threatened], 1, 603, 616, 745, 975, 976, 1434, 1766, 1930, 2027, 2051, 2257, 2796, 3434, 4545, 4548, 4856, 5018, 5031, 5248, 5336, 5385
birds, endemic [/rare], 46, 258, 1113, 1152, 1774, 1852, 2051, 2142, 2436, 2476, 2556, 2969, 3142, 3168, 3214, 3296, 3362, 3434, 3898, 3970, 4010, 4865, 5248, 5385, 5521, 5660
birds, forest [/woodland], 150, 198, 199, 202, 311, 471, 531, 632, 1177, 1271, 1852, 1996, 2208, 2557, 2575, 2716, 3184, 3317, 3421, 3768, 3970, 4037, 4120, 4122, 4141, 4420, 4421, 4476, 4495, 4889, 5042, 5043, 5044, 5045, 5047, 5224, 5283, 5540, 5615
birds: general and misc. subjects, 68, 270, 337, 443, 790, 885, 1028, 1356, 1561, 1649, 1711, 2051, 2182, 2415, 2470, 2481, 2500, 2550, 2567, 2706, 2715, 2883, 2949, 2972, 3472, 3586, 3587, 3785, 3963, 4011, 4037, 4075, 4121, 4244, 4245, 4466, 4757, 4798, 4814, 4828, 4993, 5042, 5302, 5491, 5569, 5578
birds, introduced, 2900, 2949, 2970, 3472, 4673
birds, island, 2, 46, 370, 616, 1274, 1280, 2038, 2476, 2719, 2783, 2949, 3214, 3334, 3434, 3604, 4215, 4676, 4677, 4866, 5071, 5636
birds, migratory, 200, 201, 531, 1228, 1606, 1998, 2350, 2549, 2577, 3185, 3187, 4041, 4042, 4121, 4151, 4302, 4325, 5000, 5030
birds, rare. *See* birds, endemic
birds, species density and numbers of, 1023, 1768, 2470, 2667, 2783, 3768
birds, threatened. *See* birds, endangered
birds, tropical, 259, 431, 1271, 2350, 2716, 2959, 3768, 4141, 4889, 5044, 5045
birds, woodland. *See* birds, forest
cowbirds, 601, 4325
cranes, 782, 1440, 1845, 2742, 5393

crossbills (*Loxia*), 396
finches, Darwin's, 1909
owl, northern spotted, 2790, 2796, 2823, 3423, 3500, 5652
owls, 2678
parrot, Puerto Rican, 603
quetzals, 4041, 4042
rails, Guam, 2001
raptors, 5042, 5046
robin, black, 177
seabirds, 891, 1374, 2224, 3604
songbirds, 200, 489, 601, 2689, 2935, 3630, 5031, 5141
sparrows, 5364
waterbirds, 1649, 2096, 2708, 3829, 4114, 4993
woodpecker, red-cockaded, 4856, 5336
woodpeckers, 3333, 4850

Reptiles and Amphibians

amphibian decline, 293, 374, 454, 474, 475, 476, 477, 478, 963, 1119, 1358, 1592, 2141, 2153, 2938, 3223, 3892, 3893, 3946, 4039, 4461, 5309, 5331, 5524, 5647
amphibians, 28, 373, 374, 404, 454, 478, 704, 1229, 1242, 1321, 1322, 1371, 1372, 1539, 1804, 2026, 2151, 2152, 2193, 2483, 2651, 2945, 3395, 3607, 3797, 4473, 4474, 4556, 4723, 4885, 4934, 4981, 5331, 5407
Boiga irregularis (on Guam), 4330, 4476
Bufo boreas, 476
bullfrogs (*Rana catesbeiana*), 2141, 4934
dinosaurs, 103, 104, 176, 932, 1538, 1838, 2328, 3721, 4164, 4587
frogs, 559, 1358, 1372, 1373, 2141, 2211, 2764, 2827, 2831, 4039, 4275, 4461, 4934, 5524, 5704
frogs, endemic, 2831, 4275
frogs, ranid, 476, 2141, 2211, 4934
lizards, 1235, 1373, 3463, 3952, 3953, 4330, 4511, 4514, 4556, 4840, 4919, 5290, 5484
reptiles, 28, 404, 704, 825, 826, 1229, 1321, 1322, 1371, 1539, 1804, 2651, 2945, 3797, 4175, 4186, 4281, 4473, 4474, 4661, 4885, 4981, 5585
salamanders, 963, 1507, 3937, 4574
toad, golden, 1119, 4039
tortoises, 1034
tuatara, 1185, 3196
turtles, 715, 1142, 2406, 3564, 4720

Fishes

blennies, 4523
carp, 2619
cichlids, 1861, 2269, 2999, 3345, 4239, 4564, 5577
cod, 2046
cyprinids, 5462, 5568
fish biogeography [/distribution], 2228, 3068, 3231, 3984, 4724, 5681
fish communities, 2018, 2269, 2300, 2393, 2447, 2448, 3220, 4262, 4428, 4646, 5541
fish conservation, 136, 190, 856, 2898, 3379, 3478, 3700, 4616, 4724, 5073, 5188, 5339, 5504, 5571, 5573
fish decline. *See* fish extinction
fish density. *See* fish diversity
fish distribution. *See* fish biogeography
fish diversity [/density], 62, 286, 899, 2164, 2452, 2729, 2897, 3023, 3220, 4314, 4428, 4564, 4616, 5571
fish extinction [/decline/loss], 144, 856, 1697, 3345, 3358, 3379, 3477, 3864, 4295, 5462
fish faunas, 1627, 1931, 2999, 3179, 3477, 3867, 3872, 4239, 4240, 4314, 4646, 5356, 5504, 5626, 5877
fish genetics, 3408, 3700, 5073, 5074, 5339, 5573
fish introductions [/invasions], 292, 559, 866, 868, 1109, 1062, 1358, 1559, 2508, 2619, 2902, 2999, 3095, 3345, 3724, 3981, 5577
fish loss. *See* fish extinction
fish species richness, 211, 866, 1943, 2300, 2588, 2608, 3023, 3068, 3077, 3171, 3179, 3313, 3386, 3697, 3702, 3703, 3926, 4129, 4345, 4646
fishes, endangered [/threatened], 16, 152, 426, 678, 868, 1016, 1456, 1697, 2326, 2825, 3379, 3475, 3756, 4240, 4341, 4422, 5356, 5504, 5506, 5511, 5577
fishes, endemic [/rare], 3300, 3379, 3942, 4239, 4616, 4724
fishes, freshwater, 144, 146, 190, 190, 231, 264, 286, 385, 575, 811, 866, 899, 1109, 1943, 1976, 2110, 2152, 2228, 2269, 2393, 2452, 2564, 2588, 2618, 2737, 2897, 2898, 3077, 3231, 3313, 3386, 3387, 3475, 3697, 3700, 3702, 3703, 3712, 3926, 3939, 4129, 4240, 4262, 4502, 4503, 4616, 4724, 4855, 4888, 5119, 5356, 5504, 5681, 5877
fishes: general and misc. subjects, 136, 231,

811, 952, 1016, 1062, 1142, 1213, 1870, 1976, 2046, 2152, 2411, 3023, 3027, 3068, 3077, 3180, 3805, 3906, 4024, 4035, 4133, 4262, 4345, 4412, 4413, 4463, 4502, 4732, 4818, 5541
fishes, marine, 62, 152, 160, 211, 348, 812, 850, 856, 1213, 1328, 1735, 2239, 2326, 2447, 2448, 2569, 2729, 3408, 3984, 4035, 4133, 4345, 4397, 4428, 4429, 4463, 4646, 4818
fishes, rare. *See* fishes, endemic
fishes, threatened. *See* fishes, endangered
gobies, round, 2508
ground-fish communities, 1929
herring, 2046
lamprey, sea, 899
Nile perch, 292, 868, 1880
salmon, 2352, 3088, 4455, 4738, 5339, 5340
salmonids, 939
sturgeons and paddlefishes, 5310, 5573, 5693, 5788

Arthropods

amphipods, 1655, 2260, 2439, 3137, 3508, 5060
ants, 112, 113, 582, 969, 1144, 1232, 2249, 2304, 2971, 3055, 3085, 3908, 3909, 4019, 4371, 4448, 5349, 5501, 5546, 5667
aphids, 1299, 2739
arthropod communities, 112, 128, 787, 1261, 2127, 2676, 3351, 3698, 4019, 4441, 4827, 4841
arthropods, 43, 127, 128, 132, 326, 507, 637, 969, 1108, 1346, 1369, 1484, 2680, 2754, 2823, 2824, 2977, 3192, 3911, 4019, 4204, 4331, 4656, 4827, 5049, 5202, 5369, 5563
barnacles, 1258, 4523, 5427
bees, 85, 1507, 2126, 2573, 3173, 3328, 5514, 5519
beetles, 327, 1096, 1286, 1484, 1501, 1737, 2438, 3459, 3642, 3643, 3824, 4411, 4897, 5118, 5389
beetles, tiger, 817, 818, 2678, 3891, 4332, 5294
butterflies, 300, 366, 470, 623, 817, 1159, 1160, 1221, 1246, 1247, 1261, 1421, 1504, 1569, 1778, 1985, 2033, 2062, 2070, 2072, 2129, 2203, 2676, 2751, 2825, 3101, 3498, 3609, 3613, 3615, 3845, 4008, 4060, 4100, 4303, 4622, 4679, 4772, 4829, 4841, 5051, 5052, 5053, 5157, 5357
caddisflies, 1346, 1475

chironomids, 5113
cockroaches, 1906
Coleoptera, 369, 1484, 1534, 1758, 3891, 4332
Collembola, 1108, 1230, 1928, 2977, 4149
copepods, 2161, 4133, 4223, 5131
crab, European green, 2789
crayfishes, 783, 1090, 1655
crustaceans, 5, 1325, 1326, 2260, 3698, 4947, 5809
Diptera, 1234, 1758, 3728
dragonflies, 915, 4063
Drosophila, 599, 819, 3849, 3850
empidoid flies, 1234
grasshoppers, 2611, 3791
hawkmoths, 2678
Hemiptera, 2232, 2507
Heteroptera, 369
houseflies, 5024
Hymenoptera, 1758, 2739, 2816, 3694
ichneumonids, 2426, 4679
insect abundance. *See* insect diversity
insect biogeography [/distribution/faunas], 369, 410, 1027, 1153, 1758, 1770, 2012, 2429, 2439, 3473, 4446, 4879, 5206, 5267
insect body size, 2590, 4512, 4654
insect conservation, 649, 981, 1590, 1777, 2654, 3235, 3835, 4107, 4445, 4535, 4897, 5227, 5740
insect distribution. *See* insect biogeography
insect diversity [/numbers/richness/abundance], 816, 1179, 1754, 1755, 1770, 2233, 2429, 2780, 3496, 4081, 4416, 4654, 4921, 4925, 5049, 5206, 5267, 5280, 5430, 5466, 5500, 5591, 5592, 5632, 5683
insect ecology, 1153, 1590, 1613, 2095, 3351, 4928
insect evolution [/systematics], 816, 1907, 2780, 5430, 5466
insect faunas. *See* insect biogeography
insect numbers. *See* insect diversity
insect pests, 96, 99, 922, 1613, 3344, 4920, 5330, 5367
insect richness. *See* insect diversity
insect species richness, 818, 1475, 1754, 1770, 2127, 2426, 2438, 2739, 3092, 3642, 3791, 3891, 4081, 4679, 4925, 5205, 5606, 5632
insect systematics. *See* insect evolution
insects, aquatic, 3092, 4004, 4925, 5205, 5221, 5280, 5714
insects, endangered [/threatened], 2072,

2825, 3498, 5357, 5642
insects, galling. *See* insects, parasitic
insects: general and misc. subjects, 829, 1001, 1285, 1590, 1591, 2095, 2124, 2125, 2130, 2654, 2762, 2840, 2846, 2948, 3510, 3548, 3609, 3611, 4444, 5683
insects, introduced, 845, 903, 1037, 4122, 5330
insects, parasitic [/galling], 1037, 1560, 2128, 3235, 4879, 5466, 5632
insects, threatened. *See* insects, endangered
insects, tropical, 331, 4081, 5591, 5592
isopods, 4014, 4926, 4947
leafhoppers, 5434
Lepidoptera, 2043, 2179, 2252, 3235, 3612
mites, 1108, 4655, 5333
mosquitoes, 1473
moths, 1159, 1160, 4317, 5157, 5330
mycetophilids, 3728
orthopterans, 248
ostracods, 139, 952, 1100
sawflies, 2739
seed bugs, 2507
shrimp, 923
spiders, 1319, 3114, 4007, 4441, 4513, 4840, 5711
spiny lobsters, 1492
termites, 7, 1191, 1232, 1416, 3163, 3463, 3953
Thysanoptera, 1773
treehoppers, 5606
trilobites, 47, 1628, 5426
tsetse flies, 5542
wasps, fig, 2127
wire-worms, 3903

Lower (Complex) Animals

Acanthaster planci, 974, 5268
ammonoids, 2280, 2629
Aphanoneura, 3153
ascidians, 5669
bivalves, 493, 1617, 3291, 3341, 4265, 4875, 5690
brachiopods, 529, 4265, 5215
bryozoans, 1742, 2091, 2092
corals, 5, 381, 626, 697, 1038, 1883, 2396, 2543, 2698, 2920, 4376, 4873, 5268, 5270, 5371
crinoids, 108, 2526
earthworms, 494, 629, 2169, 2170, 2836, 2843, 2977, 4260
echinoderms, 108, 3864, 4158, 5669
gastropods, 119, 2356, 3704, 4258, 4379, 5305
helminths, 385, 727, 1935, 2617, 2618, 2619, 4033, 5355
holothurians, 4305
molluscs, 235, 493, 659, 1469, 1617, 2356, 2398, 2571, 2677, 4381, 4382, 4870, 5216, 5259, 5262, 5264
mussel, zebra, 2253, 3007, 4268
mussels, 3308, 4268, 5507
nematodes, 423, 486, 525, 526, 1677, 1749, 2263
Oligochaeta, 3153
polychaetes, 1186, 5275
Radopholus, 1529
rotifers, 4588
seastars, 974, 3308, 5268
snails, 340, 1068, 1993, 2356, 2819, 3503, 3642, 4058, 5642
sponges, 2092
tapeworms (cestodes), 2224
worms, blood, 923

Plants

almond, 4774
angiosperms, 1091, 1095, 1110, 1556, 1679, 5645
apples, 38
aspen, 2925, 3317, 4494
Asteraceae, 5274
bamboo, 4185
barley, 459, 1243, 2420
beans, 2488
beans, popping, 5698
beech, 785
bryophytes, 138, 1293, 1306, 2364, 2530, 2726, 3416, 3420, 3940, 4728
cacao, 379
cacti, 4387
canola, 1798
cedar (*Widdringtonia cedarbergensis*), 5059
cereals, 1169
chestnut, American, 722
Chondrilla juncea, 4943
coca, 5670
coffee and coffee plantations, 1734, 1924, 3908, 3909, 3910, 5000
cycads, 1231, 3787, 5655

Index III 457

dipterocarp forests, 1191
Douglas-fir, 434, 4532
Epacridaceae, 2583
eucalypts, 213, 214, 1036, 1283, 1814, 3118, 3672, 4204, 5448
ferns, 5144
fig trees, 2127, 5817
grasses, 927, 1180, 2461, 2886
Haloxylon salicornicum, 630
Hepaticae, 4063
herbs, 2276, 3303, 4134
juniper, 2093
kelp, 1208, 1492
legumes, 40, 2580, 3445, 4734
lianas, 938, 5630
louseworts, 3309
macrophytes, 235, 520, 1694
mahogany, 3289, 5257
maize, 388, 1268, 2983, 3691, 4747, 4983, 5697
mangroves, 401, 1378, 1457, 1572, 2362, 2866, 3384, 3784, 4311, 4401, 5608, 5664, 5804
Mesembryanthemaceae, 2359
mesquite, honey, 3649
Miconia calvescens (Melastomataceae), 3324
moss, aquatic, 5303
neem tree, 269, 2520, 2521
oaks, 2076, 5271
orchids, 4387, 4802
palms, 4727
peas, 4877
phanerogams, 1293, 2364
pines and pine plantations, 550, 604, 1530, 1814, 2093, 3020, 3468, 4276, 4278, 4969
plant biogeography [/geography/distribution], 52, 179, 355, 881, 1102, 1712, 1783, 2146, 2196, 2513, 2766, 2814, 3315, 3705, 3706, 3928, 4088, 4675, 4727, 5068, 5083, 5144, 5334, 5442, 5610
plant collections, 777, 1365, 2231, 4933, 5069
plant [/forest] communities, 183, 604, 630, 934, 1474, 1614, 1673, 1708, 1788, 1821, 1857, 1955, 1962, 2031, 2183, 2599, 2672, 2884, 2944, 3000, 3352, 3420, 3678, 3854, 4929, 4980, 5034, 5136, 5368, 5454, 5455, 5560, 5561
plant conservation, 102, 179, 287, 566, 567, 1104, 1199, 1525, 1664, 1827, 1830, 2036, 2184, 2194, 2433, 2599, 2724, 2885, 3295, 3392, 3412, 3460, 3539, 3818, 3856, 4459, 4492, 4518, 4681, 4976, 5025, 5156, 5274, 5447
plant dispersal, 3175, 4021, 4088
plant distribution. *See* plant biogeography
plant diversification. *See* plant evolution
plant diversity, 74, 127, 287, 550, 706, 775, 1072, 1073, 1074, 1091, 1096, 1264, 1332, 1354, 1474, 1679, 1708, 1786, 1788, 1825, 1859, 2011, 2157, 2196, 2247, 2303, 2546, 2709, 2884, 3295, 3376, 3412, 3443, 3496, 3502, 3635, 3649, 3791, 3802, 4036, 4193, 4310, 4326, 4489, 4533, 4637, 4845, 4894, 4895, 4933, 5093, 5153, 5368, 5518, 5784
plant ecology, 80, 550, 1233, 2031, 2766, 3296, 3360, 4128, 4354, 4409, 5334, 5460
plant evolution [/diversification], 67, 533, 1091, 1533, 1534, 1535, 2621, 2622, 2704, 3636, 4478, 5133
plant extinction [/loss], 473, 1532, 2724, 5154, 5156
plant genetic diversity, 1523, 1974, 2047, 2983, 3468, 4068
plant geography. *See* plant biogeography
plant germplasm [/genetic] resources, 35, 36, 379, 393, 459, 625, 726, 772, 857, 859, 957, 1014, 1118, 1168, 1323, 1458, 1467, 1523, 1525, 1636, 1647, 1661, 1662, 1664, 1696, 1701, 1705, 1781, 1984, 2047, 2089, 2121, 2231, 2241, 2242, 2288, 2371, 2401, 2421, 2511, 2555, 2695, 2697, 2709, 3125, 3190, 3524, 3565, 3566, 3585, 4043, 4112, 4144, 4196, 4374, 4747, 4748, 4774, 4949, 4952, 5069, 5085, 5189, 5281, 5282, 5492, 5508, 5662, 5682, 5695, 5770
plant introductions [/invasions], 128, 212, 216, 618, 775, 855, 1104, 1180, 1260, 1609, 1923, 2220, 2886, 2919, 3182, 3183, 3427, 4109, 4241, 4276, 4278, 4515, 4886, 4908, 5023, 5424
plant litter, 1539, 4556, 5086, 5352, 5353
plant loss. *See* plant extinction
plant species abundance, 1283, 2039, 2814, 5560, 5561
plant species diversity, 986, 1072, 1250, 1659, 1787, 1857, 2031, 2093, 2464, 2944, 3000, 3012, 3244, 3801, 3854, 3857, 4136, 5028, 5455
plant species numbers, 1755, 3928, 4682, 4867, 5233
plant species richness, 183, 235, 261, 369, 520, 733, 775, 833, 938, 1070, 1138, 1283, 1329, 1340, 1565, 1576, 1793, 1821, 1833, 1891, 1900, 1962, 2076, 2157, 2158, 2190,

2238, 2578, 2650, 2719, 2979, 3638, 3639, 3642, 3705, 3706, 3949, 3950, 4009, 4021, 4409, 4419, 4675, 4726, 4828, 4980, 5120, 5203, 5429, 5442, 5560, 5584, 5632, 5709
plant systematics [/taxonomy], 1893, 2432, 3392, 4733, 4932
plants and biotechnology [/genetic engineering], 149, 246, 421, 1750, 1781, 2519, 2520, 2521, 2696, 4082, 4951
plants and climate, 316, 351, 505, 1634, 2670, 3315, 5675
plants and genetic engineering. *See* plants and biotechnology
plants, aquatic, 351, 1833, 2190, 2880, 3000, 5386
plants, arid land, 1074, 1659, 4599
plants, endangered [/threatened], 83, 138, 158, 412, 510, 528, 560, 1093, 1199, 1231, 1448, 1527, 1826, 1830, 2276, 2410, 2433, 2573, 2726, 3309, 3360, 3361, 3787, 3818, 4054, 4279, 4387, 4459, 4492, 4681, 4728, 4970, 5059, 5134, 5138, 5311
plants, endemic, 515, 1070, 1095, 1530, 1826, 1859, 2410, 2465, 2761, 3296, 3443, 4279, 4637, 5138, 5167, 5311, 5448, 5518, 5632
plants, extinct [/fossil], 118, 1091, 1679, 4193, 4255, 5083
plants, flowering, 706, 1466, 2573, 2642, 4478, 4867
plants, food, 102, 726, 1136, 2120
plants, forest [/woodland], 412, 1708, 3928, 4036, 4310, 5028, 5154, 5203
plants, fossil. *See* plants, extinct
plants, host, 1299, 2739, 4879
plants, island, 1102, 1256, 1609, 2039, 3182, 3642, 4409, 4515, 4637, 5311
plants, land, 67, 2368, 2384, 2621, 2622, 3636, 3854, 4193
plants, medicinal, 57, 246, 250, 252, 364, 384, 742, 776, 797, 851, 1135, 1449, 1532, 1729, 1792, 2187, 2505, 2797, 3880, 3992, 4111, 4134, 4309, 4529, 4847, 5117, 5387
plants: misc. subjects, 453, 505, 606, 757, 777, 1134, 1203, 1329, 1356, 1490, 1565, 1833, 1886, 2015, 2089, 2137, 2395, 2461, 2573, 2757, 2761, 2775, 2778, 2840, 2846, 2850, 2885, 2904, 2905, 2920, 3073, 3175, 3244, 3502, 3546, 3563, 3626, 3744, 3856, 4068, 4136, 4534, 4601, 4814, 5233, 5447, 5460, 5475, 5518, 5523, 5576, 5676, 5678
plants, rare, 138, 179, 332, 473, 833, 1448, 1458, 1525, 1955, 2157, 2666, 2761, 2905, 3296, 3310, 3361, 3460, 3818, 4115, 4492,
4534, 4728, 4802, 4976, 5025, 5068, 5311, 5354, 5429, 5442
plants, riparian, 1821, 1891, 3638, 3639, 4009, 4828, 4980
plants, threatened. *See* plants, endangered
plants, tropical, 777, 1673, 1708, 4036, 5154, 5738
plants, vascular, 67, 179, 833, 1256, 1939, 2157, 2158, 2578, 2650, 2704, 3244, 3360, 3361, 3420, 3502, 3628, 3636, 3928, 4115, 4419, 4420, 5068, 5153, 5311, 5442
plants, woodland. *See* plants, forest
plants, woody, 261, 445, 719, 725, 1609, 2047, 3376, 3642, 3705
poplar, 894
potatoes, 665, 672, 2121, 4116, 5697, 5702
Pteridophyta, 1826
rice, 26, 663, 859, 2400, 5249, 5281, 5282, 5367, 5695, 5801
safflower, 2709
seaweed, 2123, 3014, 4400
silverswords (Compositae), 4313
spruce trees, 1154, 4122, 4155, 4573, 5563
strangler figs, 4251
sugarcane, 4930
sweetpotato (*Ipomoea batatas*), 1015
tarweeds, 5728
taungya plantations, 4603
tree diversity, 1008, 2143, 2681, 2858, 2872, 3237, 3958, 4038, 4531, 4612, 5630
tree species diversity, 29, 568, 717, 731, 786, 1251, 1789, 2320, 2333, 2425, 3322, 3432, 3732, 3958, 4395, 4613, 4614, 4990, 5271, 5539
tree species richness, 30, 938, 1138, 1140, 1377, 1658, 1788, 2821, 4989, 5208
trees: general and misc. subjects, 29, 185, 403, 1204, 3449, 3565, 3732, 4826, 4827, 4831, 5691
trees, hardwood, 150, 471, 1453, 1496, 4506, 4573, 5136, 5473, 5769
trees, tropical, 65, 327, 717, 786, 1000, 1008, 1204, 2425, 2478, 2479, 2872, 4531, 4612, 4831, 4989, 5034
Ulmus, 3049
weeds, 216, 1298, 2220, 2840, 3054, 3649, 3822, 4943, 5023, 5529
wheat, 1168, 1798, 3019, 5659
wiregrass, 604
yew, Pacific, 364

Algae, Fungi, Microorganisms, Viruses, etc.

algae, 114, 387, 892, 929, 2025, 2301, 2497, 3000, 3674, 3699, 3945, 4125, 4376, 4662, 5167, 5583
archaea, 1706
archaebacteria, 305, 306, 837, 3161
Ascomycetes, 468, 1306
bacteria, 39, 305, 306, 543, 608, 829, 837, 1042, 1236, 1239, 1268, 1380, 1395, 1553, 1706, 1798, 1948, 2079, 2115, 2289, 2450, 2451, 2580, 2728, 2878, 2957, 3019, 3161, 3445, 3447, 3842, 3895, 3897, 4082, 5122, 5123, 5124, 5137, 5405, 5534, 5680
bacteriophages, 419
bacterioplankton, 1236, 3897
Caulerpa taxifolia, 387
ciliates, 266, 1580, 1583, 1940
coccoid picoplankton, 4029
cyanobacteria, 1948
diatoms, 2677, 3105, 4372
dinoflagellates, 1940, 3069
eubacteria, 305
foraminifera, 267, 381, 741, 946, 1871, 2523, 2589, 3655, 5004, 5596
fungi, 55, 74, 412, 444, 468, 507, 784, 878, 927, 1005, 1108, 1154, 1306, 1853, 1893, 1986, 2132, 2134, 2353, 2408, 2951, 2952, 3197, 3365, 3621, 3801, 3927, 4369, 4410, 4424, 4916, 4942, 5284, 5361, 5362, 5389, 5678, 5679, 5751, 5865
macromycetes, 5361
microorganism conservation, 1582, 4261, 4358, 4851, 4854, 4942
microorganism diversity, 305, 518, 519, 694, 1578, 2295, 2615, 3803, 3855, 4358, 4456, 4851, 4942, 5131, 5730, 5820
microorganisms: general and misc. subjects, 55, 90, 107, 132, 444, 694, 839, 989, 1024, 1108, 1462, 1583, 1635, 1680, 2131, 2135, 2270, 2295, 2353, 2673, 2674, 2807, 2977, 3714, 4656, 4916, 5012, 5131, 5277, 5342, 5680, 5730, 5749, 5837
microorganisms, marine, 305, 839, 1239, 1707, 3842
microorganisms, soil, 518, 519, 629, 1108, 1575, 2079, 2115, 2615, 2835, 2836, 2868, 3834, 4656, 4734, 4851, 5123, 5124, 5137, 5760
mushrooms, 878, 2630
nanoflagellates, 1940
Pneumocystis carinii hominis, 2822
prokaryotes, 3161, 3816, 5143
protozoa, 45, 838, 1553, 1578, 1579, 1580, 1624, 1940
shiitake (*Lentinula edodes*), 887
Trypanosoma cruzi, 2832
viruses, 419, 1042, 1522, 2849, 3348, 5578
yeasts, 22, 3691

About the Author

Charles H. Smith is Science Librarian and Associate Professor of Library Public Services at Western Kentucky University in Bowling Green. His educational background includes a B.A. (geology) from Wesleyan University, Middletown, Connecticut, an M.A. (geography) from Indiana University, Bloomington, a Ph.D. (geography) from the University of Illinois, Champaign-Urbana, and an M.L.S. from the University of Pittsburgh. In addition to his published articles in the fields of biogeography, history of science, systems theory, and bibliography and collection development, and a monographic anthology of the shorter writings of nineteenth-century naturalist Alfred Russel Wallace, Dr. Smith has developed three World Wide Web sites: "HERPFAUN," a bibliography of the distributional literature on reptiles and amphibians, "The Alfred Russel Wallace Page," an educational site reviewing the range of Wallace's work, and "The Classical Composers Navigator," an educational site providing an alternative kind of introduction to the world of classical music. Dr. Smith can be reached by email for questions at: charles.smith@wku. edu.